Lecture Notes in Mathematics

Volume 2342

This series reports on new developments in all areas of mathematics and their applications - quickly, informally and at a high level. Mathematical texts analysing new developments in modelling and numerical simulation are welcome. The type of material considered for publication includes:

1. Research monographs
2. Lectures on a new field or presentations of a new angle in a classical field
3. Summer schools and intensive courses on topics of current research.

Texts which are out of print but still in demand may also be considered if they fall within these categories. The timeliness of a manuscript is sometimes more important than its form, which may be preliminary or tentative. Please visit the LNM Editorial Policy (https://drive.google.com/file/d/19XzCzDXr0FyfcV-nwVojWYTIIhCeo2LN/view?usp=sharing)

Titles from this series are indexed by Scopus, Web of Science, Mathematical Reviews, and zbMATH.

Jennifer Johnson-Leung • Brooks Roberts •
Ralf Schmidt

Stable Klingen Vectors and Paramodular Newforms

Springer

Jennifer Johnson-Leung
Department of Mathematics
University of Idaho
Moscow, ID, USA

Brooks Roberts
Department of Mathematics
University of Idaho
Moscow, ID, USA

Ralf Schmidt
Department of Mathematics
University of North Texas
Denton, TX, USA

ISSN 0075-8434 ISSN 1617-9692 (electronic)
Lecture Notes in Mathematics
ISBN 978-3-031-45176-8 ISBN 978-3-031-45177-5 (eBook)
https://doi.org/10.1007/978-3-031-45177-5

Mathematics Subject Classification: 11F46, 11F70, 11F30, 11F60, 22E50, 11F50, 22E55

This Springer imprint is published by the registered company Springer Nature Switzerland AG
The registered company address is: Gewerbestrasse 11, 6330 Cham, Switzerland

Paper in this product is recyclable.

Preface

Elliptic modular forms are analytic functions on the complex upper half plane invariant under a certain action of the congruence subgroup $\Gamma_0(N)$ of $\mathrm{SL}(2, \mathbb{Z})$. They are connected to many topics in mathematics in a concrete fashion, and can be organized into finite-dimensional vector spaces $M_k(\Gamma_0(N))$ according to their weight k and level N. Each element f of $M_k(\Gamma_0(N))$ admits a Fourier expansion of the form $f(z) = \sum_{n=0}^{\infty} a(n)e^{2\pi inz}$ with Fourier coefficients $a(n)$. An important subspace of $M_k(\Gamma_0(N))$, characterized in terms of Fourier expansions, is the space of cusp forms $S_k(\Gamma_0(N))$.

To study these spaces, one employs natural endomorphisms $T(n)$ known as Hecke operators, first studied systematically in [54]. An important theorem states that $S_k(\Gamma_0(N))$ admits a basis consisting of simultaneous eigenvectors for the $T(n)$ for all positive integers n relatively prime to N, known as eigenforms. Each simultaneous eigenspace has a distinguished line, and if a spanning vector is not in $S_k(\Gamma_0(M))$ for any proper divisor M of N, then we call it a newform of level N. Moreover, any such newform f is a simultaneous eigenvector for the $T(n)$ for *all* positive integers n. If f has Fourier coefficients $a(n)$ and Hecke eigenvalues λ_n, then, after appropriate normalization, one can prove that $a(n) = \lambda_n$ for all n. The proof is easy because the relevant $\Gamma_0(N)$ cosets defining the Hecke operators are represented by upper triangular matrices, so that the action of the Hecke operators can be expressed in terms of Fourier expansions.

For Siegel modular forms of degree 2, which are functions on a three-dimensional complex domain, the relationship between Fourier coefficients and Hecke eigenvalues is more complicated. In the case of Siegel modular forms with respect to $\mathrm{Sp}(4, \mathbb{Z})$, Andrianov [1] has found such a relationship in terms of a rationality result for a formal power series involving certain Fourier coefficients. Similar to the case of elliptic modular forms, one ingredient in the proof is the fact that the relevant coset representatives defining the Hecke operators can be chosen to be block upper triangular. This again allows the action of the Hecke operators to be expressed in terms of Fourier expansions.

In this book we consider Siegel modular forms with level. Level can mean many things for Siegel modular forms, but for us it means that we consider Siegel modular forms with respect to the paramodular groups K(N). Such Siegel modular forms, known as paramodular forms, also admit a newforms theory, developed in [104, 105, 114], and [115]. The current interest in paramodular forms comes in part from their appearance in the generalization of the modularity theorem to the setting of abelian surfaces formulated in [21, 23]; for other connections, see the list of references on p. 51.

In contrast to the case of elliptic modular forms, one encounters a difficulty. For a prime p such that p^2 divides the level N, the paramodular Hecke operators at p involve cosets whose representatives are never block upper triangular matrices. As a consequence, their action cannot readily be calculated in terms of Fourier coefficients. This is a problem because examples of paramodular forms are typically studied via Fourier expansions. One approach to this issue, which has been successful for some examples, is the method of restriction to a modular curve; see for example [96].

The purpose of this book is to present a new approach. We consider a closely related family of congruence subgroups called the stable Klingen subgroups and denoted by $\mathrm{K}_s(N)$. For this subgroup, the relevant cosets always admit block upper triangular matrices as representatives. Consequently, they interact well with the Fourier expansion. The group $\mathrm{K}_s(N)$ is contained in K(N), so that paramodular forms may be considered as stable Klingen forms.

Perhaps surprisingly, the desired paramodular Hecke eigenvalues can be recovered using the associated stable Klingen Hecke operators. To prove this, we reinterpret Siegel modular forms as functions on the adelic group GSp(4, $\mathbb{A}$). We consider the automorphic representations generated by these adelic functions, and analyze vectors invariant under the local stable Klingen subgroups in the resulting local representations of the group GSp(4, $\mathbb{Q}_p$).

This local analysis is done in the more general context of a finite extension F of $\mathbb{Q}_p$. It turns out that stable Klingen vectors and paramodular vectors are related closely enough to allow transfer of information from the former to the latter. We describe the spaces of stable Klingen vectors in all irreducible, admissible representations of GSp(4, F). This part of the present work makes full use of the local paramodular theory of [105].

Finally, as an application of our theory, we relate Fourier coefficients of paramodular eigenforms to Hecke eigenvalues at primes p for which $p^2 \mid N$. One consequence of this connection is an efficient algorithm for computing these Hecke eigenvalues. We also provide a rationality result analogous to that of Andrianov for the p-radial Fourier coefficients. The reader who is primarily interested in Siegel modular forms from a computational point of view can directly use these results, which appear in Part II of the book; however, the details of the proofs rely on the local nonarchimedean theory of Part I. Also in Part II, we have introduced several new operators on spaces of stable Klingen forms and Jacobi forms, which may be of independent interest.

This work was supported by a Collaboration Grant for Mathematicians from the Simons Foundation (PI Ralf Schmidt). We would like to thank the anonymous referees for a careful reading of the manuscript and many helpful comments.

Moscow, ID, USA — Jennifer Johnson-Leung
Moscow, ID, USA — Brooks Roberts
Denton, TX, USA — Ralf Schmidt
June 2023

Contents

Glossary of Notations

Roman Symbols

$\hat{a}(F)$	Fourier coefficient function of F, 300
$a(\chi)$	conductor of the character χ, 73
$a(S)$	Fourier coefficient of Siegel modular form, 249
A'	conjugate-inverse-transpose of the 2×2 matrix A, 71
$\mathbb{A}$	adeles of $\mathbb{Q}$, 247
$\mathbb{A}_{\rm fin}$	finite adeles of $\mathbb{Q}$, 247
$A[B]$	${}^tBAB$ for 2×2 matrices A and B, 247
$\mathcal{A}^\circ$	cuspidal adelic automorphic forms, 9
$\mathcal{A}_k$	adelic automorphic forms of weight k, 251
$\mathcal{A}_k^\circ$	cuspidal adelic automorphic forms of weight k, 251
$\mathcal{A}_k(\mathcal{K})$	weight k adelic automorphic forms with respect to $\mathcal{K}$, 252
$\mathcal{A}_k^\circ(\mathcal{K})$	cuspidal elements of $\mathcal{A}_k(\mathcal{K})$, 252
$A(N)$	paramodular Fourier coefficient indices, 248
$A(N)^+$	cuspidal paramodular Fourier coefficient indices, 248
$A(\mathbb{Q})$	positive semi-definite 2×2 matrices with entries from $\mathbb{Q}$, 300
$A(\mathbb{Q})^+$	positive definite 2×2 matrices with entries from $\mathbb{Q}$, 300
B	Borel subgroup of $\mathrm{GSp}(4, F)$, 74
$B(N)$	stable Klingen Fourier coefficient indices, 248
$B(N)^+$	cuspidal stable Klingen Fourier coefficient indices, 248
$c_{i,j}$	value of a local newform evaluated at $\Delta_{i,j}$, 182
$c(n,r)$	Fourier coefficient of a Jacobi form, 250
$d^I y$	Haar measure on $\mathrm{GSp}(4, F)$ with $m(I) = 1$, 224
d_K	projection using the parahoric subgroup K, 229
$\mathcal{D}(k)$	discrete series representation of weight k, 10
$e_i = T_{s_i}$	element of the Iwahori-Hecke algebra defined by s_i, 224
$e = T_1$	identity element of the Iwahori-Hecke algebra, 224
$\mathcal{F}$	vector space of $\mathbb{C}$ valued functions on $A(\mathbb{Q})$, 300
$F\big|_k g$	slash action on functions on $\mathcal{H}_2$, 246

$f\big\vert_{k,m}g$	slash action on functions on $\mathcal{H}_1 \times \mathbb{C}$, 250
$f_1^{\rm para}, f_2^{\rm para}$	basis elements of $(\chi_1 \times \chi_2 \rtimes \sigma)^{\mathrm{K}(\mathfrak{p})}$, 230
f_w	certain element of $(\chi_1 \times \chi_2 \rtimes \sigma)^I$ supported on $BIwI$, 227
$g\langle Z\rangle$	action of $\mathrm{GSp}(4, \mathbb{R})^+$ on Siegel upper half-space $\mathcal{H}_2$, 246
${}^{\rm t}g$	transpose of the matrix g, 71
G^J	Jacobi group, 81
$\mathrm{GL}(2, \mathbb{R})^+$	group of elements of $\mathrm{GL}(2, \mathbb{R})$ with positive determinant, 12
$\mathrm{GSp}(4, F)$	4×4 symplectic similitudes with entries from F, 72
$\mathrm{GSp}(4, R)$	4×4 symplectic similitudes with entries from R, 245
$\mathrm{GSp}(4, \mathbb{R})^+$	group of elements of $\mathrm{GSp}(4, \mathbb{R})$ with positive multiplier, 246
$\mathcal{H}_1$	complex upper half-plane, 13, 249
$\mathcal{H}_2$	Siegel upper half-space of degree 2, 246
$\mathcal{H}_n$	Siegel upper half-space of degree n, 34
$\mathcal{H}(\mathrm{GSp}(4, F), I)$	Iwahori-Hecke algebra, 224
$H(\mathbb{R})$	Heisenberg group, 249
$\mathbb{I}$	Ideles of $\mathbb{Q}$, 4
I	Iwahori subgroup of $\mathrm{GSp}(4, F)$, 221
$I = \left[\begin{smallmatrix} i & \\ & i\end{smallmatrix}\right]$	element of the Siegel upper half-space $\mathcal{H}_2$, 246
J	matrix defining GSp(4) (particular to Parts I, II), 72, 245
$j(g, Z)$	degree 2 factor of automorphy, 246
K_∞	circle group SO(2), 9
K_∞	maximal compact subgroup of $\mathrm{Sp}(4, \mathbb{R})$, 251
$\mathcal{K}$	compact, open subgroup of $\mathrm{GSp}(4, \mathbb{A}_{\rm fin})$, 252
$\mathrm{K}(\mathfrak{p}^n)$	local paramodular subgroup of level $\mathfrak{p}^n$, 86
$\mathrm{Kl}(\mathfrak{p}^n)$	local Klingen congruence subgroup of level $\mathfrak{p}^n$, 91
$\mathrm{K}_s(\mathfrak{p}^n)$	local stable Klingen congruence subgroup of level $\mathfrak{p}^n$, 92
$\mathrm{K1}_{s,1}(\mathfrak{p}^n)$	auxiliary congruence subgroup, 212
$\mathrm{K}(N)$	paramodular congruence subgroup of level N, 247
$\mathrm{K}_s(N)$	stable Klingen congruence subgroup of level N, 248
$\mathcal{K}(N)$	adelic paramodular congruence subgroup of level N, 253
$\mathcal{K}_s(N)$	adelic stable Klingen congruence subgroup of level N, 253
K_J	standard parahoric subgroup corresponding to J, 223
L_{c^2}	index lowering operator on Jacobi forms, 268
L'_p	index lowering operator on Jacobi forms, 273
$L(s, \pi)$	L-function of π, 85
$\mathrm{M}(n)$	alternative model for local paramodular vectors of level $\mathfrak{p}^n$, 207
$\mathrm{M}_\Delta(n)$	n-th triangular subspace of $\mathrm{M}_{\infty\times\infty}(\mathbb{C})$, 208
$\mathrm{M}_s(n)$	alternative model for stable Klingen vectors of level $\mathfrak{p}^n$, 205
$\mathrm{M}_0(k)$	subspace of $\mathrm{M}_\Delta(k)$, 209
M_i	certain element of $\mathrm{GSp}(4, \mathfrak{o})$, 123
$M_k(\Gamma)$	Siegel modular forms of weight k with respect to Γ, 248
$m(W)$	paramodular vector W in the alternative model, 205
N_π	paramodular level of π, 87
$N_{\pi,s}$	stable Klingen level of π, 95

$\bar{N}_{\pi,s}$	quotient stable Klingen level of π, 95
N_s	the integer $N \prod_{p\mid N} p^{-1}$, 248
N_τ	level of the local GL(2) representation τ, 132
P	Siegel parabolic subgroup of GSp(4, F), 75
P_0	certain polynomial, 186
P_3	subgroup of GL(3, F), 81
p_n	paramodularization map, 95
Q	Klingen parabolic subgroup of GSp(4, F), 76
$\bar{Q}$	subgroup of Klingen parabolic subgroup Q of GSp(4, F), 82
$Q(\nu^{\frac{1}{2}}\pi, \nu^{-\frac{1}{2}}\sigma)$	Saito-Kurokawa representation, 81
q_w	number of left I cosets in a disjoint decomposition of IwI, 225
R_{n-1}	certain auxiliary operator, 213
$s_0 = t_1$	Weyl group element for GSp(4, F), 222
s_1, s_2	Weyl group elements for GSp(4, F), 72
$S_k(\Gamma)$	Siegel modular cusp forms of weight k with respect to Γ, 248
$S_k(\mathrm{K}(M))_{\mathrm{old}}$	subspace of oldforms in $S_k(\mathrm{K}(M))$, 255
$S_k(\mathrm{K}(M))_{\mathrm{new}}$	subspace of newforms in $S_k(\mathrm{K}(M))$, 255
s_n	trace operator on stable Klingen vectors, 107
Sp(4, F)	4×4 symplectic isometries with entries from F, 72
Sp(4, R)	symplectic isometries with entries from R, 246
$S : V \to V$	auxiliary operator, 213
St	Steinberg representation, 22
$\mathcal{S}(X)$	Schwartz functions on X, 73
T_1	local Hecke operator for GL(2), 15
$T_{0,1}, T_{1,0}$	local paramodular Hecke operators, 88
$T^s_{0,1}, T^s_{1,0}$	local stable Klingen Hecke operators, 112
$T^s_{0,1}(p)$	global stable Klingen Hecke operator at p, 279
$T^s_{1,0}(p)$	global stable Klingen Hecke operator at p, 283
$T(1, 1, p, p)$	global paramodular Hecke operator at p, 256
$T(p, 1, p, p^2)$	global paramodular Hecke operator at p, 256
$T(l, m)$	GL(2) Hecke operator, 301
$T(n)$	GL(2) Hecke operator, 301
t_n	element of $\mathrm{K}(\mathfrak{p}^n)$, level raising operator, 93, 97
U	subgroup of the Borel subgroup of GSp(4, F), 75
U_p	index raising operator on Jacobi forms, 266
U_3	subgroup of P_3, 82
u_1	extended affine Weyl group element, 222
$u_1 = T_{u_1}$	element of the Iwahori-Hecke algebra, 224
u_n	Atkin-Lehner element of GSp(4, F), 88
$V_\infty(\ell)$	weight-ℓ subspace of V_∞, 10
$V^\vee$	contragredient of V, 73
$V_{N,\chi}$	Jacquet module, 74
$v_p(N)$	exponent of the power of p that divides N exactly, 247
$V(n)$	vectors in V fixed by $\mathrm{K}(\mathfrak{p}^n)$, 86

$V_s(n)$	vectors in V fixed by $\mathrm{K}_s(\mathfrak{p}^n)$, 94
$\bar{V}_s(n)$	quotient of $V_s(n)$ by $V(n-1)+V(n)$, 94
$V(\mathfrak{p}^n)$	vectors in V fixed by $\mathrm{K}(\mathfrak{p}^n)$ (global setting notation), 247
$V_s(\mathfrak{p}^n)$	vectors in V fixed by $\mathrm{K}_s(\mathfrak{p}^n)$ (global setting notation), 247
v_s	shadow of a given newform v_{new}, 151
V_{Z^J}	P_3-quotient of V, 82
W	Weyl group, 222
W^a	affine Weyl group, 222
W^a_J	subgroup of W^a generated by J, 223
W^e	extended affine Weyl group (Iwahori-Weyl group), 222
w_p	global paramodular Atkin-Lehner operator at p, 256
$\mathcal{W}(\pi,\psi_{c_1,c_2})$	Whittaker model of π, 77
Z	center of $\mathrm{GSp}(4,F)$, 72
Z^J	center of the Jacobi group, 82
$Z(s,W)$	zeta integral, 85
$Z_N(s,W)$	simplified zeta integral, 216

Greek and Other Symbols

$\Gamma_0(\mathfrak{p}^n)$	Hecke congruence subgroup of $\mathrm{GL}(2,F)$, 132
$\Gamma_0(N)$	Hecke congruence subgroup of $\mathrm{SL}(2,\mathbb{Z})$ of level N, 300
$\Gamma'_0(N)$	Klingen congruence subgroup of level N, 247
$\gamma(s,\pi)$	gamma factor of π, 85
Δ_t^+,Δ_t^-	operators on $\mathcal{F}$, 301
$\Delta_{i,j}$	certain diagonal element of $\mathrm{GSp}(4,F)$, 72
δ_P	modular character, 73
ε_π	eigenvalue of $\pi(u_{N_\pi})$ on newform in π, 88
$\varepsilon(s,\pi)$	epsilon factor of π, 85
η	local level raising operator, 86, 98, 206
η_p	global level raising operator at p induced by η, 254, 266
θ	local level raising operator, 86, 102, 206
θ_p	global level raising operator at p induced by θ, 254, 264
θ'	local level raising operator, 86
θ'_p	global level raising operator at p induced by θ', 254
$[\lambda,\mu,\kappa]$	element of the Heisenberg group, 249
$\lambda(g)$	similitude factor, 72, 245
λ_π	eigenvalue of $T_{0,1}$ on newform in π, 88
μ_π	eigenvalue of $T_{1,0}$ on newform in π, 88
ν	absolute value character, 73
$\pi^\vee$	contragredient of π, 73
σ_n,σ	local level lowering operator, 108, 206
σ_p	global level lowering operator at p induced by σ, 270

τ_n, τ	local level raising operator, 99, 186, 206
τ_p	global level raising operator at p induced by τ, 261
$\chi_1 \times \chi_2$	local representation induced from χ_1 and χ_2, 22
χ_p	local character of $\mathbb{Q}_p^\times$ associated to adelic character χ, 5
ψ_{c_1,c_2}	character of U, 77
Ω	subgroup of W^e, 222
ω_π	central character of π, 73
$\langle \cdot, \cdot \rangle$	Petersson inner product, 255
∇_t	operator on $\mathcal{F}$, 301

Chapter 1
Introduction

This work makes a contribution to the theory of automorphic forms on the general symplectic group GSp(4). In classical language, these are known as Siegel modular forms of degree 2. We prove new results about representations of GSp(4) over a local nonarchimedean field, and we apply this theory to Siegel modular forms defined with respect to the paramodular groups.

To put our work into context, we begin by recalling the more familiar situation of automorphic forms on GL(1) and GL(2) for which the classical objects are Dirichlet characters and elliptic modular forms. We then introduce the paramodular setting and summarize the results that are developed in this work.

1.1 Dirichlet Characters

One of the themes in any course about elementary number theory is modular arithmetic. Given a positive integer N, one learns that the quotient ring $\mathbb{Z}/N\mathbb{Z}$ is a field if and only if N is a prime number. Of particular interest are the unit groups $(\mathbb{Z}/N\mathbb{Z})^\times$, represented by those residue classes $x + N\mathbb{Z}$ for which x is relatively prime to N. If $N = \prod_{p|N} p^{n_p}$ is the prime factorization of N, then the Sun Zi Theorem implies that

$$(\mathbb{Z}/N\mathbb{Z})^\times \cong \prod_{p|N} (\mathbb{Z}/p^{n_p}\mathbb{Z})^\times. \tag{1.1}$$

If $\varphi(N)$ denotes the cardinality of $(\mathbb{Z}/N\mathbb{Z})^\times$, one thus obtains a multiplicative function, known as Euler's φ-function. It is not difficult to prove that $(\mathbb{Z}/N\mathbb{Z})^\times$ is cyclic if and only if N equals 4, p^m or $2p^m$, where p is an odd prime and m a non-negative integer.

J. Johnson-Leung et al., *Stable Klingen Vectors and Paramodular Newforms*,
Lecture Notes in Mathematics 2342, https://doi.org/10.1007/978-3-031-45177-5_1

A deeper question is whether, given x relatively prime to N, the class $x + N\mathbb{Z}$ contains infinitely many prime numbers. Dirichlet [35] has proved that this is the case, a statement known as the *Dirichlet Prime Number Theorem*. In order to do so, he considered homomorphisms $\eta : (\mathbb{Z}/N\mathbb{Z})^\times \to \mathbb{C}^\times$. These *Dirichlet characters mod N* form a group isomorphic to $(\mathbb{Z}/N\mathbb{Z})^\times$ itself (in a non-canonical way).

A basic observation is that if M and N are positive integers with $M \mid N$, then there is a natural map $\alpha_M^N : (\mathbb{Z}/N\mathbb{Z})^\times \to (\mathbb{Z}/M\mathbb{Z})^\times$, given by $x + N\mathbb{Z} \mapsto x + M\mathbb{Z}$. If η is a Dirichlet character mod M, we may thus lift it to a Dirichlet character mod N by precomposing with α_M^N. Any Dirichlet character mod N thus obtained from a Dirichlet character mod M for a proper divisor M of N is "old", because in some sense it already occurred at a lower level. The "new" Dirichlet characters mod N, i.e., those who do not come via this construction for any proper divisor M of N, are known as the *primitive* Dirichlet characters mod N. If σ is a Dirichlet character mod N, then the smallest positive divisor M of N for which there exists a (necessarily primitive) Dirichlet character η mod M such that $\sigma = \eta \circ \alpha_M^N$ is called the *conductor* of σ.

A Common Domain

A somewhat unsatisfactory aspect of Dirichlet characters is that they live on different groups. It would be algebraically more appealing if we could multiply a Dirichlet character η_1 mod N_1 and a Dirichlet character η_2 mod N_2. A possible attempt would be to lift both characters to level $N := \operatorname{lcm}(N_1, N_2)$ via the maps $\alpha_{N_1}^N$ and $\alpha_{N_2}^N$, and multiply the resulting characters mod N. However this is awkward; it would be more desirable to have a universal domain of definition for all Dirichlet characters.

Such a universal domain is provided by the concept of projective limit. The groups $(\mathbb{Z}/N\mathbb{Z})^\times$ form a projective system with respect to the natural maps $\alpha_M^N : (\mathbb{Z}/N\mathbb{Z})^\times \to (\mathbb{Z}/M\mathbb{Z})^\times$ for $M \mid N$. The projective limit

$$\hat{\mathbb{Z}}^\times := \varprojlim_N (\mathbb{Z}/N\mathbb{Z})^\times \tag{1.2}$$

is the subset of the direct product $\prod_N (\mathbb{Z}/N\mathbb{Z})^\times$ consisting of all elements (x_N) for which $\alpha_M^N(x_N) = x_M$ whenever $M \mid N$. There are surjective projection maps $\alpha_N : \hat{\mathbb{Z}}^\times \to (\mathbb{Z}/N\mathbb{Z})^\times$, and by definition the diagram

$$\begin{array}{ccc} \hat{\mathbb{Z}}^\times & \xrightarrow{\alpha_N} & (\mathbb{Z}/N\mathbb{Z})^\times \\ & \searrow^{\alpha_M} & \downarrow \alpha_M^N \\ & & (\mathbb{Z}/M\mathbb{Z})^\times \end{array} \tag{1.3}$$

is commutative for $M \mid N$. We give all groups $(\mathbb{Z}/N\mathbb{Z})^\times$ the discrete topology, $\prod_N (\mathbb{Z}/N\mathbb{Z})^\times$ the product topology, and $\hat{\mathbb{Z}}^\times$ the induced (subset) topology. Then $\hat{\mathbb{Z}}^\times$ is a compact topological group.

In general, by a *character* of a topological group, we mean a continuous homomorphism of the group into $\mathbb{C}^\times$. A Dirichlet character η mod M gives rise to the character $\eta \circ \alpha_M$ of $\hat{\mathbb{Z}}^\times$. If we first lift η to the Dirichlet character $\sigma = \eta \circ \alpha_M^N$ mod N, then σ gives rise to the same character of $\hat{\mathbb{Z}}^\times$, because the diagram

$$\begin{array}{ccccc} \hat{\mathbb{Z}}^\times & \xrightarrow{\alpha_N} & (\mathbb{Z}/N\mathbb{Z})^\times & & \\ & \searrow^{\alpha_M} & \downarrow{\scriptstyle \alpha_M^N} & \searrow^{\sigma} & \\ & & (\mathbb{Z}/M\mathbb{Z})^\times & \xrightarrow[\eta]{} & \mathbb{C}^\times \end{array} \tag{1.4}$$

is commutative. Thus, all "old" Dirichlet characters define the same character of $\hat{\mathbb{Z}}^\times$ as the primitive Dirichlet character they are coming from.

It is not difficult to see that the kernels $\mathcal{K}(N)$ of the maps α_N form a neighborhood basis of the identity of $\hat{\mathbb{Z}}^\times$ consisting of open subgroups. By a standard argument, a given character of $\hat{\mathbb{Z}}^\times$ is trivial on one of these kernels, hence factors through $(\mathbb{Z}/N\mathbb{Z})^\times$ for some N. The smallest N for which this is the case is called the *conductor* of the character. We thus obtain the following result.

Theorem 1.1.1 (Correspondence Theorem for Dirichlet Characters, First Version) *The characters of $\hat{\mathbb{Z}}^\times$ are in one-to-one correspondence with the primitive Dirichlet characters. More precisely, the characters of $\hat{\mathbb{Z}}^\times$ of conductor N are in one-to-one correspondence with the primitive Dirichlet characters mod N.*

We will see shortly that it is advantageous to build an inverse into the correspondence of Theorem 1.1.1, i.e., if η is a primitive Dirichlet character mod N, then the corresponding character of $\hat{\mathbb{Z}}^\times$ is $\eta^{-1} \circ \alpha_N$.

Characters of the Ideles

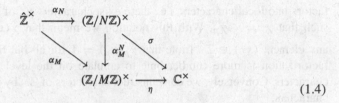

The isomorphism (1.1) may be interpreted as an isomorphism of projective systems. Taking projective limits on both sides, we obtain $\hat{\mathbb{Z}}^\times \cong \prod_p \mathbb{Z}_p^\times$ algebraically and topologically, where $\mathbb{Z}_p^\times = \varprojlim (\mathbb{Z}/p^n\mathbb{Z})^\times$ is the group of units of the p-adic integers $\mathbb{Z}_p = \varprojlim \mathbb{Z}/p^n\mathbb{Z}$. From now on we want to think of $\hat{\mathbb{Z}}^\times$ in this way, hence think of its elements as "vectors" $(x_2, x_3, x_5, \ldots)$ with $x_p \in \mathbb{Z}_p^\times$. The groups $\mathcal{K}(N)$ defined above become

$$\mathcal{K}(N) = \left(\prod_{p|N}(1 + p^{n_p}\mathbb{Z}_p)\right) \times \left(\prod_{p\nmid N}\mathbb{Z}_p^\times\right), \tag{1.5}$$

where $\prod p^{n_p}$ is the prime factorization of N.

An interesting consequence of $\hat{\mathbb{Z}}^\times = \prod_p \mathbb{Z}_p^\times$ is that any character χ of $\hat{\mathbb{Z}}^\times$ factors into local characters, i.e., there exist characters χ_p of $\mathbb{Z}_p^\times$, almost all trivial, such that $\chi = \otimes\chi_p$. With this notation we mean that $\chi(x) = \prod_p \chi_p(x_p)$ for any element $(x_p) \in \hat{\mathbb{Z}}^\times$ (note that $\chi_p(x_p) = 1$ for all but finitely many p). This factorization is more cumbersome to emulate on the level of classical Dirichlet characters. Conversely, we can construct characters of $\hat{\mathbb{Z}}^\times$ by piecing together local characters.

The transition from the multitude of groups $(\mathbb{Z}/N\mathbb{Z})^\times$ to their projective limit $\hat{\mathbb{Z}}^\times$ has thus streamlined the theory of Dirichlet characters in several ways: It has created a common domain of definition, so that characters can easily be multiplied; it has put into focus the primitive Dirichlet characters and relegated non-primitive ones into the background; and it has created a natural way to factorize characters into local objects.

To fully realize the power of this localization, it turns out to be advantageous to go one step further and introduce the concept of the *ideles*. For a prime p let $\mathbb{Q}_p$ be the field of p-adic numbers. The ring of *finite adeles* $\mathbb{A}_{\text{fin}}$ is the restricted direct product of the $\mathbb{Q}_p$ with respect to the $\mathbb{Z}_p$, meaning those elements (x_p) of the direct product for which $x_p \in \mathbb{Z}_p$ for almost all p. The topology is generated by the subsets $\prod U_p$, where U_p is open in $\mathbb{Q}_p$ and $U_p = \mathbb{Z}_p$ for almost all p. It is designed in such a way to make $\mathbb{A}_{\text{fin}}$ a locally compact topological ring. The ring of *adeles* is $\mathbb{A} = \mathbb{R} \times \mathbb{A}_{\text{fin}}$; hence, it includes the archimedean completion $\mathbb{R}$ of $\mathbb{Q}$. We also write $\mathbb{R} = \mathbb{Q}_\infty$ in this context. The rational numbers $\mathbb{Q}$ are diagonally embedded into $\mathbb{A}$ as a closed and discrete subgroup.

The *finite ideles* $\mathbb{I}_{\text{fin}}$ are, analogously, the restricted direct product of the $\mathbb{Q}_p^\times$ with respect to the $\mathbb{Z}_p^\times$. The topology is generated by the subsets $\prod U_p$, where U_p is open in $\mathbb{Q}_p^\times$ and $U_p = \mathbb{Z}_p^\times$ for almost all p, making $\mathbb{I}_{\text{fin}}$ a locally compact topological group. The group of *ideles* is $\mathbb{I} = \mathbb{R}^\times \times \mathbb{I}_{\text{fin}}$. We diagonally embed $\mathbb{Q}^\times$ into $\mathbb{I}$ as a closed and discrete subgroup.

It is not difficult to prove that there is an internal direct product decomposition

$$\mathbb{I} = \mathbb{Q}^\times \times \mathbb{R}^+ \times \hat{\mathbb{Z}}^\times, \tag{1.6}$$

where $\mathbb{R}^+$ is the group of all ideles of the form $(x, 1, 1, \ldots)$ with x a positive real number. The characters of $\hat{\mathbb{Z}}^\times$ can thus be identified with characters of $\mathbb{I}$ which are trivial on $\mathbb{Q}^\times$ and on $\mathbb{R}^+$. Characters of $\mathbb{I}$ which are trivial on $\mathbb{Q}^\times$ are called *idele class characters*. The characters of $\mathbb{R}^+$ are of the form $x \mapsto x^s$ for some $s \in \mathbb{C}$. We may thus think of the characters of $\hat{\mathbb{Z}}^\times$ as the idele class characters of finite order. Theorem 1.1.1 may therefore be reformulated as follows.

Theorem 1.1.2 (Correspondence Theorem for Dirichlet Characters, Second Version) *The idele class characters of finite order are in one-to-one correspondence with the primitive Dirichlet characters. Idele class characters of conductor N correspond to primitive Dirichlet characters mod N.*

Here, we define the *conductor* of a character of $\mathbb{I}$ as the conductor of its restriction to $\hat{\mathbb{Z}}^\times$, i.e., as the smallest positive integer N for which the character is trivial on the group $\mathcal{K}(N)$ in (1.5). If η is a primitive Dirichlet character mod N, where N has prime factorization $N = \prod p^{n_p}$, then the corresponding character χ of $\mathbb{Q}^\times\backslash\mathbb{I}$ is defined as the composition

$$\mathbb{I} \longrightarrow \hat{\mathbb{Z}}^\times \longrightarrow \prod_{p|N} \mathbb{Z}_p^\times \longrightarrow \prod_{p|N} \mathbb{Z}_p^\times/(1+p^{n_p}\mathbb{Z}_p)$$

$$\xrightarrow{\sim} \prod_{p|N} (\mathbb{Z}/p^{n_p}\mathbb{Z})^\times \xrightarrow{\sim} (\mathbb{Z}/N\mathbb{Z})^\times \xrightarrow{\eta^{-1}} \mathbb{C}^\times. \tag{1.7}$$

The reason for the inverse in the last step will become clear momentarily.

L-Functions

The idele class character χ of $\mathbb{I}$ obtained as the composition (1.7) also factors into local characters, i.e., there exist characters χ_p of $\mathbb{Q}_p^\times$ (including the case $p = \infty$) such that $\chi = \otimes\chi_p$. Specifically, $\chi_p : \mathbb{Q}_p^\times \to \mathbb{C}^\times$ is defined by $\chi_p(y) = \chi(Y)$ for $Y = (\dots, 1, y, 1, \dots) \in \mathbb{I}$ with y in the p-th place and ones elsewhere. This time the characters χ_p for primes $p \nmid N$ are not necessarily trivial but they are *unramified*, meaning trivial on $\mathbb{Z}_p^\times$, and the notation means that $\chi(x) = \prod_{p\le\infty} \chi_p(x_p)$ for $x = (x_p) \in \mathbb{I}$ (note that $\chi_p(x_p) = 1$ for all but finitely many p).

Since $\mathbb{Q}_p^\times = \langle p\rangle \times \mathbb{Z}_p^\times$, the unramified χ_p are determined by the single value $\chi_p(p) \in \mathbb{C}^\times$, known as a *Satake parameter*. Using (1.7) and $\chi(p) = 1$, where $p = (p, p, \dots)$, we see that

$$\chi_p(p) = \eta(p) \qquad \text{for primes } p \nmid N, \tag{1.8}$$

by the choice of η^{-1} and not η in (1.7). Dirichlet defined the *L-function*

$$L(s, \eta) := \prod_{p\nmid N} \frac{1}{1-\eta(p)p^{-s}}, \tag{1.9}$$

where s is a complex parameter. The product is absolutely convergent for $\mathrm{Re}(s) > 1$. Dirichlet used the analytic properties of these functions to prove his prime number theorem. We may similarly attach an L-function to the corresponding idelic character $\chi = \otimes \chi_p$ by setting

$$L(s, \chi) := \prod_{p \nmid N} \frac{1}{1 - \chi_p(p) p^{-s}}. \tag{1.10}$$

Then $L(s, \chi) = L(s, \eta)$ by (1.8). A conceptual advantage of the definition (1.10) over (1.9) is that the individual *Euler factors* only depend on the local objects χ_p. Hence, one can make a purely local definition

$$L_p(s, \chi_p) := \begin{cases} \dfrac{1}{1 - \chi_p(p) p^{-s}} & \text{if } \chi_p \text{ is unramified,} \\ 1 & \text{if } \chi_p \text{ is ramified,} \end{cases} \tag{1.11}$$

for any character χ_p of $\mathbb{Q}_p^\times$, and then $L(s, \chi) = \prod_{p<\infty} L_p(s, \chi_p)$. This possibility of giving the L-function a place-by-place definition is one of the reasons why we transitioned from $\hat{\mathbb{Z}}^\times$ to $\mathbb{I}$. It would not have been possible with the factorization of the character χ of $\hat{\mathbb{Z}}^\times$ from Theorem 1.1.1, whose local components are almost everywhere trivial. The local components outside N of the idelic character χ, on the other hand, carry the additional information given by their Satake parameters.

A further reason for the introduction of the ideles comes from the archimedean place. Since in the first step of the chain (1.7) we are forgetting the factor $\mathbb{R}^+$ appearing in (1.6), the archimedean component χ_∞ of χ is trivial on $\mathbb{R}^+$. Hence, χ_∞ can only be the trivial character or the sign character (sgn) of $\mathbb{R}^\times$. Using (1.7) and $\chi(-1) = 1$, where $-1 = (-1, -1, \ldots)$, we see that

$$\chi_\infty = \begin{cases} 1 & \text{if } \eta(-1) = 1, \\ \mathrm{sgn} & \text{if } \eta(-1) = -1. \end{cases} \tag{1.12}$$

Dirichlet characters η with $\eta(-1) = 1$ are called *even*, and those with $\eta(-1) = -1$ are called *odd*. Thus, χ_∞ carries the information about η being even or odd.

There is a local factor attached to characters at the archimedean place as well. We set

$$L_\infty(s, \chi_\infty) := \begin{cases} \pi^{-s/2} \Gamma\left(\dfrac{s}{2}\right) & \text{if } \chi_\infty = 1, \\ \pi^{-(s+1)/2} \Gamma\left(\dfrac{s+1}{2}\right) & \text{if } \chi_\infty = \mathrm{sgn}, \end{cases} \tag{1.13}$$

where Γ is the usual Γ-function. The *complete* L-function is then $\Lambda(s, \chi) = \prod_{p \le \infty} L_p(s, \chi_p)$. It is known that $\Lambda(s, \chi)$ admits a meromorphic continuation to all of $\mathbb{C}$ and satisfies the functional equation

$$\Lambda(s,\chi) = \varepsilon N^{-s+1/2} \Lambda(1-s,\bar{\chi}), \tag{1.14}$$

where ε is a constant, depending on χ, of absolute value 1. If $N = 1$, then χ is the trivial character, $L(s,\chi)$ is the Riemann zeta function, and $\Lambda(s,\chi)$ has simple poles at $s = 0$ and $s = 1$. In all other cases $\Lambda(s,\chi)$ is an entire function.

Strong Multiplicity One

The following result, whose proof is based on the Dirichlet Prime Number Theorem, states that idele class characters are determined by all but finitely many of their local components.

Theorem 1.1.3 (Strong Multiplicity One for GL(1)**)** *Let* $\chi = \otimes\chi_p$ *and* $\sigma = \otimes\sigma_p$ *be idele class characters such that* $\chi_p = \sigma_p$ *for almost all places* p*. Then* $\chi = \sigma$*.*

Proof It is enough to show that if $\chi = \otimes\chi_p$ is an idele class character such that $\chi_p = 1$ for almost all p, then $\chi = 1$. For simplicity, let us assume that χ has finite order; the extension to the general case is an exercise.

Since χ has finite order, it corresponds to a primitive Dirichlet character η mod N via (1.7). By (1.8) and the hypothesis, $\eta(p) = 1$ for almost all primes p. Since every class $x + N\mathbb{Z}$ with $\gcd(x, N) = 1$ contains infinitely many primes, it follows that $\eta = 1$. Hence $\chi = 1$, concluding the proof. □

As a consequence of Theorem 1.1.3, an idele class character $\chi = \otimes\chi_p$ is determined by its *incomplete L-function*

$$L^S(s,\chi) = \prod_{p\notin S} L_p(s,\chi_p), \tag{1.15}$$

for any finite set S of primes. By the correspondence theorem, a primitive Dirichlet character η is then determined by the incomplete L-function

$$L^S(s,\eta) = \prod_{\substack{p\nmid N\\ p\notin S}} \frac{1}{1-\eta(p)p^{-s}} \tag{1.16}$$

for any finite set S of primes. Clearly, this last statement would be wrong if we allowed non-primitive Dirichlet characters.

Given an incomplete L-function as in (1.15), there is only one way to "fill in" Euler factors at the missing primes, and to complete with appropriate Γ-factors at the archimedean place, so as to obtain a nice functional equation as in (1.14). For idele class characters of finite order this is an easy task; see (1.11) and (1.13). In particular, the Euler factors at the "bad" primes, i.e., those primes dividing the conductor, are all 1. For automorphic forms on groups of higher rank, knowing the "bad" Euler factors can be a serious issue. One of the main motivations for the present work is to address this problem for *paramodular forms*, which are automorphic forms on GSp(4).

Concluding Remarks

The above discussion shows that considering Dirichlet characters as characters of $\mathbb{I}$ brings with it several advantages, among which are the natural factorizations of the characters themselves and of their L-functions. We mention one more benefit of this point of view. Local groups like $\mathbb{R}$, $\mathbb{R}^\times$, $\mathbb{Q}_p$, $\mathbb{Q}_p^\times$, $\mathbb{Z}_p$, $\mathbb{Z}_p^\times$, as well as global groups like $\mathbb{A}$, $\mathbb{Q}\backslash\mathbb{A}$, $\mathbb{I}$, $\mathbb{Q}^\times\backslash\mathbb{I}$, are locally compact topological groups. As such they admit Haar measures and a theory of integration, opening up the possibility of Fourier analysis on these groups. This point of view was taken by Tate [126], who used harmonic analysis on local and global groups to derive functional equations like the one in (1.14).

Note that the ideles are, algebraically at least, the unit group of the ring $\mathbb{A}$ of adeles. We may thus think of $\mathbb{I}$ as $\mathrm{GL}(1,\mathbb{A})$, the *adelization* of the algebraic group GL(1). This is also true topologically if we agree to give an adelized group like $\mathrm{GL}(1,\mathbb{A})$ its natural restricted direct product topology (as opposed to the subset topology inherited from the surrounding matrix space). A character of $\mathrm{GL}(1,\mathbb{A})$ is nothing but a 1-dimensional, hence irreducible, representation of this group. The characters we considered had the special property that they were trivial on the subgroup of rational points $\mathbb{Q}^\times = \mathrm{GL}(1,\mathbb{Q})$. Functions on $\mathrm{GL}(1,\mathbb{A})$ which are invariant under $\mathrm{GL}(1,\mathbb{Q})$ and satisfy some additional regularity conditions are called *automorphic forms*. Our characters are thus *automorphic representations*, meaning representations of $\mathrm{GL}(1,\mathbb{A})$ in the space of automorphic forms. Since this turned out to be a successful theory, it is promising to consider automorphic forms on other algebraic groups. The next group to examine is GL(2). The classical objects corresponding to automorphic forms on $\mathrm{GL}(2,\mathbb{A})$ are (elliptic) modular forms, which we consider next.

1.2 Modular Forms I

In this section we start with automorphic forms on the adelic group $\mathrm{GL}(2,\mathbb{A})$, see how their "descent" to functions on the upper half plane gives rise to the notion of elliptic modular forms, and explain how this leads to a correspondence between automorphic representations and classical newforms.

Automorphic Forms on GL(2)

If p is a prime number, then $\mathrm{GL}(2, \mathbb{Z}_p)$ is a maximal compact subgroup of $\mathrm{GL}(2, \mathbb{Q}_p)$, and in analogy to the case of the ideles, we define $\mathrm{GL}(2, \mathbb{A}_{\mathrm{fin}})$ to be the restricted direct product, over all primes, of the $\mathrm{GL}(2, \mathbb{Q}_p)$ with respect to the $\mathrm{GL}(2, \mathbb{Z}_p)$. The adelized group GL(2) is $\mathrm{GL}(2, \mathbb{A}) := \mathrm{GL}(2, \mathbb{R}) \times \mathrm{GL}(2, \mathbb{A}_{\mathrm{fin}})$. We have $\mathrm{GL}(2, \mathbb{Q})$ diagonally embedded into $\mathrm{GL}(2, \mathbb{A})$ as a closed and discrete subgroup. The center $Z(\mathbb{A})$ of $\mathrm{GL}(2, \mathbb{A})$ consists of all scalar matrices, and identifies algebraically and topologically with the ideles $\mathbb{I}$.

A continuous function $\Phi : \mathrm{GL}(2, \mathbb{A}) \to \mathbb{C}$ is called an *automorphic form* if $\Phi(\rho g) = \Phi(g)$ for all $\rho \in \mathrm{GL}(2, \mathbb{Q})$ and $g \in \mathrm{GL}(2, \mathbb{A})$, and if it satisfies additional regularity conditions. (The standard reference for automorphic forms on any group is [17].) For simplicity, we will require all automorphic forms to be invariant under the center, i.e., $\Phi(zg) = \Phi(g)$ for all $z \in Z(\mathbb{A})$ and $g \in \mathrm{GL}(2, \mathbb{A})$. If the automorphic form Φ satisfies

$$\int_{\mathbb{Q}\backslash\mathbb{A}} \Phi(\left[\begin{smallmatrix} 1 & x \\ & 1 \end{smallmatrix}\right] g)\, dx = 0 \tag{1.17}$$

for all $g \in \mathrm{GL}(2, \mathbb{A})$, then Φ is called a *cuspidal automorphic form*. In essence, the cuspidality condition precludes the possibility that Φ is obtained from Dirichlet characters via a standard construction known as *Eisenstein series*. In the following we will concentrate on the space $\mathcal{A}^\circ$ of cuspidal automorphic forms.

The group $\mathrm{GL}(2, \mathbb{A}_{\mathrm{fin}})$ acts on $\mathcal{A}^\circ$ by right translations, i.e., $(g \cdot \Phi)(h) = \Phi(hg)$ for $g \in \mathrm{GL}(2, \mathbb{A}_{\mathrm{fin}})$, $h \in \mathrm{GL}(2, \mathbb{A})$ and $\Phi \in \mathcal{A}^\circ$. The group $\mathrm{GL}(2, \mathbb{R})$ does not act in the same way, since right translation by arbitrary elements of this group does not preserve all the regularity conditions. What does still act is the circle subgroup $K_\infty := \mathrm{SO}(2)$ consisting of all rotation matrices

$$r(\theta) = \left[\begin{smallmatrix} \cos\theta & \sin\theta \\ -\sin\theta & \cos\theta \end{smallmatrix}\right] \tag{1.18}$$

with $\theta \in \mathbb{R}/2\pi\mathbb{Z}$, and the Lie algebra $\mathfrak{g}$ of $\mathrm{GL}(2, \mathbb{R})$ via the derived representation. The space $\mathcal{A}^\circ$ is thus a $(\mathfrak{g}, K_\infty) \times \mathrm{GL}(2, \mathbb{A}_{\mathrm{fin}})$ module. It is common to (slightly incorrectly) refer to a complex vector space with an action of $(\mathfrak{g}, K_\infty) \times \mathrm{GL}(2, \mathbb{A}_{\mathrm{fin}})$ as a "representation of $\mathrm{GL}(2, \mathbb{A})$".

One can show that elements of $\mathcal{A}^\circ$ are square-integrable modulo $\mathrm{GL}(2, \mathbb{Q})Z(\mathbb{A})$, and there is a natural inner product on $\mathcal{A}^\circ$ given by

$$\langle \Phi_1, \Phi_2 \rangle = \int_{\mathrm{GL}(2,\mathbb{Q})Z(\mathbb{A})\backslash\mathrm{GL}(2,\mathbb{A})} \Phi_1(g)\overline{\Phi_2(g)}\, dg. \tag{1.19}$$

Evidently, the inner product is invariant with respect to right translations of the functions Φ_1 and Φ_2 by the same element.

It is known that $\mathcal{A}^\circ$ decomposes into an orthogonal direct sum of irreducible representations, each occurring with multiplicity one; see Proposition 11.1.1 of [68]. These irreducibles are called *cuspidal automorphic representations*. If π is one of them, then a general result (see [41]) states that there is an isomorphism of $(\mathfrak{g}, K_\infty) \times \mathrm{GL}(2, \mathbb{A}_{\mathrm{fin}})$ modules $\pi \cong \otimes_{p\le\infty}\pi_p$, where π_p for $p < \infty$ is an irreducible *admissible* representation of $\mathrm{GL}(2, \mathbb{Q}_p)$, and π_∞ is an irreducible $(\mathfrak{g}, K_\infty)$ module. Of course one needs to make sense out of the infinite tensor product; for many purposes one can take the notation $\pi \cong \otimes\pi_p$ to simply mean that π determines a collection of local representations π_p, with compatible actions of the local groups.

Distinguished Vectors

Let (π, V) be a cuspidal automorphic representation of $\mathrm{GL}(2, \mathbb{A})$, with factorization $\pi \cong \otimes\pi_p$. It turns out that V is always infinite-dimensional. In fact, each of the local representations (π_p, V_p) is infinite-dimensional. It would be desirable to exhibit in each π_p a distinguished vector v_p, at least up to multiples. Then $\otimes v_p$ would be a distinguished vector in $\otimes V_p$, corresponding to a distinguished automorphic form $\Phi_\pi \in V$. The hope is that the single vector Φ_π still carries essential information about the infinite-dimensional representation π.

It is indeed possible to find such distinguished $v_p \in V_p$, at each place p. Consider first the archimedean place. We decompose V_∞ according to the action of K_∞, i.e., we write $V_\infty = \bigoplus_{\ell\in\mathbb{Z}} V_\infty(\ell)$, where $V_\infty(\ell)$ is the subspace of all vectors $v \in V_\infty$ satisfying

$$\pi_\infty(r(\theta))v = e^{i\ell\theta}\, v \tag{1.20}$$

for $\theta \in \mathbb{R}$. The irreducibility of π_∞ implies that each $V_\infty(\ell)$ is at most one-dimensional; if it is one-dimensional, then we say that the *weight* ℓ occurs in π_∞. As a consequence of our assumption that the center acts trivially, only even weights can occur.

A classification of all $(\mathfrak{g}, K_\infty)$ modules shows that they come in two kinds. The first kind consists of those modules in which all even weights occur; they are called *principal series representations*, and they constitute an infinite, continuous family. The second kind consists of modules with weights $\{\pm k, \pm(k+2), \ldots\}$ for an even, positive integer k; they are called *discrete series representations*. Up to isomorphism there is exactly one discrete series representation for each k, which we denote by $\mathcal{D}(k)$.

Now if π_∞ is a principal series representation, we let v_∞ be a non-zero vector of weight 0. If $\pi_\infty = \mathcal{D}(k)$, then we let v_∞ be a non-zero vector of weight k. Thus, we have singled out a distinguished vector at the archimedean place, unique up to multiples.

The complexified Lie algebra $\mathfrak{g}_{\mathbb{C}}$ contains a particular element X_- acting as a *weight lowering operator*, meaning that if $v \in V_\infty$ has weight ℓ, then $\pi_\infty(X_-)v$ has weight $\ell - 2$. One can show that on a principal series representation $\pi_\infty(X_-)$ has no kernel. For $\pi_\infty = \mathcal{D}(k)$, a discrete series representation of lowest weight k, the weight k space is annihilated by $\pi_\infty(X_-)$, but no other weight space is. In the following let us assume that the archimedean component of our cuspidal automorphic representation $\pi \cong \otimes \pi_p$ is $\mathcal{D}(k)$. Then v_∞, chosen to be a weight k vector, satisfies

$$\pi_\infty(X_-)v_\infty = 0, \tag{1.21}$$

and this property characterizes v_∞ inside V_∞ up to multiples.

Next consider π_p for a prime number p. Let $\mathfrak{p} = p\mathbb{Z}_p$ be the maximal ideal of $\mathbb{Z}_p$. For a non-negative integer n, let

$$\Gamma_0(\mathfrak{p}^n) = \left\{\left[\begin{smallmatrix} a & b \\ c & d \end{smallmatrix}\right] \in \mathrm{GL}(2, \mathbb{Z}_p) \mid c \in \mathfrak{p}^n\right\}. \tag{1.22}$$

Let $V_p(n) = \{v \in V_p \mid \pi_p(g)v = v \text{ for all } g \in \Gamma_0(\mathfrak{p}^n)\}$. Since the groups $\Gamma_0(\mathfrak{p}^n)$ become smaller with increasing n, their fixed spaces become bigger, i.e., $V_p(0) \subset V_p(1) \subset \ldots$. It is a non-trivial fact, implied by Proposition 2.9 of [68], that $V_p(n) \neq 0$ for large enough n. Let N_p be the minimal n such that $V_p(n) \neq 0$; this is the *conductor exponent*, or *minimal level*, of π_p. Then, by the results of [27],

$$\dim V_p(n) = \begin{cases} 0 & \text{for } n < N_p, \\ n - N_p + 1 & \text{for } n \geq N_p. \end{cases} \tag{1.23}$$

In particular, $\dim V_p(N_p) = 1$, a statement known as *local multiplicity one*. Any vector spanning this one-dimensional space is known as a *local newform*. Vectors in $V_p(n)$ for $n > N_p$ are *local oldforms*. (A local newform, when viewed as an element of $V_p(n)$ for $n > N_p$, is a local oldform!) We choose our distinguished vector $v_p \in V_p$ to be a local newform.

Part of the regularity conditions of representations of $\mathrm{GL}(2, \mathbb{A})$, whose details we suppressed, is that if $\pi \cong \otimes \pi_p$, then almost every π_p must admit non-zero vectors invariant under the maximal compact subgroup $\mathrm{GL}(2, \mathbb{Z}_p)$. In other words, $N_p = 0$ for almost all p. It therefore makes sense to define

$$N_\pi = \prod_p p^{N_p}. \tag{1.24}$$

This positive integer is the (global) *conductor* of π. Local representations with conductor exponent 0 are called *spherical* or *unramified*. Hence, any such π is spherical for finite primes p not dividing N_π. Our distinguished vectors v_p for primes $p \nmid N_\pi$ are *spherical vectors*, meaning non-zero vectors invariant under $\mathrm{GL}(2, \mathbb{Z}_p)$.

Descent to the Upper Half Plane

We continue to let (π, V) be a cuspidal, automorphic representation of $\mathrm{GL}(2, \mathbb{A})$, and $\pi \cong \otimes\pi_p$ with local representations (π_p, V_p) and $\pi_\infty = \mathcal{D}(k)$. Above we singled out a distinguished vector v_p, or rather a distinguished line, in each local representation space V_p. Let $\Phi_\pi : \mathrm{GL}(2, \mathbb{A}) \to \mathbb{C}$ be the automorphic form corresponding to $\otimes v_p$. We call Φ_π, which is determined up to multiples, a *global newform*. For any integer N with prime factorization $N = \prod p^{n_p}$, let

$$\mathcal{K}_0(N) := \prod_{p<\infty} \Gamma_0(\mathfrak{p}^{n_p}). \tag{1.25}$$

Then $\mathcal{K}_0(N)$ is an open-compact subgroup of $\mathrm{GL}(2, \mathbb{A}_{\mathrm{fin}})$. Let $\mathrm{GL}(2, \mathbb{R})^+$ be the subgroup of elements of $\mathrm{GL}(2, \mathbb{R})$ with positive determinant. We define $\mathcal{A}_k^\circ(\mathcal{K}_0(N))$ to be the space of all functions $\Phi : \mathrm{GL}(2, \mathbb{A}) \to \mathbb{C}$ with the following properties.

1. Φ is an automorphic form, in particular $\Phi(\rho g) = \Phi(g)$ for all $\rho \in \mathrm{GL}(2, \mathbb{Q})$ and $g \in \mathrm{GL}(2, \mathbb{A})$;
2. $\Phi(gz) = \Phi(g)$ for all $z \in Z(\mathbb{A})$ and $g \in \mathrm{GL}(2, \mathbb{A})$;
3. For the compact, open subgroup $\mathcal{K}_0(N) := \prod_{p<\infty} \Gamma_0(\mathfrak{p}^{N_p})$ of $\mathrm{GL}(2, \mathbb{A}_{\mathrm{fin}})$, we have
$$\Phi(g\kappa) = \Phi(g) \qquad \text{for all } g \in \mathrm{GL}(2, \mathbb{A}) \text{ and } \kappa \in \mathcal{K}_0(N); \tag{1.26}$$
4. Φ has weight k, i.e.,
$$\Phi(gr(\theta)) = e^{ik\theta}\Phi(g) \qquad \text{for all } g \in \mathrm{GL}(2, \mathbb{A}) \text{ and } \theta \in \mathbb{R}/2\pi\mathbb{Z}; \tag{1.27}$$
5. For any $g_{\mathrm{fin}} \in \mathrm{GL}(2, \mathbb{A}_{\mathrm{fin}})$, the function $\mathrm{GL}(2, \mathbb{R})^+ \to \mathbb{C}$ defined by $g \mapsto \Phi(g_{\mathrm{fin}} g)$ is smooth and is annihilated by the element X_- of $\mathfrak{g}_{\mathbb{C}}$;
6. Φ satisfies the cuspidality condition (1.17).

By construction, the global newform Φ_π is an element of the space $\mathcal{A}_k^\circ(\mathcal{K}_0(N_\pi))$, where N_π is the conductor defined in (1.24).

Strong approximation for SL(2) is the theorem that $\mathrm{SL}(2, \mathbb{Q})$ is dense in $\mathrm{SL}(2, \mathbb{A}_{\mathrm{fin}})$. Together with (1.6) it implies that

$$\mathrm{GL}(2, \mathbb{A}) = \mathrm{GL}(2, \mathbb{Q}) \cdot \mathrm{GL}(2, \mathbb{R})^+ \mathcal{K}_0(N), \tag{1.28}$$

for any positive integer N. Observe that the factorization on the right-hand side is not unique:

$$\mathrm{GL}(2, \mathbb{Q}) \cap \mathrm{GL}(2, \mathbb{R})^+ \mathcal{K}_0(N) = \Gamma_0(N), \tag{1.29}$$

where

$$\Gamma_0(N) = \{\left[\begin{smallmatrix} a & b \\ c & d \end{smallmatrix}\right] \in \mathrm{SL}(2,\mathbb{Z}) \mid c \equiv 0 \bmod N\} \tag{1.30}$$

is a classical congruence subgroup of $\mathrm{SL}(2,\mathbb{Z})$.

Let $\Phi \in \mathcal{A}_k(\mathcal{K}_0(N))$. By (1.28), the left invariance of Φ under $\mathrm{GL}(2,\mathbb{Q})$, and the right invariance property (1.26), we see that Φ is determined by its restriction to $\mathrm{GL}(2,\mathbb{R})^+$. In view of the property (1.27), and because Φ is invariant under the center, Φ is in fact determined on a set of representatives for the cosets $\mathrm{GL}(2,\mathbb{R})^+/Z(\mathbb{R})\mathrm{SO}(2)$.

Recall that $\mathrm{GL}(2,\mathbb{R})^+$ acts transitively on the complex upper half plane $\mathcal{H}_1$ via

$$g\langle z\rangle = \frac{az+b}{cz+d} \tag{1.31}$$

for $g = \left[\begin{smallmatrix} a & b \\ c & d \end{smallmatrix}\right] \in \mathrm{GL}(2,\mathbb{R})^+$ and $z \in \mathcal{H}_1$. The stabilizer of the point i is $Z(\mathbb{R})\mathrm{SO}(2)$, so that $\mathrm{GL}(2,\mathbb{R})^+/Z(\mathbb{R})\mathrm{SO}(2) \cong \mathcal{H}_1$. As a set of representatives for $\mathrm{GL}(2,\mathbb{R})^+/Z(\mathbb{R})\mathrm{SO}(2) \cong \mathcal{H}_1$ we may take the elements $\left[\begin{smallmatrix} y & x \\ & 1 \end{smallmatrix}\right]$, where $x \in \mathbb{R}$ and $y \in \mathbb{R}^+$. Note that $\left[\begin{smallmatrix} y & x \\ & 1 \end{smallmatrix}\right]\langle \mathrm{i}\rangle = x + \mathrm{i}y$.

Thus Φ is determined by the function on $\mathcal{H}_1$ given by $x + \mathrm{i}y \mapsto \Phi(\left[\begin{smallmatrix} y & x \\ & 1 \end{smallmatrix}\right])$. However we work in an additional factor $y^{k/2}$ and define a function $f : \mathcal{H}_1 \to \mathbb{C}$ by

$$f(x+\mathrm{i}y) = y^{-k/2}\,\Phi(\left[\begin{smallmatrix} y & x \\ & 1 \end{smallmatrix}\right]) \tag{1.32}$$

for $x \in \mathbb{R}$ and $y \in \mathbb{R}^+$. The reason for this factor is that it leads to nicer formulas. There is a right action of $\mathrm{GL}(2,\mathbb{R})^+$ on functions $f : \mathcal{H}_1 \to \mathbb{C}$ given by

$$(f\big|_k g)(z) = \det(g)^{k/2}(cz+d)^{-k} f(g\langle z\rangle) \tag{1.33}$$

for $g = \left[\begin{smallmatrix} a & b \\ c & d \end{smallmatrix}\right] \in \mathrm{GL}(2,\mathbb{R})^+$ and $z \in \mathcal{H}_1$. The normalizing factor $\det(g)^{k/2}$ is designed in such a way that the center acts trivially. For our particular f defined in (1.32) one verifies easily that

$$\Phi(g) = (f\big|_k g)(\mathrm{i}) \tag{1.34}$$

for all $g \in \mathrm{GL}(2,\mathbb{R})^+$.

To every $\Phi \in \mathcal{A}_k(\mathcal{K}_0(N))$ thus corresponds a function f on the upper half plane. Let f_π be the function $\mathcal{H}_1 \to \mathbb{C}$ corresponding to the distinguished vector $\Phi_\pi \in V_\pi$. This f_π gives us a concrete handle on the infinite-dimensional representation π of $\mathrm{GL}(2,\mathbb{A})$. We will next investigate the properties of f_π.

Modular Forms

Let N be a positive integer, and let $\Phi \in \mathcal{A}_k(\mathcal{K}_0(N))$. Let f be the corresponding function $\mathcal{H}_1 \to \mathbb{C}$, related to Φ by the Eq. (1.34). If we take a γ from the group $\Gamma_0(N)$ defined in (1.30), replace g in (1.34) by γg, and use (1.26), we see that

$$f\big|_k\gamma = f \qquad \text{for all } \gamma \in \Gamma_0(N). \tag{1.35}$$

This is the typical transformation property of a (elliptic) *modular form of weight k and level N*. Modular forms are defined in many references; see for example [86, 118]. In order to qualify as a modular form, f also needs to be holomorphic on $\mathcal{H}_1$ and "at the cusps".

Our function f is holomorphic on $\mathcal{H}_1$ because Φ is annihilated by X_-. When transferred to the upper half plane, this condition becomes $\frac{\partial}{\partial \bar{z}} f = 0$, exhibiting f as holomorphic.

Our function f is holomorphic at the cusps because of a growth condition built into the definition of automorphic forms. The condition implies that Φ grows at most polynomially in its archimedean variables. On the upper half plane, this translates into the condition that $(f\big|_k g)(z)$ remains bounded as $\mathrm{Im}(z) \to \infty$, for all $g \in \mathrm{SL}(2, \mathbb{Z})$.

We thus see that f is indeed an element of the space $M_k(\Gamma_0(N))$ of modular forms of weight k and level N. The cuspidality condition (1.17) implies that the constant terms in the Fourier expansions of the functions $f\big|_k g$ vanish, for each $g \in \mathrm{SL}(2, \mathbb{Z})$. Hence, f "vanishes at the cusps", and is therefore an element of the subspace of *cusp forms* $S_k(\Gamma_0(N))$.

Conversely, let f be an element of $S_k(\Gamma_0(N))$. We associate with f a function $\Phi : \mathrm{GL}(2, \mathbb{A}) \to \mathbb{C}$ as follows. Write an element of $\mathrm{GL}(2, \mathbb{A})$ as $\rho g \kappa$ with $\rho \in \mathrm{GL}(2, \mathbb{Q})$, $g \in \mathrm{GL}(2, \mathbb{R})^+$ and $\kappa \in \mathcal{K}_0(N)$ according to (1.28), and set

$$\Phi(\rho g \kappa) = (f\big|_k g)(\mathrm{i}). \tag{1.36}$$

One can verify that Φ is well-defined and lies in $\mathcal{A}_k^\circ(\mathcal{K}_0(N))$.

Evidently, the maps $\Phi \mapsto f$ and $f \mapsto \Phi$ are inverse to each other and provide an isomorphism

$$\mathcal{A}_k^\circ(\mathcal{K}_0(N)) \cong S_k(\Gamma_0(N)). \tag{1.37}$$

The spaces $S_k(\Gamma_0(N))$ are known to be finite-dimensional. It follows that there are only finitely many cuspidal automorphic representations of $\mathrm{GL}(2, \mathbb{A})$ with archimedean component $\mathcal{D}(k)$ and with a given conductor.

Finally, transferring the inner product (1.19) to $S_k(\Gamma_0(N))$ via (1.37), one obtains, after appropriate normalization of measures, the *Petersson inner product*

$$\langle f_1, f_2 \rangle = \int\limits_{\Gamma_0(N)\backslash\mathcal{H}_1} f_1(z)\overline{f_2(z)}y^{k-2}\,dx\,dy, \tag{1.38}$$

where $f_1, f_2 \in S_k(\Gamma_0(N))$ and $z = x + iy$. Then the isomorphism (1.37) becomes an isometry.

Eigenforms

We return to the cuspidal, automorphic representation $\pi \cong \otimes\pi_p$ whose archimedean component is a discrete series representation $\mathcal{D}(k)$. Consider one of the representations (π_p, V_p) for a prime p not dividing the conductor N_π. Recall that π_p is spherical, i.e., contains non-zero vectors invariant under $K_p := \mathrm{GL}(2, \mathbb{Z}_p)$. The space $V_p(0)$ of K_p-invariant vectors is one-dimensional, and we chose our distinguished vector v_p to be a non-zero element of $V_p(0)$.

For an element $g \in \mathrm{GL}(2, \mathbb{Q}_p)$, the vector $\pi_p(g)v_p$ is no longer spherical, unless $g \in K_p$ (in which case $\pi_p(g)v_p = v_p$). We may transport the vector back into $V_p(0)$ by summing over a system of representatives for $K_p/(K_p \cap gK_pg^{-1})$, i.e., the vector

$$T_g v_p = \sum_{h\in K_p/(K_p\cap gK_pg^{-1})} \pi_p(h)\pi_p(g)v_p \tag{1.39}$$

is in $V_p(0)$. Note here that the coset space is finite, because K_p is open and compact. Alternatively, we can write

$$T_g v_p = \sum_{h\in K_pgK_p/K_p} \pi_p(h)v_p. \tag{1.40}$$

The map T_g thus defined is an endomorphism of the one-dimensional space $V_p(0)$, called a *Hecke operator*.

Evidently, the endomorphism T_g depends only on the double coset K_pgK_p. Such double cosets are known to be represented by diagonal matrices. Since the center acts trivially, and since units can be absorbed into K_p, we may assume $g = \begin{bmatrix} p^n & \\ & 1 \end{bmatrix}$ for some $n \geq 0$. Let us write T_n instead of T_g for $g = \begin{bmatrix} p^n & \\ & 1 \end{bmatrix}$. One can prove that T_n is a polynomial in T_1, making T_1 the only relevant Hecke operator. Let λ_p be its eigenvalue on $V_p(0)$, i.e.,

$$T_1 v_p = \lambda_p v_p. \tag{1.41}$$

This Hecke eigenvalue is a canonically defined number attached to the spherical representation π_p, and in fact determines the isomorphism class of π_p.

We have compatible Hecke operators on the space $\mathcal{A}_k^\circ(\mathcal{K}_0(N))$. For a prime $p \nmid N$ and $\Phi \in \mathcal{A}_k^\circ(\mathcal{K}_0(N))$, let

$$(T_1(p)\Phi)(g) = \sum_{h \in K_p\left[\begin{smallmatrix} p & \\ & 1\end{smallmatrix}\right]K_p/K_p} \Phi(gh) \tag{1.42}$$

for $g \in \mathrm{GL}(2, \mathbb{A})$. Then $T_1(p)\Phi$ is a well-defined element of $\mathcal{A}_k^\circ(\mathcal{K}_0(N))$, and the map $T_1(p)$ thus defined is an endomorphism of this space.

We can similarly define Hecke operators on $S_k(\Gamma_0(N))$. Let p be a prime not dividing N, and let

$$T(1, p)f = p^{\frac{k}{2}-1} \sum_{h \in \Gamma_0(N)\backslash\Gamma_0(N)\left[\begin{smallmatrix} 1 & \\ & p\end{smallmatrix}\right]\Gamma_0(N)} f\big|_k h \tag{1.43}$$

for $f \in S_k(\Gamma_0(N))$. Then $T(1, p)$ is an endomorphism of $S_k(\Gamma_0(N))$. The factor $p^{\frac{k}{2}-1}$ is useful and is already present in Hecke's foundational works; see Eq. (13) of [54] and Eq. (4) of [55].

Using $p \nmid N$, a topological argument shows that the map

$$\Gamma_0(N)\backslash\Gamma_0(N)\left[\begin{smallmatrix} 1 & \\ & p\end{smallmatrix}\right]\Gamma_0(N) \longrightarrow K_p\backslash K_p\left[\begin{smallmatrix} 1 & \\ & p\end{smallmatrix}\right]K_p \tag{1.44}$$

induced by the inclusion is surjective, and an algebraic argument shows that it is injective. Hence, if $\{h_i\}$ is a system of representatives for $\Gamma_0(N)\backslash\Gamma_0(N)\left[\begin{smallmatrix} 1 & \\ & p\end{smallmatrix}\right]\Gamma_0(N)$, then $\{\left[\begin{smallmatrix} p & \\ & p\end{smallmatrix}\right]h_i^{-1}\}$ is a system of representatives for $K_p\left[\begin{smallmatrix} p & \\ & 1\end{smallmatrix}\right]K_p/K_p$. Using this fact and (1.34), it is easy to see that the diagram

$$\begin{array}{ccc} \mathcal{A}_k^\circ(\mathcal{K}_0(N)) & \xrightarrow{\sim} & S_k(\Gamma_0(N)) \\ {\scriptstyle T_1(p)}\big\downarrow & & \big\downarrow{\scriptstyle p^{1-\frac{k}{2}}T(1,p)} \\ \mathcal{A}_k^\circ(\mathcal{K}_0(N)) & \xrightarrow{\sim} & S_k(\Gamma_0(N)) \end{array} \tag{1.45}$$

is commutative; here, the horizontal maps are the isomorphism (1.37).

The consequences for our cuspidal, automorphic representation $\pi \cong \otimes\pi_p$, its global newform Φ_π, and the associated modular form f_π are as follows. For a prime p not dividing the conductor N let v_p be the spherical vector in π_p, and let λ_p be the Hecke eigenvalue defined by (1.41). Then $T_1(p)\Phi_\pi = \lambda_p\Phi_\pi$ because of the compatibility of our isomorphism with the $\mathrm{GL}(2, \mathbb{Q}_p)$ action, and

$$T(1, p)f_\pi = p^{\frac{k}{2}-1}\lambda_p f_\pi \tag{1.46}$$

because of the commutativity of the diagram (1.45). Modular forms that are eigenvectors for the $T(1, p)$ for almost all p are called *eigenforms*.

Hence, every cuspidal automorphic representation π of $\mathrm{GL}(2, \mathbb{A})$ of conductor N_π and with archimedean component $\mathcal{D}(k)$ gives rise to an eigenform f_π of level N_π and weight k, unique up to multiples. The eigenforms thus obtained are called *newforms*. By multiplicity one for GL(2), there is no ambiguity in the definition of newforms, except for scaling.

We let $S_k(\Gamma_0(N))_{\mathrm{new}}$ be the space spanned by all the f_π for π of conductor N. To understand the rest of $S_k(\Gamma_0(N))$, we need to take a closer look at oldforms.

Oldforms

Consider a local representation (π_p, V_p) for a prime number p and the spaces $V_p(n)$ whose dimension is given by (1.23). There is a trivial linear map $\theta : V_p(n) \to V_p(n+1)$ given by the inclusion. An only slightly less trivial linear map $\theta' : V_p(n) \to V_p(n+1)$ is given by $\theta' v = \pi_p(\left[\begin{smallmatrix} 1 & \\ & p \end{smallmatrix}\right])v$. We do not include n in the notation for these *level raising operators*, so that we can conveniently consider linear maps like $\theta^i \theta'^j : V_p(n) \to V_p(n+i+j)$. Note here that θ and θ' commute in the obvious sense. Given a non-zero $v \in V_p(n)$ and a non-negative integer m, it is not difficult to see that the vectors $\theta^i \theta'^j v$ for $i + j = m$ are linearly independent. In particular, if v_p is a local newform, then, by (1.23), a basis for $V_p(n)$ is given by the vectors $\theta^i \theta'^j v_p$ for $i + j = n - N_p$. The following diagram, in which $\bullet$ represents a local newform and $\circ$ represents local oldforms, illustrates the situation.

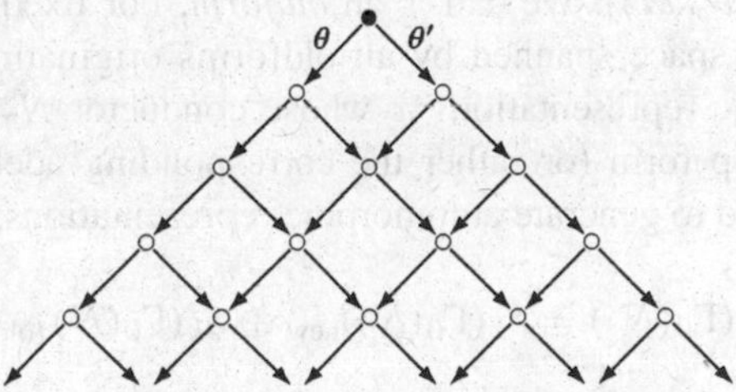

(1.47)

We refer to the fact that all local oldforms originate from a local newform via repeated application of a fixed set of level raising operators as the *oldforms principle*.

For a positive integer N and for each prime p, we have compatible operators θ_p and θ'_p from $\mathcal{A}_k^\circ(\mathcal{K}_0(N))$ to $\mathcal{A}_k^\circ(\mathcal{K}_0(Np))$. The operator θ_p is just the identity map, and θ'_p is right translation by the p-adic matrix $\left[\begin{smallmatrix} 1 & \\ & p \end{smallmatrix}\right]$.

We also have compatible operators $\theta_p, \theta'_p : S_k(\Gamma_0(N)) \to S_k(\Gamma_0(Np))$. The operator θ_p is just the inclusion map, and $(\theta'_p f)(z) = p^{k/2} f(pz)$. A calculation confirms that the diagrams

$$\begin{array}{ccc} \mathcal{A}_k^\circ(\mathcal{K}_0(N)) & \xrightarrow{\sim} & S_k(\Gamma_0(N)) \\ {\scriptstyle \theta_p,\theta'_p}\downarrow & & \downarrow{\scriptstyle \theta_p,\theta'_p} \\ \mathcal{A}_k^\circ(\mathcal{K}_0(Np)) & \xrightarrow{\sim} & S_k(\Gamma_0(Np)) \end{array} \tag{1.48}$$

are commutative; here, the horizontal maps are the isomorphisms (1.37).

Now take the cuspidal, automorphic representation $\pi \cong \otimes \pi_p$ with $\pi_\infty = \mathcal{D}(k)$, let N_π be its conductor, let Φ_π be its global newform, and let $f_\pi \in S_k(\Gamma_0(N_\pi))$ be the corresponding newform. By the oldforms principle, any *global oldform* in π is a linear combination of vectors of the form

$$\Big(\prod_{p|M} \sum_{i+j=m_p} \theta_p^i \theta_p'^j\Big) \Phi_\pi, \tag{1.49}$$

where M is a positive integer with prime factorization $\prod p^{m_p}$ such that $M > 1$. By (1.48), the element (1.49), which lies in $\mathcal{A}_k^\circ(\mathcal{K}_0(N_\pi M))$, descends to

$$f := \Big(\prod_{p|M} \sum_{i+j=m_p} \theta_p^i \theta_p'^j\Big) f_\pi, \tag{1.50}$$

which lies in $S_k(\Gamma_0(N_\pi M))$. We call f an *oldform*. For fixed $N = N_\pi M$ we let $S_k(\Gamma_0(N))_{\rm old}$ be the space spanned by all oldforms originating in this way from cuspidal, automorphic representations π whose conductor N_π is a proper divisor of N. Since any cusp form (or rather the corresponding adelic function defined by (1.36)) can be used to generate automorphic representations, it is clear that

$$S_k(\Gamma_0(N)) = S_k(\Gamma_0(N))_{\rm new} \oplus S_k(\Gamma_0(N))_{\rm old}. \tag{1.51}$$

We note one terminological peculiarity in the literature. While every element of $S_k(\Gamma_0(N))_{\rm old}$ is called an oldform, not every element of $S_k(\Gamma_0(N))_{\rm new}$ is called a newform; only the elements f_π of $S_k(\Gamma_0(N))_{\rm new}$ are called newforms.

Suppose that $f \in S_k(\Gamma_0(N))_{\rm new}$ is a newform. Hence, $f = f_\pi$ for a cuspidal, automorphic representation π of conductor N. On the other hand, any $\tilde{f} \in S_k(\Gamma_0(N))_{\rm old}$ originates from cuspidal, automorphic representations $\tilde{\pi}$ of strictly smaller conductor. Since the isomorphism (1.37) is an isometry, it follows that $\langle f, \tilde{f}\rangle = 0$. In other words, oldforms and newforms (of the same level) are orthogonal to each other with respect to the Petersson inner product.

One can therefore give the following definition of $S_k(\Gamma_0(N))_{\rm new}$ and $S_k(\Gamma_0(N))_{\rm old}$ without any reference to automorphic representations. First define

$S_k(\Gamma_0(N))_{\text{old}}$ as the space spanned by $\theta_p S_k(\Gamma_0(Np^{-1}))$ and $\theta'_p S_k(\Gamma_0(Np^{-1}))$ for all prime divisors p of N. Equivalently, $S_k(\Gamma_0(N))_{\text{old}}$ is the space spanned by all functions $z \mapsto f(dz)$, where $f \in S_k(\Gamma_0(M))$ for a proper divisor M of N, and d a divisor of N/M. Then define $S_k(\Gamma_0(N))_{\text{new}}$ as the orthogonal complement of $S_k(\Gamma_0(N))_{\text{old}}$ within $S_k(\Gamma_0(N))$ with respect to the Petersson inner product.

The Correspondence Theorem

The following result is the GL(2) analogue of Theorem 1.1.3. Its proof is based on the uniqueness of *Whittaker models* and some Fourier analysis on $\mathrm{GL}(2, \mathbb{A})$; see [92].

Theorem 1.2.1 (Strong Multiplicity One for GL(2)) *Let $\pi \cong \otimes\pi_p$ and $\tilde{\pi} \cong \otimes\tilde{\pi}_p$ be cuspidal, automorphic representations of $\mathrm{GL}(2, \mathbb{A})$ such that $\pi_p \cong \tilde{\pi}_p$ for almost all p. If V and $\tilde{V}$ are subspaces of $\mathcal{A}^\circ$ realizing π and $\tilde{\pi}$, respectively, then $V = \tilde{V}$.*

Recall how Theorem 1.1.2 established a correspondence between adelic objects (certain automorphic representations of $\mathrm{GL}(1, \mathbb{A})$) and classical objects (Dirichlet characters). We can now state an analogous result for GL(2).

Theorem 1.2.2 (Correspondence Theorem for Elliptic Cusp Forms) *Let k and N be positive integers. The map $\pi \mapsto f_\pi$ induces a bijection between the set of cuspidal, automorphic representations $\pi \cong \otimes\pi_p$ of $\mathrm{GL}(2, \mathbb{A})$ (always assumed to have trivial central character) with $\pi_\infty = \mathcal{D}(k)$ and conductor N, and the eigenforms in $S_k(\Gamma_0(N))_{\text{new}}$ modulo scalars.*

Sketch of Proof The injectivity of the map $\pi \mapsto f_\pi$ is clear: If $f_{\pi_1} = f_{\pi_2}$ (up to scalars), then we have equality of the global newforms Φ_{π_1} and Φ_{π_2} (up to scalars), because f_{π_i} determines Φ_{π_i} by (1.28) and (1.34). Hence, the spaces of automorphic forms V_{π_1} and V_{π_2} have a vector in common, making them equal by irreducibility.

We defined $S_k(\Gamma_0(N))_{\text{new}}$ to be the space spanned by all f_π for π of conductor N. Yet it is not obvious that every eigenform in $S_k(\Gamma_0(N))_{\text{new}}$ is an f_π. The issue is that it is at least conceivable that two newforms f_π and $f_{\tilde{\pi}}$ have the same Hecke eigenvalues almost everywhere. Then $f := f_\pi + f_{\tilde{\pi}}$ would be an eigenform, but it would not be an $f_{\pi'}$ for any cuspidal, automorphic representation π' (because the only irreducible subspaces of $V_\pi \oplus V_{\tilde{\pi}}$ are V_π and $V_{\tilde{\pi}}$).

However, recall that if $\pi \cong \otimes\pi_p$, and if the prime p does not divide the conductor N, then the Hecke eigenvalue of f_π determines the equivalence class of the local representation π_p. Hence if f_π and $f_{\tilde{\pi}}$ have the same Hecke eigenvalues almost everywhere, then $\pi_p \cong \tilde{\pi}_p$ for almost all p. Theorem 1.2.1 then implies $\pi = \tilde{\pi}$, which in turn implies that f_π and $f_{\tilde{\pi}}$ are multiples of each other. □

Roughly speaking, Theorem 1.2.2 asserts that newforms arise as distinguished vectors in cuspidal, automorphic representations. Just as in the case of Dirichlet characters, there are certain conceptual advantages to the representation theoretic viewpoint. One of them is a natural factorization $\pi \cong \otimes\pi_p$ of the adelic objects, for which there is no classical analogue. Hecke operators, for example, defined classically by (1.43), reveal themselves as local objects when translated into the adelic language. It is clear that Hecke operators with respect to different primes commute, that the spaces $S_k(\Gamma_0(N))_{\rm new}$ and $S_k(\Gamma_0(N))_{\rm old}$ are preserved by the $T(1, p)$, and that they are spanned by eigenforms; none of these facts is obvious if only the classical definitions are used. Another benefit of the adelic point of view is that it automatically puts the emphasis on newforms. Oldforms are still present, as non-distinguished vectors, but have been consigned to a less important role.

On the other hand, automorphic representations are not always the best setting. Automorphic representations can be unwieldy objects, being infinite tensor products of infinite-dimensional spaces. The corresponding modular forms f_π are naturally connected to Riemann surfaces and arithmetic algebraic geometry. Thus, the f_π often form a good arena for the formulation and computational investigation of applications to number theory. An example of this is given by the history of the Modularity Theorem (previously called the modularity conjecture for elliptic curves).

An Example

To illustrate how oldforms and newforms in cuspidal, automorphic representations contribute to spaces of cusp forms, we consider the weight 12 and levels dividing 20. In order to contribute to $S_{12}(\Gamma_0(20))$, a cuspidal, automorphic representation $\pi \cong \otimes\pi_p$ must have $\pi_\infty = \mathcal{D}(12)$ and conductor N_π dividing 20. It turns out that there are 13 such π: One representation π_1 has conductor 1; one representation π_4 has conductor 4; three representations $\pi_{5a}, \pi_{5b}, \pi_{5c}$ have conductor 5; five representations $\pi_{10a}, \pi_{10b}, \pi_{10c}, \pi_{10d}, \pi_{10e}$ have conductor 10; and three representations $\pi_{20a}, \pi_{20b}, \pi_{20c}$ have conductor 20.

For each of these 13 representations Table 1.1 lists the dimensions of the spaces of local fixed vectors $V_2(0)$, $V_2(1)$, $V_2(2)$ in the 2-adic component π_2, and the dimensions of the spaces $V_5(0)$, $V_5(1)$ in the 5-adic component π_5. These dimensions follow immediately from (1.23).

We can then read off the contribution of each π to the dimension of $S_{12}(\Gamma_0(N))$ for $N \mid 20$. For example, the contributions of π_1 to $S_{12}(\Gamma_0(20))$ are given by choosing a vector in the 3-dimensional space $V_2(2)$ and a vector in the 2-dimensional space $V_5(1)$; for each other prime we always choose the essentially unique spherical vector, and at the archimedean place we always choose the essentially unique lowest weight vector. Hence, π_1 contributes $\dim V_2(2) \otimes V_5(1) = 6$ to the dimension of $S_{12}(\Gamma_0(20))$. The last six columns of Table 1.1 show the dimensional contributions of each π to the spaces $S_{12}(\Gamma_0(N))$ for the divisors N of 20. The contribution of the unique newform is indicated by **1**.

If we then add up the numbers in the last six columns of Table 1.1 we obtain $\dim S_{12}(\Gamma_0(N))$ for $N \mid 20$. If we only add up the **1**'s we obtain $\dim S_{12}(\Gamma_0(N))_{\rm new}$.

Table 1.1 Automorphic representations contributing to $S_{12}(\Gamma_0(20))$

	Dimension of …					Contribution to dim $S_{12}(\Gamma_0(N))$ for $N = \dots$					
π	$V_2(0)$	$V_2(1)$	$V_2(2)$	$V_5(0)$	$V_5(1)$	1	2	4	5	10	20
π_1	1	2	3	1	2	**1**	2	3	2	4	6
π_4	0	0	1	1	2	0	0	**1**	0	0	2
π_{5a}	1	2	3	0	1	0	0	0	**1**	2	3
π_{5b}	1	2	3	0	1	0	0	0	**1**	2	3
π_{5c}	1	2	3	0	1	0	0	0	**1**	2	3
π_{10a}	0	1	2	0	1	0	0	0	0	**1**	2
π_{10b}	0	1	2	0	1	0	0	0	0	**1**	2
π_{10c}	0	1	2	0	1	0	0	0	0	**1**	2
π_{10d}	0	1	2	0	1	0	0	0	0	**1**	2
π_{10e}	0	1	2	0	1	0	0	0	0	**1**	2
π_{20a}	0	0	1	0	1	0	0	0	0	0	**1**
π_{20b}	0	0	1	0	1	0	0	0	0	0	**1**
π_{20c}	0	0	1	0	1	0	0	0	0	0	**1**
			dim $S_k(\Gamma_0(N))$			1	2	4	5	15	30
			dim $S_k(\Gamma_0(N))_{\text{new}}$			1	0	1	3	5	3

The notation is such that the cuspidal, automorphic representation π_{Nx} has conductor N

In actuality one proceeds the other way around. Dimension formulas for the $S_k(\Gamma_0(N))$ can be obtained using methods of Riemann surfaces; see for example Theorem 2.24 of [118], or Theorem 2.5.2 of [86]. From these one can determine oldform dimensions, and then newform dimensions. Hence, we get the number of cuspidal, automorphic representations of a given conductor by the correspondence theorem. This is an example of how the concrete functions f_π help us learn something about automorphic representations.

1.3 Modular Forms II

Recall how to each idele class character $\chi = \otimes\chi_p$ local factors $L_p(s, \chi_p)$ were attached, using the definitions (1.11) and (1.13), such that the resulting global L-function $L(s, \chi) = \prod_{p\le\infty} L_p(s, \chi_p)$ satisfies a functional equation as in (1.14). It was proven in Theorem 11.1 of [68] that something similar can be done for cuspidal, automorphic representations $\pi \cong \otimes\pi_p$ of $\mathrm{GL}(2, \mathbb{A})$. More precisely, there are local factors $L_p(s, \pi_p)$ and $\varepsilon_p(s, \pi_p)$, depending only on the equivalence class of π_p, such that $L(s, \pi) := \prod_{p\le\infty} L_p(s, \pi_p)$, convergent for $\mathrm{Re}(s) > 1$, has analytic continuation to an entire function and satisfies

$$L(s, \pi) = \varepsilon(s, \pi)L(1 - s, \pi), \qquad (1.52)$$

where $\varepsilon(s,\pi)=\prod_{p\le\infty}\varepsilon_p(s,\pi_p)$. Here, we work under the simplifying assumption that the center acts trivially; otherwise the contragredient of π would appear on the right-hand side of (1.52), and also it would not be entirely true that the local ε-factors only depend on the equivalence class of π_p. In any case, for almost all primes p we have $\varepsilon_p(s,\pi_p)=1$, and $L_p(s,\pi_p)$ is the reciprocal of a polynomial of degree 2 in p^{-s}.

In the following we will review the definition of the local factors $L_p(s,\pi_p)$ and $\varepsilon_p(s,\pi_p)$, and then make the connection with modular forms.

Local Representations

Let p be a prime number. The question of classifying irreducible representations of $\mathrm{GL}(1,\mathbb{Q}_p)$ did not come up, since these are just the characters of $\mathbb{Q}_p^\times$. In the GL(2) case the question of writing down all irreducible representations becomes more interesting. The basic classification we will now review can be done for representations of $\mathrm{GL}(2,F)$, where F is a finite extension of $\mathbb{Q}_p$. Since there are no additional difficulties, we will work in this more general context. Let $\mathfrak{o}$ be the ring of integers of F, let $\mathfrak{p}$ be the maximal ideal of $\mathfrak{o}$, and let ϖ be a generator of $\mathfrak{p}$. Let q be the cardinality of the residue class field $\mathfrak{o}/\mathfrak{p}$. We denote by $|\cdot|$ the absolute value on F, normalized such that $|\varpi|=q^{-1}$. The restriction of $|\cdot|$ to $F^\times$ will be denoted by ν. Every unramified character of $F^\times$ (i.e., a character which is trivial on $\mathfrak{o}^\times$) is of the form ν^s for some $s\in\mathbb{C}$.

The representations of $\mathrm{GL}(2,F)$ to be considered will be *smooth* and *admissible*. Here, a representation is smooth, if every vector is stabilized by some compact-open subgroup of $\mathrm{GL}(2,F)$, and it is admissible, if in addition the subspace fixed by any compact-open subgroup is finite-dimensional.

A standard way to construct representations of $\mathrm{GL}(2,F)$ is via *parabolic induction*. For this we take two characters χ_1,χ_2 of $F^\times$ and consider the space of all functions $\varphi:\mathrm{GL}(2,F)\to\mathbb{C}$ with the transformation property

$$\varphi(\left[\begin{smallmatrix}a&b\\0&d\end{smallmatrix}\right]g)=|a/d|^{1/2}\chi_1(a)\chi_2(d)\varphi(g) \tag{1.53}$$

for $a,d\in F^\times$, $b\in F$ and $g\in\mathrm{GL}(2,F)$. We also require that φ is smooth, i.e., that there exists a compact-open subgroup H of $\mathrm{GL}(2,F)$ such that $\varphi(gh)=\varphi(g)$ for $g\in\mathrm{GL}(2,F)$ and $h\in H$. Let $\chi_1\times\chi_2$ be the representation of $\mathrm{GL}(2,F)$ on the space of all such φ given by right translation. It is proven in Theorem 3.3 of [68] that $\chi_1\times\chi_2$ is irreducible unless $\chi_1\chi_2^{-1}=\nu^{\pm1}$. The irreducible $\chi_1\times\chi_2$ are called *principal series representations*; here, we also call them *group I representations*.

Group II representations will be those arising from reducible $\chi_1\times\chi_2$. Up to equivalence we may assume $\chi_1=\nu^{1/2}\chi$ and $\chi_2=\nu^{-1/2}\chi$. It turns out that $\nu^{1/2}\chi\times\nu^{-1/2}\chi$ has an irreducible subspace of codimension 1. The representation of $\mathrm{GL}(2,F)$ on this subspace is denoted by $\chi\mathrm{St}_{\mathrm{GL}(2)}$; it is the χ-twist of the *Steinberg representation*. The representation of $\mathrm{GL}(2,F)$ on the 1-dimensional quotient is

Table 1.2 Non-supercuspidal representations of GL(2, F)

Constituent of	Group		Representation	Tempered	ess. L^2	Generic
$\chi_1 \times \chi_2$ (irreducible)	I		$\chi_1 \times \chi_2$	χ_1, χ_2 unitary		•
$\nu^{1/2}\chi \times \nu^{-1/2}\chi$	II	a	$\chi \mathrm{St}_{\mathrm{GL}(2)}$	χ unitary	•	•
		b	$\chi 1_{\mathrm{GL}(2)}$			

denoted by $\chi 1_{\mathrm{GL}(2)}$; it is the χ-twist of the trivial representation. Thus, there is an exact sequence

$$0 \longrightarrow \chi \mathrm{St}_{\mathrm{GL}(2)} \longrightarrow \nu^{1/2}\chi \times \nu^{-1/2}\chi \longrightarrow \chi 1_{\mathrm{GL}(2)} \longrightarrow 0. \tag{1.54}$$

We say that representations of type $\chi \mathrm{St}_{\mathrm{GL}(2)}$ are in *group IIa*, and those of type $\chi 1_{\mathrm{GL}(2)}$ are in *group IIb*.

The representations in group I, IIa and IIb are those irreducibles which can be obtained via parabolic induction. There are many others, called *supercuspidal representations*. For our purposes it is not necessary to further classify the supercuspidals.

Table 1.2 lists the non-supercuspidal, irreducible representations of GL(2, F), and lists the conditions for these being *tempered*, *essentially square integrable*, and *generic*. If π_p is a representation of GL(2, $\mathbb{Q}_p$) occurring as the local component of a cuspidal, automorphic representation of GL(2, $\mathbb{A}$) of the type that we are considering, then π_p must be *generic*; it is also true that π_p must be *tempered*, thanks to Deligne's proof of the Ramanujan-Petersson Conjecture (see [32]). Thus, the one-dimensional representations of type IIb are irrelevant for our current purposes.

Recall that we imposed on our automorphic forms the condition that the center of GL(2, $\mathbb{A}$) acts trivially. As a consequence, only local representations for which the center of GL(2, F) acts trivially will be relevant. Among the non-supercuspidals, these are the group I representations of the form $\chi \times \chi^{-1}$, and the group IIa and IIb representations $\chi \mathrm{St}_{\mathrm{GL}(2)}$ and $\chi 1_{\mathrm{GL}(2)}$ for which $\chi^2 = 1$.

Three Invariants Derived from the Local Newform

Except for those of type IIb, irreducible, admissible representations of GL(2, F) are infinite-dimensional. A standard method to understand a given representation is to restrict the action to compact subgroups. Generalizing (1.22), let

$$\Gamma_0(\mathfrak{p}^n) = \{\left[\begin{smallmatrix} a & b \\ c & d \end{smallmatrix}\right] \in \mathrm{GL}(2, \mathfrak{o}) \mid c \in \mathfrak{p}^n\} \tag{1.55}$$

for a non-negative integer n. The group $\Gamma_0(\mathfrak{p}^0) = \mathrm{GL}(2, \mathfrak{o})$ represents the unique conjugacy class of maximal compact subgroups of GL(2, F).

If (π, V) is an admissible representation of GL(2, F) and n is a non-negative integer, then let $V(n)$ be the subspace of V stabilized by $\Gamma_0(\mathfrak{p}^n)$. By admissibility,

$V(n)$ is finite-dimensional. Clearly, if $\pi = \chi 1_{\mathrm{GL}(2)}$, then $\dim V(n) = 1$ for all n if χ is unramified, and $\dim V(n) = 0$ for all n if χ is ramified.

Now assume that (π, V) is irreducible, infinite-dimensional, and that the center acts trivially. Then the structure of the spaces $V(n)$ is just like in the special case of $F = \mathbb{Q}_p$; see (1.23). In particular, there exists a smallest non-negative integer N_π, called the *conductor exponent* or *minimal level* of π, such that $V(N_\pi) \neq 0$. We have $\dim V(N_\pi) = 1$, and any non-zero vector v_0 in $V(N_\pi)$ is called a *local newform.*

Every canonically defined endomorphism of the 1-dimensional space $V(N_\pi)$ acts via an eigenvalue, which may be viewed as an invariant attached to the representation π. For example, for any $g \in \mathrm{GL}(2, F)$ we have a *Hecke operator* given by

$$T_g v_0 = \sum_{h \in \Gamma_0(\mathfrak{p}^{N_\pi}) g \Gamma_0(\mathfrak{p}^{N_\pi}) / \Gamma_0(\mathfrak{p}^{N_\pi})} \pi(h) v_0. \tag{1.56}$$

Of particular interest is the case $g = \left[\begin{smallmatrix} \varpi & \\ & 1 \end{smallmatrix}\right]$, for which we denote the corresponding Hecke operator by T_1. Let λ_π be its eigenvalue, i.e.,

$$T_1 v_0 = \lambda_\pi v_0. \tag{1.57}$$

Then λ_π is an invariant, which we may attempt to calculate for each π. Working in local models, this is in fact not difficult to do. The results are shown in the last column of Table 1.3.

Another important Hecke operator is defined by the *Atkin-Lehner element* u_{N_π}, where

$$u_n = \left[\begin{smallmatrix} & 1 \\ -\varpi^n & \end{smallmatrix}\right]. \tag{1.58}$$

Note that u_n normalizes $\Gamma_0(\mathfrak{p}^n)$, so that the corresponding Hecke operator on $V(N_\pi)$ is simply $\pi(u_{N_\pi})$. Since $u_{N_\pi}^2$ is in the center, the resulting eigenvalue on $V(N_\pi)$ is ± 1. We denote this sign by ε_π and call it the *Atkin-Lehner eigenvalue* of π. The ε_π can also be calculated using local models, and the results are shown in Table 1.3.

Table 1.3 Minimal levels, Atkin-Lehner eigenvalues and Hecke eigenvalues

Group		Representation	Inducing data	N_π	ε_π	λ_π
I		$\chi \times \chi^{-1}$	χ unramified	0	1	$q^{1/2}(\chi(\varpi) + \chi(\varpi)^{-1})$
			χ ramified	$2a(\chi)$	$\chi(-1)$	0
II	a	$\chi \mathrm{St}_{\mathrm{GL}(2)}$	χ unramified	1	$-\chi(\varpi)$	$\chi(\varpi)$
			χ ramified	$2a(\chi)$	$\chi(-1)$	0
	b	$\chi 1_{\mathrm{GL}(2)}$	χ unramified	0	1	$(q+1)\chi(\varpi)$
			χ ramified	no $\Gamma_0(\mathfrak{p}^n)$ invariant vectors for any n		
Supercuspidal				≥ 2	± 1	0

Local Factors

We now present the definition of local factors $L(s, \pi)$ and $\varepsilon(s, \pi)$ for each irreducible, admissible representation of GL(2, F). Theorem 2.18 of [68] provides an abstract definition of both quantities in terms of zeta integrals and the local functional equation. From the latter one derives the explicit form

$$\varepsilon(s, \pi) = \varepsilon_\pi q^{-N_\pi(s-1/2)}, \tag{1.59}$$

where N_π and ε_π are two of the quantities listed in Table 1.3. Here, we have to assume that π is not of the form $\chi 1_{\mathrm{GL}(2)}$ with ramified χ. (The same formula would hold if we set $N_\pi = 2a(\chi)$ and $\varepsilon_\pi = \chi(-1)$ for such π.)

For a character χ of $F^\times$, let

$$L(s, \chi) := \begin{cases} \dfrac{1}{1 - \chi(\varpi)q^{-s}} & \text{if } \chi \text{ is unramified,} \\ 1 & \text{if } \chi \text{ is ramified,} \end{cases} \tag{1.60}$$

as in (1.11) for the case $F = \mathbb{Q}_p$. Then Propositions 3.5 and 3.6 of [68] give the explicit form of $L(s, \pi)$ as follows,

$$L(s, \pi) = \begin{cases} L(s, \chi_1)L(s, \chi_2) & \text{if } \pi = \chi_1 \times \chi_2 \text{ (group I),} \\ L(s, \nu^{1/2}\chi) & \text{if } \pi = \chi \mathrm{St}_{\mathrm{GL}(2)} \text{ (group IIa),} \\ L(s, \nu^{1/2}\chi)L(s, \nu^{-1/2}\chi) & \text{if } \pi = \chi 1_{\mathrm{GL}(2)} \text{ (group IIb),} \\ 1 & \text{if } \pi \text{ is supercuspidal.} \end{cases} \tag{1.61}$$

Thus $L(s, \pi)$ is "compatible with inducing data". We now observe from Table 1.3 that $L(s, \pi)$ can be calculated from the local invariants N_π and λ_π, namely,

$$L(s, \pi) = \begin{cases} \dfrac{1}{1 - \lambda_\pi q^{-1/2}q^{-s} + q^{-2s}} & \text{if } N_\pi = 0, \\ \dfrac{1}{1 - \lambda_\pi q^{-1/2}q^{-s}} & \text{if } N_\pi > 0. \end{cases} \tag{1.62}$$

Here again we have to assume that $L(s, \pi)$ is not of the form $\chi 1_{\mathrm{GL}(2)}$ with a ramified character χ. (However (1.62) would hold in this case as well if we set $\lambda_\pi = 0$.)

The important point here is that the local factors $L(s, \pi)$ and $\varepsilon(s, \pi)$ can be calculated from the invariants given in Table 1.3. In this sense the local newform carries all the necessary information about the local factors.

L-Functions

We return to the global setting, with $\pi \cong \otimes\pi_p$ a cuspidal, automorphic representation of $\mathrm{GL}(2, \mathbb{A})$. As in previous sections, we will assume that $\pi(z) = 1$ for $z \in Z(\mathbb{A})$, i.e., the center acts trivially. For each finite prime p let $N_p := N_{\pi_p}$ be the conductor exponent of π_p. Let $N_\pi = \prod p^{N_p}$ be the global conductor of π. (There will be no danger of confusing N_π with the minimal level of a local representation π, for which we used the same symbol.)

Above we have assigned local factors $L_p(s, \pi_p)$ and $\varepsilon_p(s, \pi_p)$ for all finite places p. We also need archimedean factors $L_\infty(s, \pi_\infty)$ and $\varepsilon_\infty(s, \pi_\infty)$. These are given in Sect. 1.5 of [68]. If π_∞ is a discrete series representation $\mathcal{D}(k)$ for a positive, even integer k, which we assume, then

$$L_\infty(s, \pi_\infty) = (2\pi)^{-s-\frac{k-1}{2}}\,\Gamma\left(s + \frac{k-1}{2}\right), \qquad \varepsilon_\infty(s, \pi_\infty) = (-1)^{\frac{k}{2}}. \tag{1.63}$$

Having thus defined local factors at all places, let

$$L(s, \pi) := \prod_{p \le \infty} L_p(s, \pi_p) \tag{1.64}$$

be the global L-function of π. The product is absolutely convergent for $\mathrm{Re}(s) > 1$. By Theorem 11.1 of [68], $L(s, \pi)$ admits analytic continuation to all of $\mathbb{C}$ and satisfies the functional Eq. (1.52). Substituting the ε-factors from (1.59) and (1.63), the functional equation can be written as

$$L(s, \pi) = (-1)^{\frac{k}{2}}\Big(\prod_{p \mid N_\pi} \varepsilon_p\Big) N_\pi^{-(s-1/2)} L(1 - s, \pi), \tag{1.65}$$

where $\varepsilon_p := \varepsilon_{\pi_p}$ is the Atkin-Lehner eigenvalue of π_p. We see that "everything factors", including the sign appearing in the functional equation.

For any finite set S of places, the *partial L-function* $L^S(s, \pi)$ is defined as

$$L^S(s, \pi) := \prod_{p \notin S} L_p(s, \pi_p). \tag{1.66}$$

For almost every prime, π_p is a type I representation $\chi \times \chi^{-1}$ with unramified χ. For such representations, a look at (1.61) shows that $L_p(s, \pi_p)$ determines π_p. The strong multiplicity one property, Theorem 1.2.1, thus implies that $L^S(s, \pi)$ determines π, for any S. In view of (1.62), this can also be stated as saying that the Hecke eigenvalues $\lambda_p := \lambda_{\pi_p}$, for almost all p, determine π. However to get a nice functional equation as in (1.65), we need to "fill in" the correct Euler factors at all places.

Fourier Coefficients

Let $f \in S_k(\Gamma_0(N))$. Since f is holomorphic, $f(z) = f(z+1)$ for all $z \in \mathcal{H}_1$, and f vanishes at the cusps, we have a *Fourier expansion*

$$f(z) = \sum_{n=1}^{\infty} a(n) e^{2\pi i n z} \tag{1.67}$$

with certain *Fourier coefficients* $a(n) \in \mathbb{C}$. Suppose that f is a newform, so that $f = f_\pi$ for a cuspidal, automorphic representation $\pi \cong \otimes \pi_p$ with conductor $N = N_\pi$. Since f determines π, it should be possible to calculate the Euler factors of $L(s, \pi)$ from f alone. How can this be done in practical terms?

By (1.62), it suffices to know the Hecke eigenvalues λ_p of π_p for each prime p. Recall from (1.45) and (1.46) that, for a good place at least, and up to a factor depending only on the weight, λ_p is also the Hecke eigenvalue of f:

$$T(1, p) f = p^{\frac{k}{2}-1} \lambda_p f. \tag{1.68}$$

But the same argument used to show the compatibility of the classical and adelic Hecke operators applies to all places, so that (1.68) holds in fact for all primes. Our task is then to calculate the λ_p from the $a(n)$.

This is not difficult to do using the coset decomposition

$$\Gamma_0(N)\left[\begin{smallmatrix}1 & \\ & p\end{smallmatrix}\right]\Gamma_0(N) = \begin{cases} \displaystyle\bigsqcup_{m \in \mathbb{Z}/p\mathbb{Z}} \Gamma_0(N)\left[\begin{smallmatrix}1 & m \\ & p\end{smallmatrix}\right] \sqcup \Gamma_0(N)\left[\begin{smallmatrix}p & \\ & 1\end{smallmatrix}\right] & \text{for } p \nmid N, \\ \displaystyle\bigsqcup_{m \in \mathbb{Z}/p\mathbb{Z}} \Gamma_0(N)\left[\begin{smallmatrix}1 & m \\ & p\end{smallmatrix}\right] & \text{for } p \mid N. \end{cases} \tag{1.69}$$

Using (1.68), (1.69) and the definition (1.43), a straightforward calculation shows that

$$\lambda_p a(n) = \begin{cases} p^{1-\frac{k}{2}} a(np) + p^{\frac{k}{2}} a(n/p) & \text{for } p \nmid N, \\ p^{1-\frac{k}{2}} a(np) & \text{for } p \mid N, \end{cases} \tag{1.70}$$

for all positive integers n, with the understanding that $a(n/p) = 0$ for $p \nmid n$. For $n = 1$, we get

$$\lambda_p a(1) = p^{1-\frac{k}{2}} a(p). \tag{1.71}$$

If $a(1)$ would be 0, then $a(p^t)$ would be 0 for all $t \geq 0$. Since the $a(n)$ can also be shown to be multiplicative, it follows that $a(1) \neq 0$. Hence, the λ_p can easily be read off the Fourier coefficients.

We note that the recursion formulas (1.70) are equivalent to the formal identity

$$\sum_{t=0}^{\infty} a(np^t)X^t = \begin{cases} \dfrac{a(n) - p^{k-1}a(n/p)X}{1 - \lambda_p p^{\frac{k}{2}-1}X + p^{k-1}X^2} & \text{for } p \nmid N, \\ \dfrac{a(n)}{1 - \lambda_p p^{\frac{k}{2}-1}X} & \text{for } p \mid N. \end{cases} \tag{1.72}$$

Setting $X = p^{-s}$ in (1.72), we get

$$\sum_{t=0}^{\infty} \frac{a(np^t)}{p^{ts}} = N(p^{-s}, n)L_p(s, f), \tag{1.73}$$

where

$$N(X, n) = \begin{cases} a(n) - p^{k-1}a(n/p)X & \text{for } p \nmid N, \\ a(n) & \text{for } p \mid N, \end{cases} \tag{1.74}$$

and

$$L_p(s, f) = \begin{cases} \dfrac{1}{1 - \lambda_p p^{\frac{k}{2}-1} p^{-s} + p^{k-1} p^{-2s}} & \text{for } p \nmid N, \\ \dfrac{1}{1 - \lambda_p p^{\frac{k}{2}-1} p^{-s}} & \text{for } p \mid N, \end{cases} \tag{1.75}$$

is the *classical Euler factor* of f at p. Comparison of (1.75) with (1.62) shows that

$$L_p(s, f) = L_p\left(s - \frac{k-1}{2}, \pi_p\right) \tag{1.76}$$

for all primes p. Hence, up to a translation, the Dirichlet series (1.73) involving the *radial* Fourier coefficients $a(np^t)$, $t \geq 0$, calculates the local Euler factor $L(s, \pi_p)$.

The Classical L-Function

In view of (1.76) it is natural to introduce the shifted L-function

$$L(s, f) := L\left(s - \frac{k-1}{2}, \pi\right). \tag{1.77}$$

By (1.65), it satisfies the functional equation

$$L(s, f) = (-1)^{\frac{k}{2}}\Big(\prod_{p|N} \varepsilon_p\Big) N^{-(s-k/2)} L(k-s, f). \tag{1.78}$$

We also say that $L(s, f)$ is given in *arithmetic normalization* (functional equation relates s and $k-s$), whereas $L(s, \pi)$ is given in *analytic normalization* (functional equation relates s and $1-s$).

There is a direct way of calculating $L(s, f)$ that avoids Hecke operators altogether. Assume that $a(1) = 1$, which can always be achieved by normalization. Setting $n = 1$ in (1.73) then gives

$$\sum_{t=0}^{\infty} \frac{a(p^t)}{p^{ts}} = L_p(s, f) \tag{1.79}$$

for all primes p. Taking the product over all primes, and using the multiplicativity of the $a(n)$, we see that

$$\sum_{n=1}^{\infty} \frac{a(n)}{n^s} = \prod_{p<\infty} L_p(s, f). \tag{1.80}$$

It follows that

$$\int_0^{\infty} f(iy) y^{s-1}\, dy = (2\pi)^{-s}\,\Gamma(s) \prod_{p<\infty} L_p(s, f). \tag{1.81}$$

Note from (1.63) that $(2\pi)^{-s}\Gamma(s) = L_\infty(s - \frac{k-1}{2}, \pi_\infty)$. Thus the integral on the left-hand side of (1.81), known as a *Mellin transform*, calculates the complete L-function $L(s, f) = L(s - \frac{k-1}{2}, \pi)$. Of course, the functional equation for $L(s, f)$ was first proved using the direct connection (1.81) with the modular form f; see [53].

Atkin-Lehner Operators

We saw that the L-function $L(s, f)$ appearing in (1.78) can be defined purely in terms of the cusp form f, even though its origin is representation-theoretic. The functional Eq. (1.78) still has the "flaw" however that it contains the numbers ε_p, which are the eigenvalues of the local Atkin-Lehner operators on the newforms in the representations π_p. It is desirable to express these in terms of f as well, without reference to the underlying automorphic representation.

To do so, consider a prime dividing N, and let p^n be the exact power of p dividing N. Since the Atkin-Lehner element $u_n = \begin{bmatrix} & 1 \\ -p^n & \end{bmatrix}$ (see (1.58)) normalizes the local group $\Gamma_0(\mathfrak{p}^n)$, right translation by the p-adic matrix u_n defines an endomorphism of $\mathcal{A}_k^\circ(\mathcal{K}_0(N))$. We seek an operator w_p on $S_k(\Gamma_0(N))$ for which the diagram

$$\begin{array}{ccc} \mathcal{A}_k^\circ(\mathcal{K}_0(N)) & \xrightarrow{\sim} & S_k(\Gamma_0(N)) \\ {\scriptstyle u_{p^n}}\downarrow & & \downarrow{\scriptstyle w_p} \\ \mathcal{A}_k^\circ(\mathcal{K}_0(N)) & \xrightarrow{\sim} & S_k(\Gamma_0(N)) \end{array} \tag{1.82}$$

is commutative; here, the horizontal maps are the isomorphism (1.37).

To define w_p, we choose a $\gamma \in \mathrm{SL}(2, \mathbb{Z})$ such that

$$\gamma \equiv \begin{bmatrix} & 1 \\ -1 & \end{bmatrix} \pmod{p^n}, \qquad \gamma \equiv \begin{bmatrix} 1 & \\ & 1 \end{bmatrix} \pmod{Np^{-n}}, \tag{1.83}$$

and let $u = \begin{bmatrix} 1 & \\ & p^n \end{bmatrix}\gamma$. Then u normalizes $\Gamma_0(N)$ and $u^2 \in p^n\Gamma_0(N)$. We now define the map $w_p : S_k(\Gamma_0(N)) \to S_k(\Gamma_0(N))$ by

$$w_p f = f\big|_k u. \tag{1.84}$$

Then w_p, referred to as an *Atkin-Lehner operator*, is a well-defined involution of $S_k(\Gamma_0(N))$.

A standard verification shows that the diagram (1.82), with w_p thus defined, is commutative. Therefore, if $f = f_\pi$ is a newform belonging to the cuspidal, automorphic representation $\pi \cong \otimes\pi_p$, we have

$$w_p f = \varepsilon_p f \tag{1.85}$$

for all primes p, where ε_p is the Atkin-Lehner eigenvalue of π_p. Clearly, if $p \nmid N$, then w_p is the identity and $\varepsilon_p = 1$.

While (1.85) provides a way to access the ε_p without reference to the underlying representation, the calculation of $w_p f$ can be tricky. We observe from (1.78) that in order to know the sign in the functional equation, one does not need to know the individual ε_p, but only their product. We therefore consider the endomorphism

$$w_N := \prod_{p|N} w_p. \tag{1.86}$$

Note here that the w_p commute with each other, a fact which is not obvious from their definition, but from the commutativity of the diagram (1.82). The endomorphism w_N of $S_k(\Gamma_0(N))$ is called the *Fricke involution*. It is simply given by

$$w_N f = f\big|_k \begin{bmatrix} & 1 \\ N & \end{bmatrix}, \tag{1.87}$$

which follows from the commutativity of the diagram

$$\begin{array}{ccc} \mathcal{A}_k^\circ(\mathcal{K}_0(N)) & \xrightarrow{\ \sim\ } & S_k(\Gamma_0(N)) \\ \Big\downarrow{\scriptstyle \begin{bmatrix} & 1 \\ N & \end{bmatrix}} & & \Big\downarrow{\scriptstyle w_N} \\ \mathcal{A}_k^\circ(\mathcal{K}_0(N)) & \xrightarrow{\ \sim\ } & S_k(\Gamma_0(N)) \end{array} \tag{1.88}$$

Here, the map on the left-hand side is right translation by $\begin{bmatrix} & 1 \\ N & \end{bmatrix} \in \mathrm{GL}(2, \mathbb{A}_{\mathrm{fin}})$. We see that if $w_N f = \varepsilon_N f$ with $\varepsilon_N \in \{\pm 1\}$, then the sign in the functional equation of $L(s, f)$ is $(-1)^{k/2}\varepsilon_N$.

Summary

Classical modular forms in $S_k(\Gamma_0(N))$ appear in many contexts. The point of view we have taken here is that they arise as certain vectors in cuspidal, automorphic representations $\pi \cong \otimes \pi_p$. It makes obvious a number of factorizations, like the Euler product for the L-functions of classical newforms or the sign in the functional equation of these L-functions. Looking at a particular π_p provides a certain localization of the corresponding newform f_π. The $\Gamma_0(\mathfrak{p}^n)$-invariant vectors in the space of π_p may be viewed as "local modular forms". By the oldforms principle, they all arise from the local newform via level raising operators, explaining the classical decomposition of $S_k(\Gamma_0(N))$ into oldforms and newforms.

Given π, for each prime p there is a "frequency" λ_p attached to π, namely the eigenvalue of the Hecke operator T_1 on the local newform in π_p. While λ_p does not determine π_p in all cases, it does determine the Euler factor $L_p(s, \pi_p)$. Using classical Hecke operators, it is possible to calculate the λ_p directly from f_π without reference to the underlying representation. If the Fourier expansion of f is known, then the λ_p can be directly read off the Fourier coefficients; see (1.71). The sign in the functional equation can be determined from f_π as well via the Fricke involution.

As we will explain in the following section, many aspects of this theory still work for cuspidal, automorphic representations of the group $\mathrm{GSp}(4, \mathbb{A})$, and the corresponding classical objects, which are Siegel modular forms with respect to paramodular groups. One aspect is considerably more difficult though, and this is the connection between Hecke eigenvalues and Fourier coefficients. A major motivation for the present work is to provide theorems clarifying this connection. Our main tool will be local nonarchimedean representation theory.

1.4 Paramodular Forms I

We saw in Sect. 1.1 how reinterpreting Dirichlet characters as idele class characters, or automorphic representations of $\mathrm{GL}(1, \mathbb{A})$, gives a number of conceptual insights. Similarly, reinterpreting elliptic cusp forms as vectors in cuspidal, automorphic representations of $\mathrm{GL}(2, \mathbb{A})$ gives structure to the spaces $S_k(\Gamma_0(N))$ and naturally produces a number of results from the classical theory.

As elliptic modular forms are *Siegel modular forms of degree* 1, it is natural to suspect that higher degree Siegel modular forms can also be understood with the help of automorphic representation theory. In general this program is in its infancy, but much progress has been made in degree 2. We will review what is known and what is not known.

Automorphic Forms

Standard references for Siegel modular forms are [42, 76] and [3]. We will however start with the adelic language. For an integer n such that $n > 0$, let

$$J_n = \begin{bmatrix} & 1_n \\ -1_n & \end{bmatrix}, \tag{1.89}$$

where 1_n is the $n \times n$ identity matrix. The *general symplectic group* $\mathrm{GSp}(2n)$ of rank n is the algebraic group whose R-points are

$$\mathrm{GSp}(2n, R) = \{g \in \mathrm{GL}(2n, R) \mid {}^t g J_n g = \lambda(g) J_n \text{ for some } \lambda(g) \in R^\times\} \tag{1.90}$$

for any commutative ring R. The map $\lambda : \mathrm{GSp}(2n, R) \to R^\times$ is called the *multiplier homomorphism*. Its kernel is the *symplectic group* $\mathrm{Sp}(2n, R)$. Note that $\mathrm{GSp}(2) = \mathrm{GL}(2)$ and $\mathrm{Sp}(2) = \mathrm{SL}(2)$.

The group $\mathrm{GSp}(2n, \mathbb{A}_{\mathrm{fin}})$ is defined as the restricted directed product of the $\mathrm{GSp}(2n, \mathbb{Q}_p)$ with respect to the compact groups $\mathrm{GSp}(2n, \mathbb{Z}_p)$, for all prime numbers p. The adelization of $\mathrm{GSp}(2n)$ is $\mathrm{GSp}(2n, \mathbb{A}) := \mathrm{GSp}(2n, \mathbb{R}) \times \mathrm{GSp}(2n, \mathbb{A}_{\mathrm{fin}})$. The center $Z(\mathbb{A})$ of $\mathrm{GSp}(2, \mathbb{A})$ consists of all scalar matrices and is isomorphic to the ideles.

Just as in the $\mathrm{GL}(2)$ case, an *automorphic form* on $\mathrm{GSp}(2n, \mathbb{A})$ is a continuous function $\Phi : \mathrm{GSp}(2n, \mathbb{A}) \to \mathbb{C}$ satisfying $\Phi(\rho g) = \Phi(g)$ for all $\rho \in \mathrm{GSp}(2n, \mathbb{Q})$ and $g \in \mathrm{GSp}(2n, \mathbb{A})$, plus some additional regularity conditions. To keep things simple, we will require all automorphic forms to be invariant under the center $Z(\mathbb{A})$. The cuspidality condition (cf. (1.17)) now is

$$\int\limits_{N(\mathbb{Q})\backslash N(\mathbb{A})} \Phi(ng)\, dn = 0 \tag{1.91}$$

for all $g \in \mathrm{GSp}(2n, \mathbb{A})$ and the unipotent radicals N of all parabolic subgroups. We let $\mathcal{A}$ be the space of all automorphic forms on $\mathrm{GSp}(2n, \mathbb{A})$, and $\mathcal{A}^\circ$ be the subspace of all cuspidal automorphic forms. As in the GL(2) case our focus will be on $\mathcal{A}^\circ$.

The group $\mathrm{GSp}(2n, \mathbb{A}_{\mathrm{fin}})$ acts on $\mathcal{A}^\circ$ by right translations. At the archimedean place we have an action of the compact group K_∞ consisting of all matrices of the form $\left[\begin{smallmatrix} A & B \\ -B & A \end{smallmatrix}\right]$ for which $A + \mathrm{i}B$ is in $\mathrm{U}(n)$. We also have an action of the Lie algebra of $\mathrm{GSp}(2n, \mathbb{R})$ via the derived representation. This makes $\mathcal{A}^\circ$ a $(\mathfrak{g}, K_\infty) \times \mathrm{GSp}(2n, \mathbb{A}_{\mathrm{fin}})$ module. We will adopt a slightly incorrect terminology and refer to any such module as a "representation of $\mathrm{GSp}(2n, \mathbb{A})$".

There is an inner product on $\mathcal{A}^\circ$ defined analogously to (1.19). By general principles, $\mathcal{A}^\circ$ decomposes into a discrete direct sum of irreducible representations, each occurring with finite multiplicity. Any such irreducible is called a *cuspidal, automorphic representation.* If π is one of them, then in a certain technical sense $\pi \cong \otimes \pi_p$ with local representations π_p. More precisely, for finite primes, π_p is an irreducible, admissible representation of $\mathrm{GSp}(2n, \mathbb{Q}_p)$, and at the archimedean place π_∞ is an irreducible, admissible $(\mathfrak{g}, K_\infty)$ module.

Siegel Modular Forms

Let $\pi \cong \otimes \pi_p$ be a cuspidal, automorphic representation of $\mathrm{GSp}(2n, \mathbb{A})$, acting on the irreducible subspace V of $\mathcal{A}^\circ$. In the GL(2) case we were able to find in V a distinguished vector Φ_π and descend to the upper half plane to obtain a modular form f_π of weight k. In order for this to work, we needed to assume that the archimedean component π_∞ is a discrete series representation $\mathcal{D}(k)$ of some weight k.

We will make a similar assumption in the $\mathrm{GSp}(2n)$ case. The symbol $\mathcal{D}(k)$ will now denote a certain *lowest weight module* of lowest weight k. These exist for all positive k. (This time there is no assumption that k must be even.) For $k > n$ they will again be discrete series representations, but for $k \leq n$ they are not. In either case there exists a distinguished vector v_∞ in $\mathcal{D}(k)$, unique up to scalars, and characterized by the property that it is annihilated by a certain subalgebra $\mathfrak{p}_\mathbb{C}^-$ of the complexified Lie algebra $\mathfrak{g}_\mathbb{C}$. (See Sect. 3.5 of [6] for the definition of $\mathfrak{p}_\mathbb{C}^-$.) If $n = 2$, then the vector v_∞ has weight (k, k); here, we say a vector v has weight $(k, \ell) \in \mathbb{Z}^2$ if

$$\pi_\infty\left(\begin{bmatrix} \cos\theta_1 & & \sin\theta_1 & \\ & \cos\theta_2 & & \sin\theta_2 \\ -\sin\theta_1 & & \cos\theta_1 & \\ & -\sin\theta_2 & & \cos\theta_2 \end{bmatrix}\right)v = e^{ik\theta_1 + i\ell\theta_2}v \tag{1.92}$$

for all $\theta_1, \theta_2 \in \mathbb{R}$.

Next we consider finite primes p. For almost all such the representation π_p will be *spherical* (or *unramified*), meaning π_p contains a non-zero vector v_p, unique up to multiples, which is invariant under the maximal compact subgroup $\mathrm{GSp}(2n, \mathbb{Z}_p)$. This will be our distinguished vector in π_p.

To find distinguished vectors in π_p if π_p is not spherical is in general an unsolved problem. For now let v_p be *any* non-zero vector in π_p. By smoothness, v_p is stabilized by the *principal congruence subgroup*

$$\Gamma(\mathfrak{p}^n) := \{g \in \mathrm{GSp}(2n, \mathbb{Q}_p) \mid g \equiv 1 \ (\mathrm{mod}\ \mathfrak{p}^n)\} \tag{1.93}$$

for some non-negative n.

So now every v_p is stabilized by some open-compact subgroup $\mathcal{K}_p$, with $\mathcal{K}_p = \mathrm{GSp}(2n, \mathbb{Z}_p)$ for almost all p. Let $\mathcal{K}_{\mathrm{fin}} = \prod_{p<\infty} \mathcal{K}_p$, and

$$\Gamma := \mathrm{GSp}(2n, \mathbb{Q}) \cap \mathrm{GSp}(2n, \mathbb{R})^+ \mathcal{K}_{\mathrm{fin}}, \tag{1.94}$$

where $\mathrm{GSp}(2n, \mathbb{R})^+$ is the subgroup of $\mathrm{GSp}(2n, \mathbb{R})$ consisting of elements with positive multiplier. Then Γ is a subgroup of $\mathrm{Sp}(2n, \mathbb{Q})$ which, for some positive integer N, contains a principal congruence subgroup

$$\Gamma(N) := \{g \in \mathrm{Sp}(2n, \mathbb{Z}) \mid g \equiv 1 \ (\mathrm{mod}\ N)\} \tag{1.95}$$

with finite index.

Having chosen a distinguished vector v_∞ in π_∞, spherical vectors v_p for almost all finite p, and random vectors v_p for the remaining places, let Φ be the automorphic form in the space of π corresponding to $\prod_{p\le\infty} v_p$ via the isomorphism $\pi \cong \otimes\pi_p$. If the multiplier map $\lambda : \mathcal{K}_p \to \mathbb{Z}_p^\times$ is surjective for all finite p, then

$$\mathrm{GSp}(2n, \mathbb{A}) = \mathrm{GSp}(2n, \mathbb{Q}) \cdot \mathrm{GSp}(2n, \mathbb{R})^+ \mathcal{K}_{\mathrm{fin}} \tag{1.96}$$

by *strong approximation* for the group $\mathrm{Sp}(2n)$. In this case Φ is determined by its values on $\mathrm{GSp}(2n, \mathbb{R})^+$, or even on $\mathrm{Sp}(2n, \mathbb{R})$, since we assumed the center acts trivially. In either event we may consider the restriction of Φ to $\mathrm{Sp}(2n, \mathbb{R})$.

Now $\mathrm{GSp}(2n, \mathbb{R})^+$ acts on the *Siegel upper half space* $\mathcal{H}_n$ of degree n, consisting of all complex, symmetric $n\times n$ matrices Z whose imaginary part is positive definite, via

$$g\langle Z\rangle = (AZ + B)(CZ + D)^{-1} \tag{1.97}$$

for $g = \left[\begin{smallmatrix} A & B \\ C & D \end{smallmatrix}\right]$. This action is transitive, and the stabilizer of the point $I_n := \mathrm{i}1_n$ is K_∞. There is a right action of $\mathrm{GSp}(2n, \mathbb{R})^+$ on functions $F : \mathcal{H}_n \to \mathbb{C}$ given by

$$(F\big|_k g)(Z) = \lambda(g)^{nk/2} \det(CZ + D)^{-k} F(g\langle Z\rangle) \tag{1.98}$$

for $g = \left[\begin{smallmatrix} A & B \\ C & D \end{smallmatrix}\right]$. The effect of the factor $\lambda(g)^{nk/2}$ is to make the center act trivially.

Generalizing (1.32), we now attach to our automorphic form Φ a function $F : \mathcal{H}_n \to \mathbb{C}$ by setting

$$F(X + \mathrm{i}Y) = \det(Y)^{-k/2}\,\Phi(\begin{bmatrix} 1 & X \\ & 1 \end{bmatrix}\begin{bmatrix} Y^{1/2} & \\ & Y^{-1/2} \end{bmatrix}), \tag{1.99}$$

where $Y^{1/2}$ is the square root of the positive definite matrix Y. Then one can verify that

$$\Phi(g) = (F\big|_k g)(I_n) \tag{1.100}$$

for all $g \in \mathrm{GSp}(2n, \mathbb{R})^+$ and $I_n = \mathrm{i}1_n$. The fact that Φ is annihilated by $\mathfrak{p}_{\mathbb{C}}^-$ translates into F being holomorphic. The relationship (1.100) and the invariance properties of Φ imply that

$$F\big|_k \gamma = F \tag{1.101}$$

for all γ in the group Γ defined in (1.94). A holomorphic function F on $\mathcal{H}_n$ with the property (1.101) is called a *Siegel modular form* of weight k and degree n with respect to Γ. In the case $n = 1$, which is the case of elliptic modular forms discussed earlier, one has to also require that the function is holomorphic at the cusps, which it turns out is automatic for $n \geq 2$ by the *Koecher principle* proved in [78].

Our function F *vanishes at the cusps* since Φ satisfies the cuspidality condition (1.91); we say that F is a *cusp form*. Let $S_k(\Gamma)$ be the space of all cusp forms of weight k with respect to Γ. (We do not include the degree n in the notation, hoping it will be clear from the context.) Since one can reverse the above process and use cusp forms to generate automorphic representations, every element of $S_k(\Gamma)$ is a linear combination of cusp forms originating as above from a vector in a cuspidal, automorphic representation.

GSp(4)

In the GL(2) case Theorem 1.2.2 established a correspondence between cuspidal, automorphic representations and newforms in $S_k(\Gamma_0(N))$. A key ingredient was the existence of local newforms, i.e., distinguished vectors in nonarchimedean representations. If there is any hope for a correspondence theorem in degree n, then we need a local newform theory for irreducible, admissible representations of $\mathrm{GSp}(2n, \mathbb{Q}_p)$. Unfortunately, no such theory is currently known for $n \geq 3$. However for $n = 2$ a local newform theory does exist; see [105].

Besides this essential local ingredient, the correspondence theorem for GL(2) also required the strong multiplicity one result of Theorem 1.2.1. Strong multiplicity one is no longer true for cuspidal, automorphic representations of $\mathrm{GSp}(2n, \mathbb{A})$ if $n \geq 2$. Hence, an important global ingredient for a correspondence theorem is also not available in the higher degree case. However, in the case $n = 2$ the work of Arthur [4, 5] has clarified to what extent strong multiplicity one fails, thanks to a complete description of the *packet structure* of cuspidal, automorphic representations of $\mathrm{GSp}(4, \mathbb{A})$.

It turns out that for $n = 2$ the existing local newform theory for representations of $\mathrm{GSp}(4, \mathbb{Q}_p)$ and the global packet structure combine to allow for a correspondence theorem involving "most" cuspidal, automorphic representations of $\mathrm{GSp}(4, \mathbb{A})$, and the class of Siegel modular forms of degree 2 known as *paramodular forms*. From now on we will restrict exclusively to the case of GSp(4).

Local Paramodular New- and Oldforms

We temporarily switch to a local setting, letting F be a finite extension of $\mathbb{Q}_p$ for some prime number p. Let $\mathfrak{p}$ be the maximal ideal in the ring of integers $\mathfrak{o}$ of F. When considering representations of $\mathrm{GL}(2, F)$, we had success by looking at vectors fixed under the congruence subgroup $\Gamma_0(\mathfrak{p}^n)$ defined in (1.55). Now let (π, V) be an irreducible, admissible representation of $\mathrm{GSp}(4, F)$ for which the center acts trivially. It is tempting to again consider fixed vectors under

$$\Gamma_0(\mathfrak{p}^n) = \{\left[\begin{smallmatrix} A & B \\ C & D \end{smallmatrix}\right] \in \mathrm{GSp}(4, \mathfrak{o}) \mid C \equiv 0 \ (\mathrm{mod}\ \mathfrak{p}^n)\}. \tag{1.102}$$

However, examples show that this sequence of congruence subgroups does not lead to a good newforms theory. For example, there is in general no uniqueness at the minimal level, and there is no oldforms principle.

Instead, the right setting is provided by the paramodular groups. In the global context, these groups arise naturally in the consideration of abelian surfaces. For a non-negative integer n let

$$J(\varpi^n) = \begin{bmatrix} & & 1 & \\ & & & \varpi^n \\ -1 & & & \\ & -\varpi^n & & \end{bmatrix}, \tag{1.103}$$

and if R is a subring of F define

$$\begin{aligned} &\mathrm{GSp}(J(\varpi^n), R) \\ &\quad = \{g \in \mathrm{GL}(4, R) \mid {}^t g J(\varpi^n) g = \lambda(g) J(\varpi^n) \text{ for some } \lambda(g) \in R^\times\}. \end{aligned} \tag{1.104}$$

Then $\mathrm{GSp}(J(\varpi^0), F) = \mathrm{GSp}(4, F)$ and $\mathrm{GSp}(J(\varpi^0), \mathfrak{o}) = \mathrm{GSp}(4, \mathfrak{o})$. More generally, since all non-degenerate four-dimensional symplectic forms over F

are equivalent, we have $\mathrm{GSp}(J(\varpi^n), F) \cong \mathrm{GSp}(4, F)$ so that we may regard $\mathrm{GSp}(J(\varpi^n), \mathfrak{o})$ as a subgroup of $\mathrm{GSp}(4, F)$. Concretely, if

$$h = \begin{bmatrix} 1 & & & \\ & 1 & & \\ & & 1 & \\ & & & \varpi^n \end{bmatrix}, \tag{1.105}$$

then $h\mathrm{GSp}(J(\varpi^n), F)h^{-1} = \mathrm{GSp}(4, F)$, and we define the *paramodular group* $\mathrm{K}(\mathfrak{p}^n)$ to be $h\mathrm{GSp}(J(\varpi^n), \mathfrak{o})h^{-1}$. Explicitly, we have

$$\mathrm{K}(\mathfrak{p}^n) = \{g \in \mathrm{GSp}(4, F) \mid \lambda(g) \in \mathfrak{o}^\times\} \cap \begin{bmatrix} \mathfrak{o} & \mathfrak{p}^n & \mathfrak{o} & \mathfrak{o} \\ \mathfrak{o} & \mathfrak{o} & \mathfrak{o} & \mathfrak{p}^{-n} \\ \mathfrak{o} & \mathfrak{p}^n & \mathfrak{o} & \mathfrak{o} \\ \mathfrak{p}^n & \mathfrak{p}^n & \mathfrak{p}^n & \mathfrak{o} \end{bmatrix}. \tag{1.106}$$

It is interesting to note that in Sect. 7.1 of [127], Sect. 1 of [21], Sect. 5 of [52] and Sect. 7 of [128] sequences of congruence subgroups in split orthogonal groups $\mathrm{SO}(2m+1, F)$ are defined for any positive integer m. Using the exceptional isomorphisms $\mathrm{PGL}(2) \cong \mathrm{SO}(3)$ and $\mathrm{PGSp}(4) \cong \mathrm{SO}(5)$, the images of $\Gamma_0(\mathfrak{p}^n)$ in $\mathrm{SO}(3, F)$ and $\mathrm{K}(\mathfrak{p}^n)$ in $\mathrm{SO}(5, F)$ are conjugate to the congruence subgroups defined in these references.

We now let

$$V(n) = \{v \in V \mid \pi(k)v = v \text{ for all } k \in \mathrm{K}(\mathfrak{p}^n)\}. \tag{1.107}$$

If $V(n)$ is non-zero for some n, then we say that π is *paramodular*. While not every representation is paramodular, most representations are, in particular all *generic* representations. If π is paramodular, we let N_π be the minimal n for which $V(n) \neq 0$, and call this number the *minimal level* or *conductor exponent* of π.

One of the main results of [105] is that $\dim V(N_\pi) = 1$. Hence, π admits a *local newform* v_{new}, unique up to scalars.

As for oldforms, there are two level raising operators $\theta, \theta' : V(n) \to V(n+1)$, similar to the GL(2) case. In addition, there is a third operator $\eta : V(n) \to V(n+2)$. All these operators commute with each other. The oldform principle proven in [105] states that the spaces $V(n)$ for $n > N_\pi$ are spanned by the vectors $\theta'^i \theta^j \eta^k v$ for $i, j, k \geq 0$ and $i+j+2k = n-N_\pi$ where v spans the one-dimensional space $V(N_\pi)$. For generic representations these vectors are all linearly independent, producing the following picture

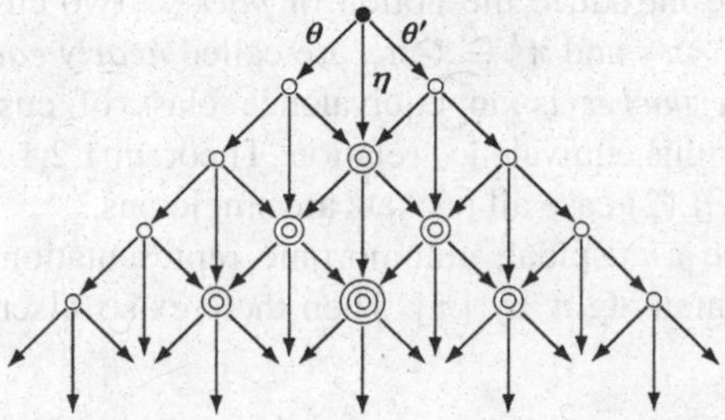

(1.108)

similar to (1.47) but with the additional η operator. Instead of the linear growth of dimensions given by (1.23) we now have the quadratic growth

$$\dim V(n) = \begin{cases} 0 & \text{for } n < N_p, \\ \left\lfloor \dfrac{(n - N_p + 2)^2}{4} \right\rfloor & \text{for } n \geq N_p. \end{cases} \tag{1.109}$$

For paramodular non-generic representations there are linear relations between θ, θ' and η, leading the picture (1.108) to collapse. For example, for the class of *Saito-Kurokawa representations*, defined in Proposition 5.5.1 of [105], the operators θ and θ' are linearly dependent, and we get

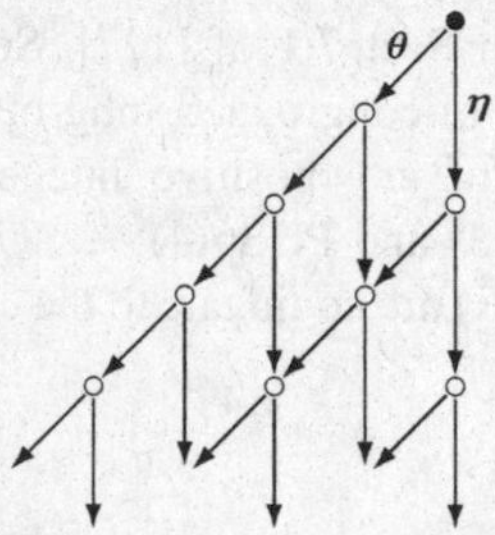

(1.110)

with the linear dimension formula

$$\dim V(n) = \begin{cases} 0 & \text{for } n < N_p, \\ \left\lfloor \dfrac{n - N_p + 2}{2} \right\rfloor & \text{for } n \geq N_p. \end{cases} \tag{1.111}$$

We next need to understand which cuspidal, automorphic representations have local components that are paramodular at all finite places.

Five Types of Cuspidal, Automorphic Representations

We review some results of [4]. While in the GL(2) case there is only one type of cuspidal, automorphic representation, in the GSp(4) case there are five types. To describe these, we introduce the notion of *packet*. Two cuspidal, automorphic representations $\pi \cong \otimes\pi_p$ and $\pi' \cong \otimes\pi'_p$ are called *nearly equivalent* if $\pi_p \cong \pi'_p$ for almost all p. A *packet* is an equivalence class of cuspidal, automorphic representations under this equivalence relation. Theorem 1.2.1 can be reformulated as saying that in the GL(2) case all packets are singletons.

Let $\pi \cong \otimes\pi_p$ be a cuspidal, automorphic representation of GSp(4, $\mathbb{A}$). We denote the packet containing π by $[\pi]$. Then there exists a set S of places, which may be finite or infinite, and for each $p \in S$ a *local packet* consisting of π_p and another representation π'_p, such that each other element of $[\pi]$ is of the form

$$\Big(\bigotimes_{p\in T}\pi'_p\Big)\otimes\Big(\bigotimes_{p\notin T}\pi_p\Big) \tag{1.112}$$

for a finite subset T of S. If T can be any finite subset of S, then we say that $[\pi]$ is *stable*, otherwise *unstable*. In the unstable case, there is a parity condition for T: If S is finite, then T must have the same parity as S, and if S is infinite, then T must have even parity.

We say a packet is *tempered* if for each of its members $\pi \cong \otimes\pi_p$ the π_p are tempered representations for all places p. We say a packet is *generic* if it contains a globally generic member.

According to [4], the packets for GSp(4) come in five types. The *general type* **(G)** consists of π's for which the L-function $L(s,\pi)$ is *primitive*, i.e., not a product of L-functions for GL(1) and GL(2). The packets of general type are finite, stable, tempered and generic. The *Yoshida type* **(Y)** consists of π's for which $L(s,\pi) = L(s,\pi_1)L(s,\pi_2)$ with cuspidal automorphic representations π_1, π_2 of $\mathrm{GL}(2,\mathbb{A})$. The packets of type **(Y)** are finite, unstable, tempered and generic.

The remaining three types of packets are associated with the three parabolic subgroups B, P and Q; see Sect. 2.2 for the definition of the parabolic subgroups. All these packets are non-tempered and non-generic; in fact, the local components for any π in any of these packets are non-generic at every place. Packets of type **(Q)**, also known as the *Soudry type*, are infinite and stable. Packets of type **(P)**, also known as the *Saito-Kurokawa type*, are finite and unstable. Packets of type **(B)**, also known as the *Howe-Piatetski-Shapiro type*, are infinite and unstable. The names refer to places in the literature where examples of these representations, or corresponding Siegel modular forms, have first been exhibited; see [58, 81, 122, 131].

Table 1.4 gives an overview of the five different types of packets of cuspidal, automorphic representations of $\mathrm{GSp}(4,\mathbb{A})$. We mention that [4] lists a sixth type consisting of 1-dimensional representations. While these are in the discrete spectrum, they are not cuspidal, and so are not relevant for our purposes.

It remains to explain the last column of Table 1.4. We say a cuspidal, automorphic representation $\pi \cong \otimes\pi_p$ is *paramodular of weight k* if $\pi_\infty = \mathcal{D}(k)$ for some $k > 0$

Table 1.4 Arthur packets for GSp(4)

Type	Name	Finite	Stable	Tempered	Generic	Paramodular
(G)	General	•	•	•	•	•
(Y)	Yoshida	•		•	•	
(Q)	Soudry		•			
(P)	Saito-Kurokawa	•				•
(B)	Howe-Piatetski-Shapiro					

and π_p is paramodular for all finite p. It is such π that contribute to spaces of Siegel modular forms of weight k with respect to the paramodular group. We say a packet is paramodular if it contains a paramodular member of some weight; such a member is unique if it exists. The last column in Table 1.4 indicates which types admit paramodular packets; see [114, 115] for proofs. It is the paramodular packets for which we will be able to formulate a correspondence theorem.

Paramodular Forms

Let $\pi \cong \otimes \pi_p$ be a cuspidal, automorphic representation of $\mathrm{GSp}(4, \mathbb{A})$ which is paramodular of weight k. Let v_∞ be the lowest weight vector in the space of π_∞. For finite primes, let v_p be the local newform in the space of π_p. Let Φ_π be the automorphic form corresponding to $\otimes_{p\le\infty} v_p$. Then Φ_π is a distinguished vector in the space of π; we call it a *global newform*. Let N_p be the minimal level of π_p. The *conductor* of π is the positive integer $N_\pi := \prod_p p^{N_p}$. Let F_π be the function corresponding to Φ_π via (1.100). Then $F_\pi \in S_k(\mathrm{K}(N_\pi))$, where

$$\mathrm{K}(N) = \mathrm{Sp}(4, \mathbb{Q}) \cap \begin{bmatrix} \mathbb{Z} & N\mathbb{Z} & \mathbb{Z} & \mathbb{Z} \\ \mathbb{Z} & \mathbb{Z} & \mathbb{Z} & N^{-1}\mathbb{Z} \\ \mathbb{Z} & N\mathbb{Z} & \mathbb{Z} & \mathbb{Z} \\ N\mathbb{Z} & N\mathbb{Z} & N\mathbb{Z} & \mathbb{Z} \end{bmatrix}, \qquad N \in \mathbb{Z}_{>0}, \tag{1.113}$$

is the *paramodular group* of level N. Similar to the local setting, the paramodular group $\mathrm{K}(N)$ is $h\mathrm{Sp}(J(N), \mathbb{Z})h^{-1}$ where

$$J(N) = \begin{bmatrix} & & 1 & \\ & & & N \\ -1 & & & \\ & -N & & \end{bmatrix}, \qquad h = \begin{bmatrix} 1 & & & \\ & 1 & & \\ & & 1 & \\ & & & N \end{bmatrix}, \tag{1.114}$$

and

$$\mathrm{Sp}(J(N), \mathbb{Z}) = \{g \in \mathrm{GL}(4, \mathbb{Z}) \mid {}^t g J(N) g = J(N)\}. \tag{1.115}$$

As mentioned earlier, the group $\mathrm{Sp}(J(N), \mathbb{Z})$ is connected to abelian surfaces. In fact, the isomorphism classes of polarized abelian surfaces of type $(1, N)$ are in natural bijection with $\mathrm{Sp}(J(N), \mathbb{Z})\backslash\mathcal{H}_2$ (see [66], p. 118–120). A further connection to abelian surfaces is provided by the Paramodular Conjecture of Brumer and Kramer which we will briefly discuss below. Historically, $\mathrm{K}(N)$ is a special case of one of the groups $\Delta(\mathfrak{G}, \mathfrak{H})$ considered by Siegel in [120]; see [29] for a list of some early references. The paramodular group is also defined on p. 124 of [31]. Perhaps the earliest use of the term "paramodular group" in the literature occurs in [117].

Siegel modular forms of degree 2 with respect to $\mathrm{K}(N)$ for some N are referred to as *paramodular forms*. Hence, we are able to associate to a cuspidal, automorphic

π which is paramodular of weight k and has conductor N_π a paramodular cusp form F_π of weight k, determined up to scalars.

For any integer N with prime factorization $N = \prod p^{n_p}$, let

$$\mathcal{K}(N) := \prod_{p<\infty} \mathrm{K}(\mathfrak{p}^{n_p}). \tag{1.116}$$

Then $\mathcal{K}(N)$ is an open-compact subgroup of $\mathrm{GSp}(4, \mathbb{A}_{\rm fin})$. We can define a space $\mathcal{A}_k^\circ(\mathcal{K}(N))$ analogously to the space $\mathcal{A}_k^\circ(\mathcal{K}_0(N))$ in the GL(2) case; see page 12, and Sect. 10.2 for more details. If we associate to every $\Phi \in \mathcal{A}_k^\circ(\mathcal{K}(N))$ the function $F : \mathcal{H}_2 \to \mathbb{C}$ with the property (1.100), then we obtain an isomorphism

$$\mathcal{A}_k^\circ(\mathcal{K}(N)) \cong S_k(\mathrm{K}(N)). \tag{1.117}$$

Our Φ_π above is an element of $\mathcal{A}_k^\circ(\mathcal{K}(N_\pi))$, and F_π is the element of $S_k(\mathrm{K}(N_\pi))$ corresponding to Φ_π via the isomorphism (1.117). We call F_π a *paramodular newform.*

For a fixed N let $S_k(\mathrm{K}(N))_{\rm new}$ be the space spanned by the newforms of level N, i.e., spanned by the F_π for all π's that are paramodular of weight k and conductor N. Just like on page 18 we define the space $S_k(\mathrm{K}(N))_{\rm old}$ of *paramodular oldforms.* Since every Siegel modular form is of adelic origin, it is clear that

$$S_k(\mathrm{K}(N)) = S_k(\mathrm{K}(N))_{\rm new} \oplus S_k(\mathrm{K}(N))_{\rm old}. \tag{1.118}$$

Like we did in (1.48), the local paramodular level raising operators θ_p, θ'_p and η_p translate into compatible versions on spaces of paramodular forms, so that the diagrams

$$\begin{array}{ccc} \mathcal{A}_k^\circ(\mathcal{K}(N)) & \xrightarrow{\sim} & S_k(\mathrm{K}(N)) \\ {\scriptstyle \theta_p,\theta'_p}\downarrow & & \downarrow{\scriptstyle \theta_p,\theta'_p} \\ \mathcal{A}_k^\circ(\mathcal{K}(Np)) & \xrightarrow{\sim} & S_k(\mathrm{K}(Np)) \end{array} \tag{1.119}$$

and

$$\begin{array}{ccc} \mathcal{A}_k^\circ(\mathcal{K}(N)) & \xrightarrow{\sim} & S_k(\mathrm{K}(N)) \\ {\scriptstyle \eta_p}\downarrow & & \downarrow{\scriptstyle \eta_p} \\ \mathcal{A}_k^\circ(\mathcal{K}(Np^2)) & \xrightarrow{\sim} & S_k(\mathrm{K}(Np^2)) \end{array} \tag{1.120}$$

are commutative. (See Lemma 10.2.2 for the formulas defining the classical operators.) By the oldforms principle, we may characterize $S_k(\mathrm{K}(N))_{\rm old}$ as the space spanned by $\theta_p S_k(\mathrm{K}(Np^{-1}))$ and $\theta'_p S_k(\mathrm{K}(Np^{-1}))$ for $p \mid N$ and $\eta_p S_k(\mathrm{K}(Np^{-2}))$

for $p^2 \mid N$. The space $S_k(\mathrm{K}(N))_{\mathrm{new}}$ is then the orthogonal complement of $S_k(\mathrm{K}(N))_{\mathrm{old}}$ with respect to the Petersson inner product. It is thus possible to give definitions of $S_k(\mathrm{K}(N))_{\mathrm{old}}$ and $S_k(\mathrm{K}(N))_{\mathrm{new}}$ without reference to automorphic representations.

Eigenforms

Let N be a positive integer and p a prime not dividing N. In the GL(2) case we saw that there is essentially only one relevant Hecke operator at the place p on the space $\mathcal{A}_k^\circ(\mathcal{K}_0(N))$; this was the operator T_1 given by the formula (1.42). The compatible Hecke operator on $S_k(\Gamma_0(N))$ was denoted by $T(1, p)$ and is given by the formula (1.43).

Similar arguments show that there are essentially only two relevant Hecke operators on the space $\mathcal{A}_k^\circ(\mathcal{K}(N))$. They are denoted by $T_{0,1}(p)$ and $T_{1,0}(p)$, and are given by

$$(T_{0,1}(p)\Phi)(g) = \sum_{h \in K_p \left[\begin{smallmatrix} p &&& \\ & p && \\ && 1 & \\ &&& 1 \end{smallmatrix}\right] K_p/K_p} \Phi(gh) \tag{1.121}$$

and

$$(T_{1,0}(p)\Phi)(g) = \sum_{h \in K_p \left[\begin{smallmatrix} p &&& \\ & p^2 && \\ && p & \\ &&& 1 \end{smallmatrix}\right] K_p/K_p} \Phi(gh) \tag{1.122}$$

for $g \in \mathrm{GSp}(4, \mathbb{A})$; here, K_p abbreviates $\mathrm{GSp}(4, \mathbb{Z}_p)$. The classical versions of these operators will be denoted by $T(1, 1, p, p)$ and $T(p, 1, p, p^2)$. See (10.37) for their precise definitions. By Lemma 10.2.3 we have commutative diagrams similar to (1.45).

An element of $S_k(\mathrm{K}(N))$ is called an *eigenform* if it is an eigenvector for $T(1, 1, p, p)$ and $T(p, 1, p, p^2)$ for almost all p. The newform $F_\pi \in S_k(\mathrm{K}(N_\pi))$, coming from a paramodular representation $\pi \cong \otimes \pi_p$ of conductor N_π, has this property. More precisely, if

$$T_{0,1} v_p = \lambda_p v_p, \qquad T_{1,0} v_p = \mu_p v_p \tag{1.123}$$

with $\lambda_p, \mu_p \in \mathbb{C}$, where v_p is the spherical vector in the space of π_p, then

$$T(1, 1, p, p) F_\pi = p^{k-3} \lambda_p F_\pi, \qquad T(p, 1, p, p^2) F_\pi = p^{2(k-3)} \mu_p F_\pi \tag{1.124}$$

by Lemma 10.2.3.

The Correspondence Theorem

The map $\pi \mapsto F_\pi$ establishes a correspondence between cuspidal, automorphic representations of $\mathrm{GSp}(4,\mathbb{A})$, paramodular of weight k and with conductor N, and newforms in $S_k(\mathrm{K}(N))_{\mathrm{new}}$ up to multiples. Here, we adopt the convention that only the F_π are called newforms, making the statement trivially true. We would like the statement still to be true if "newforms" is replaced by "eigenforms". As we saw in the proof of Theorem 1.2.2, this requires a strong multiplicity one result. Now strong multiplicity one is not true for GSp(4), since the packets of cuspidal, automorphic representations generally have more than one element. However, as already mentioned, an investigation of local packets (see Theorem 1.1 of [114] and Sect. 5 of [115]) shows that a global packet can have at most one paramodular element. In other words, if we insist on paramodular representations only, then we restore the strong multiplicity one property. Thus, we obtain the following result.

Theorem 1.4.1 (Correspondence Theorem for Paramodular Forms) *Let k and N be positive integers. The map $\pi \mapsto F_\pi$ induces a bijection between the set of cuspidal, automorphic representations $\pi \cong \otimes\pi_p$ of $\mathrm{GSp}(4,\mathbb{A})$ (always assumed to have trivial central character) which are paramodular of weight k and have conductor N, and the eigenforms in $S_k(\mathrm{K}(N))_{\mathrm{new}}$ modulo scalars.*

There is a refinement of Theorem 1.4.1 with respect to Arthur types. Recall from Table 1.4 that only packets of type **(G)** or **(P)** can be paramodular. Let **X** be **G** or **P**. Let $S_k(\mathrm{K}(N))_{\mathrm{new}}^{(\mathbf{X})}$ be the subspace of $S_k(\mathrm{K}(N))_{\mathrm{new}}$ spanned by newforms originating from paramodular π of type **(X)**. We define $S_k(\mathrm{K}(N))_{\mathrm{old}}^{(\mathbf{X})}$ similarly. Then there are orthogonal decompositions

$$S_k(\mathrm{K}(N))_{\mathrm{new}} = S_k(\mathrm{K}(N))_{\mathrm{new}}^{(\mathbf{G})} \oplus S_k(\mathrm{K}(N))_{\mathrm{new}}^{(\mathbf{P})}, \tag{1.125}$$

$$S_k(\mathrm{K}(N))_{\mathrm{old}} = S_k(\mathrm{K}(N))_{\mathrm{old}}^{(\mathbf{G})} \oplus S_k(\mathrm{K}(N))_{\mathrm{old}}^{(\mathbf{P})}, \tag{1.126}$$

$$S_k(\mathrm{K}(N)) = S_k(\mathrm{K}(N))^{(\mathbf{G})} \oplus S_k(\mathrm{K}(N))^{(\mathbf{P})}, \tag{1.127}$$

where $S_k(\mathrm{K}(N))^{(\mathbf{X})} := S_k(\mathrm{K}(N))_{\mathrm{new}}^{(\mathbf{X})} \oplus S_k(\mathrm{K}(N))_{\mathrm{old}}^{(\mathbf{X})}$. The refined version of Theorem 1.4.1 is then as follows.

Theorem 1.4.2 (Refined Correspondence Theorem for Paramodular Forms) *Let* **X** *be* **G** *or* **P**. *Let k and N be positive integers. The map $\pi \mapsto F_\pi$ induces a bijection between the set of cuspidal, automorphic representations $\pi \cong \otimes\pi_p$ of $\mathrm{GSp}(4,\mathbb{A})$ (always assumed to have trivial central character) of type* **(X)** *which are paramodular of weight k and have conductor N, and the eigenforms in $S_k(\mathrm{K}(N))_{\mathrm{new}}^{(\mathbf{X})}$ modulo scalars.*

An Example

We will use the example of $S_{12}(\mathrm{K}(N))$ for $N \mid 16$ to illustrate how new- and oldforms in cuspidal, automorphic representations contribute to spaces of paramodular cusp forms. The required data comes from the work [99]. We note that, in contrast to the case of elliptic modular forms, there is no known general algorithm for computing the spaces $S_k(\mathrm{K}(N))$.

There are 35 cuspidal, automorphic representations of $\mathrm{GSp}(4, \mathbb{A})$ that are paramodular of weight 12 and with conductor dividing 16. Some of these are grouped together into Galois orbits. Table 1.5 gives a name to each orbit and shows the size of the orbits. There are 26 representations of type **(G)** and 9 representations of type **(P)**.

Among the 35 representations there is only one of conductor 1. It is of type **(P)** and comprises the orbit π_1' in Table 1.5. The newform corresponding to π_1' is the

Table 1.5 Paramodular forms of weight 12 and level dividing 16

			Contribution to dim $S_{12}(\mathrm{K}(N))$ for $N = \ldots$				
Type	Orbit	Size	1	2	4	8	16
(G)	π_4	1	0	0	**1**	2	4
	π_{8a}	1	0	0	0	**1**	2
	π_{8b}	1	0	0	0	**1**	2
	π_{8c}	2	0	0	0	**2**	4
	π_{16a}	1	0	0	0	0	**1**
	π_{16b}	1	0	0	0	0	**1**
	π_{16c}	2	0	0	0	0	**2**
	π_{16d}	2	0	0	0	0	**2**
	π_{16e}	2	0	0	0	0	**2**
	π_{16f}	5	0	0	0	0	**5**
	π_{16g}	8	0	0	0	0	**8**
(P)	π_1'	1	**1**	1	2	2	3
	π_2'	1	0	**1**	1	2	2
	π_8'	2	0	0	0	**2**	2
	π_{16a}'	2	0	0	0	0	**2**
	π_{16b}'	3	0	0	0	0	**3**
dim $S_{12}(\mathrm{K}(N))$			1	2	4	12	45
dim $S_{12}(\mathrm{K}(N))_{\mathrm{new}}$			1	1	1	6	26
dim $S_{12}(\mathrm{K}(N))^{(\mathbf{G})}$			0	0	1	6	33
dim $S_{12}(\mathrm{K}(N))^{(\mathbf{G})}_{\mathrm{new}}$			0	0	1	4	21
dim $S_{12}(\mathrm{K}(N))^{(\mathbf{P})}$			1	2	3	6	12
dim $S_{12}(\mathrm{K}(N))^{(\mathbf{P})}_{\mathrm{new}}$			1	1	0	2	5

The notation is such that the cuspidal, automorphic representations π_{Nx} and π'_{Nx} have conductor N

cusp form χ_{12} defined on page 195 of [65]. (It is a *Saito-Kurokawa lift* of the unique elliptic cusp form of weight 22 with respect to $\mathrm{SL}(2,\mathbb{Z})$.)

The numbers in the last five columns of Table 1.5 show the dimensions of the spaces $V(n)$ defined in (1.107) for the local component at $p = 2$ for all the automorphic representations in the given orbit combined. For example, for π_1', we see the beginning of the linear growth $1, 1, 2, 2, 3, 3 \ldots$ typical for Saito-Kurokawa representations; see (1.111). For π_4 we see the beginning of the quadratic growth $1, 2, 4, 6, 9, \ldots$ typical for generic representations; see (1.109). In each case the local newforms are indicated in bold.

Adding up the numbers in the columns of Table 1.5 for each orbit gives the dimension of $S_{12}(\mathrm{K}(N))$ for N dividing 16. Adding up the numbers for only the newforms gives $\dim S_{12}(\mathrm{K}(N))_{\mathrm{new}}$. We also see the refined dimensions for the spaces $S_k(\mathrm{K}(N))^{(\mathbf{G})}$ and $S_k(\mathrm{K}(N))^{(\mathbf{P})}$ and their newform subspaces.

1.5 Paramodular Forms II

In the previous section we were able to formulate a correspondence between cuspidal, automorphic representations π which are paramodular of weight k and have conductor N, and newforms F_π in $S_k(\mathrm{K}(N))$. In this section we consider the L-function of π, and how to calculate its Euler factors from the Fourier coefficients of F_π. As we saw in Sect. 1.3, this is a rather straightforward exercise in the $\mathrm{GL}(2)$ case. In the $\mathrm{GSp}(4)$ case however, it will lead us to the new results contained in this monograph.

We remark that the L-function $L(s,\pi)$ considered in this work is always the degree 4, or *spin* L-function. We do not consider the degree 5 L-function, which, perhaps confusingly, is known as the *standard* L-function for automorphic representations of $\mathrm{GSp}(4)$.

Local Representations

As before let F be a finite extension of $\mathbb{Q}_p$ for some prime number p. Let $\mathfrak{o}$ be the ring of integers of F, let $\mathfrak{p}$ be the maximal ideal of $\mathfrak{o}$, and let ϖ be a generator of $\mathfrak{p}$. Let q be the cardinality of the residue class field $\mathfrak{o}/\mathfrak{p}$.

Recall the Table 1.2 of irreducible, admissible, non-supercuspidal representations of $\mathrm{GL}(2,F)$. Thanks to the work [109] it is possible to make a similar table for $\mathrm{GSp}(4)$; see Table A.1. Instead of only 2 groups we now have 11 groups. We refer to Sect. 2.2 for the notations used in Table A.1. Prominent examples of representations in this table are the twists of the Steinberg representation $\sigma\mathrm{St}_{\mathrm{GSp}(4)}$ (type IVa) and the twists of the trivial representation $\sigma 1_{\mathrm{GSp}(4)}$ (type IVd). Some of the theorems of [105], as well as of the present work, require case by case considerations according to the classification of Table A.1. Hence, the classification takes on a more conspicuous role than in the GL(2) case.

Four Invariants Derived from the Local Newform

Recall that an irreducible, admissible representation (π, V) of $\mathrm{GSp}(4, F)$ with trivial central character is called *paramodular* if it admits non-zero vectors fixed by one of the paramodular groups $\mathrm{K}(\mathfrak{p}^n)$ defined in (1.106). Not every representation is paramodular; Table A.2 contains a complete list of those which are not.

Now assume that π is paramodular. We defined the minimal level (or conductor exponent) N_π to be the smallest n for which the fixed space $V(n)$ defined in (1.107) is not zero. Then $\dim V(N_\pi) = 1$ by Theorem 7.5.1 of [105]. This allows us to define invariants attached to π as the eigenvalues of certain Hecke operators on this one-dimensional space.

There will be three relevant Hecke operators, which we will define in more detail when we review the local paramodular theory in Sect. 2.3. The first two, called $T_{0,1}$ and $T_{1,0}$, look similar to the unramified Hecke operators defined in (1.121) and (1.122), but with $\mathrm{K}(\mathfrak{p}^{N_\pi})$ as the compact subgroup. We let λ_π be the eigenvalue of $T_{0,1}$ on $V(N_\pi)$, and μ_π be the eigenvalue of $T_{1,0}$ on $V(N_\pi)$. Working in certain models, Theorem 7.5 of [105] calculates λ_π and μ_π for all paramodular, irreducible, admissible representations π of $\mathrm{GSp}(4, F)$ with trivial central character.

The third relevant Hecke operator is the *Atkin-Lehner involution* defined by the GSp(4) analogue of the Atkin-Lehner element (1.58); see (2.31) for details. Its action on $V(N_\pi)$ produces a sign ε_π.

The four invariants N_π, λ_π, μ_π and ε_π were calculated in [105] for each paramodular, irreducible, admissible, paramodular representation of $\mathrm{GSp}(4, F)$ with trivial central character (except that for supercuspidals ε_π was left undetermined). Here, we list these quantities in Table A.4.

Local Factors

There are several ways to attach L- and ε-factors to irreducible, admissible representations π of $\mathrm{GSp}(4, F)$. For generic representations, there is the method of greatest common divisor of zeta integrals, which was carried out in [125] and is analogous to the method of [68] for GL(2). For non-supercuspidal representations, the desiderata of the local Langlands correspondence determine the L- and ε-factors, as explained in Sect. 2.4 of [105]. Since every paramodular representation is either generic or non-supercuspidal, we thus have factors $L(s, \pi)$ and $\varepsilon(s, \pi)$ attached at least to each paramodular representation. Tables A.8 and A.9 of [105] contain a complete list of these factors. They are designed to fit into a global (spin) L-function with analytic continuation and functional equation.

Remarkably, it turns out that $L(s,\pi)$ and $\varepsilon(s,\pi)$ can be expressed entirely in terms of the four invariants N_π, λ_π, μ_π and ε_π. For the ε-factor we have

$$\varepsilon(s,\pi)=\varepsilon_\pi q^{-N_\pi(s-1/2)}, \tag{1.128}$$

just like in the GL(2) case. For the L-factor see Theorems 2.3.4 and 2.3.5. The formulas for the L-factor given in Theorem 2.3.4 are analogous to those for GL(2) in (1.62). Indeed, the topic of higher level Hecke operators seems to be largely unexplored outside the setting of GL(2) and GSp(4).

As for the formula for $L(s,\pi)$ given in (2) of Theorem 2.3.4 in the case $N_\pi=1$, we mention that in this case there is a relation $\lambda_\pi\varepsilon_\pi+\mu_\pi+q+1=0$; see Sect. 7.2 of [105]. Furthermore, $\lambda_\pi\neq 0$, so that ε_π can be eliminated from the formula for $L(s,\pi)$. Hence, in all cases does $L(s,\pi)$ only depend on N_π, λ_π and μ_π, just like in the GL(2) case the L-factor only depended on N_π and λ_π.

L-Functions

Now let $\pi\cong\otimes\pi_p$ be a cuspidal, automorphic representation of GSp(4, $\mathbb{A}$) for which the center acts trivially, assumed to be paramodular of weight k. For each finite prime p let $N_p:=N_{\pi_p}$ be the minimal level of π_p. Let $N_\pi=\prod p^{N_p}$ be the global conductor of π. (The context will always make clear whether N_π is the global conductor or a local conductor exponent.)

Since π is assumed to be paramodular at every finite place, we have the local factors $L_p(s,\pi_p)$ and $\varepsilon_p(s,\pi_p)$ as defined above. We also need archimedean factors $L_\infty(s,\pi_\infty)$ and $\varepsilon_\infty(s,\pi_\infty)$. These follow from the archimedean local Langlands correspondence and are given by

$$L_\infty(s,\pi_\infty)=(2\pi)^{-2s-k+1}\,\Gamma\Big(s+k-\frac{3}{2}\Big)\Gamma\Big(s+\frac{1}{2}\Big),\quad \varepsilon_\infty(s,\pi_\infty)=(-1)^k; \tag{1.129}$$

see Table 5 of [113].

Having thus defined local factors at all places, let

$$L(s,\pi):=\prod_{p\le\infty}L_p(s,\pi_p) \tag{1.130}$$

be the global (spin) L-function of π. The product is absolutely convergent for $\mathrm{Re}(s)>3/2$. The L-function admits meromorphic continuation to all of $\mathbb{C}$ and satisfies the functional equation

$$L(s,\pi)=\varepsilon(s,\pi)L(1-s,\pi), \tag{1.131}$$

where $\varepsilon(s,\pi)=\prod_{p\le\infty}\varepsilon_p(s,\pi_p)$. This can be seen as follows. If π is of type **(G)**, then π can be transferred to a cuspidal, automorphic representation of $\mathrm{GL}(4,\mathbb{A})$ by [5]; in this case one can use results about L-functions on general linear groups. If π is of type **(P)**, then $L(s,\pi)$ is (essentially) a product of an L-function for GL(2) and shifted Riemann zeta functions (see Theorem 5.3 of [115]); in this case one can use results about GL(1) and GL(2). It follows also that $L(s,\pi)$ is entire if π is of type **(G)**, but has a pole at $s=3/2$ if π is of type **(P)**. The observation that the Arthur type of π is reflected in the analytic properties of $L(s,\pi)$ has been made in various places, for example [39, 90, 93].

Substituting the ε-factors from (1.128) and (1.129), the functional equation can be written as

$$L(s,\pi)=(-1)^k\Big(\prod_{p|N_\pi}\varepsilon_p\Big)N_\pi^{-(s-1/2)}L(1-s,\pi), \tag{1.132}$$

where $\varepsilon_p:=\varepsilon_{\pi_p}$ is the Atkin-Lehner eigenvalue of π_p.

Fourier Coefficients

Let $F\in S_k(\mathrm{K}(N))$ be a paramodular cusp form of level N. Since F is holomorphic and $F(\left[\begin{smallmatrix}\tau&z\\z&\tau'\end{smallmatrix}\right])=F(\left[\begin{smallmatrix}\tau+n&z+r\\z+r&\tau'+m\end{smallmatrix}\right])$ for all $Z=\left[\begin{smallmatrix}\tau&z\\z&\tau'\end{smallmatrix}\right]\in\mathcal{H}_2$ and $n,r,m\in\mathbb{Z}$, we have a *Fourier expansion*

$$F(Z)=\sum_S a(S)\mathrm{e}^{2\pi\mathrm{i}\mathrm{Tr}(SZ)} \tag{1.133}$$

with certain *Fourier coefficients* $a(S)\in\mathbb{C}$, where S runs over symmetric matrices $\left[\begin{smallmatrix}\alpha&\beta\\\beta&\gamma\end{smallmatrix}\right]$ with $\alpha,2\beta,\gamma\in\mathbb{Z}$. By the Koecher principle, and since F is a cusp form, the summation may be restricted to positive definite S. Since $F(\left[\begin{smallmatrix}\tau&z\\z&\tau'\end{smallmatrix}\right])=F(\left[\begin{smallmatrix}\tau&z\\z&\tau'+N^{-1}\end{smallmatrix}\right])$, we may further assume that S comes from the set $A(N)$ defined in (10.13).

Now assume that $F=F_\pi$ is a paramodular newform, corresponding to the automorphic representation $\pi\cong\otimes\pi_p$ of conductor $N=N_\pi$ via Theorem 1.4.1. Since F determines π, it must be possible to calculate the local factors $L(s,\pi_p)$ from the Fourier coefficients $a(S)$. By the formulas in Theorem 2.3.4, we only need to determine $\lambda_p:=\lambda_{\pi_p}$ and $\mu_p:=\mu_{\pi_p}$ from the $a(S)$. It turns out that it is much more difficult to do so than it was in the GL(2) case. Andrianov [1] writes: "The principal gap in the theory as compared with Hecke's theory for $n=1$ is the lack of any analytical connection between the Euler product and the corresponding modular form."

Andrianov then goes on to provide such an "analytical connection" in the case $N = 1$ (and for degree 2). It is expressed as a formal identity

$$\sum_{t=0}^{\infty} a(p^t S)X^t \tag{1.134}$$

$$= \frac{N(X, S)}{1 - \lambda_p p^{k-3}X + (p^{-2}\mu_p + 1 + p^{-2})p^{2k-3}X^2 + \lambda_p p^{3k-6}X^3 + p^{4k-6}X^4},$$

for $S = \begin{bmatrix} \alpha & \beta \\ \beta & \gamma \end{bmatrix}$ with $\gcd(\alpha, 2\beta, \gamma, p) = 1$, where $N(X, s)$ is a certain polynomial of degree 2. Comparison of the denominator in (1.134) with the formula for $L_p(s, \pi_p)$ in (1) of Theorem 2.3.4 shows that

$$\sum_{t=0}^{\infty} \frac{a(p^t S)}{p^{ts}} = N(p^{-s}, S)L_p(s, F), \tag{1.135}$$

where

$$L_p(s, F) := L_p\Big(s - k + \frac{3}{2}, \pi_p\Big) \tag{1.136}$$

is the *classical Euler factor* at p for the eigenform F. We refer to the elaborations in [121] for a practical way of calculating λ_p and μ_p from the Fourier coefficients in the case $N = 1$.

For cases where $p \mid N$ the calculation of λ_p and μ_p from the Fourier coefficients is trickier. It has been accomplished for some examples in [96] and [99], but in a complicated and non-systematic way.

The main application of our work will be to provide relations between the Hecke eigenvalues λ_p and μ_p and the Fourier coefficients $a(S)$ in the "bad" case $p^2 \mid N$. In particular, we will prove (Theorem 12.5.2) that a radial Fourier coefficient formula as in (1.134) holds in this case as well. We will also prove formulas (Corollary 12.1.3) which provide a practical way of calculating Hecke eigenvalues from Fourier coefficients when $p^2 \mid N$. In Sect. 12.3 we will test these formulas on the examples from [99].

The Classical L-Function

Let $F = F_\pi$ as above. In view of (1.136) we introduce the shifted L-function $L(s, F) := L(s - k + \frac{3}{2}, \pi)$. By (1.132), it satisfies the functional equation

$$L(s, F) = (-1)^k\Big(\prod_{p \mid N} \varepsilon_p\Big)N^{-(s-k+1)}L(2k - 2 - s, F) \tag{1.137}$$

with $N = N_\pi$. This is the L-function given in *arithmetic normalization* (functional equation relates s and $2k-2-s$), whereas $L(s,\pi)$ is given in *analytic normalization* (functional equation relates s and $1-s$). The functional equation (1.137) was first proved by Andrianov in [1] for the case $N=1$.

Using the references cited above, including this work, the Hecke eigenvalues, and hence the Euler factors of $L(s,F)$, can be obtained from the Fourier coefficients of F without consideration of the underlying representations. Similarly, the Atkin-Lehner signs ε_p appearing in (1.137) can be obtained from F via an operator similar to w_p in (1.82); see Lemma 10.2.3. There is also a Fricke involution w_N calculating the product $\prod_{p|N_\pi}\varepsilon_p$ all at once, given by

$$w_N F = F\big|_k \begin{bmatrix} & & & 1 \\ & & 1 & \\ & N & & \\ N & & & \end{bmatrix}. \tag{1.138}$$

If $w_N F = \varepsilon_N F$ with $\varepsilon_N F \in \{\pm 1\}$, then the sign in the functional equation is $(-1)^k \varepsilon_N$.

A Difficulty with Paramodular Hecke Operators

The formulas (1.70) relating Hecke eigenvalues and Fourier coefficients for elliptic modular forms are easily obtained, thanks to the fact that the coset representatives in (1.69) are upper triangular. The situation is different for the double cosets

$$\mathrm{K}(N)\begin{bmatrix} 1 & & & \\ & 1 & & \\ & & p & \\ & & & p \end{bmatrix}\mathrm{K}(N), \qquad \mathrm{K}(N)\begin{bmatrix} p & & & \\ & 1 & & \\ & & p & \\ & & & p^2 \end{bmatrix}\mathrm{K}(N), \tag{1.139}$$

defining the basic paramodular Hecke operators $T(1,1,p,p)$ and $T(p,1,p,p^2)$. In general, the coset representatives for these Hecke operators contain lower triangular matrices (see Proposition 2.10 of [114]), which do not interact well with the Fourier expansion (1.133). Only if p divides N at most once is it possible to find coset representatives with block upper triangular form (see Sect. 2.3 of [114]). Hence, only in these cases can the Fourier expansion of $T(1,1,p,p)F$ and $T(p,1,p,p^2)F$ be easily calculated from the Fourier expansion of F.

The Present Work

To overcome the difficulty for $p^2 \mid N$ just mentioned, we will replace $\mathrm{K}(N)$ by a slightly smaller congruence subgroup $\mathrm{K}_s(N)$, which we call the *stable Klingen congruence subgroup* of level N; see (1.140) for the definition. As it turns out, the Hecke operators for $K_s(N)$ are block upper triangular, allowing us to calculate Fourier expansions; see Lemmas 11.4.2 and 11.4.3.

The question is if this is useful. Can the action of the stable Klingen Hecke operators still be used to calculate paramodular Hecke eigenvalues? The answer is yes, and we present it in the form of an algorithm in Sect. 12.4.

To understand the relationship between the paramodular theory and the stable Klingen theory, we rely heavily on the localization principle provided by the adelic method. Our paramodular eigenforms correspond to cuspidal, automorphic representations $\pi \cong \otimes\pi_p$, and we will analyze stable Klingen vectors in each local representation π_p, in particular for primes p with $p^2 \mid N$. In fact, we will more generally analyze stable Klingen vectors in all irreducible, admissible representations of $\mathrm{GSp}(4, F)$ with trivial central character, where F is a finite extension of $\mathbb{Q}_p$.

Our work thus has a local and a global part, which we describe in more detail in Sects. 1.6 and 1.7. But before doing so, we will provide further context by pointing out some of the uses of paramodular forms in the literature.

Applications and Significance of Paramodular Forms

Paramodular forms are of significant current interest, due in part to the Paramodular Conjecture of Brumer and Kramer [21, 23]. This conjecture asserts that for every abelian surface A defined over $\mathbb{Q}$ with trivial endomorphism ring and conductor N there exists a suitable paramodular newform F of weight two and level N such that $L(s, A) = L(s, F)$. This is an attractive conjecture because it is specific enough to be checked for examples (see [13, 14, 19, 22, 24, 25, 30, 33, 51, 94, 95, 101]). On the modular forms side, such examples F are often investigated via their Fourier coefficients, and the local factors $L_p(s, F)$ are calculated by determining the Hecke eigenvalues of F at the prime p. The Paramodular Conjecture has been generalized by Benedict Gross to a conjecture involving symplectic motives and the split orthogonal group $\mathrm{SO}(2n + 1)$ (see p. 2467 of [21]).

Some other active areas of investigation concerning paramodular forms include Eichler-Jacquet-Langlands type correspondences ([7, 37, 57, 60–64, 82, 102]), new- and oldforms ([104, 105, 114, 115]), Borcherds products and lifting ([34, 46–51, 56, 80, 88, 97, 100, 112, 129]), twisting ([72–74]), congruences ([20, 37, 43]), theta and Eisenstein series ([10, 116, 119, 124]), the Böcherer conjecture ([36, 107, 108]), Fourier coefficients and Bessel models ([83], [84, 85]), and relations to physics ([9, 11, 12, 89]).

1.6 Local Results

In this section we summarize our local results. We begin with some definitions. As above, let F be a nonarchimedean local field of characteristic zero with ring of integers $\mathfrak{o}$. Let $\mathfrak{p}$ be the maximal ideal of $\mathfrak{o}$, let ϖ be a generator of $\mathfrak{p}$, and let q be the order of $\mathfrak{o}/\mathfrak{p}$. For this introduction we continue to define $\mathrm{GSp}(4, F)$ with respect to $\begin{bmatrix} & 1_2 \\ -1_2 & \end{bmatrix}$ (though we will use a more convenient form in Part I of this work). Let n be a non-negative integer. We define the *Klingen congruence subgroup of level* $\mathfrak{p}^n$ to be

$$\mathrm{Kl}(\mathfrak{p}^n) = \mathrm{GSp}(4, \mathfrak{o}) \cap \begin{bmatrix} \mathfrak{o} & \mathfrak{p}^n & \mathfrak{o} & \mathfrak{o} \\ \mathfrak{o} & \mathfrak{o} & \mathfrak{o} & \mathfrak{o} \\ \mathfrak{o} & \mathfrak{p}^n & \mathfrak{o} & \mathfrak{o} \\ \mathfrak{p}^n & \mathfrak{p}^n & \mathfrak{p}^n & \mathfrak{o} \end{bmatrix},$$

and we define the *stable Klingen congruence subgroup of level* $\mathfrak{p}^n$ to be the subgroup $\mathrm{K}_s(\mathfrak{p}^n)$ of $\mathrm{GSp}(4, F)$ generated by $\mathrm{Kl}(\mathfrak{p}^n)$ and the subgroup

$$\begin{bmatrix} 1 & & & \\ & 1 & \mathfrak{p}^{-n+1} & \\ & & 1 & \\ & & & 1 \end{bmatrix}.$$

We have $\mathrm{K}_s(\mathfrak{p}^0) = \mathrm{GSp}(4, \mathfrak{o})$, $\mathrm{K}_s(\mathfrak{p}^1) = \mathrm{Kl}(\mathfrak{p})$, and if n is positive, then

$$\mathrm{K}_s(\mathfrak{p}^n) = \{g \in \mathrm{GSp}(4, F) \mid \lambda(g) \in \mathfrak{o}^\times\} \cap \begin{bmatrix} \mathfrak{o} & \mathfrak{p}^n & \mathfrak{o} & \mathfrak{o} \\ \mathfrak{o} & \mathfrak{o} & \mathfrak{o} & \mathfrak{p}^{-n+1} \\ \mathfrak{o} & \mathfrak{p}^n & \mathfrak{o} & \mathfrak{o} \\ \mathfrak{p}^n & \mathfrak{p}^n & \mathfrak{p}^n & \mathfrak{o} \end{bmatrix}.$$

Evidently, the stable Klingen subgroup $\mathrm{K}_s(\mathfrak{p}^n)$ is contained in the paramodular subgroup $\mathrm{K}(\mathfrak{p}^n)$. Let (π, V) be a smooth representation of $\mathrm{GSp}(4, F)$ for which the center acts trivially. We define

$$V_s(n) = \{v \in V \mid \pi(g)v = v \text{ for all } g \in \mathrm{K}_s(\mathfrak{p}^n)\},$$

and refer to the elements of $V_s(n)$ as *stable Klingen vectors of level* $\mathfrak{p}^n$. If $V_s(n)$ is non-zero for some $n \geq 0$, then we let $N_{\pi,s}$ be the smallest such n and say that π admits non-zero stable Klingen vectors and refer to $N_{\pi,s}$ as the *stable Klingen level* of π. As mentioned above, the main idea of this work is to investigate stable Klingen vectors with a view toward applications to the paramodular theory. We were motivated to consider stable Klingen vectors because of the shadow of a newform. The shadow of a newform is contained in $V_s(N_\pi - 1)$ when π is paramodular and was introduced, but not deeply studied, in [105]. The shadow of a newform, defined below, plays a significant role in this work.

A Partition

Our first major local result establishes a fundamental connection between the spaces $V_s(n)$ and $V(n)$ and a consequent partition of the set of paramodular representations. Since $V(n-1)+V(n)$ is contained $V_s(n)$ for all n, we see that if π is paramodular, then π admits non-zero stable Klingen vectors. The following theorem implies that the converse also holds when π is irreducible.

Theorem 5.1.5 *Let n be an integer such that $n \geq 0$. Let (π, V) be an irreducible, admissible representation of* GSp(4, F) *with trivial central character. Then*

$$\dim V_s(n) \leq \dim V(n) + \dim V(n+1).$$

The proof of this theorem uses the P_3-quotient of π (see Sect. 2.2) along with detailed results about stable Klingen vectors in representations that are induced from the Siegel parabolic subgroup of GSp(4, F) (see Chap. 4). Theorem 5.1.5 has the following corollary.

Corollary 5.1.6 *Let (π, V) be an irreducible, admissible representation of* GSp(4, F) *with trivial central character. Then π admits non-zero paramodular vectors if and only if π admits non-zero stable Klingen vectors. If π is paramodular, then $N_{\pi,s} = N_\pi - 1$ or $N_{\pi,s} = N_\pi$.*

This corollary divides the set of paramodular representations into two classes. Assume that π is a paramodular, irreducible, admissible representation of GSp(4, F) with trivial central character. We will say that π is a *category 1 paramodular representation* if $N_{\pi,s} = N_\pi - 1$; if $N_{\pi,s} = N_\pi$, then we will say that π is a *category 2 paramodular representation*. Interestingly, one can prove that this dichotomy admits a characterization independent of the concept of stable Klingen vectors. We prove the following result.

Corollary 5.7.1 *Let (π, V) be an irreducible, admissible representation of* GSp(4, F) *with trivial central character. Assume that π is paramodular. Then the following are equivalent.*

(1) *π is a category 1 paramodular representation, i.e., $N_{\pi,s} \doteq N_\pi - 1$.*
(2) *The decomposition of the L-parameter of π into indecomposable representations contains no unramified one-dimensional factors.*

The partition of the set of paramodular representations into category 1 and category 2 representations plays an important role in the results described below.

Structure of the $V_s(n)$

Our second major local result is the determination of the structure of the spaces $V_s(n)$. A cornerstone for this is the calculation of the dimensions of the spaces $V_s(n)$

and certain related vector spaces for all irreducible, admissible representations of $\mathrm{GSp}(4, F)$ with trivial central character and all non-negative integers n. Again let (π, V) be a smooth representation of $\mathrm{GSp}(4, F)$ for which the center acts trivially. For n a non-negative integer the vector space $V_s(n)$ contains $V(n-1) + V(n)$, and we define the quotient of $V_s(n)$ by this subspace to be

$$\bar{V}_s(n) = V_s(n)/(V(n-1) + V(n)).$$

The quotient $\bar{V}_s(n)$ measures the difference between $V_s(n)$ and the subspace of paramodular vectors $V(n-1)+V(n)$; when V is infinite-dimensional this subspace is actually the direct sum of $V(n-1)$ and $V(n)$, and has structure as determined in [105]. If $\bar{V}_s(n)$ is non-zero for some n, then we let $\bar{N}_{\pi,s}$ be the smallest such n and refer to $\bar{N}_{\pi,s}$ as the *quotient stable Klingen level* of π. We prove the following theorem about the dimensions of the $V_s(n)$ and $\bar{V}_s(n)$.

Theorem 5.6.1 *For every irreducible, admissible representations* (π, V) *of the group* $\mathrm{GSp}(4, F)$ *with trivial central character, the stable Klingen level* $N_{\pi,s}$*, the dimensions of the spaces* $V_s(n)$*, the quotient stable Klingen level* $\bar{N}_{\pi,s}$*, and the dimensions of the spaces* $\bar{V}_s(n)$ *are given in Table A.3.*

The proof of this theorem uses the inequality of Theorem 5.1.5, the already mentioned results about stable Klingen vectors in representations induced from the Siegel parabolic subgroup from Chap. 4, zeta integrals, and results from [105].

Theorem 5.6.1 completely determines the partition of paramodular representations into category 1 and category 2 representations; additionally, Theorem 5.6.1 reveals dimensional growth patterns that are explained by the following result. To state the theorem we need to introduce two level raising operators. Let (π, V) be a smooth representation for which the center acts trivially, and let n be a non-negative integer. We define

$$\tau, \theta : V_s(n) \longrightarrow V_s(n+1)$$

by

$$\tau v = q^{-1} \sum_{z \in \mathfrak{o}/\mathfrak{p}} \pi(\begin{bmatrix} 1 &&& \\ & 1 && z\varpi^{-n} \\ && 1 & \\ &&& 1 \end{bmatrix})v$$

and

$$\theta v = \pi(\begin{bmatrix} 1 &&& \\ & 1 && \\ && \varpi & \\ &&& \varpi \end{bmatrix})v + \sum_{x \in \mathfrak{o}/\mathfrak{p}} \pi(\begin{bmatrix} 1 && x & \\ & 1 && \\ && 1 & \\ &&& 1 \end{bmatrix}\begin{bmatrix} \varpi &&& \\ & 1 && \\ && 1 & \\ &&& \varpi \end{bmatrix})$$

for $v \in V_s(n)$. Then τ and θ are well-defined, commute, and induce maps from $\bar{V}_s(n)$ to $\bar{V}_s(n+1)$ (see Sects. 3.4 and 3.5). The following theorem determines the structure of the spaces $V_s(n)$ for $n \geq N_{\pi,s}$.

Theorem 7.4.2 *Let (π, V) be an infinite-dimensional, irreducible, admissible representation of* GSp(4, *F*) *with trivial central character. Assume that π is paramodular. Define $v_0 \in V$ as follows. If $\bar{N}_{\pi,s}$ is not defined, set $v_0 = 0$; if $\bar{N}_{\pi,s}$ is defined, let v_0 be an element of $V_s(\bar{N}_{\pi,s})$ that is not contained in $V(\bar{N}_{\pi,s}-1)+V(\bar{N}_{\pi,s})$. For integers n such that $n \geq N_{\pi,s}$ let $E(n)$ be the subspace of $V_s(n)$ spanned by the vectors*

$$\tau^i\theta^j v_0, \qquad i, j \geq 0, \quad i+j = n - \bar{N}_{\pi,s}.$$

Then

$$V_s(n) = V(n-1) \oplus V(n) \oplus E(n)$$

for integers n such that $n \geq N_{\pi,s}$.

The main ingredients for the proof of this theorem are zeta integrals and Theorem 5.6.1. If π is generic, then the integer $\bar{N}_{\pi,s}$ is defined and $V(n-1)+V(n)$ is a proper subspace of $V_s(n)$ for $n \geq \max(N_\pi, 1)$; in this case there is a natural choice for the vector v_0 and the $\tau^i\theta^j v_0$ form a basis for $E(n)$ (see Sect. 7.2 and Fig. 7.1 on p. 189 for a visualization). If π is non-generic, and does not belong to subgroup IVb or IVc, then $V_s(n) = V(n-1) + V(n)$ for $n \geq 0$. Thus, for almost all non-generic π, including all Saito-Kurokawa representations, we have $V_s(n) = V(n-1)+V(n)$ for $n \geq 0$ so that $\bar{N}_{\pi,s}$ is not defined.

As an interesting corollary of Theorem 7.4.2 we prove that the quotients $\bar{V}_s(n)$ exhibit a regular structure reminiscent of $\Gamma_0(\mathfrak{p}^n)$-invariant vectors in irreducible, admissible representations of GL(2, *F*) with trivial central character.

Corollary 7.4.3 *Let (π, V) be an irreducible, admissible representation of* GSp(4, *F*) *with trivial central character. Assume that π is paramodular. Assume that $\bar{N}_{\pi,s}$ is defined. Then*

$$\bar{N}_{\pi,s} = \begin{cases} 1 & \text{if } N_\pi = 0 \text{ or } N_\pi = 1, \\ N_\pi - 1 & \text{if } N_\pi \geq 2 \text{ and } \pi \text{ is a category 1 representation,} \\ N_\pi & \text{if } N_\pi \geq 2 \text{ and } \pi \text{ is a category 2 representation} \end{cases}$$

and

$$\dim \bar{V}_s(\bar{N}_{\pi,s}) = 1.$$

Let $v_{s,\mathrm{new}}$ be a non-zero element of the one-dimensional vector space $\bar{V}_s(\bar{N}_{\pi,s})$. Then the vectors

$$\tau^i\theta^j v_{s,\mathrm{new}}, \qquad i, j \geq 0, \quad i+j = n - \bar{N}_{\pi,s}$$

span $\bar{V}_s(n)$ for $n \geq \bar{N}_{\pi,s}$.

Paramodular Hecke Eigenvalues

Our third major local result relates stable Klingen vectors to paramodular Hecke eigenvalues. To explain the connection we need some definitions. Let (π, V) be a smooth representation of $\mathrm{GSp}(4, F)$ for which the center acts trivially, and let n be a non-negative integer. In analogy to the paramodular Hecke operators $T_{0,1}$ and $T_{1,0}$, let

$$\mathrm{K}_s(\mathfrak{p}^n)\begin{bmatrix} \varpi &&& \\ & \varpi && \\ && 1 & \\ &&& 1 \end{bmatrix}\mathrm{K}_s(\mathfrak{p}^n) = \sqcup_i g_i \mathrm{K}_s(\mathfrak{p}^n),$$

$$\mathrm{K}_s(\mathfrak{p}^n)\begin{bmatrix} \varpi &&& \\ & \varpi^2 && \\ && \varpi & \\ &&& 1 \end{bmatrix}\mathrm{K}_s(\mathfrak{p}^n) = \sqcup_i h_j \mathrm{K}_s(\mathfrak{p}^n)$$

be disjoint decompositions, and define

$$T_{0,1}^s, T_{1,0}^s : V_s(n) \longrightarrow V_s(n)$$

by

$$T_{0,1}^s v = \sum_i \pi(g_i)v, \qquad T_{1,0}^s v = \sum_j \pi(h_j)v$$

for $v \in V_s(n)$. We refer to $T_{0,1}^s$ and $T_{1,0}^s$ as *stable Klingen Hecke operators*. In contrast to the paramodular case, if $n \geq 1$, then $T_{0,1}^s$ and $T_{1,0}^s$ admit simple upper block formulas. We prove that if $n \geq 1$, then

$$\begin{aligned} T_{0,1}^s(v) = &\sum_{y,z\in\mathfrak{o}/\mathfrak{p}} \pi(\begin{bmatrix} 1 &&& \\ y & 1 && z\varpi^{-n+1} \\ && 1 & -y \\ &&& 1 \end{bmatrix}\begin{bmatrix} 1 &&& \\ & \varpi && \\ && \varpi & \\ &&& 1 \end{bmatrix})v \\ &+ \sum_{c,y,z\in\mathfrak{o}/\mathfrak{p}} \pi(\begin{bmatrix} 1 && c & y \\ & 1 & y & z\varpi^{-n+1} \\ && 1 & \\ &&& 1 \end{bmatrix}\begin{bmatrix} \varpi &&& \\ & \varpi && \\ && 1 & \\ &&& 1 \end{bmatrix})v, \end{aligned}$$

$$T_{1,0}^s(v) = \sum_{\substack{x,y\in\mathfrak{o}/\mathfrak{p} \\ z\in\mathfrak{o}/\mathfrak{p}^2}} \pi(\begin{bmatrix} 1 &&& y \\ x & 1 & y & z\varpi^{-n+1} \\ && 1 & -x \\ &&& 1 \end{bmatrix}\begin{bmatrix} \varpi &&& \\ & \varpi^2 && \\ && \varpi & \\ &&& 1 \end{bmatrix})v$$

for $v \in V_s(n)$ (see Sect. 3.8). If $n \geq 2$, then we also introduce a level lowering operator

$$\sigma_n : V_s(n+1) \longrightarrow V_s(n)$$

that is given by the upper block formula

$$\sigma_n v = q^{-3} \sum_{x,y,z \in \mathfrak{o}/\mathfrak{p}} \pi(\begin{bmatrix} 1 & & y & \\ x & 1 & y & z\varpi^{-n+1} \\ & & 1 & -x \\ & & & 1 \end{bmatrix} \begin{bmatrix} 1 & & & \\ & \varpi & & \\ & & 1 & \\ & & & \varpi^{-1} \end{bmatrix}) v$$

for $v \in V_s(n+1)$ (see Sect. 3.7). Finally, assume that π is irreducible, admissible, and paramodular. Let $v_{\text{new}} \in V(N_\pi)$ be a newform, i.e., a non-zero element of $V(N_\pi)$. When $N_\pi \geq 2$, we define the *shadow* of v_{new} to be the vector

$$v_s = \sum_{x,y,z \in \mathfrak{o}/\mathfrak{p}} \pi(\begin{bmatrix} 1 & x\varpi^{N_\pi - 1} & & \\ & 1 & & \\ & y\varpi^{N_\pi - 1} & 1 & \\ y\varpi^{N_\pi - 1} & z\varpi^{N_\pi - 1} & -x\varpi^{N_\pi - 1} & 1 \end{bmatrix}) v_{\text{new}}.$$

The vector v_s is contained in $V_s(N_\pi - 1)$. The next proposition proves that π being a category 1 paramodular representation is equivalent to a number of conditions involving the just introduced concepts. These equivalences are used in subsequent parts of this work. For (3) and (4) note that v_{new} is contained in both $V_s(N_\pi)$ and $V_s(N_\pi + 1)$.

Proposition 5.7.3 *Let (π, V) be an irreducible, admissible representation of* GSp(4, *F*) *with trivial central character. Assume that π is paramodular, and that $N_\pi \geq 2$. Let $v_{\text{new}} \in V(N_\pi)$ be a newform, i.e., a non-zero element of $V(N_\pi)$. Let $v_s \in V_s(N_\pi - 1)$ be the shadow of v_{new} as defined in Lemma 5.2.1. Then the following are equivalent:*

(1) *π is a category 1 paramodular representation, i.e., $N_{\pi,s} = N_\pi - 1$.*
(2) *The $T_{1,0}$-eigenvalue μ_π on v_{new} is non-zero.*
(3) $\sigma_{N_\pi - 1} v_{\text{new}} \neq 0$.
(4) $\sigma_{N_\pi} v_{\text{new}} \neq 0$.
(5) $T^s_{1,0} v_{\text{new}} \neq 0$.
(6) $v_s \neq 0$.

The next theorem ties together stable Klingen operators and paramodular Hecke eigenvalues for irreducible π. As a consequence of this theorem the paramodular Hecke eigenvalues λ_π and μ_π may be calculated using only upper block operators. We apply this theorem to Siegel modular forms in the second part of this work.

Theorem 6.1.1 *Let (π, V) be an irreducible, admissible representation of the group* GSp(4, *F*) *with trivial central character. Assume that π is paramodular and that $N_\pi \geq 2$. Let $v_{\text{new}} \in V(N_\pi)$ be a newform, and let v_s be the shadow of v_{new}.*

(1) *Assume that π is a category 1 representation, so that $N_{\pi,s} = N_\pi - 1$ and $\mu_\pi \neq 0$. Then $V_s(N_{\pi,s}) = V_s(N_\pi - 1)$ is one-dimensional and*

$$T^s_{0,1} v_s = \lambda_\pi v_s \quad \textit{and} \quad T^s_{1,0} v_s = (\mu_\pi + q^2) v_s.$$

(2) *Assume that* π *is a category 2 representation, so that* $N_{\pi,s} = N_\pi$ *and* $\mu_\pi = 0$. *Let* $v_{\text{new}} \in V(N_\pi)$ *be a newform. The vector space* $V_s(N_{\pi,s})$ *is spanned by the vectors* v_{new} *and* $T^s_{0,1}v_{\text{new}}$. *If* v_{new} *is not an eigenvector for* $T^s_{0,1}$, *so that* $V_s(N_{\pi,s})$ *is two-dimensional, then* π *is generic, and the matrix of* $T^s_{0,1}$ *in the ordered basis* $v_{\text{new}}, T^s_{0,1}v_{\text{new}}$ *is*

$$\begin{bmatrix} 0 & -q^3 \\ 1 & \lambda_\pi \end{bmatrix},$$

so that

$$q^3 v_{\text{new}} + (T^s_{0,1})^2 v_{\text{new}} = \lambda_\pi T^s_{0,1} v_{\text{new}}.$$

If v_{new} *is an eigenvector for* $T^s_{0,1}$, *so that* $V_s(N_{\pi,s})$ *is one-dimensional, then* π *is non-generic, and*

$$T^s_{0,1} v_{\text{new}} = (1+q^{-1})^{-1}\lambda_\pi v_{\text{new}}.$$

Further Local Results

In this work we also prove a number of further local results about stable Klingen vectors. The first collection of results is about the case when π is generic. Let (π, V) be a generic, irreducible, admissible representation of GSp(4, F) with trivial central character. For the first result, assume that $V = \mathcal{W}(\pi, \psi_{c_1,c_2})$ is a Whittaker model of π with $c_1, c_2 \in \mathfrak{o}^\times$ (see Sect. 2.2). We prove that if $n \geq 0$ and $W \in V_s(n)$, then $W \neq 0$ if and only if W does not vanish on the diagonal subgroup of GSp(4, F). This generalizes a foundational result from the paramodular theory. Our second result about generic π is about the kernel of $\sigma_{n-1} : V_s(n) \to V_s(n-1)$. We prove that

$$V_s(n) = V(n-1) \oplus V(n) + \ker(\sigma_{n-1})$$

for integers $n \geq \max(N_{\pi,s}, 2)$. It follows that σ_{n-1} induces an isomorphism

$$(V(n-1) \oplus V(n))/\mathcal{K}_n \xrightarrow{\sim} V_s(n-1)$$

for integers $n \geq \max(N_{\pi,s}, 2)$, where $\mathcal{K}_n$ is the intersection of $V(n-1) \oplus V(n)$ with $\ker(\sigma_{n-1})$. We determine $\mathcal{K}_n$ for $n \geq \max(N_{\pi,s}, 2)$ and show that this subspace is at most two-dimensional. To prove these statements we use the result about the non-vanishing of stable Klingen vectors on the diagonal of GSp(4, F). Third, assuming that $L(s, \pi) = 1$ (this includes all supercuspidal π), we define a useful graphical model for $V_s(n)$ for integers $n \geq N_{\pi,s}$. The existence of this model also uses the just mentioned nonvanishing result. In this model, our level changing operators have simple and visual interpretations. Lastly, still under the hypothesis that $L(s, \pi) = 1$, we prove that $N_\pi \geq 4$. For these results see Chap. 8.

Finally, for all irreducible, admissible representations (π, V) of $\mathrm{GSp}(4, F)$ with trivial central character for which $V_s(1)$ is non-zero, we determine the characteristic polynomials of $T^s_{0,1}$ and $T^s_{1,0}$ on $V_s(1)$. If π is such a representation, then π is Iwahori spherical, and thus may be realized as a subquotient of a representation induced from an unramified character of the Borel subgroup of $\mathrm{GSp}(4, F)$; moreover, the action of the Iwahori-Hecke algebra of $\mathrm{GSp}(4, F)$ on such induced representations may be explicitly realized. We use this observation, via the expression of $T^s_{0,1}$ and $T^s_{1,0}$ in the Iwahori-Hecke algebra of $\mathrm{GSp}(4, F)$, to calculate the desired characteristic polynomials. These results appear in Chap. 9.

1.7 Results About Siegel Modular Forms

In (10.37) we define paramodular Hecke operators $T(1, 1, q, q)$ and $T(q, 1, q, q^2)$, and in (10.39) we define the Atkin-Lehner involution w_q. Let $F \in S_k(\mathrm{K}(N))_{\mathrm{new}}$ be an eigenvector of $T(1, 1, q, q)$ and $T(q, 1, q, q^2)$ for all but finitely many primes q of $\mathbb{Z}$ with $q \nmid N$; by Schmidt [115] F is an eigenvector of $T(1, 1, q, q)$ and $T(q, 1, q, q^2)$ for all primes q of $\mathbb{Z}$. Let $T(1, 1, q, q)F = q^{k-3}\lambda_q F$ and $T(q, 1, q, q^2)F = q^{2(k-3)}\mu_q F$ for each prime q of $\mathbb{Z}$. In the second part of this work we apply some of the above local results to solve the problem of finding formulas relating λ_p and μ_p to the Fourier coefficients of F for the primes p of $\mathbb{Z}$ such that $p^2 \mid N$.

To make the application, we consider Siegel modular forms defined with respect to the global analogues of the local stable Klingen congruence subgroups from the previous section. We define

$$\mathrm{K}_s(N) = \mathrm{Sp}(4, \mathbb{Q}) \cap \begin{bmatrix} \mathbb{Z} & N\mathbb{Z} & \mathbb{Z} & \mathbb{Z} \\ \mathbb{Z} & \mathbb{Z} & \mathbb{Z} & N_s^{-1}\mathbb{Z} \\ \mathbb{Z} & N\mathbb{Z} & \mathbb{Z} & \mathbb{Z} \\ N\mathbb{Z} & N\mathbb{Z} & N\mathbb{Z} & \mathbb{Z} \end{bmatrix}, \qquad N \in \mathbb{Z}_{>0}, \tag{1.140}$$

where $N_s = N \prod_{p \mid N} p^{-1}$ and p runs over the primes dividing N. We refer to $\mathrm{K}_s(N)$ as the *stable Klingen congruence subgroup* of level N. Evidently, $\mathrm{K}_s(N)$ is contained in $\mathrm{K}(N)$. As usual, we let $M_k(\mathrm{K}_s(N))$ denote the vector space of Siegel modular forms of degree two and weight k defined with respect to $\mathrm{K}_s(N)$; the subspace of cusp forms is denoted by $S_k(\mathrm{K}_s(N))$. If $F \in M_k(\mathrm{K}_s(N))$, then F has a Fourier expansion of the form

$$F(Z) = \sum_{S \in B(N)} a(S) e^{2\pi i \mathrm{Tr}(SZ)}.$$

Here, $B(N)$ consists of the half-integral positive semi-definite matrices $S = \begin{bmatrix} \alpha & \beta \\ \beta & \gamma \end{bmatrix}$ such that $N_s \mid \gamma$; if $F \in S_k(\mathrm{K}_s(N))$, then the sum runs over the subset $B(N)^+$ of positive definite S. We note that the space $M_k(\mathrm{K}(N))$ is contained in $M_k(\mathrm{K}_s(N))$. If p is a prime dividing N, then the local stable Klingen Hecke operators from the previous section induce operators

$$T^s_{0,1}(p), T^s_{1,0}(p) : M_k(\mathrm{K}_s(N)) \longrightarrow M_k(\mathrm{K}_s(N)),$$

and if p^2 divides N, then the local map σ induces a level lowering operator

$$\sigma_p : M_k(\mathrm{K}_s(N)) \longrightarrow M_k(\mathrm{K}_s(Np^{-1})).$$

These maps take cusp forms to cusp forms. We prove that these operators are given by slash formulas involving only upper block matrices. As a consequence, we show that if F has the above Fourier expansion, then the Fourier expansions of $T^s_{0,1}(p)F$, $T^s_{1,0}(p)F$, and $\sigma_p F$ are given by

$$\begin{aligned}(T^s_{0,1}(p)F)(Z) = &\sum_{S=\left[\begin{smallmatrix} \alpha & \beta \\ \beta & \gamma \end{smallmatrix}\right]\in B(N)} \Big(p^{3-k}a(pS) \\ &+ \sum_{\substack{y\in\mathbb{Z}/p\mathbb{Z}\\ p\mid(\alpha+2\beta y+\gamma y^2)}} pa(p^{-1}S[\left[\begin{smallmatrix} 1 & \\ y & p \end{smallmatrix}\right]])\Big)e^{2\pi i \mathrm{Tr}(SZ)},\\
(T^s_{1,0}(p)F)(Z) = &\sum_{S\in B(N)}\sum_{a\in\mathbb{Z}/p\mathbb{Z}} p^{3-k}a(S[\left[\begin{smallmatrix} 1 & \\ a & p \end{smallmatrix}\right]])e^{2\pi i \mathrm{Tr}(SZ)},\\
(\sigma_p F)(Z) = &\sum_{S\in B(Np^{-1})}\sum_{a\in\mathbb{Z}/p\mathbb{Z}} p^{-k-1}a(S[\left[\begin{smallmatrix} 1 & \\ a & p \end{smallmatrix}\right]])e^{2\pi i \mathrm{Tr}(SZ)}.\end{aligned}$$

In these formulas if S and A are 2×2 matrices, then $S[A] = {}^tASA$. We also calculate the Fourier-Jacobi expansions of $T^s_{0,1}(p)F$, $T^s_{1,0}(p)F$, and $\sigma_p F$. These formulas involve new operators on Jacobi forms that may be of independent interest. Similarly, we derive formulas for the operators induced by the other local maps discussed in the previous section; see Chap. 11.

The Main Theorem

Let $F \in S_k(\mathrm{K}(N))_{\mathrm{new}}$ be an eigenvector of $T(1,1,q,q)$ and $T(q,1,q,q^2)$ for all but finitely primes q of $\mathbb{Z}$ with $q \nmid N$; as mentioned above, by Schmidt [115] F is an eigenvector of $T(1,1,q,q)$ and $T(q,1,q,q^2)$ for all primes q of $\mathbb{Z}$. Let $T(1,1,q,q)F = q^{k-3}\lambda_q F$ and $T(q,1,q,q^2)F = q^{2(k-3)}\mu_q F$ for each prime q of

$\mathbb{Z}$. Using Theorem 6.1.1 and other local results we obtain the following result, which relates λ_p and μ_p to the action of the upper block operators $T^s_{0,1}(p)$, $T^s_{1,0}(p)$, and σ_p on F for primes p such that $p^2 \mid N$. This theorem mentions the representation π generated by the adelization of F; see Theorem 12.1.1 (based on [114] and [115]) for a summary of the properties of π.

Theorem 12.1.2 *Let N and k be integers such that $N > 0$ and $k > 0$, and let $F \in S_k(\mathrm{K}(N))$. Assume that F is a newform and is an eigenvector for the Hecke operators $T(1,1,q,q)$ and $T(q,1,q,q^2)$ for all but finitely many of the primes q of $\mathbb{Z}$ such that $q \nmid N$; then by Theorem 12.1.1, F is an eigenvector for $T(1,1,q,q)$ and $T(q,1,q,q^2)$ for all primes q of $\mathbb{Z}$. Let $\otimes_{v\le\infty}\pi_v$ be as in Theorem 12.1.1. For every prime q of $\mathbb{Z}$ let $\lambda_q, \mu_q \in \mathbb{C}$ be such that*

$$T(1,1,q,q)F = q^{k-3}\lambda_q F, \tag{12.1}$$

$$T(q,1,q,q^2)F = q^{2(k-3)}\mu_q F. \tag{12.2}$$

Let p be a prime of $\mathbb{Z}$ with $v_p(N) \ge 2$. Then

$$\mu_p = 0 \iff \sigma_p F = 0 \iff T^s_{1,0}(p)F = 0. \tag{12.3}$$

Moreover:

(1) *If $v_p(N) \ge 3$, then $\sigma_p^2 F = 0$.*
(2) *We have*

$$\mu_p F = p^4\tau_p^2\sigma_p F - p^2\eta_p\sigma_p F. \tag{12.4}$$

(3) *Assume that $\mu_p \ne 0$. Then $\sigma_p F \ne 0$, and*

$$T^s_{0,1}(p)(\sigma_p F) = \lambda_p(\sigma_p F), \tag{12.5}$$

$$T^s_{1,0}(p)(\sigma_p F) = (\mu_p + p^2)(\sigma_p F), \tag{12.6}$$

and the representation π_p is generic.
(4) *Assume that $\mu_p = 0$. Then*

$$T^s_{1,0}(p)F = 0. \tag{12.7}$$

If F is not an eigenvector for $T^s_{0,1}(p)$, then π_p is generic and

$$T^s_{0,1}(p)^2 F = -p^3 F + \lambda_p T^s_{0,1}(p)F. \tag{12.8}$$

The newform F is an eigenvector for $T^s_{0,1}(p)$ if and only if

$$T^s_{0,1}(p)F = (1+p^{-1})^{-1}\lambda_p F; \tag{12.9}$$

in this case π_p is non-generic.

Fourier Coefficients

The next corollary translates the assertions of Theorem 12.1.2 into identities involving the Fourier coefficients of F. This is a natural application of this theorem since the operators $T^s_{0,1}(p)$, $T^s_{1,0}(p)$, and σ_p are upper block operators. These relations in turn allow for efficient numerical calculations, at least in the examples that we discuss after the statement of the corollary.

Corollary 12.1.3 *Let N and k be integers such that $N > 0$ and $k > 0$, and let $F \in S_k(\mathrm{K}(N))$. Assume that F is a newform and is an eigenvector for the Hecke operators $T(1,1,q,q)$ and $T(q,1,q,q^2)$ for all but finitely many of the primes q of $\mathbb{Z}$ such that $q \nmid N$; by Theorem 12.1.1, F is an eigenvector for $T(1,1,q,q)$, $T(q,1,q,q^2)$ and w_q for all primes q of $\mathbb{Z}$. Let $\lambda_q, \mu_q \in \mathbb{C}$ and $\varepsilon_q \in \{\pm 1\}$ be such that*

$$T(1,1,q,q)F = q^{k-3}\lambda_q F,$$
$$T(q,1,q,q^2)F = q^{2(k-3)}\mu_q F,$$
$$w_q F = \varepsilon_q F$$

for all primes q of $\mathbb{Z}$. Regard F as an element of $S_k(\mathrm{K}_s(N))$, and let

$$F(Z) = \sum_{S \in B(N)^+} a(S) e^{2\pi i \mathrm{Tr}(SZ)}$$

be the Fourier expansion of F. Let $\pi \cong \otimes_{v \le \infty} \pi_v$ be as in Theorem 12.1.2. Let p be a prime of $\mathbb{Z}$ with $v_p(N) \ge 2$. Then

$$\sum_{a \in \mathbb{Z}/p\mathbb{Z}} a(S[\left[\begin{smallmatrix} 1 & \\ a & p \end{smallmatrix}\right]]) = 0 \qquad \text{for } S \in B(Np^{-1})^+$$

$$\Updownarrow \tag{12.22}$$

$$\mu_p = 0$$

$$\Updownarrow \tag{12.23}$$

$$\sum_{a \in \mathbb{Z}/p\mathbb{Z}} a(S[\left[\begin{smallmatrix} 1 & \\ a & p \end{smallmatrix}\right]]) = 0 \qquad \text{for } S \in B(N)^+.$$

Moreover:

(1) *If* $v_p(N) \geq 3$ *and* $S \in B(Np^{-2})^+$, *then*

$$\sum_{z \in \mathbb{Z}/p^2\mathbb{Z}} a(S[\left[\begin{smallmatrix} 1 & \\ z & p^2 \end{smallmatrix}\right]]) = 0. \tag{12.24}$$

(2) *If* $S = \left[\begin{smallmatrix} \alpha & \beta \\ \beta & \gamma \end{smallmatrix}\right] \in B(Np)^+$, *then*

$$\mu_p a(S) = \begin{cases} \sum\limits_{x \in \mathbb{Z}/p\mathbb{Z}} p^{3-k} a(S[\left[\begin{smallmatrix} 1 & \\ x & p \end{smallmatrix}\right]]) & \text{if } p \nmid 2\beta, \\ \sum\limits_{x \in \mathbb{Z}/p\mathbb{Z}} p^{3-k} a(S[\left[\begin{smallmatrix} 1 & \\ x & p \end{smallmatrix}\right]]) - \sum\limits_{x \in \mathbb{Z}/p\mathbb{Z}} pa(S[\left[\begin{smallmatrix} 1 & \\ xp^{-1} & 1 \end{smallmatrix}\right]]) & \text{if } p \mid 2\beta. \end{cases} \tag{12.25}$$

(3) *Assume that* $\mu_p \neq 0$. *If* $S = \left[\begin{smallmatrix} \alpha & \beta \\ \beta & \gamma \end{smallmatrix}\right] \in B(Np^{-1})^+$, *then*

$$\lambda_p \sum_{x \in \mathbb{Z}/p\mathbb{Z}} a(S[\left[\begin{smallmatrix} 1 & \\ x & p \end{smallmatrix}\right]]) = \sum_{x \in \mathbb{Z}/p\mathbb{Z}} p^{3-k} a(pS[\left[\begin{smallmatrix} 1 & \\ x & p \end{smallmatrix}\right]]) + \sum_{\substack{z \in \mathbb{Z}/p^2\mathbb{Z} \\ p \mid (\alpha + 2\beta z + \gamma z^2)}} pa(p^{-1} S[\left[\begin{smallmatrix} 1 & \\ z & p^2 \end{smallmatrix}\right]]) \tag{12.26}$$

and

$$\sum_{y \in \mathbb{Z}/p^2\mathbb{Z}} a(S[\left[\begin{smallmatrix} 1 & \\ y & p^2 \end{smallmatrix}\right]]) = \begin{cases} \varepsilon_p \sum\limits_{x \in \mathbb{Z}/p\mathbb{Z}} p^{k-2} a(S[\left[\begin{smallmatrix} 1 & \\ x & p \end{smallmatrix}\right]]) & \text{if } v_p(N) = 2, \\ 0 & \text{if } v_p(N) > 2. \end{cases} \tag{12.27}$$

The representation π_p *is generic.*

(4) *Assume that* $\mu_p = 0$. *Then*

$$\sum_{a \in \mathbb{Z}/p\mathbb{Z}} a(S[\left[\begin{smallmatrix} 1 & \\ a & p \end{smallmatrix}\right]]) = 0 \qquad \text{for } S \in B(Np^{-1})^+. \tag{12.28}$$

We have $T^s_{0,1}(p)F = \sum_{S \in B(N)^+} c(S) e^{2\pi i \mathrm{Tr}(SZ)}$ *where*

$$c(S) = p^{3-k} a(pS) + \sum_{\substack{x \in \mathbb{Z}/p\mathbb{Z} \\ p \mid (\alpha + 2\beta x)}} pa(p^{-1} S[\left[\begin{smallmatrix} 1 & \\ x & p \end{smallmatrix}\right]]) \tag{12.29}$$

for $S = \left[\begin{smallmatrix} \alpha & \beta \\ \beta & \gamma \end{smallmatrix}\right] \in B(N)^+$. *If* F *is not an eigenvector for* $T^s_{0,1}(p)$, *then* π_p *is generic,*

$$\lambda_p c(S) = p^3 a(S) + p^{6-2k} a(p^2 S) + \sum_{\substack{y \in \mathbb{Z}/p\mathbb{Z} \\ p \mid (\alpha+2\beta y)}} p^{4-k} a(S[\left[\begin{smallmatrix} 1 & \\ y & p \end{smallmatrix}\right]])$$

$$+ \sum_{\substack{z \in \mathbb{Z}/p^2\mathbb{Z} \\ p^2 \mid (\alpha+2\beta z+\gamma z^2)}} p^2 a(p^{-2} S[\left[\begin{smallmatrix} 1 & \\ z & p^2 \end{smallmatrix}\right]]) \tag{12.30}$$

for $S = \left[\begin{smallmatrix} \alpha & \beta \\ \beta & \gamma \end{smallmatrix}\right] \in B(N)^+$, *and* $c(S) \neq 0$ *for some* $S \in B(N)^+$. *The newform* F *is an eigenvector for* $T^s_{0,1}(p)$ *if and only if*

$$\lambda_p a(S) = (1+p^{-1})c(S) = (1+p)p^{2-k} a(pS) + \sum_{\substack{x \in \mathbb{Z}/p\mathbb{Z} \\ p \mid (\alpha+2\beta x)}} (1+p) a(p^{-1} S[\left[\begin{smallmatrix} 1 & \\ x & p \end{smallmatrix}\right]]) \tag{12.31}$$

for $S = \left[\begin{smallmatrix} \alpha & \beta \\ \beta & \gamma \end{smallmatrix}\right] \in B(N)^+$; *in this case* π_p *is non-generic.*

The formulas of Corollary 12.1.3 provide a solution to the problem of finding formulas that relate λ_p and μ_p to the Fourier coefficients of F when F is as in Corollary 12.1.3 and $p^2 \mid N$. As mentioned earlier, when $p^2 \mid N$, the Hecke operators $T(1,1,p,p)$ and $T(p,1,p,p^2)$ do not admit formulas involving only upper block matrices; consequently, a direct calculation is not possible. Instead, our solution uses the new theory of stable Klingen vectors to relate these Hecke eigenvalues to the new upper block operators $T^s_{0,1}(p)$, $T^s_{1,0}(p)$, and σ_p.

The reader may recognize the sums occurring in this corollary as involving GL(2) Hecke operators. In Sect. 12.2 we use this observation to reformulate the results of Corollary 12.1.3 into briefer conceptual statements (see Theorem 12.2.2).

Calculations

The formulas of Corollary 12.1.3 seem to be of practical value. Using a computer algebra program, we verified that the identities of Corollary 12.1.3 do indeed hold for the case $N = 16$, $p = 2$, $k = 6, \ldots, 14$ for the examples presented in [99] and [98] (see Sect. 12.3 for a discussion of this verification). Besides providing a welcome check, our calculations also showed that the formulas of Corollary 12.1.3 can be used to quickly determine λ_2 and μ_2 from the Fourier coefficients of the examples of [99] and [98]. See Sect. 12.4 for a description of how to use Corollary 12.1.3 to compute Hecke eigenvalues, along with some illustrative examples.

A Recurrence Relation

Finally, as an application of Corollary 12.1.3 we prove that the Fourier coefficients of F as in Corollary 12.1.3 satisfy a recurrence relation. The following theorem shows that the Fourier coefficients $a(p^t S)$ of F are determined by $a(S)$, $a(pS)$, $a(p^2 S)$ and the Hecke eigenvalues λ_p and μ_p for p such that $v_p(N) \geq 2$.

Theorem 12.5.2 *Let N and k be integers such that $N > 0$ and $k > 0$, and let $F \in S_k(\mathrm{K}(N))$. Assume that F is a newform and is an eigenvector for the Hecke operators $T(1,1,q,q)$ and $T(q,1,q,q^2)$ for all but finitely many of the primes q of $\mathbb{Z}$ such that $q \nmid N$; by Theorem 12.1.1, F is an eigenvector for $T(1,1,q,q)$, $T(q,1,q,q^2)$ and w_q for all primes q of $\mathbb{Z}$. Let $\lambda_q, \mu_q \in \mathbb{C}$ be such that*

$$T(1,1,q,q)F = q^{k-3}\lambda_q F,$$

$$T(q,1,q,q^2)F = q^{2(k-3)}\mu_q F$$

for all primes q of $\mathbb{Z}$. Regard F as an element of $S_k(\mathrm{K}_s(N))$, and let

$$F(Z) = \sum_{S \in B(N)^+} a(S) e^{2\pi i \mathrm{Tr}(SZ)}$$

be the Fourier expansion of F. Let p be a prime of $\mathbb{Z}$ with $v_p(N) \geq 2$. If $S \in B(N)^+$, then there is formal identity of power series in p^{-s}

$$\sum_{t=0}^{\infty} \frac{a(p^t S)}{p^{ts}} = N(p^{-s}, S) L_p(s, F) \tag{12.96}$$

where

$$N(p^{-s}, S) = a(S) + \big(a(pS) - p^{k-3}\lambda_p a(S)\big)p^{-s} + \big(a(p^2 S) - p^{k-3}\lambda_p a(pS) + p^{2k-5}(\mu_p + p^2)a(S)\big)p^{-2s} \tag{12.97}$$

and

$$L_p(s, F) = \frac{1}{1 - p^{k-3}\lambda_p p^{-s} + p^{2k-5}(\mu_p + p^2)p^{-2s}} \tag{12.98}$$

is the spin L-factor of F at p (e.g., see [71], p. 547).

The equation (12.96) is analogous to (1.135), and a similar statement when $\mathrm{K}(N)$ is replaced by $\Gamma_0(N)$ and $p \nmid N$ can be found in Sect. 4.3.2 of [2]. Finally, we point out that, as a consequence of (12.96), in certain cases the Hecke eigenvalues λ_p, μ_p can be computed directly from the radial Fourier coefficients $a(pS), a(p^2S), a(p^3S), a(p^4S)$; see Corollary 12.5.3. We thank an anonymous referee for this observation.

1.8 Further Directions

The theory of local paramodular new- and oldforms was developed in [105]. The present work considers invariant vectors with respect to stable Klingen subgroups, which can be regarded as a natural extension of the paramodular theory. (In fact, one genesis of the stable Klingen theory is contained in the shadow vector from Sect. 7.4 of [105].) Thus, the paramodular and stable Klingen theory form a natural unit. At the same time, the considerations of this theory lead to some broader questions, which we state below.

Invariants and Eigenvalues Let F be a p-adic field, and let (π, V) be an irreducible, admissible representation of $\mathrm{GSp}(4, F)$ with trivial central character. Suppose that π admits non-zero paramodular vectors, and let $V(n)$ be the subspace of V consisting of vectors invariant under $\mathrm{K}(\mathfrak{p}^n)$; see (2.26). Let N_π be the smallest n for which $V(n)$ is non-zero. Then $V(N_\pi)$ is 1-dimensional and spanned by a local newform v_π. Let $T_{0,1}$ and $T_{1,0}$ be the Hecke operators defined in (2.30), and let u_n be the Atkin-Lehner element defined in (2.31). Let λ_π, μ_π, and ε_π be defined by

$$T_{0,1}v_\pi = \lambda_\pi v_\pi, \qquad T_{1,0}v_\pi = \mu_\pi v_\pi, \qquad \pi(u_{N_\pi})v_\pi = \varepsilon_\pi v_\pi.$$

As we already mentioned on p. 47, the four quantities N_π, λ_π, μ_π, and ε_π determine the L-factor $L(s, \pi)$ and ε-factor $\varepsilon(s, \pi)$. Counterexamples show that the four quantities do not determine the degree-5 L- and ε-factors. A similar phenomenon holds for $\mathrm{GL}(2, F)$: the minimal level, Hecke eigenvalue and Atkin-Lehner eigenvalue determine the standard L- and ε-factors, but not the symmetric square factors. Is there a conceptual explanation for the fact that N_π, λ_π, μ_π, and ε_π determine the invariants $L(s, \pi)$ and $\varepsilon(s, \pi)$?

Structure of the Paramodular Hecke Algebra An answer to the above question could involve a better understanding of Hecke operators. Currently, the structure of the general Hecke algebra $\mathcal{H}_n := \mathcal{H}(\mathrm{GSp}(4, F), \mathrm{K}(\mathfrak{p}^n))$ is unknown for $n \geq 2$. (For $n = 1$, see [44, 75].) By Proposition 6.2.1 of [105], the subalgebra $\mathcal{H}'_n$ generated by $T_{0,1}$, $T_{1,0}$ and u_n is non-commutative for $n > 0$. For $n = 0$ and $n = 1$ we have $\mathcal{H}_n = \mathcal{H}'_n$. For $n \geq 2$, how much bigger is $\mathcal{H}_n$ compared to $\mathcal{H}'_n$? Also, a relation between $\mathcal{H}_n$ and $\mathcal{H}_{n+1}$ would be desirable. Some connections between $\mathcal{H}_n$ and $\mathcal{H}_{n+1}$ were made in Proposition 6.3.1 of [105].

Generalization to SO(2n + 1) As already mentioned after (1.106), the works [21, 52, 127], and [128] define a sequence of congruence subgroups in certain orthogonal groups $\mathrm{SO}(2m+1, F)$ and partially generalize the newforms theory for the groups $\mathrm{GL}(2, F)$ and $\mathrm{GSp}(4, F)$. This setting still has a number of open problems, for example the one-dimensionality of the space of newforms for non-supercuspidal representations, or the oldforms principle as in Theorem 5.6.1 of [105].

The Theory of Newforms The literature now contains a number of instances for which a newforms theory exists for representations of $G(F)$ with respect to a family of compact, open subgroups of $G(F)$ indexed by a totally ordered set; besides this work, see [27] for $G = \mathrm{GL}(2)$, [69] for $G = \mathrm{GL}(n)$ (generic representations), [87] for G the unramified $U(2, 1)$, [128] for $G = \mathrm{SO}(2n+1)$, [91] for $G = \mathrm{GSp}(4)$ (with arbitrary central character), and [8] for $G = \mathrm{GL}(n)$ (including non-generic representations). Can one provide a systematic characterization of the groups G and the involved compact, open subgroups that lead to a satisfactory newforms theory?

$\Gamma_0(\mathfrak{p}^n)$-invariant vectors The theta series of degree 2 attached to an even-dimensional quadratic form transforms according to the group $\Gamma_0(N) = \{\left[\begin{smallmatrix} A & B \\ C & D \end{smallmatrix}\right] \in \mathrm{Sp}(4, \mathbb{Z}) \mid C \equiv 0 \bmod N\}$, where N is the level of the quadratic form. For this and other reasons, it is natural to consider Siegel modular forms of degree 2 with respect to $\Gamma_0(N)$. For a local analysis of such Siegel modular forms analogous to what we carry out in this book for paramodular forms one would require an understanding of the spaces of fixed vectors $V_0(n) := V^{\Gamma_0(\mathfrak{p}^n)}$ with respect to the group $\Gamma_0(\mathfrak{p}^n) = \{\left[\begin{smallmatrix} A & B \\ C & D \end{smallmatrix}\right] \in \mathrm{GSp}(4, \mathfrak{o}) \mid C \equiv 0 \bmod \mathfrak{p}^n\}$ in an irreducible, admissible representation (π, V) of $\mathrm{GSp}(4, F)$ with trivial central character. In general these spaces are more complicated than spaces of paramodular fixed vectors. For example, for the smallest n such that $V_0(n) \neq 0$ it can happen that $\dim V_0(n) > 1$. Also, it seems that the dimensions of the spaces $V_0(n)$ grow more quickly than the dimensions of the spaces of paramodular vectors. Nevertheless, a better understanding of the spaces $V_0(n)$ seems essential for applications to important Siegel modular forms.

As a final point, it is interesting to observe that some non-generic representations, for example non-generic supercuspidal π, do not admit paramodular vectors; at the same time, such representations often have $\Gamma_0(\mathfrak{p}^n)$-invariant vectors. One might wonder whether $\Gamma_0(\mathfrak{p}^n)$-invariant vectors might serve as a substitute for the missing newforms theory for such representations.

Part I
Local Theory

Chapter 2
Background

In this chapter we recall some essential definitions and results concerning the group GSp(4) over a nonarchimedean local field of characteristic zero and its representation theory. We also review the theory of paramodular vectors.

2.1 Some Definitions

The following objects and definitions will be fixed for Part I of this work.

The Base Field and Matrices

We let F be a nonarchimedean local field of characteristic zero. Let $\mathfrak{o}$ denote the ring of integers of F, and let $\mathfrak{p}$ be the maximal proper ideal of $\mathfrak{o}$. Once and for all, we fix a generator ϖ of $\mathfrak{p}$, so that $\mathfrak{p} = \mathfrak{o}\varpi$. If $x \in F^\times$, then there exists a unique integer $v(x)$ and $u \in \mathfrak{o}^\times$ such that $x = u\varpi^{v(x)}$. The order of $\mathfrak{o}/\mathfrak{p}$ is denoted by q, and we denote by $|\cdot|$ the absolute value on F such that $|\varpi| = q^{-1}$. We will use the Haar measure dx on F that gives $\mathfrak{o}$ volume 1. Occasionally, we will use the Haar measure on $F^\times$ that is given by $d^\times x = dx/|\cdot|$. We let $\psi : F \to \mathbb{C}^\times$ denote a fixed continuous homomorphism such that $\psi(\mathfrak{o}) = 1$ but $\psi(\mathfrak{p}^{-1}) \neq 1$. If g is a matrix, then we denote the *transpose* of g by ${}^tg$. In this work, if a matrix is specified, then a blank entry means a zero entry. If $A = \left[\begin{smallmatrix} a_1 & a_2 \\ a_3 & a_4 \end{smallmatrix}\right]$ is in GL(2, F), then we define

$$A' = \left[\begin{smallmatrix} & 1 \\ 1 & \end{smallmatrix}\right] {}^tA^{-1} \left[\begin{smallmatrix} & 1 \\ 1 & \end{smallmatrix}\right] = \det(A)^{-1} \left[\begin{smallmatrix} a_1 & -a_2 \\ -a_3 & a_4 \end{smallmatrix}\right]. \tag{2.1}$$

It is useful to use the asterisk symbol $*$ as a wild card to define a subset of a given set of matrices. For example, when working with GL(2, F), the phrase "let $B = \left[\begin{smallmatrix} * & * \\ & * \end{smallmatrix}\right]$"

J. Johnson-Leung et al., *Stable Klingen Vectors and Paramodular Newforms*,
Lecture Notes in Mathematics 2342, https://doi.org/10.1007/978-3-031-45177-5_2

would define the subgroup of all upper triangular elements of GL(2, F). It is useful to note that if $x \in F^\times$, then

$$\begin{bmatrix} 1 & \\ x & 1 \end{bmatrix} = \begin{bmatrix} 1 & x^{-1} \\ & 1 \end{bmatrix} \begin{bmatrix} -x^{-1} & \\ & -x \end{bmatrix} \begin{bmatrix} & 1 \\ -1 & \end{bmatrix} \begin{bmatrix} 1 & x^{-1} \\ & 1 \end{bmatrix}. \tag{2.2}$$

The Symplectic Similitude Group

In this first part we define GSp(4, F) to be the set of g in GL(4, F) such that ${}^t g J g = \lambda J$ for some $\lambda \in F^\times$, where

$$J = \begin{bmatrix} & & & 1 \\ & & 1 & \\ & -1 & & \\ -1 & & & \end{bmatrix}. \tag{2.3}$$

In the second part of this work, which concerns Siegel modular forms, we will use the traditional form of J to define GSp(4). If $g \in$ GSp(4, F), then the unit λ such that ${}^t g J g = \lambda J$ is unique, and will be denoted by $\lambda(g)$. It is easy to see that GSp(4, F) is a subgroup of GL(4, F). If $g = \begin{bmatrix} A & B \\ C & D \end{bmatrix} \in$ GSp(4, F) with $A = \begin{bmatrix} a_1 & a_2 \\ a_3 & a_4 \end{bmatrix}$, $B = \begin{bmatrix} b_1 & b_2 \\ b_3 & b_4 \end{bmatrix}$, $C = \begin{bmatrix} c_1 & c_2 \\ c_3 & c_4 \end{bmatrix}$, and $D = \begin{bmatrix} d_1 & d_2 \\ d_3 & d_4 \end{bmatrix}$ in M(2, F), then

$$g^{-1} = \lambda(g)^{-1} \begin{bmatrix} d_4 & d_2 & -b_4 & -b_2 \\ d_3 & d_1 & -b_3 & -b_1 \\ -c_4 & -c_2 & a_4 & a_2 \\ -c_3 & -c_1 & a_3 & a_1 \end{bmatrix}. \tag{2.4}$$

We define Sp(4, F) to be the subgroup of g in GSp(4, F) such that $\lambda(g) = 1$, i.e., ${}^t g J g = J$. The center Z of the group GSp(4, F) consists of all the matrices of the form

$$\begin{bmatrix} z & & & \\ & z & & \\ & & z & \\ & & & z \end{bmatrix}, \qquad z \in F^\times.$$

We give GSp(4, F) $\subset$ GL(4, F) the relative topology. The groups GSp(4, F) and Sp(4, F) are unimodular. We let

$$s_1 = \begin{bmatrix} & 1 & & \\ 1 & & & \\ & & & 1 \\ & & 1 & \end{bmatrix} \quad \text{and} \quad s_2 = \begin{bmatrix} 1 & & & \\ & & 1 & \\ & -1 & & \\ & & & 1 \end{bmatrix}, \tag{2.5}$$

and

$$\Delta_{i,j} = \begin{bmatrix} \varpi^{2i+j} & & & \\ & \varpi^{i+j} & & \\ & & \varpi^{i} & \\ & & & 1 \end{bmatrix} \tag{2.6}$$

for integers i and j.

Characters and Representations

Let G be a group of td-type, as defined in [26], and assume that the topology of G has a countable basis.

Characters A *character* of G is a continuous homomorphism $G \to \mathbb{C}^\times$. The *trivial character* of G is denoted by 1_G. If $\chi : F^\times \to \mathbb{C}^\times$ is a character, then we let $a(\chi)$ be the smallest non-negative integer n such that $\chi(1+\mathfrak{p}^n) = 1$; here, we use the convention that $1+\mathfrak{p}^0 = \mathfrak{o}^\times$. We refer to $a(\chi)$ as the *conductor* of χ. We will use the continuous character $\nu : F^\times \to \mathbb{C}^\times$ defined by $\nu(x) = |x|$ for $x \in F^\times$; we have $a(\nu) = 0$.

Schwartz Functions If X is a closed subset of G, then the complex vector space of locally constant and compactly supported functions on X will be denoted by $\mathcal{S}(X)$.

Representations A *representation* of G is a pair (π, V), where V is a complex vector space, and $\pi : G \to \mathrm{GL}_{\mathbb{C}}(V)$ is a homomorphism. Let (π, V) be a representation of G. We say that π is *smooth* if for every $v \in V$, the stabilizer of v in G is open. Evidently, characters of G can be identified with one-dimensional smooth representations of G. We say that π is *admissible* if π is smooth and, for every compact open subgroup K of G, the subspace of vectors in V fixed by every element of K is finite-dimensional. Let (π, V) be a smooth representation of G. We say that π is *irreducible* if the only G subspaces of V are 0 and V. An *irreducible constituent*, or *irreducible subquotient*, of π is an irreducible representation of G that is isomorphic to W/W', where W and W' are G subspaces of V with $W' \subset W$. The group G acts on the complex vector space of linear functionals $\lambda : V \to \mathbb{C}$ via the formula $(g \cdot \lambda)(v) = \lambda(\pi(g^{-1})v)$ for $v \in V$ and $g \in G$; we say that λ is *smooth* if there exists a compact, open subgroup K of G such that $k \cdot \lambda = \lambda$ for $k \in K$. The complex vector space of all smooth linear functionals on V will be denoted by $V^\vee$ and is a smooth representation $\pi^\vee$ of G called the *contragredient* of π. If π admits a *central character*, then we denote it by ω_π.

Induction Let H be a closed subgroup of G, and let (π, V) be a smooth representation of H. Let $\text{c-Ind}_H^G \pi$ be the complex vector space of all functions $f : G \to V$ such that $f(hg) = \pi(h)f(g)$ for $h \in H$ and $g \in G$, $f(gk) = f(g)$ for all k in a compact, open subgroup of G and $g \in G$, and there exists a compact subset X of G such that $f(g) = 0$ for g not in HX. Then with the right translation action defined by $(g' \cdot f)(g) = f(gg')$ for $g, g' \in G$ and f in $\text{c-Ind}_H^G \pi$, the space $\text{c-Ind}_H^G \pi$ is a smooth representation of G. Next, assume that G is unimodular, and let M and U be closed subgroups of G such that $M \cap U = 1$, U is unimodular, M normalizes U, the group $P = MU$ is closed in G, and $P\backslash G$ is compact. We define the *modular character* $\delta_P : P \to \mathbb{C}^\times$ of P in the following way. Fix a Haar measure on U. For p in P we let $\delta_P(p)$ be the unique positive number such that

$$\int_U f(p^{-1}up)\,du = \delta_P(p)\int_U f(u)\,du \tag{2.7}$$

for f in $\mathcal{S}(U)$. Further assume that (π, V) is a smooth representation of M. The complex vector space of all functions $f : G \to V$ such that $f(mug) = \delta_P(m)^{1/2}\pi(m)f(g)$ for $m \in M$, $u \in U$ and $g \in G$, and $f(gk) = f(g)$ for all k in some compact, open subgroup of G and $g \in G$, will be denoted by $\mathrm{Ind}_P^G\pi$. With the right translation action, $\mathrm{Ind}_P^G\pi$ is a smooth representation of G called the representation obtained from π by *normalized induction*.

Jacquet Modules Let N be a closed subgroup of G, let χ be a character of N, and let (π, V) be a smooth representation of G. We define $V(N, \chi)$ to be the $\mathbb{C}$ subspace of V spanned by all the vectors of the form $\chi(n)v - \pi(n)v$ for $n \in N$ and $v \in V$, and we set $V_{N,\chi} = V/V(N, \chi)$. The subspace $V(N, \chi)$ is a N subspace of V, and N acts on the quotient $V_{N,\chi}$ by the character χ. If H is a closed subgroup of G that normalizes N and is such that $\chi(hnh^{-1}) = \chi(n)$ for $n \in N$ and $h \in H$, then $V(N, \chi)$ is also an H subspace, and H acts on $V_{N,\chi}$. We refer to $V_{N,\chi}$ as a *Jacquet module*.

2.2 Representations

In this section we review some needed definitions and results about representations of GSp(4, F).

Parabolic Induction

In this work we will often consider representations of GSp(4, F) that are constructed via parabolic induction from the three standard proper parabolic subgroups. We refer to Sect. 30 of [59] for general facts on parabolic subgroups, and to Sect. 2.3 of [105] for specifics regarding GSp(4).

The first, called the *Borel subgroup* of GSp(4, F), is the subgroup of all upper triangular elements, i.e.

$$B = \begin{bmatrix} * & * & * & * \\ & * & * & * \\ & & * & * \\ & & & * \end{bmatrix}.$$

We also define the following subgroups of B:

$$T = \begin{bmatrix} * & & & \\ & * & & \\ & & * & \\ & & & * \end{bmatrix}, \qquad U = \begin{bmatrix} 1 & * & * & * \\ & 1 & * & * \\ & & 1 & * \\ & & & 1 \end{bmatrix}. \tag{2.8}$$

Then T normalizes U, and $B = TU = UT$. Every element h of B can be written in the form

$$h = tu, \qquad \text{where} \quad t = \begin{bmatrix} a & & & \\ & b & & \\ & & cb^{-1} & \\ & & & ca^{-1} \end{bmatrix} \quad \text{and} \quad u = \begin{bmatrix} 1 & & & \\ & 1 & w & \\ & & 1 & \\ & & & 1 \end{bmatrix} \begin{bmatrix} 1 & x & y & z \\ & 1 & & y \\ & & 1 & -x \\ & & & 1 \end{bmatrix},$$

with $a, b, c \in F^\times$ and $w, x, y, z \in F$. With this notation, $\lambda(h) = c$. Let χ_1, χ_2 and σ be characters of $F^\times$. The representation of $\mathrm{GSp}(4, F)$ obtained from the character of T defined by $t \mapsto \chi_1(a)\chi_2(b)\sigma(c)$ for t as above via normalized induction as in Sect. 2.1 is denoted by $\chi_1 \times \chi_2 \rtimes \sigma$. The standard model for $\chi_1 \times \chi_2 \rtimes \sigma$ is the complex vector space all the locally constant functions $f : \mathrm{GSp}(4, F) \to \mathbb{C}$ that satisfy

$$f(hg) = |a^2 b||c|^{-\frac{3}{2}} \chi_1(a)\chi_2(b)\sigma(c) f(g)$$

for g in $\mathrm{GSp}(4, F)$ and h in B as above; the action of $\mathrm{GSp}(4, F)$ is by right translation. We remark that the modular character (see (2.1)) of B is given by $\delta_B(h) = |a|^4 |b|^2 |c|^{-3}$ for h as above, and that $\chi_1 \times \chi_2 \rtimes \sigma$ admits a central character, which is $\chi_1 \chi_2 \sigma^2$.

The *Siegel parabolic subgroup* of $\mathrm{GSp}(4, F)$ is defined to be

$$P = \begin{bmatrix} * & * & * & * \\ * & * & * & * \\ & & * & * \\ & & * & * \end{bmatrix}.$$

We also define the following subgroups of P:

$$M_P = \begin{bmatrix} * & * & & \\ * & * & & \\ & & * & * \\ & & * & * \end{bmatrix}, \qquad N_P = \begin{bmatrix} 1 & & * & * \\ & 1 & * & * \\ & & 1 & \\ & & & 1 \end{bmatrix}. \tag{2.9}$$

Then M_P normalizes N_P, and we have $P = M_P N_P = N_P M_P$. An element p of P has a decomposition of the form

$$p = mn, \qquad \text{where} \quad m = \begin{bmatrix} a_1 & a_2 & & \\ a_3 & a_4 & & \\ & & cd_1 & cd_2 \\ & & cd_3 & cd_4 \end{bmatrix} \quad \text{and} \quad n = \begin{bmatrix} 1 & & y & z \\ & 1 & x & y \\ & & 1 & \\ & & & 1 \end{bmatrix}$$

with $A = \left[\begin{smallmatrix} a_1 & a_2 \\ a_3 & a_4 \end{smallmatrix}\right] \in \mathrm{GL}(2, F)$, $\left[\begin{smallmatrix} d_1 & d_2 \\ d_3 & d_4 \end{smallmatrix}\right] = A'$ (see (2.1)), $c \in F^\times$, and $x, y, z \in F$. We note that $\lambda(p) = c$. Let (π, V) be an admissible representation of $\mathrm{GL}(2, F)$, and let σ be a character of $F^\times$. The representation of $\mathrm{GSp}(4, F)$ obtained from the representation of M_P on V defined by $m \mapsto \sigma(c)\pi(A)$ for m as above via normalized induction as in Sect. 2.1 is denoted by $\pi \rtimes \sigma$. The modular character of P (see (2.1)) is given by $\delta_P(p) = |\det(A)|^3|c|^{-3}$ for p as above. Thus, the standard model of $\pi \rtimes \sigma$ is comprised of all the locally constant functions $f : \mathrm{GSp}(4, F) \to V$ such that

$$f(pg) = |c^{-1}\det(A)|^{\frac{3}{2}}\sigma(c)\pi(A)f(g)$$

for p as above and $g \in \mathrm{GSp}(4, F)$. If the representation π admits a central character ω_π, then $\pi \rtimes \sigma$ admits $\omega_\pi\sigma^2$ as a central character.

Finally, we define the *Klingen parabolic subgroup* of $\mathrm{GSp}(4, F)$ to be

$$Q = \begin{bmatrix} * & * & * & * \\ & * & * & * \\ & * & * & * \\ & & & * \end{bmatrix}.$$

We also define the following subgroups of Q:

$$M_Q = \begin{bmatrix} * & & & \\ & * & * & \\ & * & * & \\ & & & * \end{bmatrix}, \qquad N_Q = \begin{bmatrix} 1 & * & * & * \\ & 1 & & * \\ & & 1 & * \\ & & & 1 \end{bmatrix}. \tag{2.10}$$

If q is in Q, then q has a decomposition

$$q = mn, \qquad \text{where} \quad m = \begin{bmatrix} t & & & \\ & a & b & \\ & c & d & \\ & & & \lambda t^{-1} \end{bmatrix} \quad \text{and} \quad n = \begin{bmatrix} 1 & x & y & z \\ & 1 & & y \\ & & 1 & -x \\ & & & 1 \end{bmatrix},$$

with $\left[\begin{smallmatrix} a & b \\ c & d \end{smallmatrix}\right] \in \mathrm{GL}(2, F)$, $t \in F^\times$, $x, y, z \in F$, and $\lambda = \det(\left[\begin{smallmatrix} a & b \\ c & d \end{smallmatrix}\right])$. We note that $\lambda(q) = \lambda$. Let (π, V) be an admissible representation of $\mathrm{GSp}(2, F) = \mathrm{GL}(2, F)$, and let χ be a character of $F^\times$. The representation of $\mathrm{GSp}(4, F)$ obtained by normalized induction as in Sect. 2.1 from the representation of M_Q on V defined by $m \mapsto \chi(t)\pi(\left[\begin{smallmatrix} a & b \\ c & d \end{smallmatrix}\right])$ for m as above is denoted by $\chi \rtimes \pi$. The modular character of Q is given by $\delta_Q(q) = |t|^4|ad - bc|^{-2}$ for q as above. The standard model of $\chi \rtimes \pi$ consists of all the locally constant functions $f : \mathrm{GSp}(4, F) \to V$ such that

$$f(qg) = |t^2(ad - bc)^{-1}|\chi(t)\pi(\left[\begin{smallmatrix} a & b \\ c & d \end{smallmatrix}\right])f(g)$$

for q as above and $g \in \mathrm{GSp}(4, F)$.

Generic Representations

Let U be the subgroup of B defined in (2.8). Let $c_1, c_2 \in F$, and let $\psi : F \to \mathbb{C}^\times$ be the character fixed in Sect. 2.1. We define a character $\psi_{c_1,c_2} : U \to \mathbb{C}^\times$ by

$$\psi_{c_1,c_2}(\begin{bmatrix} 1 & x & * & * \\ & 1 & y & * \\ & & 1 & -x \\ & & & 1 \end{bmatrix}) = \psi(c_1 x + c_2 y), \qquad x, y \in F. \tag{2.11}$$

Let (π, V) be an irreducible, admissible representation of $\mathrm{GSp}(4, F)$. We say that π is *generic* if $\mathrm{Hom}_U(V, \psi_{c_1,c_2})$ is non-zero for some $c_1, c_2 \in F^\times$; we note that this does not depend on the choice of c_1 and c_2 in $F^\times$. Assume that π is generic. Then π has a Whittaker model. This means that π is isomorphic to a subspace of the complex vector space of all functions $W : \mathrm{GSp}(4, F) \to \mathbb{C}$ such that

$$W(\begin{bmatrix} 1 & x & * & * \\ & 1 & y & * \\ & & 1 & -x \\ & & & 1 \end{bmatrix} g) = \psi(c_1 x + c_2 y) W(g), \qquad x, y \in F, \quad g \in \mathrm{GSp}(4, F).$$

Here, $\mathrm{GSp}(4, F)$ acts on the space of all such functions by right translation. The subspace isomorphic to π is unique, and will be denoted by $\mathcal{W}(\pi, \psi_{c_1,c_2})$. We call $\mathcal{W}(\pi, \psi_{c_1,c_2})$ the *Whittaker model* of π with respect to ψ_{c_1,c_2}. In the remainder of this work, if we consider a generic representation π and its Whittaker model $\mathcal{W}(\pi, \psi_{c_1,c_2})$, then we always assume that $c_1, c_2 \in \mathfrak{o}^\times$.

The List of Non-supercuspidal Representations

An irreducible, admissible representation of $\mathrm{GSp}(4, F)$ is called *supercuspidal*, if it is not a subquotient of a representation obtained via parabolic induction from one of the three standard parabolic subgroups; this is equivalent to the vanishing of all Jacquet modules with respect to the unipotent radicals of all proper parabolic subgroup (see [28]). Thanks to [109], it is possible to provide a convenient list of all the non-supercuspidal, irreducible, admissible representations of $\mathrm{GSp}(4, F)$. This list is described in [105]; here, we provide a brief summary. The list also appears in Table A.1. The list consists of eleven groups of representations, denoted by capital Roman numerals. These groups are further divided into (sometimes only one, and at most four) subgroups, denoted by small Roman letters (e.g., IIa and IIb); in the case that a group admits only one subgroup, the letter is omitted. Every non-supercuspidal, irreducible, admissible representation of $\mathrm{GSp}(4, F)$ is isomorphic to a representation that is a member of a group and a subgroup. Representations from distinct subgroups are not isomorphic, with the exception that subgroup Vb equals subgroup Vc if the parameters are allowed to run through all possibilities. The notation is such that representations of an "a" subgroup (e.g., IIa) are generic (if a group consists only of one subgroup, then all the representations in that group are generic).

Group I The members of this group are the irreducible representations of the form $\chi_1 \times \chi_2 \rtimes \sigma$, where χ_1, χ_2 and σ are characters of $F^\times$. A representation of this form is irreducible if and only if $\chi_1 \neq \nu^{\pm 1}$, $\chi_2 \neq \nu^{\pm 1}$, and $\chi_2 \neq \chi_1^{\pm 1}\nu^{\pm 1}$. This group of representations has only one subgroup.

Group II Let χ be a character of $F^\times$ such that $\chi \neq \nu^{\pm\frac{3}{2}}$ and $\chi^2 \neq \nu^{\pm 1}$, and let σ be a character of $F^\times$. The members of Group II are the irreducible constituents of the representations of the form $\nu^{\frac{1}{2}}\chi \times \nu^{-\frac{1}{2}}\chi \rtimes \sigma$. The representation $\nu^{\frac{1}{2}}\chi \times \nu^{-\frac{1}{2}}\chi \rtimes \sigma$ has two irreducible constituents. One of these irreducible constituents, the subrepresentation, is $\chi \mathrm{St}_{\mathrm{GL}(2)} \rtimes \sigma$; here $\mathrm{St}_{\mathrm{GL}(2)}$ is the Steinberg representation of $\mathrm{GL}(2, F)$. The representations of the form $\chi \mathrm{St}_{\mathrm{GL}(2)} \rtimes \sigma$ comprise subgroup IIa. The other irreducible constituent, the quotient, of $\nu^{\frac{1}{2}}\chi \times \nu^{-\frac{1}{2}}\chi \rtimes \sigma$ is $\chi 1_{\mathrm{GL}(2)} \rtimes \sigma$. The representations of the form $\chi 1_{\mathrm{GL}(2)} \rtimes \sigma$ make up subgroup IIb.

Group III Let χ be a character of $F^\times$ such that $\chi \neq 1_{F^\times}$ and $\chi \neq \nu^{\pm 2}$, and let σ be a character of $F^\times$. The elements of Group III are the irreducible constituents of the representations of the form $\chi \times \nu \rtimes \nu^{-\frac{1}{2}}\sigma$. The representation $\chi \times \nu \rtimes \nu^{-\frac{1}{2}}\sigma$ has two irreducible constituents. The irreducible representation $\chi \rtimes \sigma \mathrm{St}_{\mathrm{GSp}(2)}$ is a subrepresentation of $\chi \times \nu \rtimes \nu^{-\frac{1}{2}}\sigma$, and representations of the form $\chi \rtimes \sigma \mathrm{St}_{\mathrm{GSp}(2)}$ make up subgroup IIIa. The irreducible representation $\chi \rtimes \sigma 1_{\mathrm{GSp}(2)}$ is a quotient of $\chi \times \nu \rtimes \nu^{-\frac{1}{2}}\sigma$, and representations of the form $\chi \rtimes \sigma 1_{\mathrm{GSp}(2)}$ comprise subgroup IIIb.

Group IV Let σ be a character of $F^\times$. The irreducible constituents of the representations of the form $\nu^2 \times \nu \rtimes \nu^{-\frac{3}{2}}\sigma$ make up Group IV. The representation $\nu^2 \times \nu \rtimes \nu^{-\frac{3}{2}}\sigma$ has four irreducible constituents, which can be described as follows. There are two exact sequences

$$0 \to \nu^{\frac{3}{2}}\mathrm{St}_{\mathrm{GL}(2)} \rtimes \nu^{-\frac{3}{2}}\sigma \to \nu^2 \times \nu \rtimes \nu^{-\frac{3}{2}}\sigma \to \nu^{\frac{3}{2}} 1_{\mathrm{GL}(2)} \rtimes \nu^{-\frac{3}{2}}\sigma \to 0 \tag{2.12}$$

and

$$0 \to \nu^2 \times \nu^{-1}\sigma \mathrm{St}_{\mathrm{GSp}(2)} \to \nu^2 \times \nu \rtimes \nu^{-\frac{3}{2}}\sigma \to \nu^2 \rtimes \nu^{-1}\sigma 1_{\mathrm{GSp}(2)} \to 0. \tag{2.13}$$

In these exact sequences, the second and fourth representations are reducible, and each has two irreducible constituents as in the following table. The subrepresentations are on the top and left, and the quotients are on the bottom and right.

$$\begin{array}{c|cc} & \nu^{\frac{3}{2}}\mathrm{St}_{\mathrm{GL}(2)} \rtimes \nu^{-\frac{3}{2}}\sigma & \nu^{\frac{3}{2}}1_{\mathrm{GL}(2)} \rtimes \nu^{-\frac{3}{2}}\sigma \\ \hline \nu^2 \rtimes \nu^{-1}\sigma\mathrm{St}_{\mathrm{GSp}(2)} & \sigma\mathrm{St}_{\mathrm{GSp}(4)} & L(\nu^2, \nu^{-1}\sigma\mathrm{St}_{\mathrm{GSp}(2)}) \\ \nu^2 \rtimes \nu^{-1}\sigma 1_{\mathrm{GSp}(2)} & L(\nu^{\frac{3}{2}}\mathrm{St}_{\mathrm{GL}(2)}, \nu^{-\frac{3}{2}}\sigma) & \sigma 1_{\mathrm{GSp}(4)} \end{array} \tag{2.14}$$

Thus, $\sigma\mathrm{St}_{\mathrm{GSp}(4)}$, $L(\nu^2, \nu^{-1}\sigma\mathrm{St}_{\mathrm{GSp}(2)})$, $L(\nu^{\frac{3}{2}}\mathrm{St}_{\mathrm{GL}(2)}, \nu^{-\frac{3}{2}}\sigma)$, and $\sigma 1_{\mathrm{GSp}(4)}$ are the irreducible constituents of $\nu^2 \times \nu \rtimes \nu^{-\frac{3}{2}}\sigma$, and representations of this form constitute, respectively, subgroups IVa, IVb, IVc and IVd.

Group V Let ξ be a non-trivial quadratic character of $F^\times$, and let σ be a character of $F^\times$. Then the irreducible constituents of representations of the form $\nu\xi \rtimes \xi \rtimes \nu^{-\frac{1}{2}}\sigma$ make up Group V. The representation $\nu\xi \rtimes \xi \rtimes \nu^{-\frac{1}{2}}\sigma$ has four irreducible constituents which are delineated as follows. There are two exact sequences

$$0 \to \nu^{\frac{1}{2}}\xi\mathrm{St}_{\mathrm{GL}(2)} \rtimes \nu^{-\frac{1}{2}}\sigma \to \nu\xi \times \xi \rtimes \nu^{-\frac{1}{2}}\sigma \to \nu^{\frac{1}{2}}\xi 1_{\mathrm{GL}(2)} \rtimes \nu^{-\frac{1}{2}}\sigma \to 0 \tag{2.15}$$

and

$$0 \to \nu^{\frac{1}{2}}\xi\mathrm{St}_{\mathrm{GL}(2)} \rtimes \xi\nu^{-\frac{1}{2}}\sigma \to \nu\xi \times \xi \rtimes \nu^{-\frac{1}{2}}\sigma \to \nu^{\frac{1}{2}}\xi 1_{\mathrm{GL}(2)} \rtimes \xi\nu^{-\frac{1}{2}}\sigma \to 0. \tag{2.16}$$

In these exact sequences, the second and fourth representations are reducible, and each has two irreducible constituents as in the following table. The subrepresentations are on the top and left, and the quotients are on the bottom and right.

$$\begin{array}{c|cc} & \nu^{\frac{1}{2}}\xi\mathrm{St}_{\mathrm{GL}(2)} \rtimes \nu^{-\frac{1}{2}}\xi\sigma & \nu^{\frac{1}{2}}\xi 1_{\mathrm{GL}(2)} \rtimes \xi\nu^{-\frac{1}{2}}\sigma \\ \hline \nu^{\frac{1}{2}}\xi\mathrm{St}_{\mathrm{GL}(2)} \rtimes \nu^{-\frac{1}{2}}\sigma & \delta([\xi, \nu\xi], \nu^{-\frac{1}{2}}\sigma) & L(\nu^{\frac{1}{2}}\xi\mathrm{St}_{\mathrm{GL}(2)}, \nu^{-\frac{1}{2}}\sigma) \\ \nu^{\frac{1}{2}}\xi 1_{\mathrm{GL}(2)} \rtimes \nu^{-\frac{1}{2}}\sigma & L(\nu^{\frac{1}{2}}\xi\mathrm{St}_{\mathrm{GL}(2)}, \nu^{-\frac{1}{2}}\xi\sigma) & L(\nu\xi, \xi \rtimes \nu^{-\frac{1}{2}}\sigma) \end{array} \tag{2.17}$$

The representation $\nu\xi \rtimes \xi \rtimes \nu^{-\frac{1}{2}}\sigma$ has as irreducible constituents $\delta([\xi, \nu\xi], \nu^{-\frac{1}{2}}\sigma)$, $L(\nu^{\frac{1}{2}}\xi\mathrm{St}_{\mathrm{GL}(2)}, \nu^{-\frac{1}{2}}\sigma)$, $L(\nu^{\frac{1}{2}}\xi\mathrm{St}_{\mathrm{GL}(2)}, \nu^{-\frac{1}{2}}\xi\sigma)$, and $L(\nu\xi, \xi \rtimes \nu^{-\frac{1}{2}}\sigma)$, and representations of this form make up subgroups Va, Vb, Vc and Vd, respectively. We note that the Va representation $\delta([\xi, \nu\xi], \nu^{-\frac{1}{2}}\sigma)$ is essentially square integrable, and while the Vb representation $L(\nu^{\frac{1}{2}}\xi\mathrm{St}_{\mathrm{GL}(2)}, \nu^{-\frac{1}{2}}\sigma)$ and the Vc representation $L(\nu^{\frac{1}{2}}\xi\mathrm{St}_{\mathrm{GL}(2)}, \nu^{-\frac{1}{2}}\xi\sigma)$ are twists of each other, these representations are not isomorphic.

Group VI Let σ be a character of $F^\times$. The irreducible constituents of representations of the form $\nu \times 1_{F^\times} \rtimes \nu^{-\frac{1}{2}}\sigma$ make up Group VI. The representation $\nu \times 1_{F^\times} \rtimes \nu^{-\frac{1}{2}}\sigma$ has four irreducible constituents which can be described as follows.

There are two exact sequences

$$0 \to \nu^{\frac{1}{2}}\mathrm{St}_{\mathrm{GL}(2)} \rtimes \nu^{-\frac{1}{2}}\sigma \to \nu \times 1_{F^\times} \rtimes \nu^{-\frac{1}{2}}\sigma \to \nu^{\frac{1}{2}}1_{\mathrm{GL}(2)} \rtimes \nu^{-\frac{1}{2}}\sigma \to 0 \qquad (2.18)$$

and

$$0 \to 1_{F^\times} \rtimes \sigma\mathrm{St}_{\mathrm{GSp}(2)} \to \nu \times 1_{F^\times} \rtimes \nu^{-\frac{1}{2}}\sigma \to 1_{F^\times} \rtimes \sigma 1_{\mathrm{GSp}(2)} \to 0. \qquad (2.19)$$

In these exact sequences, the second and fourth representations are reducible, and each has two irreducible constituents as in the following table. The subrepresentations are on the top and left, and the quotients are on the bottom and right.

(2.20)

	$\nu^{\frac{1}{2}}\mathrm{St}_{\mathrm{GL}(2)} \rtimes \nu^{-\frac{1}{2}}\sigma$	$\nu^{\frac{1}{2}}1_{\mathrm{GL}(2)} \rtimes \nu^{-\frac{1}{2}}\sigma$
$1_{F^\times} \rtimes \sigma\mathrm{St}_{\mathrm{GSp}(2)}$	$\tau(S, \nu^{-\frac{1}{2}}\sigma)$	$\tau(T, \nu^{-\frac{1}{2}}\sigma)$
$1_{F^\times} \rtimes \sigma 1_{\mathrm{GSp}(2)}$	$L(\nu^{\frac{1}{2}}\mathrm{St}_{\mathrm{GL}(2)}, \nu^{-\frac{1}{2}}\sigma)$	$L(\nu, 1_{F^\times} \rtimes \nu^{-\frac{1}{2}}\sigma)$

The representation $\nu \times 1_{F^\times} \rtimes \nu^{-\frac{1}{2}}\sigma$ has the representations $\tau(S, \nu^{-\frac{1}{2}}\sigma)$, $\tau(T, \nu^{-\frac{1}{2}}\sigma)$, $L(\nu^{\frac{1}{2}}\mathrm{St}_{\mathrm{GL}(2)}, \nu^{-\frac{1}{2}}\sigma)$, and $L(\nu, 1_{F^\times} \rtimes \nu^{-\frac{1}{2}}\sigma)$ as irreducible constituents, and representations of this form constitute subgroups VIa, VIb, VIc, and VId, respectively. We note that the representations $\tau(S, \nu^{-\frac{1}{2}}\sigma)$ and $\tau(T, \nu^{-\frac{1}{2}}\sigma)$ are essentially tempered but not essentially square integrable.

Group VII The members of this group are the irreducible representations of the form $\chi \rtimes \pi$, where χ is a character of $F^\times$, and π is a supercuspidal, irreducible, admissible representation of $\mathrm{GSp}(2, F) = \mathrm{GL}(2, F)$. A representation of this form is irreducible if and only if $\chi \neq 1_{F^\times}$ and $\chi \neq \xi\nu^{\pm 1}$ for every character ξ of $F^\times$ of order two such that $\xi\pi \cong \pi$. This group of representations has only one subgroup.

Group VIII Let π be a supercuspidal, irreducible, admissible representation of $\mathrm{GSp}(2, F) = \mathrm{GL}(2, F)$. The members of Group VIII are the irreducible constituents of representations of the form $1 \rtimes \pi$. The representation $1 \rtimes \pi$ is the direct sum of two essentially tempered, irreducible, admissible representations $\tau(S, \pi)$ and $\tau(T, \pi)$. Representations of the form $\tau(S, \pi)$ and $\tau(T, \pi)$ make up subgroups VIIIa and VIIIb, respectively.

Group IX Let ξ be a non-trivial quadratic character of $F^\times$, and let π be a supercuspidal, irreducible, admissible representation of $\mathrm{GSp}(2, F) = \mathrm{GL}(2, F)$ such that $\xi\pi \cong \pi$. The members of group IX are the irreducible constituents of the representations of the form $\nu\xi \rtimes \nu^{-\frac{1}{2}}\pi$. The representation $\nu\xi \rtimes \nu^{-\frac{1}{2}}\pi$ has two irreducible constituents. One of these irreducible constituents, the subrepresentation $\delta(\nu\xi, \nu^{-\frac{1}{2}}\pi)$, is essentially square-integrable; the quotient $L(\nu\xi, \nu^{-\frac{1}{2}}\pi)$ is non-tempered. Representations of the form $\delta(\nu\xi, \nu^{-\frac{1}{2}}\pi)$ make up subgroup IXa, and representations of the form $L(\nu\xi, \nu^{-\frac{1}{2}}\pi)$ make up subgroup IXb.

Group X The members of this group are the irreducible representations of the form $\pi \rtimes \sigma$ where π is a supercuspidal, irreducible, admissible representation of GL(2, F), and σ is a character of $F^\times$. A representation of this form is irreducible if and only if π is not of the form $\nu^{\pm\frac{1}{2}}\rho$ where ρ is a supercuspidal, irreducible, admissible representation of GL(2, F) with trivial central character. This group has only one subgroup.

Group XI Let π be a supercuspidal, irreducible, admissible representation of GL(2, F) with trivial central character, and let σ be a character of $F^\times$. The members of group XI are the irreducible constituents of representations of the form $\nu^{\frac{1}{2}}\pi \rtimes \nu^{-\frac{1}{2}}\sigma$. The representation $\nu^{\frac{1}{2}}\pi \rtimes \nu^{-\frac{1}{2}}\sigma$ has two irreducible constituents. One of these constituents, the subrepresentation $\delta(\nu^{\frac{1}{2}}\pi, \nu^{-\frac{1}{2}}\sigma)$, is essentially square integrable; the quotient $L(\nu^{\frac{1}{2}}\pi, \nu^{-\frac{1}{2}}\sigma)$ is non-tempered. Representations of the form $\delta(\nu^{\frac{1}{2}}\pi, \nu^{-\frac{1}{2}}\sigma)$ make up subgroup XIa, and representations of the form $L(\nu^{\frac{1}{2}}\pi, \nu^{-\frac{1}{2}}\sigma)$ make up subgroup XIb.

Saito-Kurokawa Representations

Let π be an infinite-dimensional, irreducible, admissible representation of GL(2, F) with trivial central character, and let σ be a character of $F^\times$. Assume that $\pi \not\cong \nu^{\frac{3}{2}} \times \nu^{-\frac{3}{2}}$. Then the representation $\nu^{\frac{1}{2}}\pi \rtimes \nu^{-\frac{1}{2}}\sigma$ of GSp(4, F) has a unique irreducible quotient $Q(\nu^{\frac{1}{2}}\pi, \nu^{-\frac{1}{2}}\sigma)$; we say that $Q(\nu^{\frac{1}{2}}\pi, \nu^{-\frac{1}{2}}\sigma)$ is a *Saito-Kurokawa* representation. The central character of $Q(\nu^{\frac{1}{2}}\pi, \nu^{-\frac{1}{2}}\sigma)$ is σ^2. The Saito-Kurokawa representations are all the representations in groups IIb, Vb, Vc, VIc, and XIb, and form an important family of non-generic representations of GSp(4, F). See Sect. 5.5 of [105] for more information.

The P_3-Quotient

As developed in [105] and [106], there is a useful map of representations of GSp(4, F) to representations of the group P_3. Here, P_3 is the subgroup of GL(3, F) defined as

$$P_3 = \begin{bmatrix} * & * & * \\ * & * & * \\ & & 1 \end{bmatrix}. \tag{2.21}$$

To explain the map we require some more definitions. The *Jacobi subgroup* of GSp(4, F) is defined to be

$$G^J = \begin{bmatrix} 1 & * & * & * \\ & * & * & * \\ & * & * & * \\ & & & 1 \end{bmatrix}.$$

The Jacobi subgroup G^J is a normal subgroup of the Klingen parabolic subgroup Q. The center of the Jacobi group is

$$Z^J = \begin{bmatrix} 1 & & & * \\ & 1 & & \\ & & 1 & \\ & & & 1 \end{bmatrix}. \tag{2.22}$$

The group Z^J is also a normal subgroup of Q. Next, let $\bar{Q}$ be the following subgroup of Q:

$$\bar{Q} = \begin{bmatrix} * & * & * & * \\ & * & * & * \\ & * & * & * \\ & & & 1 \end{bmatrix}.$$

Then Z^J is a normal subgroup of $\bar{Q}$. Every element of $\bar{Q}$ can be uniquely written in the form

$$q = \begin{bmatrix} ad-bc & & & \\ & a & b & \\ & c & d & \\ & & & 1 \end{bmatrix} \begin{bmatrix} 1 & -y & x & z \\ & 1 & & x \\ & & 1 & y \\ & & & 1 \end{bmatrix}$$

with $\left[\begin{smallmatrix} a & b \\ c & d \end{smallmatrix}\right] \in \mathrm{GL}(2, F)$ and $x, y, z \in F$. Define $i : \bar{Q} \to P_3$ by

$$i(q) = \begin{bmatrix} a & b & \\ c & d & \\ & & 1 \end{bmatrix} \begin{bmatrix} 1 & & x \\ & 1 & y \\ & & 1 \end{bmatrix}$$

for q as above. Then i is a surjective homomorphism with kernel Z^J, and thus induces an isomorphism $\bar{Q}/Z^J \cong P_3$. Now suppose that (π, V) is a smooth representation of $\mathrm{GSp}(4, F)$. Let $V(Z^J)$ be the $\mathbb{C}$ subspace of V spanned by the vectors $v - \pi(z)v$ for $v \in V$ and $z \in Z^J$. Since $\bar{Q}$ normalizes Z^J, the subspace $V(Z^J)$ is a $\bar{Q}$ subspace of V. It follows that the quotient $V_{Z^J} = V/V(Z^J)$ is also a representation of $\bar{Q}$. Since Z^J acts trivially on V_{Z^J}, the group $\bar{Q}/Z^J \cong P_3$ acts on V_{Z^J}, so that V_{Z^J} may be regarded as a representation of P_3. We refer to V_{Z^J} as the P_3*-quotient* of V. We will write $p : V \to V_{Z^J}$ for the projection from V to V_{Z^J}. We have $p(\pi(q)v) = i(q)v$ for $v \in V$ and $q \in \bar{Q}$.

To state the main result about the P_3-quotient of an irreducible, admissible representation of $\mathrm{GSp}(4, F)$ we need to first recall some facts about representations of P_3, as explained in [15]. Define a subgroup U_3 of P_3 by

$$U_3 = \begin{bmatrix} 1 & * & * \\ & 1 & * \\ & & 1 \end{bmatrix}.$$

Define characters Θ and Θ' of U_3 by

$$\Theta(\begin{bmatrix} 1 & u_{12} & * \\ & 1 & u_{23} \\ & & 1 \end{bmatrix}) = \psi(u_{12} + u_{23}), \qquad \Theta'(\begin{bmatrix} 1 & u_{12} & * \\ & 1 & u_{23} \\ & & 1 \end{bmatrix}) = \psi(u_{23})$$

where $u_{12}, u_{23} \in F$. There are three families of representations of P_3, obtained from representations of the groups $\mathrm{GL}(0, F) = 1$, $\mathrm{GL}(1, F) = F^\times$ and $\mathrm{GL}(2, F)$ by an induction. First, we define

$$\tau^{P_3}_{\mathrm{GL}(0)}(1) = \text{c-Ind}^{P_3}_{U_3}(\Theta).$$

The representation $\tau^{P_3}_{\mathrm{GL}(0)}(1)$ is smooth and irreducible. Second, let (χ, V) be smooth representation of $F^\times$. Define a smooth representation $\Theta' \otimes \chi$ of the subgroup $\begin{bmatrix} * & * & * \\ & 1 & * \\ & & 1 \end{bmatrix}$ of P_3 on the space V of χ by

$$(\Theta' \otimes \chi)(\begin{bmatrix} a & * & * \\ & 1 & y \\ & & 1 \end{bmatrix})v = \Theta'(\begin{bmatrix} a & * & * \\ & 1 & y \\ & & 1 \end{bmatrix})\chi(a)v = \psi(y)\chi(a)v$$

for $a \in F^\times$, $y \in F$, and $v \in V$. Then we define

$$\tau^{P_3}_{\mathrm{GL}(1)}(\chi) = \text{c-Ind}^{P_3}_{\begin{bmatrix} * & * & * \\ & 1 & * \\ & & 1 \end{bmatrix}}(\Theta' \otimes \chi)$$

The representation $\tau^{P_3}_{\mathrm{GL}(1)}(\chi)$ is smooth, and if χ is irreducible (i.e., a character of $F^\times$), then $\tau^{P_3}_{\mathrm{GL}(1)}(\chi)$ is irreducible. Third, let ρ be a smooth representation of $\mathrm{GL}(2, F)$. We define

$$\tau^{P_3}_{\mathrm{GL}(2)}(\rho)$$

to have the same space as ρ, with action given by

$$\tau^{P_3}_{\mathrm{GL}(2)}(\rho)(\begin{bmatrix} a & b & * \\ c & d & * \\ & & 1 \end{bmatrix}) = \rho(\begin{bmatrix} a & b \\ c & d \end{bmatrix}),$$

where $\begin{bmatrix} a & b \\ c & d \end{bmatrix} \in \mathrm{GL}(2, F)$. Clearly, $\tau^{P_3}_{\mathrm{GL}(2)}(\rho)$ is smooth, and if ρ is irreducible, then $\tau^{P_3}_{\mathrm{GL}(2)}(\rho)$ is irreducible. Let η be an irreducible, smooth representation of P_3. Then η is isomorphic to a representation of the form $\tau^{P_3}_{\mathrm{GL}(0)}(1)$, $\tau^{P_3}_{\mathrm{GL}(1)}(\chi)$, where χ is a character of $F^\times$, or $\tau^{P_3}_{\mathrm{GL}(2)}(\rho)$ where ρ is an irreducible, admissible representation of P_3; moreover, if $\eta \cong \tau^{P_3}_{\mathrm{GL}(k_1)}(\pi_1)$ and $\eta \cong \tau^{P_3}_{\mathrm{GL}(k_2)}(\pi_2)$ where $k_1, k_2 \in \{0, 1, 2\}$ and π_1 and π_2 are of the types just described, then $k_1 = k_2$ and $\pi_1 \cong \pi_2$.

Theorem 2.2.1 *Let (π, V) be an irreducible, admissible representation of the group $\mathrm{GSp}(4, F)$. The quotient $V_{Z^J} = V/V(Z^J)$ is a smooth representation of $\bar{Q}/Z^J$, and hence defines a smooth representation of $P_3 \cong \bar{Q}/Z^J$. As a representation of P_3, V_{Z^J} has a finite filtration by P_3 subspaces such that the successive quotients are irreducible and of the form $\tau^{P_3}_{\mathrm{GL}(0)}(1)$, $\tau^{P_3}_{\mathrm{GL}(1)}(\chi)$, or $\tau^{P_3}_{\mathrm{GL}(2)}(\rho)$, where χ is a character of $F^\times$, and ρ is an irreducible, admissible representation of* $\mathrm{GL}(2, F)$. *Moreover:*

(1) *There exists a chain of P_3 subspaces*

$$0 \subset V_2 \subset V_1 \subset V_0 = V_{Z^J}$$

such that

$$V_2 \cong \dim \mathrm{Hom}_U(V, \psi_{-1,1}) \cdot \tau_{\mathrm{GL}(0)}^{P_3}(1),$$
$$V_1/V_2 \cong \tau_{\mathrm{GL}(1)}^{P_3}(V_{U,\psi_{-1,0}}),$$
$$V_0/V_1 \cong \tau_{\mathrm{GL}(2)}^{P_3}(V_{N_Q}).$$

Here, the vector space $V_{U,\psi_{-1,0}}$ admits a smooth action of $\mathrm{GL}(1,F) \cong F^\times$ *induced by the operators*

$$\pi(\begin{bmatrix} a & & & \\ & a & & \\ & & 1 & \\ & & & 1 \end{bmatrix}), \qquad a \in F^\times,$$

and V_{N_Q} admits a smooth action of $\mathrm{GL}(2,F)$ *induced by the operators*

$$\pi(\begin{bmatrix} \det(g) & & \\ & g & \\ & & 1 \end{bmatrix}), \qquad g \in \mathrm{GL}(2,F).$$

(2) *The representation π is generic if and only if $V_2 \neq 0$, and if π is generic, then $V_2 \cong \tau_{\mathrm{GL}(0)}^{P_3}(1)$.*
(3) *We have $V_2 \cong V_{Z^J}$ if and only if π is supercuspidal. If π is supercuspidal and generic, then $V_{Z^J} = V_2 \cong \tau_{\mathrm{GL}(0)}^{P_3}(1)$ is non-zero and irreducible. If π is supercuspidal and non-generic, then $V_{Z^J} = V_2 = 0$.*

Proof See Sect. 2.5 of [105] and Sect. 3 of [106]. □

The semisimplifications of the quotients V_0/V_1 and V_1/V_2 from Theorem 2.2.1 for all irreducible, admissible representations (π, V) of $\mathrm{GSp}(4,F)$ are listed in Appendix A.4 of [105]. Note that this reference assumed that π has trivial central character; however, as pointed out in [106], the results are exactly the same without this assumption. Note also that there is typo in Table A.5 of [105]: The entry for Vd in the "s.s.(V_0/V_1)" column should be $\tau_{\mathrm{GL}(2)}^{P_3}(\nu(\nu^{-\frac{1}{2}}\sigma \times \nu^{-\frac{1}{2}}\xi\sigma))$.

Zeta Integrals

Let π be a generic, irreducible, admissible representation of $\mathrm{GSp}(4,F)$ with trivial central character. Let $c_1, c_2 \in \mathfrak{o}^\times$, and let $\mathcal{W}(\pi, \psi_{c_1,c_2})$ be the Whittaker model of π defined with respect to ψ_{c_1,c_2}, as defined above. For $W \in \mathcal{W}(\pi, \psi_{c_1,c_2})$ and $s \in \mathbb{C}$, we define the *zeta integral* $Z(s, W)$ by

$$Z(s, W) = \int\limits_{F^\times}\int\limits_{F} W(\begin{bmatrix} a & & & \\ & a & & \\ & x & 1 & \\ & & & 1 \end{bmatrix})|a|^{s-\frac{3}{2}}\, dx\, d^\times a.$$

The following results were proven in [105]. There exists a real number s_0 such that, for all $W \in \mathcal{W}(\pi, \psi_{c_1,c_2})$, $Z(s, W)$ converges for $s \in \mathbb{C}$ with $\mathrm{Re}(s) > s_0$ to an element of $\mathbb{C}(q^{-s})$. Thus, $Z(s, W)$ for $W \in \mathcal{W}(\pi, \psi_{c_1,c_2})$ has a meromorphic continuation to $\mathbb{C}$. Let $I(\pi)$ be the $\mathbb{C}$ vector subspace of $\mathbb{C}(q^{-s})$ spanned by the $Z(s, W)$ for $W \in \mathcal{W}(\pi, \psi_{c_1,c_2})$. Then $I(\pi)$ is independent of the choice of $c_1, c_2 \in \mathfrak{o}^\times$. Moreover, $I(\pi)$ is a non-zero $\mathbb{C}[q^{-s}, q^s]$ module containing $\mathbb{C}$, and there exists $R(X) \in \mathbb{C}[X]$ such that $R(q^{-s})I(\pi) \subset \mathbb{C}[q^{-s}, q^s]$, so that $I(\pi)$ is a fractional ideal of the principal ideal domain $\mathbb{C}[q^{-s}, q^s]$ whose quotient field is $\mathbb{C}(q^{-s})$. The fractional ideal $I(\pi)$ admits a generator of the form $1/Q(q^{-s})$ with $Q(0) = 1$, where $Q(X) \in \mathbb{C}[X]$. We define

$$L(s, \pi) = \frac{1}{Q(q^{-s})}, \tag{2.23}$$

and call $L(s, \pi)$ the *L-function* of π. Define

$$w = \begin{bmatrix} & & 1 & \\ & & & -1 \\ 1 & & & \\ & -1 & & \end{bmatrix}. \tag{2.24}$$

Then there exists $\gamma(s, \pi, \psi_{c_1,c_2}) \in \mathbb{C}(q^{-s})$ such that

$$Z(1 - s, \pi(w)W) = \gamma(s, \pi, \psi_{c_1,c_2})Z(s, W)$$

for $W \in \mathcal{W}(\pi, \psi_{c_1,c_2})$. The factor $\gamma(s, \pi, \psi_{c_1,c_2})$ does not depend on the choice of $c_1, c_2 \in \mathfrak{o}^\times$. Further define

$$\varepsilon(s, \pi, \psi_{c_1,c_2}) = \gamma(s, \pi, \psi_{c_1,c_2})\frac{L(s, \pi)}{L(1 - s, \pi)}.$$

Then there exists $\varepsilon \in \{\pm 1\}$ and an integer N such that $\varepsilon(s, \pi, \psi_{c_1,c_2}) = \varepsilon q^{-N(s-\frac{1}{2})}$. We will write $\gamma(s, \pi)$ for $\gamma(s, \pi, \psi_{c_1,c_2})$ and $\varepsilon(s, \pi)$ for $\varepsilon(s, \pi, \psi_{c_1,c_2})$, it being understood that we have fixed ψ as on p. 71 and that $c_1, c_2 \in \mathfrak{o}^\times$.

2.3 The Paramodular Theory

In this section we recall some facts about paramodular vectors from [105]. Let n be a non-negative integer. We define

$$\mathrm{K}(\mathfrak{p}^n) = \{g \in \mathrm{GSp}(4, F) \mid \lambda(g) \in \mathfrak{o}^\times\} \cap \begin{bmatrix} \mathfrak{o} & \mathfrak{o} & \mathfrak{o} & \mathfrak{p}^{-n} \\ \mathfrak{p}^n & \mathfrak{o} & \mathfrak{o} & \mathfrak{o} \\ \mathfrak{p}^n & \mathfrak{o} & \mathfrak{o} & \mathfrak{o} \\ \mathfrak{p}^n & \mathfrak{p}^n & \mathfrak{p}^n & \mathfrak{o} \end{bmatrix}. \tag{2.25}$$

Using (2.4), it is easy to verify that $\mathrm{K}(\mathfrak{p}^n)$ is a compact subgroup of $\mathrm{GSp}(4, F)$. We refer to $\mathrm{K}(\mathfrak{p}^n)$ as the *paramodular subgroup of level* $\mathfrak{p}^n$. We have $\mathrm{K}(\mathfrak{p}^0) = \mathrm{GSp}(4, \mathfrak{o})$. Let (π, V) be a smooth representation of the group $\mathrm{GSp}(4, F)$ for which the center of $\mathrm{GSp}(4, F)$ acts trivially. We define

$$V(n) = \{v \in V \mid \pi(k)v = v \text{ for all } k \in \mathrm{K}(\mathfrak{p}^n)\}. \tag{2.26}$$

We refer to the non-zero elements of $V(n)$ as *paramodular vectors*. Paramodular vectors have the following important property.

Theorem 2.3.1 *Let (π, V) be a smooth representation of* $\mathrm{GSp}(4, F)$ *for which the center acts trivially. Assume that the subspace of vectors fixed by* $\mathrm{Sp}(4, F)$ *is trivial. Then paramodular vectors from different levels are linearly independent, i.e., $v_1 \in V(n_1), \dots, v_t \in V(n_t)$ with $0 \le n_1 < \dots < n_t$ and $v_1 + \dots + v_t = 0$ implies that $v_1 = \dots = v_t = 0$*

Proof This is Theorem 3.1.3 of [105]. □

We note that the assumption in Theorem 2.3.1 is satisfied if π is an infinite-dimensional, irreducible, admissible representation of $\mathrm{GSp}(4, F)$ with trivial central character.

Let (π, V) be a smooth representation of $\mathrm{GSp}(4, F)$ for which the center acts trivially. We define three operators

$$\eta, \theta, \theta' : V \longrightarrow V$$

by

$$\eta v = \pi(\begin{bmatrix} \varpi^{-1} & & & \\ & 1 & & \\ & & 1 & \\ & & & \varpi \end{bmatrix})v, \tag{2.27}$$

$$\theta v = \pi(\begin{bmatrix} 1 & & & \\ & 1 & & \\ & & \varpi & \\ & & & \varpi \end{bmatrix})v + q \int\limits_{\mathfrak{o}} \pi(\begin{bmatrix} 1 & & & \\ & 1 & c & \\ & & 1 & \\ & & & 1 \end{bmatrix}\begin{bmatrix} 1 & & & \\ & \varpi & & \\ & & 1 & \\ & & & \varpi \end{bmatrix})v\,dc, \tag{2.28}$$

$$\theta' v = \eta v + q \int\limits_{\mathfrak{o}} \pi(\begin{bmatrix} 1 & & & c\varpi^{-(n+1)} \\ & 1 & & \\ & & 1 & \\ & & & 1 \end{bmatrix})v\,dc \tag{2.29}$$

for $v \in V$. If $v \in V(n)$, then $\eta v \in V(n+2)$ and $\theta v, \theta' v \in V(n+1)$. Further assume that π is admissible and irreducible. If $V(n) \neq 0$ for some non-negative integer n, then we say that π is *paramodular*, and we define N_π to be the smallest non-negative integer n such that $V(n) \neq 0$; we call N_π the *paramodular level* of π. The following theorem is the main structural result about paramodular vectors. See Table A.2 for the list of irreducible, admissible representations of $\mathrm{GSp}(4, F)$ with trivial central character that are *not* paramodular.

Theorem 2.3.2 *Let (π, V) be an irreducible, admissible representation of the group* $\mathrm{GSp}(4, F)$ *with trivial central character. Assume that π is paramodular. Then* $\dim V(N_\pi) = 1$. *Let v be a non-zero element of the one-dimensional space $V(N_\pi)$. If $n \geq N_\pi$, then the space $V(n)$ is spanned by the vectors*

$$\theta'^i \theta^j \eta^k v, \qquad i, j, k \geq 0, \quad i + j + 2k = n - N_\pi.$$

Proof This is Theorem 7.5.1 and Theorem 7.5.7 of [105]. □

Any non-zero element of the one-dimensional vector space $V(N_\pi)$ is called a *paramodular newform.*

The next theorem shows that all generic representations are paramodular, and gives some sense of the place of paramodular representations among all representations. For a schematic summary of this result see Fig. 2.1. More generally, the work [105] determines exactly which representations π of $\mathrm{GSp}(4, F)$ with trivial central character are paramodular and calculates N_π for all paramodular representations.

Theorem 2.3.3 *Let (π, V) be an irreducible, admissible representation of the group* $\mathrm{GSp}(4, F)$ *with trivial central character. If π is generic, then π is paramodular. If π is tempered, then π is paramodular if and only if π is generic.*

Proof This follows from Theorem 7.5.4 and Theorem 7.5.8 of [105]. □

As regards non-generic representations, "most" of the non-generic, irreducible, admissible representations of the group $\mathrm{GSp}(4, F)$ with trivial central character

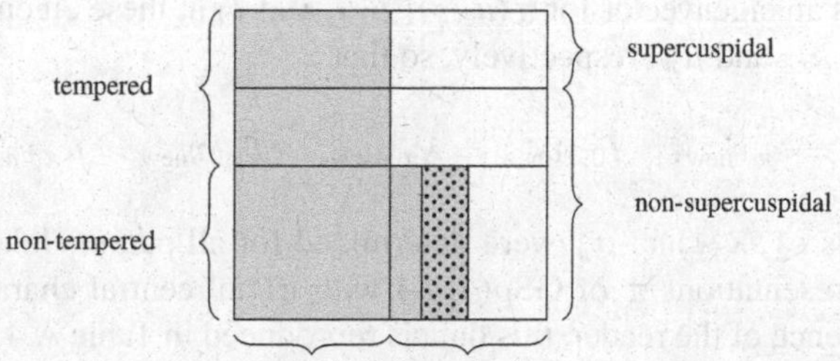

Fig. 2.1 Paramodular representations. A schematic diagram of the paramodular representations among all irreducible, admissible representations of GSp(4, F) with trivial central character. Paramodular representations are in gray, and paramodular Saito-Kurokawa representations are dotted

that are paramodular are Saito-Kurokawa representations (see Sect. 2.2). If $Q(\nu^{\frac{1}{2}}\pi, \nu^{-\frac{1}{2}}\sigma)$ is a Saito-Kurokawa representation with $\sigma^2 = 1$, then $Q(\nu^{\frac{1}{2}}\pi, \nu^{-\frac{1}{2}}\sigma)$ is paramodular if and only if σ is unramified.

Again let (π, V) be a smooth representation of $\mathrm{GSp}(4, F)$ for which the center of $\mathrm{GSp}(4, F)$ acts trivially. Let n be a non-negative integer. Let

$$\mathrm{K}(\mathfrak{p}^n)\begin{bmatrix} \varpi &&& \\ & \varpi && \\ && 1 & \\ &&& 1 \end{bmatrix}\mathrm{K}(\mathfrak{p}^n) = \bigsqcup_{i\in I} g_i\mathrm{K}(\mathfrak{p}^n),$$

$$\mathrm{K}(\mathfrak{p}^n)\begin{bmatrix} \varpi^2 &&& \\ & \varpi && \\ && \varpi & \\ &&& 1 \end{bmatrix}\mathrm{K}(\mathfrak{p}^n) = \bigsqcup_{j\in J} h_j\mathrm{K}(\mathfrak{p}^n)$$

be disjoint decompositions. We define endomorphisms

$$T_{0,1} : V(n) \longrightarrow V(n) \quad \text{and} \quad T_{1,0} : V(n) \longrightarrow V(n) \tag{2.30}$$

by

$$T_{0,1}v = \sum_{i\in I}\pi(g_i)v \quad \text{and} \quad T_{1,0}v = \sum_{j\in J}\pi(h_j)v$$

for $v \in V(n)$. We refer to $T_{0,1}$ and $T_{1,0}$ as the *paramodular Hecke operators of level* n. Also, let

$$u_n = \begin{bmatrix} &&1& \\ &&&-1 \\ \varpi^n &&& \\ & -\varpi^n && \end{bmatrix}. \tag{2.31}$$

The element u_n normalizes $\mathrm{K}(\mathfrak{p}^n)$, and the operator $\pi(u_n)$ maps the space $V(n)$ into itself; we call $\pi(u_n)$ the *paramodular Atkin-Lehner operator of level* n. If π is a paramodular, irreducible, admissible representation of $\mathrm{GSp}(4, F)$ with trivial central character, then any non-zero element v_{new} of the one-dimensional vector space $V(N_\pi)$ is an eigenvector for $\pi(u_{N_\pi})$, $T_{0,1}$, and $T_{1,0}$; these eigenvalues will be denoted by ε_π, λ_π, and μ_π, respectively, so that

$$\pi(u_{N_\pi})v_{\mathrm{new}} = \varepsilon_\pi v_{\mathrm{new}}, \quad T_{0,1}v_{\mathrm{new}} = \lambda_\pi v_{\mathrm{new}}, \quad T_{1,0}v_{\mathrm{new}} = \mu_\pi v_{\mathrm{new}}. \tag{2.32}$$

The eigenvalues ε_π, λ_π, and μ_π were determined for all paramodular, irreducible, admissible representations π of $\mathrm{GSp}(4, F)$ with trivial central character in [105]; for the convenience of the reader, this data is reproduced in Table A.4. The meaning of these eigenvalues for generic representations is given by the following theorem.

Theorem 2.3.4 *Let (π, V) be a generic, irreducible, admissible representation of $\mathrm{GSp}(4, F)$ with trivial central character; π is paramodular by Theorem 2.3.3. Let $W \in V(N_\pi)$ be a newform. Let ε_π be the Atkin-Lehner eigenvalue of W, i.e.,*

$\pi(u_n)W = \varepsilon_\pi W$, and let λ_π and μ_π be the Hecke eigenvalues of W, defined by $T_{0,1}W = \lambda_\pi W$ and $T_{1,0}W = \mu_\pi W$. We may choose W such that

$$Z(s, W) = L(s, \pi),$$

and we have

$$\varepsilon(s, \pi) = \varepsilon_\pi q^{-N_\pi(s-1/2)}.$$

Here, the zeta integral $Z(s, W)$, $L(s, \pi)$ and $\varepsilon(s, \pi)$ are defined as on p. 84. Moreover:

(1) *If $N_\pi = 0$, so that π is unramified, then*

$$L(s, \pi) = \frac{1}{1 - q^{-\frac{3}{2}}\lambda_\pi q^{-s} + (q^{-2}\mu_\pi + 1 + q^{-2})q^{-2s} - q^{-\frac{3}{2}}\lambda_\pi q^{-3s} + q^{-4s}}.$$

(2) *If $N_\pi = 1$, then,*

$$L(s, \pi) = \frac{1}{1 - q^{-\frac{3}{2}}(\lambda_\pi + \varepsilon_\pi)q^{-s} + (q^{-2}\mu_\pi + 1)q^{-2s} + \varepsilon_\pi q^{-\frac{1}{2}}q^{-3s}}.$$

(3) *If $N_\pi \geq 2$, then*

$$L(s, \pi) = \frac{1}{1 - q^{-\frac{3}{2}}\lambda_\pi q^{-s} + (q^{-2}\mu_\pi + 1)q^{-2s}}.$$

Proof See Theorem 7.5.4, Corollary 7.5.5, and Theorem 7.5.3 of [105]. □

A similar result holds for all paramodular representations.

Theorem 2.3.5 *Let (π, V) be a paramodular, irreducible, admissible representation of* GSp(4, F) *with trivial central character. Let $\varphi_\pi : W'_F \to$* GSp(4, $\mathbb{C}$) *be the L-parameter of π. Let $v \in V(N_\pi)$ be a non-zero vector. Let ε_π be the Atkin-Lehner eigenvalue of v, i.e., $\pi(u_n)v = \varepsilon_\pi v$, and let λ_π and μ_π be the Hecke eigenvalues of v, defined by $T_{0,1}v = \lambda_\pi v$ and $T_{1,0}v = \mu_\pi v$. Then*

$$\varepsilon(s, \varphi_\pi) = \varepsilon_\pi q^{-N_\pi(s-1/2)}.$$

Moreover, $L(s, \varphi_\pi)$ is given by exactly the same formulas as in (1), (2), and (3) of Theorem 2.3.4.

Proof The work [45] assigns an L-parameter φ_π to π. As remarked in [45], the L-parameter assigned to π by [45] coincides with the L-parameter assigned to π in [105] when π is non-supercuspidal; in this case, the L-parameter of π is determined by the desiderata of the local Langlands conjecture. The statement of the theorem when π is non-supercuspidal is Theorem 7.5.9 of [105]. Assume that π is supercuspidal. By Theorem 2.3.3, π is generic. By [70, 123] and [45] one has $L(s, \varphi_\pi) = L(s, \pi)$ and $\varepsilon(s, \varphi_\pi) = \varepsilon(s, \pi)$, where $L(s, \pi)$ and $\varepsilon(s, \pi)$ are as on p. 84. The theorem follows now from Theorem 2.3.4. □

In Chap. 10 we will use the following result about oldforms in smooth representations. Let (π, V) be a smooth representation of $\mathrm{GSp}(4, F)$ for which the center acts trivially, and let n be a non-negative integer. We let $V(n)_{\mathrm{old}}$ be the subspace of $V(n)$ spanned by the vectors of the form θv and $\theta' v$ for $v \in V(n-1)$ and the vectors of the form ηv for $v \in V(n-2)$ (for this definition we take $V(k) = 0$ if $k < 0$). We note that the next lemma follows trivially from Theorem 2.3.2 in case π is an irreducible, admissible representation.

Lemma 2.3.6 *Let (π, V) be a smooth representation of* $\mathrm{GSp}(4, F)$ *for which the center acts trivially, and let n be a non-negative integer. The endomorphisms $T_{0,1}$, $T_{1,0}$, and $\pi(u_n)$ of $V(n)$ map $V(n)_{\mathrm{old}}$ into $V(n)_{\mathrm{old}}$.*

Proof The statement is trivial if $n = 0$ because $V(0)_{\mathrm{old}} = 0$. Assume that $n \geq 1$. Let $w \in V(n)_{\mathrm{old}}$. We need to prove that $T_{0,1}w, T_{1,0}w, \pi(u_n)w \in V(n)_{\mathrm{old}}$. One can prove that

$$T_{0,1}\eta v = q\theta'\theta v, \qquad T_{1,0}\eta v = q^2\theta'^2 v - q^2(q+1)\eta v$$

for $v \in V(n-1)$, and

$$\pi(u_n)\eta v = \eta\pi(u_{n-2})v$$

for $v \in V(n-2)$ (again, if $n-2 < 0$ we take $V(n-2) = 0$). It follows that we may assume that w is of the form θv or $\theta' v$ for some $v \in V(n-1)$. If $n \geq 2$, then $T_{0,1}w, T_{1,0}w, \pi(u_n)w \in V(n)_{\mathrm{old}}$ by Proposition 6.3 of [105]. If $n = 1$, then $n - 1 = 0$. One may prove that if $v \in V(0)$ then

$$\begin{aligned}
T_{0,1}\theta v &= \theta T_{0,1}v + (q^2-1)\theta' v,\\
T_{0,1}\theta' v &= \theta' T_{0,1}v + (q^2-1)\theta v,\\
T_{1,0}\theta v &= q\theta' T_{0,1}v - q(q+1)\theta v,\\
T_{1,0}\theta' v &= \theta' T_{1,0}v + (q^3+1)\theta' v - \theta T_{0,1}v.
\end{aligned}$$

This proves the lemma. □

Chapter 3
Stable Klingen Vectors

Let (π, V) be a smooth representation of $\mathrm{GSp}(4, F)$ with trivial central character. In this chapter, for non-negative integers n, we introduce the stable Klingen congruence subgroups $\mathrm{K}_s(\mathfrak{p}^n)$ of $\mathrm{GSp}(4, F)$ and begin the consideration of the subspaces $V_s(n)$ of vectors in V fixed by $\mathrm{K}_s(\mathfrak{p}^n)$. We refer to such vectors as stable Klingen vectors. Since $\mathrm{K}_s(\mathfrak{p}^n)$ is contained in the paramodular group $\mathrm{K}(\mathfrak{p}^n)$, every paramodular vector is a stable Klingen vector, and we will use this connection throughout this work. In this chapter we will also introduce and study various operators between the spaces $V_s(n)$ and $V_s(m)$ for non-negative integers n and m. Similar level changing operators are important in the paramodular theory. However, in contrast to the paramodular theory, we will find that level changing operators in the stable Klingen theory often admit upper block formulas. As we will see in the second part of this work, such formulas are useful for effective calculations and applications. The results of this chapter are algebraic in the sense that we only assume that π is smooth; subsequent chapters will assume that π is admissible and irreducible.

3.1 The Stable Klingen Subgroup

In this section we define the stable Klingen subgroups of $\mathrm{GSp}(4, F)$ and describe some of their basic properties. We will make reference to the paramodular theory; see Sect. 2.3 for a summary. Let n be an integer such that $n \geq 0$. We define

$$\mathrm{Kl}(\mathfrak{p}^n) = \mathrm{GSp}(4, \mathfrak{o}) \cap \begin{bmatrix} \mathfrak{o} & \mathfrak{o} & \mathfrak{o} & \mathfrak{o} \\ \mathfrak{p}^n & \mathfrak{o} & \mathfrak{o} & \mathfrak{o} \\ \mathfrak{p}^n & \mathfrak{o} & \mathfrak{o} & \mathfrak{o} \\ \mathfrak{p}^n & \mathfrak{p}^n & \mathfrak{p}^n & \mathfrak{o} \end{bmatrix}. \tag{3.1}$$

Using (2.4), it is easy to see that $\mathrm{Kl}(\mathfrak{p}^n)$ is a subgroup of $\mathrm{GSp}(4, \mathfrak{o})$. We refer to $\mathrm{Kl}(\mathfrak{p}^n)$ as the *Klingen congruence subgroup of level* $\mathfrak{p}^n$. Evidently, $\mathrm{Kl}(\mathfrak{p}^n)$ is

J. Johnson-Leung et al., *Stable Klingen Vectors and Paramodular Newforms*,
Lecture Notes in Mathematics 2342, https://doi.org/10.1007/978-3-031-45177-5_3

contained in the paramodular subgroup $\mathrm{K}(\mathfrak{p}^n)$. Also, $\mathrm{Kl}(\mathfrak{p}^0) = \mathrm{GSp}(4, \mathfrak{o})$. Next, we let $\mathrm{K}_s(\mathfrak{p}^n)$ to be the subgroup of $\mathrm{GSp}(4, F)$ generated by $\mathrm{Kl}(\mathfrak{p}^n)$ and the subgroup

$$\begin{bmatrix} 1 & & \mathfrak{p}^{-n+1} & \\ & 1 & & \\ & & 1 & \\ & & & 1 \end{bmatrix}.$$

We refer to $\mathrm{K}_s(\mathfrak{p}^n)$ as the *stable Klingen subgroup of level* $\mathfrak{p}^n$. Thus,

$$\mathrm{K}_s(\mathfrak{p}^n) = \langle \mathrm{Kl}(\mathfrak{p}^n), \begin{bmatrix} 1 & & \mathfrak{p}^{-n+1} & \\ & 1 & & \\ & & 1 & \\ & & & 1 \end{bmatrix} \rangle. \tag{3.2}$$

We note that $\mathrm{K}_s(\mathfrak{p}^0) = \mathrm{GSp}(4, \mathfrak{o})$ and $\mathrm{K}_s(\mathfrak{p}^1) = \mathrm{Kl}(\mathfrak{p})$. It is useful to note that the element s_2 from (2.5) is contained in $\mathrm{Kl}(\mathfrak{p}^n)$, and hence in $\mathrm{K}_s(\mathfrak{p}^n)$ and $\mathrm{K}(\mathfrak{p}^n)$, for all integers n such that $n \geq 0$.

Lemma 3.1.1 *Let n be an integer such that $n \geq 1$. Then*

$$\mathrm{K}_s(\mathfrak{p}^n) = \{g \in \mathrm{GSp}(4, F) \mid \lambda(g) \in \mathfrak{o}^\times\} \cap \begin{bmatrix} \mathfrak{o} & \mathfrak{o} & \mathfrak{o} & \mathfrak{p}^{-n+1} \\ \mathfrak{p}^n & \mathfrak{o} & \mathfrak{o} & \mathfrak{o} \\ \mathfrak{p}^n & \mathfrak{o} & \mathfrak{o} & \mathfrak{o} \\ \mathfrak{p}^n & \mathfrak{p}^n & \mathfrak{p}^n & \mathfrak{o} \end{bmatrix}. \tag{3.3}$$

Moreover, if $g = (g_{ij}) \in \mathrm{K}_s(\mathfrak{p}^n)$, then

$$g_{11}, g_{44} \in \mathfrak{o}^\times. \tag{3.4}$$

Proof It is obvious that the left-hand side is contained in the right-hand side. Conversely, let $g = (g_{ij})$ be an element of the right-hand side. Then $\det(g)$ is in $\mathfrak{o}^\times$. This implies that g_{11} and g_{44} are also in $\mathfrak{o}^\times$. Consequently,

$$\begin{bmatrix} 1 & & & -g_{14}g_{44}^{-1} \\ & 1 & & \\ & & 1 & \\ & & & 1 \end{bmatrix} g \in \mathrm{Kl}(\mathfrak{p}^n). \tag{3.5}$$

This implies that $g \in \mathrm{K}_s(\mathfrak{p}^n)$. □

Let n be an integer such that $n \geq 1$. Then it is possible to conjugate $\mathrm{K}_s(\mathfrak{p}^n)$ into $\mathrm{GSp}(4, \mathfrak{o})$. If n is even, then

$$\begin{bmatrix} \varpi^{\frac{n}{2}} & & & \\ & 1 & & \\ & & 1 & \\ & & & \varpi^{-\frac{n}{2}} \end{bmatrix} \mathrm{K}_s(\mathfrak{p}^n) \begin{bmatrix} \varpi^{-\frac{n}{2}} & & & \\ & 1 & & \\ & & 1 & \\ & & & \varpi^{\frac{n}{2}} \end{bmatrix} \subset \begin{bmatrix} \mathfrak{o} & \mathfrak{p}^{\frac{n}{2}} & \mathfrak{p}^{\frac{n}{2}} & \mathfrak{p} \\ \mathfrak{p}^{\frac{n}{2}} & \mathfrak{o} & \mathfrak{o} & \mathfrak{p}^{\frac{n}{2}} \\ \mathfrak{p}^{\frac{n}{2}} & \mathfrak{o} & \mathfrak{o} & \mathfrak{p}^{\frac{n}{2}} \\ \mathfrak{o} & \mathfrak{p}^{\frac{n}{2}} & \mathfrak{p}^{\frac{n}{2}} & \mathfrak{o} \end{bmatrix}, \tag{3.6}$$

and if n is odd, then

$$\begin{bmatrix} \varpi^{\frac{n-1}{2}} &&& \\ & 1 && \\ && 1 & \\ &&& \varpi^{\frac{1-n}{2}} \end{bmatrix} \mathrm{K}_s(\mathfrak{p}^n) \begin{bmatrix} \varpi^{\frac{1-n}{2}} &&& \\ & 1 && \\ && 1 & \\ &&& \varpi^{\frac{n-1}{2}} \end{bmatrix} \subset \begin{bmatrix} \mathfrak{o} & \mathfrak{p}^{\frac{n-1}{2}} & \mathfrak{p}^{\frac{n-1}{2}} & \mathfrak{o} \\ \mathfrak{p}^{\frac{n+1}{2}} & \mathfrak{o} & \mathfrak{o} & \mathfrak{p}^{\frac{n-1}{2}} \\ \mathfrak{p}^{\frac{n+1}{2}} & \mathfrak{o} & \mathfrak{o} & \mathfrak{p}^{\frac{n-1}{2}} \\ \mathfrak{p} & \mathfrak{p}^{\frac{n+1}{2}} & \mathfrak{p}^{\frac{n+1}{2}} & \mathfrak{o} \end{bmatrix}. \tag{3.7}$$

For the paramodular group, such a conjugation is only possible if n is even.

In the next lemma we describe the coset decompositions of $\mathrm{Kl}(\mathfrak{p}^n)$ in $\mathrm{K}_s(\mathfrak{p}^n)$ and of $\mathrm{K}_s(\mathfrak{p}^n)$ in $\mathrm{K}(\mathfrak{p}^n)$. Let n be an integer such that $n \geq 0$. We define

$$t_n = \begin{bmatrix} &&& -\varpi^{-n} \\ & 1 && \\ && 1 & \\ \varpi^n &&& \end{bmatrix}. \tag{3.8}$$

Clearly, t_n is contained in the paramodular group $\mathrm{K}(\mathfrak{p}^n)$ of level $\mathfrak{p}^n$.

Lemma 3.1.2 *Let n be an integer such that $n \geq 1$. Then*

$$\mathrm{K}_s(\mathfrak{p}^n) = \bigsqcup_{x \in \mathfrak{o}/\mathfrak{p}^{n-1}} \begin{bmatrix} 1 &&& x\varpi^{-n+1} \\ & 1 && \\ && 1 & \\ &&& 1 \end{bmatrix} \mathrm{Kl}(\mathfrak{p}^n) \tag{3.9}$$

and

$$\mathrm{K}(\mathfrak{p}^n) = \bigsqcup_{x \in \mathfrak{o}/\mathfrak{p}} \begin{bmatrix} 1 &&& x\varpi^{-n} \\ & 1 && \\ && 1 & \\ &&& 1 \end{bmatrix} \mathrm{K}_s(\mathfrak{p}^n) \sqcup t_n \mathrm{K}_s(\mathfrak{p}^n). \tag{3.10}$$

Moreover,

$$\mathrm{K}_s(\mathfrak{p}^n) = \mathrm{K}(\mathfrak{p}^{n-1}) \cap \mathrm{K}(\mathfrak{p}^n). \tag{3.11}$$

Proof It is clear that the right-hand side of (3.9) is contained in the left-hand side. The opposite inclusion follows from the proof of Lemma 3.1.1. It is straightforward to verify that the decomposition is disjoint. Next, by Lemma 3.3.1 of [105],

$$\mathrm{K}(\mathfrak{p}^n) = \bigsqcup_{x \in \mathfrak{o}/\mathfrak{p}^n} \begin{bmatrix} 1 &&& x\varpi^{-n} \\ & 1 && \\ && 1 & \\ &&& 1 \end{bmatrix} \mathrm{Kl}(\mathfrak{p}^n) \sqcup \bigsqcup_{x \in \mathfrak{o}/\mathfrak{p}^{n-1}} t_n \begin{bmatrix} 1 &&& x\varpi^{-n+1} \\ & 1 && \\ && 1 & \\ &&& 1 \end{bmatrix} \mathrm{Kl}(\mathfrak{p}^n). \tag{3.12}$$

Multiplying (3.12) from the right by $\mathrm{K}_s(\mathfrak{p}^n)$, we get

$$\mathrm{K}(\mathfrak{p}^n) = \bigcup_{x \in \mathfrak{o}/\mathfrak{p}} \begin{bmatrix} 1 & & x\varpi^{-n} & \\ & 1 & & \\ & & 1 & \\ & & & 1 \end{bmatrix} \mathrm{K}_s(\mathfrak{p}^n) \cup t_n \mathrm{K}_s(\mathfrak{p}^n). \tag{3.13}$$

It is easy to see that this decomposition is disjoint. To prove (3.11), we note first that $\mathrm{K}_s(\mathfrak{p}^n)$ is contained in $\mathrm{K}(\mathfrak{p}^{n-1}) \cap \mathrm{K}(\mathfrak{p}^n)$. Conversely, let $g \in \mathrm{K}(\mathfrak{p}^{n-1}) \cap \mathrm{K}(\mathfrak{p}^n)$. Then g is in the right-hand side of (3.3). By (3.3), $g \in \mathrm{K}_s(\mathfrak{p}^n)$. □

Since the Klingen subgroup has an Iwahori decomposition, it follows from (3.9) that the stable Klingen subgroup has as well. Let n be an integer such that $n \geq 1$. Then $\mathrm{K}_s(\mathfrak{p}^n)$ is equal to

$$\left(\begin{bmatrix} 1 & \mathfrak{o} & \mathfrak{o} & \mathfrak{p}^{-n+1} \\ & 1 & & \mathfrak{o} \\ & & 1 & \mathfrak{o} \\ & & & 1 \end{bmatrix} \cap \mathrm{K}_s(\mathfrak{p}^n)\right)\left(\begin{bmatrix} \mathfrak{o}^\times & & & \\ & \mathfrak{o} & \mathfrak{o} & \\ & \mathfrak{o} & \mathfrak{o} & \\ & & & \mathfrak{o}^\times \end{bmatrix} \cap \mathrm{K}_s(\mathfrak{p}^n)\right)\left(\begin{bmatrix} 1 & & & \\ \mathfrak{p}^n & 1 & & \\ \mathfrak{p}^n & & 1 & \\ \mathfrak{p}^n & \mathfrak{p}^n & \mathfrak{p}^n & 1 \end{bmatrix} \cap \mathrm{K}_s(\mathfrak{p}^n)\right) \tag{3.14}$$

and is also equal to

$$\left(\begin{bmatrix} 1 & & & \\ \mathfrak{p}^n & 1 & & \\ \mathfrak{p}^n & & 1 & \\ \mathfrak{p}^n & \mathfrak{p}^n & \mathfrak{p}^n & 1 \end{bmatrix} \cap \mathrm{K}_s(\mathfrak{p}^n)\right)\left(\begin{bmatrix} \mathfrak{o}^\times & & & \\ & \mathfrak{o} & \mathfrak{o} & \\ & \mathfrak{o} & \mathfrak{o} & \\ & & & \mathfrak{o}^\times \end{bmatrix} \cap \mathrm{K}_s(\mathfrak{p}^n)\right)\left(\begin{bmatrix} 1 & \mathfrak{o} & \mathfrak{o} & \mathfrak{p}^{-n+1} \\ & 1 & & \mathfrak{o} \\ & & 1 & \mathfrak{o} \\ & & & 1 \end{bmatrix} \cap \mathrm{K}_s(\mathfrak{p}^n)\right). \tag{3.15}$$

3.2 Stable Klingen Vectors

We now introduce the spaces of vectors which play a central role in this work. Let (π, V) be a smooth representation of $\mathrm{GSp}(4, F)$ for which the center acts trivially, and let n be an integer such that $n \geq 0$. Let $\mathrm{K}_s(\mathfrak{p}^n)$ be the stable Klingen subgroup defined in (3.2), and let

$$V_s(n) = \{v \in V \mid \pi(g)v = v \text{ for all } g \in \mathrm{K}_s(\mathfrak{p}^n)\} \tag{3.16}$$

be the corresponding space of vectors in V fixed by $\mathrm{K}_s(\mathfrak{p}^n)$. We refer to the elements of $V_s(n)$ as *stable Klingen vectors of level* $\mathfrak{p}^n$. If $n \geq 1$, then since $\mathrm{K}_s(\mathfrak{p}^n) = \mathrm{K}(\mathfrak{p}^{n-1}) \cap \mathrm{K}(\mathfrak{p}^n)$ by Lemma 3.1.2, we have $V(n-1) + V(n) \subset V_s(n)$. In this work we define $V(-1) = 0$; with this definition, we also have $V(n-1) + V(n) \subset V_s(n)$ if $n = 0$. We refer to $V(n-1) + V(n)$ as the *paramodular subspace* of $V_s(n)$. We define

$$\bar{V}_s(n) = V_s(n)/(V(n-1) + V(n)). \tag{3.17}$$

Since $V_s(0) = V(0)$, we have $\bar{V}_s(0) = 0$. We refer to the elements of $\bar{V}_s(n)$ as *quotient stable Klingen vectors*. If $V_s(n) \neq 0$ for some integer such that $n \geq 0$, then we define $N_{\pi,s}$ to be the smallest such integer, and we call $N_{\pi,s}$ the *stable Klingen level* of π. If there exists no integer $n \geq 0$ such that $V_s(n) \neq 0$, then we will say that $N_{\pi,s}$ is not defined. If $\bar{V}_s(n) \neq 0$ for some integer such that $n \geq 0$, then we define $\bar{N}_{\pi,s}$ to be the smallest such integer, and we call $\bar{N}_{\pi,s}$ the *quotient stable Klingen level* of π; if there exists no integer $n \geq 0$ such that $\bar{V}_s(n) \neq 0$, then we will say that $\bar{N}_{\pi,s}$ is not defined. As a point of orientation, we note that it is trivial that if π admits non-zero paramodular vectors, then π admits non-zero stable Klingen vectors; in Corollary 5.1.6 we will see that the converse also holds when π is irreducible.

3.3 Paramodularization

Let (π, V) be a smooth representation of $\mathrm{GSp}(4, F)$ for which the center acts trivially. Let n be an integer such that $n \geq 0$, and let dk be a Haar measure on $\mathrm{GSp}(4, F)$. We define the *paramodularization map* $p_n : V_s(n) \to V(n)$ by

$$p_n v = \frac{1}{\mathrm{vol}(\mathrm{K}(\mathfrak{p}^n))} \int\limits_{\mathrm{K}(\mathfrak{p}^n)} \pi(k)v\,dk, \qquad \text{for } v \in V_s(n). \tag{3.18}$$

Lemma 3.3.1 *Let (π, V) be a smooth representation of* $\mathrm{GSp}(4, F)$ *for which the center acts trivially. Let n be an integer such that $n \geq 0$. Then*

$$p_n v = \frac{1}{q+1}\Big(\pi(t_n)v + \sum_{x \in \mathfrak{o}/\mathfrak{p}} \pi(\begin{bmatrix} 1 & & x\varpi^{-n} & \\ & 1 & & \\ & & 1 & \\ & & & 1 \end{bmatrix})v\Big) \tag{3.19}$$

$$= \frac{1}{q+1}\Big(\pi(t_n)v + q\int\limits_{\mathfrak{o}} \pi(\begin{bmatrix} 1 & & x\varpi^{-n} & \\ & 1 & & \\ & & 1 & \\ & & & 1 \end{bmatrix})v\,dx\Big) \tag{3.20}$$

for $v \in V_s(n)$. The map $p_n : V_s(n) \to V(n)$ is a linear projection, so that $p_n^2 = p_n$ and

$$V_s(n) = V(n) \oplus \ker(p_n). \tag{3.21}$$

Proof The assertion (3.19) is clear if $n = 0$. If n is an integer such that $n \geq 1$, then (3.19) follows from (3.10). It is clear that p_n is a linear projection. □

3.4 Operators on Stable Klingen Vectors

In this work we will use a variety of linear operators to investigate stable Klingen vectors. These linear maps are instances of a general definition. To describe this concept, let (π, V) be a smooth representation of $\mathrm{GSp}(4, F)$ for which the center acts trivially, let n and m be integers such that $n, m \geq 0$, and let $g \in \mathrm{GSp}(4, F)$. We define a linear map

$$T_g : V_s(n) \longrightarrow V_s(m)$$

by

$$T_g v = \frac{1}{\mathrm{vol}(\mathrm{K}_s(\mathfrak{p}^m) \cap g\mathrm{K}_s(\mathfrak{p}^n)g^{-1})} \int\limits_{\mathrm{K}_s(\mathfrak{p}^m)} \pi(k)\pi(g)v\, dk \tag{3.22}$$

for $v \in V_s(n)$. Here, dk is a fixed Haar measure on $\mathrm{GSp}(4, F)$; we note that the definition of T_g does not depend on the choice of Haar measure. The operator T_g has a simpler expression. Let $J = \mathrm{K}_s(\mathfrak{p}^m) \cap g\mathrm{K}_s(\mathfrak{p}^n)g^{-1}$, and let

$$\mathrm{K}_s(\mathfrak{p}^m) = \bigsqcup_{i \in I} g_i J \tag{3.23}$$

be a disjoint decomposition, where $g_i \in \mathrm{K}_s(\mathfrak{p}^m)$ for $i \in I$. The set I is finite since J is open and $\mathrm{K}_s(\mathfrak{p}^m)$ is compact. A calculation shows that

$$T_g v = \sum_{i \in I} \pi(g_i)\pi(g)v \tag{3.24}$$

for $v \in V_s(n)$. We will say that $T_g : V_s(n) \to V_s(m)$ is a *level raising operator* if $n < m$, that T_g is a *level lowering operator* if $m < n$, and that T_g is a *Hecke operator* if $n = m$. We will also say that T_g is an *upper block operator* if there exists a disjoint decomposition (3.23) such that $g_i g$ lie in the Siegel parabolic subgroup P for all $i \in I$. In the following sections we will define some important examples of level raising, level lowering, and Hecke operators and investigate the basic properties of these linear maps.

We close this section with a lemma showing that the above definition of a Hecke operator agrees with the traditional definition.

Lemma 3.4.1 *Let* (π, V) *be a smooth representation of* $\mathrm{GSp}(4, F)$ *for which the center acts trivially, and let* n *be an integer such that* $n \geq 0$. *Let* $g \in \mathrm{GSp}(4, F)$, *and let*

$$\mathrm{K}_s(\mathfrak{p}^n) g \mathrm{K}_s(\mathfrak{p}^n) = \bigsqcup_{i \in I} g_i \mathrm{K}_s(\mathfrak{p}^n)$$

be a disjoint decomposition. Then

$$T_g v = \sum_{i\in I} \pi(g_i)v$$

for $v \in V_s(n)$.

Proof For each $i \in I$, let $k_i, k'_i \in \mathrm{K}_s(\mathfrak{p}^n)$ be such that $g_i = k_i g k'_i$. Define $g'_i = g_i k_i'^{-1}$ for $i \in I$. Then $g'_i = k_i g$ for $i \in I$, $\mathrm{K}_s(\mathfrak{p}^n) g \mathrm{K}_s(\mathfrak{p}^n) = \sqcup_{i\in I} g'_i \mathrm{K}_s(\mathfrak{p}^n)$, and $\sum_{i\in I} \pi(g'_i)v = \sum_{i\in I} \pi(g_i)v$ for $v \in V_s(n)$. An argument shows that $\mathrm{K}_s(\mathfrak{p}^n) = \sqcup_{i\in I} g'_i g^{-1} J$, where $J = \mathrm{K}_s(\mathfrak{p}^n) \cap g\mathrm{K}_s(\mathfrak{p}^n)g^{-1}$. Hence, if $v \in V_s(n)$, then $T_g v = \sum_{i\in I} \pi(g'_i g^{-1})\pi(g)v = \sum_{i\in I} \pi(g'_i)v = \sum_{i\in I} \pi(g_i)v$. □

3.5 Level Raising Operators

Let (π, V) be a smooth representation of $\mathrm{GSp}(4, F)$ for which the center acts trivially, and let n be an integer such that $n \geq 0$. In this section we will consider the three level raising operators $T_g : V_s(n) \to V_s(n+1)$ determined by $g = t_n$, $g = 1$, and

$$g = \begin{bmatrix} 1 & & & \\ & 1 & & \\ & & \varpi & \\ & & & \varpi \end{bmatrix}. \tag{3.25}$$

The level raising operator determined by t_n is particularly simple.

Lemma 3.5.1 *Let* (π, V) *be a smooth representation of* $\mathrm{GSp}(4, F)$ *for which the center acts trivially, and let* n *be an integer such that* $n \geq 0$. *We have*

$$\mathrm{K}_s(\mathfrak{p}^{n+1}) \cap t_n \mathrm{K}_s(\mathfrak{p}^n) t_n^{-1} = \mathrm{K}_s(\mathfrak{p}^{n+1}). \tag{3.26}$$

Thus, the level raising operator $T_g : V_s(n) \to V_s(n+1)$ *determined by* $g = t_n$ *is given by* $T_g(v) = \pi(t_n)v$ *for* $v \in V_s(n)$.

Proof The equation (3.26) is equivalent to $\mathrm{K}_s(\mathfrak{p}^{n+1}) \subset t_n \mathrm{K}_s(\mathfrak{p}^n) t_n^{-1}$, which is equivalent to $t_n^{-1} \mathrm{K}_s(\mathfrak{p}^{n+1}) t_n \subset \mathrm{K}_s(\mathfrak{p}^n)$. This is clear if $n = 0$. Assume that $n > 0$. Then $t_n^{-1} \mathrm{K}_s(\mathfrak{p}^{n+1}) t_n \subset \mathrm{K}_s(\mathfrak{p}^n)$ follows by a calculation using the criterion (3.3). □

With the notation as in Lemma 3.5.1, this lemma asserts that the level raising operator $T_{t_n} : V_s(n) \to V_s(n+1)$ is given by applying $\pi(t_n)$; consequently, we will write

$$t_n : V_s(n) \longrightarrow V_s(n+1) \tag{3.27}$$

for this level raising operator. Using, for example, (3.4), it can be verified that $t_n : V_s(n) \to V_s(n+1)$ is not an upper block operator. Further properties of the operator t_n are given in the following lemma.

Lemma 3.5.2 *Let* (π, V) *be a smooth representation of* GSp(4, *F*) *for which the center acts trivially, and let n be an integer such that* $n \geq 0$.

(1) *The level raising operator* $t_n : V_s(n) \to V_s(n+1)$ *is injective.*
(2) *The level raising operator* $t_n : V_s(n) \to V_s(n+1)$ *induces a linear map*

$$t_n : \bar{V}_s(n) \longrightarrow \bar{V}_s(n+1) \tag{3.28}$$

which is also injective.
(3) *The map* $t_{n+1} \circ t_n : V_s(n) \to V_s(n+2)$ *is given by applying*

$$\eta = \pi(\begin{bmatrix} \varpi^{-1} & & & \\ & 1 & & \\ & & 1 & \\ & & & \varpi \end{bmatrix}). \tag{3.29}$$

Proof (1). This is obvious.

(2). Because the element t_n is in $\mathrm{K}(\mathfrak{p}^n)$, we have $t_n V(n) = V(n)$. This proves (2) in the case $n = 0$; assume that n is an integer such that $n \geq 1$. We have $t_n V(n-1) = \eta t_{n-1} V(n-1) = \eta V(n-1) \subset V(n+1)$ (see Sect. 2.3). It follows that t_n maps $V(n-1) + V(n)$ into $V(n) + V(n+1)$, and therefore induces a map $\bar{V}_s(n) \longrightarrow \bar{V}_s(n+1)$. To prove injectivity, assume that $v \in V_s(n)$, and that $t_n v$ lies in the subspace $V(n) + V(n+1)$ of $V_s(n+1)$. We will show that v lies in the subspace $V(n-1) + V(n)$ of $V_s(n)$, thus proving the injectivity of (3.28). Write $t_n v = v_1 + v_2$ with $v_1 \in V(n)$ and $v_2 \in V(n+1)$. We have

$$v_2 \text{ is invariant under } \begin{bmatrix} 1 & & & \\ \mathfrak{p}^n & 1 & & \\ \mathfrak{p}^n & & 1 & \\ \mathfrak{p}^{n+1} & \mathfrak{p}^n & \mathfrak{p}^n & 1 \end{bmatrix}, \tag{3.30}$$

because both $t_n v$ and v_1 have this property. Then also

$$v_2 = t_{n+1} v_2 \text{ is invariant under } \begin{bmatrix} 1 & \mathfrak{p}^{-1} & \mathfrak{p}^{-1} & \mathfrak{p}^{-n-1} \\ & 1 & & \mathfrak{p}^{-1} \\ & & 1 & \mathfrak{p}^{-1} \\ & & & 1 \end{bmatrix}. \tag{3.31}$$

We see from (3.30) and (3.31) that $\eta^{-1} v_2 \in V(n-1)$. Applying t_n to $t_n v = v_1 + v_2$, we get

$$v = t_n v_1 + t_n v_2 = v_1 + \eta^{-1} t_{n+1} v_2 = v_1 + \eta^{-1} v_2 \in V(n) + V(n-1),$$

as claimed.

(3). This is obvious. □

Let (π, V) be a smooth representation of $\mathrm{GSp}(4, F)$ for which the center acts trivially, and let n be an integer such that $n \geq 0$. We describe the level raising operator $T_1 : V_s(n) \to V_s(n+1)$ induced by $g = 1$. If $n = 0$, then the map $T_1 : V_s(0) \to V_s(1)$ is just inclusion, since $\mathrm{K}_s(\mathfrak{p}^0) = \mathrm{GSp}(4, \mathfrak{o})$ and $\mathrm{K}_s(\mathfrak{p}) = \mathrm{Kl}(\mathfrak{p})$. We need the following lemma for the case $n \geq 1$.

Lemma 3.5.3 *Let n be an integer such that $n \geq 1$. There is a disjoint decomposition*

$$\mathrm{K}_s(\mathfrak{p}^{n+1}) = \bigsqcup_{x \in \mathfrak{o}/\mathfrak{p}} \begin{bmatrix} 1 & & x\varpi^{-n} & \\ & 1 & & \\ & & 1 & \\ & & & 1 \end{bmatrix} \Big(\mathrm{K}_s(\mathfrak{p}^{n+1}) \cap \mathrm{K}_s(\mathfrak{p}^n)\Big). \tag{3.32}$$

Proof It is clear that the union of the cosets on the right-hand side of (3.32) is contained in $\mathrm{K}_s(\mathfrak{p}^{n+1})$. Using (3.14) (with n replaced by $n+1$), an argument shows that $\mathrm{K}_s(\mathfrak{p}^{n+1})$ is contained in this union. Finally, it is easy to see that the cosets on the right-hand side of (3.32) are disjoint. □

Again let the notation be as in the paragraph preceding Lemma 3.5.3, and assume that $n \geq 1$. Then by Lemma 3.5.3 the level raising operator $T_1 : V_s(n) \to V_s(n+1)$ is given by

$$T_1(v) = \sum_{x \in \mathfrak{o}/\mathfrak{p}} \pi(\begin{bmatrix} 1 & & x\varpi^{-n} & \\ & 1 & & \\ & & 1 & \\ & & & 1 \end{bmatrix})v$$

for $v \in V_s(n)$. In fact, we will work with a slight modification of T_1 when $n \geq 1$. For any integer n such that $n \geq 0$, define

$$\tau_n : V_s(n) \longrightarrow V_s(n+1) \tag{3.33}$$

by

$$\tau_n v = \int_{\mathfrak{o}} \pi(\begin{bmatrix} 1 & & z\varpi^{-n} & \\ & 1 & & \\ & & 1 & \\ & & & 1 \end{bmatrix})v\, dz \tag{3.34}$$

for $v \in V_s(n)$. We see that if $n = 0$, then $\tau_n = T_1$; however, if $n \geq 1$, then $\tau_n = q^{-1}T_1$. The operator $\tau_n : V_s(n) \to V_s(n+1)$ is clearly an upper block operator. We also note that using τ_n, equation (3.20) for the paramodularization operator p_n can be written as

$$(q+1)p_n v = q\tau_n v + t_n v \qquad \text{for } v \in V_s(n). \tag{3.35}$$

Lemma 3.5.4 *Let (π, V) be a smooth representation of* $\mathrm{GSp}(4, F)$ *for which the center acts trivially, and let n be an integer such that $n \geq 0$.*

(1) *The level raising operator $\tau_n : V_s(n) \to V_s(n+1)$ is injective.*

(2) *The level raising operator* $\tau_n : V_s(n) \to V_s(n+1)$ *induces a linear map*

$$\tau_n : \bar{V}_s(n) \longrightarrow \bar{V}_s(n+1), \tag{3.36}$$

which is also injective.

(3) *We have* $\tau_n = -q^{-1} t_n$ *as operators from* $\bar{V}_s(n)$ *to* $\bar{V}_s(n+1)$.

Proof (1). To prove injectivity, let $v \in V_s(n)$ and assume that $\tau_n v = 0$. Then, from (3.35) we get

$$p_n v = \frac{1}{q+1} t_n v \qquad \text{for } v \in V_s(n). \tag{3.37}$$

Applying t_n to both sides of this equation, we see that $v \in V(n)$. It now follows that $v = \tau_n v = 0$.

(2) and (3). It follows from (3.35) that

$$q\,\tau_n v + V(n) = -t_n v + V(n) \qquad \text{for } v \in V_s(n). \tag{3.38}$$

Since t_n maps $V(n-1) + V(n)$ to $V(n) + V(n+1)$, so does τ_n. Thus τ_n induces a map $\bar{V}_s(n) \longrightarrow \bar{V}_s(n+1)$. It follows from (3.38) that this map coincides with $-q^{-1} t_n$. Since (3.28) is injective, so is (3.36). □

Let the assumptions and notation be as in Lemma 3.5.4. Two other expressions for τ_n will be useful. First, if $v \in V_s(n)$, then τ_n also can be written as a sum

$$\tau_n v = q^{-1} \sum_{z \in \mathfrak{o}/\mathfrak{p}} \pi(\begin{bmatrix} 1 & & z\varpi^{-n} & \\ & 1 & & \\ & & 1 & \\ & & & 1 \end{bmatrix})v.$$

Second, let dk be the Haar measure on $\mathrm{K}_s(\mathfrak{p}^{n+1})$ that assigns $\mathrm{K}_s(\mathfrak{p}^{n+1})$ volume 1. If $v \in V_s(n)$, then $\tau_n v$ can be simply written as

$$\tau_n v = \int\limits_{\mathrm{K}_s(\mathfrak{p}^{n+1})} \pi(k) v \, dk. \tag{3.39}$$

We extend $\tau_n : V_s(n) \to V_s(n+1)$ to all of V by defining $\tau_n v$ for $v \in V$ to be as in (3.34). This extension simplifies formulas involving iterations. That is, let m be an integer such that $m \geq n$. Then we have

$$\tau_m \tau_{m-1} \cdots \tau_n v = \tau_m v$$

for $v \in V_s(n)$. Finally, if n is an integer such that $n < 0$, then we will also define $\tau_n : V \to V$ by (3.34).

Again let (π, V) be a smooth representation of $\mathrm{GSp}(4, F)$ for which the center acts trivially, and let n be an integer such that $n \geq 0$. Our third level raising operator is $T_g : V_s(n) \to V_s(n+1)$ with g as in (3.25). The following lemma provides the required coset decomposition.

Lemma 3.5.5 *Let n be an integer such that $n \geq 0$, and let*

$$J = \mathrm{K}_s(\mathfrak{p}^{n+1}) \cap \begin{bmatrix} 1 &&& \\ & 1 && \\ && \varpi & \\ &&& \varpi \end{bmatrix} \mathrm{K}_s(\mathfrak{p}^n) \begin{bmatrix} 1 &&& \\ & 1 && \\ && \varpi & \\ &&& \varpi \end{bmatrix}^{-1}.$$

There is a disjoint decomposition

$$\mathrm{K}_s(\mathfrak{p}^{n+1}) = s_2 J \sqcup \bigsqcup_{x \in \mathfrak{o}/\mathfrak{p}} \begin{bmatrix} 1 &&& \\ & 1 && \\ & x & 1 & \\ &&& 1 \end{bmatrix} J. \tag{3.40}$$

Proof Let $k \in \mathrm{GSp}(4, F)$. Using (3.3), a calculation shows that $k \in J$ if and only if $\lambda(k) \in \mathfrak{o}^\times$ and

$$k \in \begin{bmatrix} \mathfrak{o} & \mathfrak{o} & \mathfrak{o} & \mathfrak{p}^{-n} \\ \mathfrak{p}^{n+1} & \mathfrak{o} & \mathfrak{o} & \mathfrak{o} \\ \mathfrak{p}^{n+1} & \mathfrak{p} & \mathfrak{o} & \mathfrak{o} \\ \mathfrak{p}^{n+1} & \mathfrak{p}^{n+1} & \mathfrak{p}^{n+1} & \mathfrak{o} \end{bmatrix}.$$

Again using (3.3), it follows that the union of the cosets on the right-hand side of (3.40) is contained in $\mathrm{K}_s(\mathfrak{p}^{n+1})$. Conversely, let $k \in \mathrm{K}_s(\mathfrak{p}^{n+1})$, and write $k = (k_{ij})_{1 \leq i,j \leq 4}$. Assume that $k_{22} \in \mathfrak{o}^\times$. Then

$$\begin{bmatrix} 1 &&& \\ & 1 && \\ & -k_{32}k_{22}^{-1} & 1 & \\ &&& 1 \end{bmatrix} k \in \begin{bmatrix} \mathfrak{o} & \mathfrak{o} & \mathfrak{o} & \mathfrak{p}^{-n} \\ \mathfrak{p}^{n+1} & \mathfrak{o} & \mathfrak{o} & \mathfrak{o} \\ \mathfrak{p}^{n+1} & \mathfrak{p} & \mathfrak{o} & \mathfrak{o} \\ \mathfrak{p}^{n+1} & \mathfrak{p}^{n+1} & \mathfrak{p}^{n+1} & \mathfrak{o} \end{bmatrix}.$$

This implies that k is in the union. Assume that $k_{22} \in \mathfrak{p}$. Then

$$s_2 k \in \begin{bmatrix} \mathfrak{o} & \mathfrak{o} & \mathfrak{o} & \mathfrak{p}^{-n} \\ \mathfrak{p}^{n+1} & \mathfrak{o} & \mathfrak{o} & \mathfrak{o} \\ \mathfrak{p}^{n+1} & \mathfrak{p} & \mathfrak{o} & \mathfrak{o} \\ \mathfrak{p}^{n+1} & \mathfrak{p}^{n+1} & \mathfrak{p}^{n+1} & \mathfrak{o} \end{bmatrix}.$$

This again implies that k is in the union. It is straightforward to verify that the cosets on the right-hand side of (3.40) are disjoint. □

By Lemma 3.5.5, with the notation preceding this lemma, we see that T_g sends $v \in V_s(n)$ to the following element of $V_s(n+1)$:

$$\pi(s_2)\pi\left(\begin{bmatrix} 1 &&& \\ & 1 && \\ && \varpi & \\ &&& \varpi \end{bmatrix}\right)v + \sum_{x \in \mathfrak{o}/\mathfrak{p}} \pi\left(\begin{bmatrix} 1 &&& \\ & 1 && \\ & x & 1 & \\ &&& 1 \end{bmatrix}\right)\pi\left(\begin{bmatrix} 1 &&& \\ & 1 && \\ && \varpi & \\ &&& \varpi \end{bmatrix}\right)v.$$

This vector and v are invariant under s_2; applying s_2, we see that this vector is equal to

$$\pi(\begin{bmatrix}1&&&\\&1&&\\&&\varpi&\\&&&\varpi\end{bmatrix})v+\sum_{x\in\mathfrak{o}/\mathfrak{p}}\pi(\begin{bmatrix}1&&&\\&1&x&\\&&1&\\&&&1\end{bmatrix})\pi(\begin{bmatrix}1&&&\\&\varpi&&\\&&1&\\&&&\varpi\end{bmatrix})v.$$

Thus, if $v\in V(n)$, then $T_g(v)=\theta(v)$, where $\theta: V(n)\to V(n+1)$ is the paramodular level raising operator from Sect. 2.3. Consequently, we also will write θ for T_g, so that

$$\theta: V_s(n)\longrightarrow V_s(n+1) \tag{3.41}$$

is given by

$$\theta v=\pi(s_2)\pi(\begin{bmatrix}1&&&\\&1&&\\&&\varpi&\\&&&\varpi\end{bmatrix})v+\sum_{x\in\mathfrak{o}/\mathfrak{p}}\pi(\begin{bmatrix}1&&&\\&1&&\\&x&1&\\&&&1\end{bmatrix})\pi(\begin{bmatrix}1&&&\\&1&&\\&&\varpi&\\&&&\varpi\end{bmatrix})v$$

for $v\in V_s(n)$. Some alternative expressions for θ are

$$\theta v=\pi(\begin{bmatrix}1&&&\\&1&&\\&&\varpi&\\&&&\varpi\end{bmatrix})v+\sum_{x\in\mathfrak{o}/\mathfrak{p}}\pi(\begin{bmatrix}1&&&\\&1&x&\\&&1&\\&&&1\end{bmatrix}\begin{bmatrix}1&&&\\&\varpi&&\\&&1&\\&&&\varpi\end{bmatrix})v,$$

$$\theta v=\pi(\begin{bmatrix}1&&&\\&\varpi&&\\&&1&\\&&&\varpi\end{bmatrix})v+\sum_{x\in\mathfrak{o}/\mathfrak{p}}\pi(\begin{bmatrix}1&&&\\&1&&\\&x&1&\\&&&1\end{bmatrix}\begin{bmatrix}1&&&\\&1&&\\&&\varpi&\\&&&\varpi\end{bmatrix})v$$

for $v\in V_s(n)$. Just as for τ_n, we may express θ using an integral over $\mathfrak{o}$. Thus,

$$\theta v=\pi(\begin{bmatrix}1&&&\\&1&&\\&&\varpi&\\&&&\varpi\end{bmatrix})v+q\int\limits_{\mathfrak{o}}\pi(\begin{bmatrix}1&&&\\&1&c&\\&&1&\\&&&1\end{bmatrix}\begin{bmatrix}1&&&\\&\varpi&&\\&&1&\\&&&\varpi\end{bmatrix})v\,dc, \tag{3.42}$$

$$\theta v=\pi(\begin{bmatrix}1&&&\\&\varpi&&\\&&1&\\&&&\varpi\end{bmatrix})v+q\int\limits_{\mathfrak{o}}\pi(\begin{bmatrix}1&&&\\&1&&\\&c&1&\\&&&1\end{bmatrix}\begin{bmatrix}1&&&\\&1&&\\&&\varpi&\\&&&\varpi\end{bmatrix})v\,dc \tag{3.43}$$

for $v\in V_s(n)$. The operator $\theta: V_s(n)\to V_s(n+1)$ is clearly an upper block operator.

Lemma 3.5.6 *Let (π, V) be a smooth representation of* GSp(4, F) *for which the center acts trivially, and let n be an integer such that $n\geq 0$. The level raising operator $\theta: V_s(n)\to V_s(n+1)$ induces a linear map*

$$\theta: \bar{V}_s(n)\longrightarrow \bar{V}_s(n+1). \tag{3.44}$$

Proof This follows from $\theta V(m) \subset V(m+1)$ for integers m such that $m \geq 0$ (see Sect. 2.3). □

Finally, we point out yet another useful expression for θ. Let the notation be as in Lemma 3.5.6. Let dk be the Haar measure on $\mathrm{GSp}(4, F)$ that assigns $\mathrm{K}_s(\mathfrak{p}^{n+1})$ volume 1. Then

$$\theta v = (q+1) \int\limits_{\mathrm{K}_s(\mathfrak{p}^{n+1})} \pi(k \begin{bmatrix} 1 &&& \\ & 1 && \\ && \varpi & \\ &&& \varpi \end{bmatrix}) v \, dk = (q+1) \int\limits_{\mathrm{K}_s(\mathfrak{p}^{n+1})} \pi(k \begin{bmatrix} 1 &&& \\ & \varpi && \\ && 1 & \\ &&& \varpi \end{bmatrix}) v \, dk \tag{3.45}$$

for $v \in V_s(n)$.

3.6 Four Conditions

In this section we prove the equivalence of four conditions on a stable Klingen vector. In Chap. 5, these conditions will play an important role in proving that an irreducible, admissible representation π of $\mathrm{GSp}(4, F)$ with trivial central character admits non-zero stable Klingen vectors if and only if π is paramodular, and in establishing the partition of paramodular representations into two classes. We begin by introducing a variant of paramodularization.

Lemma 3.6.1 *Let (π, V) be a smooth representation of* $\mathrm{GSp}(4, F)$ *for which the center acts trivially, and let n be an integer such that $n \geq 0$. Define*

$$\rho'_n : V_s(n) \longrightarrow V(n+1) \tag{3.46}$$

by

$$\rho'_n v = (1+q^{-1}) p_{n+1} t_n v \tag{3.47}$$

for $v \in V_s(n)$.

(1) *The linear map ρ'_n has the alternative form*

$$\rho'_n v = q^{-1} \eta v + \tau_{n+1} t_n v \tag{3.48}$$

for $v \in V_s(n)$.

(2) *Assume that $v \in V(n)$. Then*

$$\rho'_n v = q^{-1} \eta v + \tau_{n+1} v = q^{-1} \theta' v. \tag{3.49}$$

(3) *Assume that* $n \geq 1$ *and* $v \in V(n-1)$. *Then*

$$\rho_n' v = (1 + q^{-1})\eta v. \tag{3.50}$$

(4) *Assume that* $v \in V_s(n)$ *and that the paramodularization* $p_n v$ *is zero. Then*

$$\rho_n' v = q^{-1}\eta v - q\tau_{n+1} v. \tag{3.51}$$

Proof (1). Let $v \in V_s(n)$. We have, using (3.35) and (3) of Lemma 3.5.2,

$$\begin{aligned} \rho_n' v &= q^{-1}(q+1)p_{n+1}t_n v \\ &= q^{-1}(t_{n+1}t_n v + q\tau_{n+1}t_n v) \\ &= q^{-1}\eta v + \tau_{n+1}t_n v. \end{aligned}$$

(2). Let $v \in V(n)$. Then $t_n v = v$, and hence by (1), $\rho_n' v = q^{-1}\eta v + \tau_{n+1} v$. This is $q^{-1}\theta' v$ by (2.29).

(3). Let $v \in V(n-1)$. Then by (1),

$$\begin{aligned} \rho_n' v &= q^{-1}\eta v + \tau_{n+1}t_n v \\ &= q^{-1}\eta v + \eta\tau_{n-1}t_{n-1} v \\ &= q^{-1}\eta v + \eta v \\ &= (1 + q^{-1})\eta v. \end{aligned}$$

(4). Let $v \in V_s(n)$ and assume that $p_n v = 0$. By (3.35) we have $t_n v = -q\tau_n v$. Hence, by (1),

$$\begin{aligned} \rho_n' v &= q^{-1}\eta v + \tau_{n+1}(-q\tau_n v) \\ &= q^{-1}\eta v - q\tau_{n+1}\tau_n v \\ &= q^{-1}\eta v - q\tau_{n+1} v. \end{aligned}$$

This completes the proof. □

We will now prove the main result of this section using ρ_n'.

Lemma 3.6.2 *Let* (π, V) *be a smooth representation of* GSp(4, F) *for which the center acts trivially. Assume that the subspace of vectors of* V *fixed by* Sp(4, F) *is trivial. Let* n *be an integer such that* $n \geq 0$, *and let* $v \in V_s(n)$. *The following are equivalent:*

(1) *The vector v satisfies*

$$\tau_{n+1}v = q^{-2}\eta v. \tag{3.52}$$

(2) *For all integers $m \geq n$, the paramodular projection $p_m(\tau_{m-1}v)$ from (3.19) is zero, or equivalently,*

$$-q\tau_m v = t_m \tau_{m-1} v. \tag{3.53}$$

(3) *The paramodular projections $p_n(v)$ and $p_{n+1}(\tau_n v)$ from (3.19) are zero, or equivalently,*

$$-q\tau_n v = t_n v \quad \textit{and} \quad -q\tau_{n+1} v = t_{n+1}\tau_n v. \tag{3.54}$$

(4) *The paramodular projection $p_n(v)$ from (3.19) is zero and $\rho_n'(v) = 0$.*

Proof (1) ⇒ (2). Write $v = v_1 + v_2$ with $v_1 \in V(n)$ and $v_2 \in V_s(n)$ such that $p_n(v_2) = 0$ (see (3.21)). Then

$$\begin{aligned}
\rho_n' v &= \rho_n' v_1 + \rho_n' v_2 \\
&= q^{-1}\eta v_1 + \tau_{n+1} v_1 + q^{-1}\eta v_2 - q\tau_{n+1} v_2 \quad \text{(see (3.49) and (3.51))} \\
&= q^{-1}\eta(v_1 + v_2) + (q+1)\tau_{n+1} v_1 - q\tau_{n+1}(v_1 + v_2) \\
&= q^{-1}\eta v - q\tau_{n+1} v + (q+1)\tau_{n+1} v_1 \\
&= (q+1)\tau_{n+1} v_1 \quad \text{(see (3.52))} \\
&= (q+1)q^{-1} \sum_{z \in \mathfrak{o}/\mathfrak{p}} \pi(\begin{bmatrix} 1 & & z\varpi^{-n-1} & \\ & 1 & & \\ & & 1 & \\ & & & 1 \end{bmatrix}) v_1,
\end{aligned}$$

so that, by (2.29),

$$\rho_n' v = (q+1)q^{-1}\theta' v_1 - (q+1)q^{-1}\eta v_1.$$

Observe that $\rho_n' v \in V(n+1)$ by (3.46). By the linear independence of paramodular vectors at different levels, Theorem 2.3.1 (this theorem uses the hypothesis that the subspace of vectors fixed by $\mathrm{Sp}(4, F)$ is trivial), we get $\eta v_1 = 0$, and hence $v_1 = 0$. Thus $v = v_2$, so that $p_n v = 0$. Using (3.35), we obtain (3.53) for $m = n$.

Now let m be an integer such that $m \geq n+1$ and set $v' = \tau_{m-1} v$. Then $v' \in V_s(m)$. Moreover,

$$\begin{aligned}
\tau_{n+1} v &= q^{-2}\eta v \\
\tau_{m+1}\tau_{n+1} v &= q^{-2}\tau_{m+1}\eta v
\end{aligned}$$

$$\tau_{m+1}v = q^{-2}\eta\tau_{m-1}v$$
$$\tau_{m+1}\tau_{m-1}v = q^{-2}\eta\tau_{m-1}v$$
$$\tau_{m+1}v' = q^{-2}\eta v'.$$

Arguing as in the last paragraph with v' replacing v shows that $-q\tau_m v' = t_m v'$, or equivalently, $-q\tau_m v = t_m\tau_{m-1}v$; this is (3.53).

(2) $\Rightarrow$ (3). This is trivial.

(3) $\Rightarrow$ (1). We have $\tau_{n+1}v = -q^{-1}t_{n+1}\tau_n v = q^{-2}t_{n+1}t_n v = q^{-2}\eta v$.

(1) $\Rightarrow$ (4). See the proof of (1) $\Rightarrow$ (2).

(4) $\Rightarrow$ (1). This follows from (4) of Lemma 3.6.1. □

3.7 Level Lowering Operators

Let (π, V) be a smooth representation of $\mathrm{GSp}(4, F)$ for which the center acts trivially, and let n be an integer such that $n \geq 1$. In this work we will use the two level lowering operators $T_g : V_s(n+1) \to V_s(n)$ determined by $g = 1$ and $g = \eta^{-1}$.

Lemma 3.7.1 *Let n be an integer such that $n \geq 1$. There is a disjoint decomposition*

$$\mathrm{K}_s(\mathfrak{p}^n) = \bigsqcup_{x,y,z\in\mathfrak{o}/\mathfrak{p}} \begin{bmatrix} 1 &&& \\ x\varpi^n & 1 && \\ y\varpi^n && 1 & \\ z\varpi^n & y\varpi^n & -x\varpi^n & 1 \end{bmatrix} \Big(\mathrm{K}_s(\mathfrak{p}^n) \cap \mathrm{K}_s(\mathfrak{p}^{n+1})\Big). \tag{3.55}$$

Proof It is clear that the union of the cosets on the right-hand side of (3.55) is contained in $\mathrm{K}_s(\mathfrak{p}^n)$. Conversely, let $k \in \mathrm{K}_s(\mathfrak{p}^n)$. By (3.15), we may write $k = k_1k_2k_3$ where $k_1, k_2, k_3 \in \mathrm{K}_s(\mathfrak{p}^n)$ and

$$k_1 \in \begin{bmatrix} 1 &&& \\ \mathfrak{p}^n & 1 && \\ \mathfrak{p}^n && 1 & \\ \mathfrak{p}^n & \mathfrak{p}^n & \mathfrak{p}^n & 1 \end{bmatrix}, \quad k_2 \in \begin{bmatrix} \mathfrak{o}^\times &&& \\ & \mathfrak{o} & \mathfrak{o} & \\ & \mathfrak{o} & \mathfrak{o} & \\ &&& \mathfrak{o}^\times \end{bmatrix}, \quad \text{and} \quad k_3 \in \begin{bmatrix} 1 & \mathfrak{o} & \mathfrak{o} & \mathfrak{p}^{-n+1} \\ & 1 && \mathfrak{o} \\ && 1 & \mathfrak{o} \\ &&& 1 \end{bmatrix}.$$

Since $k_2k_3 \in \mathrm{K}_s(\mathfrak{p}^n) \cap \mathrm{K}_s(\mathfrak{p}^{n+1})$, we see that k is in the union. It is straightforward to see that the cosets in (3.55) are disjoint. □

Let the notation be as in the paragraph preceding Lemma 3.7.1. By (3.24) and Lemma 3.7.1, the level lowering operator $T_1 : V_s(n+1) \to V_s(n)$ sends $v \in V_s(n+1)$ to

$$\sum_{x,y,z\in\mathfrak{o}/\mathfrak{p}} \pi(\begin{bmatrix} 1 &&& \\ x\varpi^n & 1 && \\ y\varpi^n && 1 & \\ z\varpi^n & y\varpi^n & -x\varpi^n & 1 \end{bmatrix})v.$$

In fact, we will work with a multiple of T_1. Define

$$s_n : V_s(n+1) \longrightarrow V_s(n) \tag{3.56}$$

by

$$s_n v = \int\limits_{\mathfrak{o}}\int\limits_{\mathfrak{o}}\int\limits_{\mathfrak{o}} \pi(\begin{bmatrix} 1 &&& \\ x\varpi^n & 1 && \\ y\varpi^n && 1 & \\ z\varpi^n & y\varpi^n & -x\varpi^n & 1 \end{bmatrix})v\,dx\,dy\,dz \tag{3.57}$$

for $v \in V_s(n+1)$. Evidently, $s_n v = q^{-3}T_1 v$ for $v \in V_s(n+1)$. It is easy to see that $s_n : V_s(n+1) \to V_s(n)$ is not an upper block operator.

Next, we consider the level lowering operator $T_{\eta^{-1}} : V_s(n+1) \to V_s(n)$. As usual, we need a coset decomposition.

Lemma 3.7.2 *Let n be an integer such that $n \geq 1$, and let*

$$J = \mathrm{K}_s(\mathfrak{p}^n) \cap \eta^{-1}\mathrm{K}_s(\mathfrak{p}^{n+1})\eta.$$

There is a disjoint decomposition

$$\mathrm{K}_s(\mathfrak{p}^n) = \bigsqcup_{x,y,z\in\mathfrak{o}/\mathfrak{p}} \begin{bmatrix} 1 & x & y & z\varpi^{-n+1} \\ & 1 && y \\ && 1 & -x \\ &&& 1 \end{bmatrix} J. \tag{3.58}$$

Proof Using (3.3), we see that if $k \in \mathrm{GSp}(4, F)$, then $k \in J$ if and only if $\lambda(k) \in \mathfrak{o}^\times$ and

$$k \in \begin{bmatrix} \mathfrak{o} & \mathfrak{p} & \mathfrak{p} & \mathfrak{p}^{-n+2} \\ \mathfrak{p}^n & \mathfrak{o} & \mathfrak{o} & \mathfrak{p} \\ \mathfrak{p}^n & \mathfrak{o} & \mathfrak{o} & \mathfrak{p} \\ \mathfrak{p}^n & \mathfrak{p}^n & \mathfrak{p}^n & \mathfrak{o} \end{bmatrix}.$$

It follows that the union of the cosets on the right-hand side of (3.58) is contained in $\mathrm{K}_s(\mathfrak{p}^n)$. Conversely, let $k \in \mathrm{K}_s(\mathfrak{p}^n)$. By (3.14), we may write $k = k_1k_2k_3$ where $k_1, k_2, k_3 \in \mathrm{K}_s(\mathfrak{p}^n)$ and

$$k_1 \in \begin{bmatrix} 1 & \mathfrak{o} & \mathfrak{o} & \mathfrak{p}^{-n+1} \\ & 1 && \mathfrak{o} \\ && 1 & \mathfrak{o} \\ &&& 1 \end{bmatrix}, \quad k_2 \in \begin{bmatrix} \mathfrak{o}^\times &&& \\ & \mathfrak{o} & \mathfrak{o} & \\ & \mathfrak{o} & \mathfrak{o} & \\ &&& \mathfrak{o}^\times \end{bmatrix}, \quad \text{and} \quad k_3 \in \begin{bmatrix} 1 &&& \\ \mathfrak{p}^n & 1 && \\ \mathfrak{p}^n && 1 & \\ \mathfrak{p}^n & \mathfrak{p}^n & \mathfrak{p}^n & 1 \end{bmatrix}.$$

Since $k_2k_3 \in J$, we see that k is in the union. It is straightforward to see that the cosets in (3.58) are disjoint. □

Let the notation be as at the beginning of this section. By (3.24) and Lemma 3.7.2, the level lowering operator $T_{\eta^{-1}} : V_s(n+1) \to V_s(n)$ sends $v \in V_s(n+1)$ to

$$\sum_{x,y,z\in\mathfrak{o}/\mathfrak{p}} \pi(\begin{bmatrix} 1 & x & y & z\varpi^{-n+1} \\ & 1 & & y \\ & & 1 & -x \\ & & & 1 \end{bmatrix}\begin{bmatrix} \varpi & & & \\ & 1 & & \\ & & 1 & \\ & & & \varpi^{-1} \end{bmatrix})v.$$

Again, we prefer to work with a multiple of $T_{\eta^{-1}}$. Define

$$\sigma_n : V_s(n+1) \longrightarrow V_s(n) \tag{3.59}$$

by

$$\sigma_n v = \int\limits_{\mathfrak{o}}\int\limits_{\mathfrak{o}}\int\limits_{\mathfrak{o}} \pi(\begin{bmatrix} 1 & x & y & z\varpi^{-n+1} \\ & 1 & & y \\ & & 1 & -x \\ & & & 1 \end{bmatrix}\begin{bmatrix} \varpi & & & \\ & 1 & & \\ & & 1 & \\ & & & \varpi^{-1} \end{bmatrix})v\,dx\,dy\,dz \tag{3.60}$$

for $v \in V_s(n)$. Evidently, $\sigma_n v = q^{-3}T_{\eta^{-1}}v$ for $v \in V_s(n+1)$. Clearly, the map $\sigma_n : V_s(n+1) \to V_s(n)$ is an upper block operator.

Lemma 3.7.3 *Let (π, V) be a smooth representation of* GSp(4, F) *for which the center acts trivially, and let n be an integer such that $n \geq 1$.*

(1) *If $w \in V(n-1)$, then $\sigma_n\eta w = w$.*
(2) *If $n \geq 2$ and $w \in V(n-2)$, then $\sigma_n\eta w = \tau_{n-1}w$.*
(3) *If $w \in V(n)$, then $\sigma_n\theta' w = w + q\sigma_n w$, so that $w = \sigma_n(\theta' w - qw)$.*
(4) *If $w \in V_s(n)$ with $p_n(w) = 0$, then $w = q\sigma_n(\rho_n' w + q\tau_n w)$.*
(5) *If $n \geq 2$ and $w \in V_s(n)$, then $t_{n-1}s_{n-1}w = \sigma_n t_n w$.*
(6) *If $n \geq 2$ and $w \in V_s(n)$, then $\sigma_n\tau_n w = \tau_{n-1}\sigma_{n-1}w$.*

Proof (1) and (2). These statements are straightforward.

(3). This follows by a calculation using the formula (2.29) for θ'.

(4). Let $w \in V_s(n)$ with $p_n(w) = 0$. By (4) of Lemma 3.6.1 we have $\rho_n' w = q^{-1}\eta w - q\tau_{n+1}w$. We calculate

$$\begin{aligned}
\sigma_n\rho_n' w &= \int\limits_{\mathfrak{o}}\int\limits_{\mathfrak{o}}\int\limits_{\mathfrak{o}} \pi(\begin{bmatrix} 1 & x & y & z\varpi^{-n+1} \\ & 1 & & y \\ & & 1 & -x \\ & & & 1 \end{bmatrix}\begin{bmatrix} \varpi & & & \\ & 1 & & \\ & & 1 & \\ & & & \varpi^{-1} \end{bmatrix})(\rho_n' w)\,dz\,dy\,dx \\
&= \int\limits_{\mathfrak{o}}\int\limits_{\mathfrak{o}}\int\limits_{\mathfrak{o}} \pi(\begin{bmatrix} 1 & x & y & z\varpi^{-n+1} \\ & 1 & & y \\ & & 1 & -x \\ & & & 1 \end{bmatrix}\begin{bmatrix} \varpi & & & \\ & 1 & & \\ & & 1 & \\ & & & \varpi^{-1} \end{bmatrix})(q^{-1}\eta w)\,dz\,dy\,dx \\
&\quad - \int\limits_{\mathfrak{o}}\int\limits_{\mathfrak{o}}\int\limits_{\mathfrak{o}} \pi(\begin{bmatrix} 1 & x & y & z\varpi^{-n+1} \\ & 1 & & y \\ & & 1 & -x \\ & & & 1 \end{bmatrix}\begin{bmatrix} \varpi & & & \\ & 1 & & \\ & & 1 & \\ & & & \varpi^{-1} \end{bmatrix})(q\tau_{n+1}w)\,dz\,dy\,dx
\end{aligned}$$

$$= q^{-1} \int_{\mathfrak{o}} \int_{\mathfrak{o}} \int_{\mathfrak{o}} \pi(\begin{bmatrix} 1 & x & y & z\varpi^{-n+1} \\ & 1 & & y \\ & & 1 & -x \\ & & & 1 \end{bmatrix}) w \, dz \, dy \, dx$$

$$- q \int_{\mathfrak{o}} \int_{\mathfrak{o}} \int_{\mathfrak{o}} \pi(\begin{bmatrix} 1 & x & y & z\varpi^{-n+1} \\ & 1 & & y \\ & & 1 & -x \\ & & & 1 \end{bmatrix} \begin{bmatrix} \varpi & & & \\ & 1 & & \\ & & 1 & \\ & & & \varpi^{-1} \end{bmatrix}) w \, dz \, dy \, dx$$

$$= q^{-1} w - q \int_{\mathfrak{o}} \int_{\mathfrak{o}} \int_{\mathfrak{o}} \pi(\begin{bmatrix} 1 & x & y & z\varpi^{-n+1} \\ & 1 & & y \\ & & 1 & -x \\ & & & 1 \end{bmatrix} \begin{bmatrix} \varpi & & & \\ & 1 & & \\ & & 1 & \\ & & & \varpi^{-1} \end{bmatrix})(\tau_n w) \, dz \, dy \, dx$$

$$= q^{-1} w - q \sigma_n \tau_n w.$$

Hence, $w = q\sigma_n(\rho'_n w + q\tau_n w)$, as claimed.

(5) and (6). These statements follow by direct calculations. □

Again, other expressions for σ_n will be useful. Let the notation be as in Lemma 3.7.3. If $v \in V_s(n+1)$, then $\sigma_n v$ can be written as a sum:

$$\sigma_n v = q^{-3} \sum_{x,y,z \in \mathfrak{o}/\mathfrak{p}} \pi(\begin{bmatrix} 1 & x & y & z\varpi^{-n+1} \\ & 1 & & y \\ & & 1 & -x \\ & & & 1 \end{bmatrix} \begin{bmatrix} \varpi & & & \\ & 1 & & \\ & & 1 & \\ & & & \varpi^{-1} \end{bmatrix}) v. \tag{3.61}$$

Also, let dk be the Haar measure on $\mathrm{GSp}(4, F)$ that assigns $\mathrm{K}_s(\mathfrak{p}^n)$ volume 1. If $v \in V_s(n+1)$, then

$$\sigma_n v = \int_{\mathrm{K}_s(\mathfrak{p}^n)} \pi(k \begin{bmatrix} \varpi & & & \\ & 1 & & \\ & & 1 & \\ & & & \varpi^{-1} \end{bmatrix}) v \, dk. \tag{3.62}$$

Proposition 3.7.4 *Let (π, V) be a smooth representation of* $\mathrm{GSp}(4, F)$ *for which the center acts trivially. Let n be an integer such that $n \geq 1$. Then the level lowering operator*

$$\sigma_n : V_s(n+1) \longrightarrow V_s(n) \tag{3.63}$$

is surjective. Moreover,

$$\sigma_n(V(n) + V(n+1)) \supset V(n-1) + V(n). \tag{3.64}$$

Proof By Lemma 3.3.1 we have $V_s(n) = V(n) \oplus \ker(p_n)$. If $v \in V(n)$, then v lies in the image of σ_n by (3) of Lemma 3.7.3. If $v \in \ker(p_n)$, then v lies in the image of σ_n by (4) of Lemma 3.7.3. Hence, the map (3.63) is surjective. Next, (1) of Lemma 3.7.3 shows that $V(n-1) \subset \sigma_n(V(n+1))$, and (3) of Lemma 3.7.3 shows that $V(n) \subset \sigma_n(V(n) + V(n+1))$. Hence, we get (3.64). □

Let (π, V) be a smooth representation of $\mathrm{GSp}(4, F)$ for which the center acts trivially. Let n be an integer such that $n \geq 1$. We may consider the composition of $\tau_n : V_s(n) \to V_s(n+1)$ and $\sigma_n : V_s(n+1) \to V_s(n)$, which defines an endomorphism

$$\sigma_n \circ \tau_n : V_s(n) \longrightarrow V_s(n). \tag{3.65}$$

Lemma 3.7.5 *Let (π, V) be a smooth representation of $\mathrm{GSp}(4, F)$ for which the center acts trivially. For $n \geq 2$, the kernel of the endomorphism $\sigma_n \circ \tau_n$ of $V_s(n)$ coincides with the kernel of the map $\sigma_{n-1} : V_s(n) \to V_s(n-1)$.*

Proof By (6) of Lemma 3.7.3 we have $(\sigma_n \circ \tau_n)v = \tau_{n-1}(\sigma_{n-1}v)$ for $v \in V_s(n)$. Since the map $\tau_{n-1} : V_s(n-1) \to V_s(n)$ is injective by (1) of Lemma 3.5.4, the assertion follows. □

3.8 Stable Hecke Operators

Let (π, V) be a smooth representation for which the center acts trivially, and let n be an integer such that $n \geq 1$. In this section we will consider the Hecke operators $T_g : V_s(n) \to V_s(n)$ determined by

$$g = \Delta_{0,1} = \begin{bmatrix} \varpi &&& \\ & \varpi && \\ && 1 & \\ &&& 1 \end{bmatrix} \quad \text{and} \quad g = \Delta_{1,0} = \begin{bmatrix} \varpi^2 &&& \\ & \varpi && \\ && \varpi & \\ &&& 1 \end{bmatrix}.$$

The following lemma provides the required coset decompositions; see also Sect. 3.4 and Lemma 3.4.1.

Lemma 3.8.1 *Let n be an integer such that $n \geq 1$.*

(1) *There is a disjoint decomposition*

$$\begin{aligned} \mathrm{K}_s(\mathfrak{p}^n) \begin{bmatrix} \varpi &&& \\ & \varpi && \\ && 1 & \\ &&& 1 \end{bmatrix} \mathrm{K}_s(\mathfrak{p}^n) = & \bigsqcup_{y,z \in \mathfrak{o}/\mathfrak{p}} \begin{bmatrix} 1 & y && z\varpi^{-n+1} \\ & 1 && \\ && 1 & -y \\ &&& 1 \end{bmatrix} \begin{bmatrix} \varpi &&& \\ & 1 && \\ && \varpi & \\ &&& 1 \end{bmatrix} \mathrm{K}_s(\mathfrak{p}^n) \\ & \sqcup \bigsqcup_{c,y,z \in \mathfrak{o}/\mathfrak{p}} \begin{bmatrix} 1 && y & z\varpi^{-n+1} \\ & 1 & c & y \\ && 1 & \\ &&& 1 \end{bmatrix} \begin{bmatrix} \varpi &&& \\ & \varpi && \\ && 1 & \\ &&& 1 \end{bmatrix} \mathrm{K}_s(\mathfrak{p}^n). \end{aligned} \tag{3.66}$$

(2) *There is a disjoint decomposition*

$$\mathrm{K}_s(\mathfrak{p}^n)\begin{bmatrix}\varpi^2&&&\\&\varpi&&\\&&\varpi&\\&&&1\end{bmatrix}\mathrm{K}_s(\mathfrak{p}^n)=\bigsqcup_{\substack{x,y\in\mathfrak{o}/\mathfrak{p}\\ z\in\mathfrak{o}/\mathfrak{p}^2}}\begin{bmatrix}1&x&y&z\varpi^{-n+1}\\&1&&y\\&&1&-x\\&&&1\end{bmatrix}\begin{bmatrix}\varpi^2&&&\\&\varpi&&\\&&\varpi&\\&&&1\end{bmatrix}\mathrm{K}_s(\mathfrak{p}^n). \tag{3.67}$$

Proof (1). The asserted identity is equivalent to

$$\begin{aligned}\mathrm{K}_s(\mathfrak{p}^n)\begin{bmatrix}\varpi&&&\\&\varpi&&\\&&1&\\&&&1\end{bmatrix}\mathrm{K}_s(\mathfrak{p}^n)=&\bigsqcup_{y,z\in\mathfrak{o}/\mathfrak{p}}s_2\begin{bmatrix}1&&y&z\varpi^{-n+1}\\&1&&y\\&&1&\\&&&1\end{bmatrix}\begin{bmatrix}\varpi&&&\\&\varpi&&\\&&1&\\&&&1\end{bmatrix}\mathrm{K}_s(\mathfrak{p}^n)\\ &\sqcup\bigsqcup_{c,y,z\in\mathfrak{o}/\mathfrak{p}}\begin{bmatrix}1&&y&z\varpi^{-n+1}\\&1&c&y\\&&1&\\&&&1\end{bmatrix}\begin{bmatrix}\varpi&&&\\&\varpi&&\\&&1&\\&&&1\end{bmatrix}\mathrm{K}_s(\mathfrak{p}^n),\end{aligned} \tag{3.68}$$

where s_2 is defined in (2.5). Clearly, the right-hand side is contained in the left-hand side. By the Iwahori factorization (3.14), the left-hand side is contained in

$$\begin{bmatrix}1&&&\\&\mathfrak{o}&\mathfrak{o}&\\&\mathfrak{o}&\mathfrak{o}&\\&&&1\end{bmatrix}\begin{bmatrix}1&&\mathfrak{o}&\mathfrak{p}^{-n+1}\\&1&&\mathfrak{o}\\&&1&\\&&&1\end{bmatrix}\begin{bmatrix}\varpi&&&\\&\varpi&&\\&&1&\\&&&1\end{bmatrix}\mathrm{K}_s(\mathfrak{p}^n).$$

Here, and below, when we write, for example, $\begin{bmatrix}1&&&\\&\mathfrak{o}&\mathfrak{o}&\\&\mathfrak{o}&\mathfrak{o}&\\&&&1\end{bmatrix}$, we actually mean the intersection of this set with $\mathrm{GSp}(4,F)$. We have

$$\mathrm{SL}(2,\mathfrak{o})=\bigsqcup_{c\in\mathfrak{o}/\mathfrak{p}}\left[\begin{smallmatrix}1&c\\&1\end{smallmatrix}\right]{}^t\Gamma_0(\mathfrak{p})\sqcup\left[\begin{smallmatrix}&1\\-1&\end{smallmatrix}\right]{}^t\Gamma_0(\mathfrak{p}),$$

where $\Gamma_0(\mathfrak{p})=\{\left[\begin{smallmatrix}a&b\\c&d\end{smallmatrix}\right]\in\mathrm{SL}(2,\mathfrak{o})\mid c\equiv0\pmod{\mathfrak{p}}\}$. Hence the left-hand side of (3.68) is contained in

$$\begin{aligned}&\bigcup_{c\in\mathfrak{o}/\mathfrak{p}}\begin{bmatrix}1&&&\\&1&c&\\&&1&\\&&&1\end{bmatrix}\begin{bmatrix}1&&\mathfrak{o}&\mathfrak{p}^{-n+1}\\&1&&\mathfrak{o}\\&&1&\\&&&1\end{bmatrix}\begin{bmatrix}\varpi&&&\\&\varpi&&\\&&1&\\&&&1\end{bmatrix}\mathrm{K}_s(\mathfrak{p}^n)\\ &\cup s_2\begin{bmatrix}1&&\mathfrak{o}&\mathfrak{p}^{-n+1}\\&1&&\mathfrak{o}\\&&1&\\&&&1\end{bmatrix}\begin{bmatrix}\varpi&&&\\&\varpi&&\\&&1&\\&&&1\end{bmatrix}\mathrm{K}_s(\mathfrak{p}^n).\end{aligned}$$

It follows that the left-hand side of (3.68) is contained in the right-hand side of (3.68). The disjointness of the right-hand side is easy to see.

(2). This follows by a similar argument. □

Let (π, V) be a smooth representation of $\mathrm{GSp}(4, F)$ for which the center acts trivially, and let n be an integer such that $n \geq 1$. By Lemma 3.8.1, there are endomorphisms $T^s_{0,1}$ and $T^s_{1,0}$ of $V_s(n)$ given by

$$\begin{aligned} T^s_{0,1}v = & \sum_{y,z\in\mathfrak{O}/\mathfrak{p}} \pi(\begin{bmatrix} 1 & y & & z\varpi^{-n+1} \\ & 1 & & \\ & & 1 & -y \\ & & & 1 \end{bmatrix}\begin{bmatrix} \varpi & & & \\ & 1 & & \\ & & \varpi & \\ & & & 1 \end{bmatrix})v \\ & + \sum_{c,y,z\in\mathfrak{O}/\mathfrak{p}} \pi(\begin{bmatrix} 1 & & y & z\varpi^{-n+1} \\ & 1 & c & y \\ & & 1 & \\ & & & 1 \end{bmatrix}\begin{bmatrix} \varpi & & & \\ & \varpi & & \\ & & 1 & \\ & & & 1 \end{bmatrix})v \end{aligned} \tag{3.69}$$

and

$$T^s_{1,0}v = \sum_{\substack{x,y\in\mathfrak{O}/\mathfrak{p} \\ z\in\mathfrak{O}/\mathfrak{p}^2}} \pi(\begin{bmatrix} 1 & x & y & z\varpi^{-n+1} \\ & 1 & & y \\ & & 1 & -x \\ & & & 1 \end{bmatrix}\begin{bmatrix} \varpi^2 & & & \\ & \varpi & & \\ & & \varpi & \\ & & & 1 \end{bmatrix})v \tag{3.70}$$

for $v \in V_s(n)$. We refer to the endomorphisms $T^s_{0,1}$ and $T^s_{1,0}$ of $V_s(n)$ as the *stable Klingen Hecke operators*, or simply *stable Hecke operators*, of level n. We note that $T^s_{0,1}$ and $T^s_{1,0}$ are upper block operators.

Lemma 3.8.2 *Let (π, V) be a smooth representation of* $\mathrm{GSp}(4, F)$ *for which the center acts trivially, and let n be an integer such that $n \geq 1$. The endomorphisms $T^s_{0,1}$ and $T^s_{1,0}$ of $V_s(n)$ commute.*

Proof Let $K = \mathrm{K}_s(\mathfrak{p}^n)$, and dk be the Haar measure on $\mathrm{GSp}(4, F)$ that assigns K volume 1. Let $v \in V_s(n)$. We have

$$\begin{aligned} T^s_{0,1}v &= \int_K \pi(k)T^s_{0,1}v\,dk \\ &= \sum_{y,z\in\mathfrak{O}/\mathfrak{p}} \int_K \pi(k\begin{bmatrix} 1 & y & & z\varpi^{-n+1} \\ & 1 & & \\ & & 1 & -y \\ & & & 1 \end{bmatrix}\begin{bmatrix} \varpi & & & \\ & 1 & & \\ & & \varpi & \\ & & & 1 \end{bmatrix})v\,dk \\ &\quad + \sum_{c,y,z\in\mathfrak{O}/\mathfrak{p}} \int_K \pi(k\begin{bmatrix} 1 & & y & z\varpi^{-n+1} \\ & 1 & c & y \\ & & 1 & \\ & & & 1 \end{bmatrix}\begin{bmatrix} \varpi & & & \\ & \varpi & & \\ & & 1 & \\ & & & 1 \end{bmatrix})v\,dk \\ &= q^2 \int_K \pi(k\begin{bmatrix} \varpi & & & \\ & 1 & & \\ & & \varpi & \\ & & & 1 \end{bmatrix})v\,dk + q^3 \int_K \pi(k\begin{bmatrix} \varpi & & & \\ & \varpi & & \\ & & 1 & \\ & & & 1 \end{bmatrix})v\,dk, \end{aligned}$$

so that

$$T^s_{0,1}v = (q^2+q^3)\int\limits_K \pi(k\begin{bmatrix}\varpi&&&\\&\varpi&&\\&&1&\\&&&1\end{bmatrix})v\,dk. \tag{3.71}$$

For the last step we recall that $s_2 \in K = \mathrm{K}_s(\mathfrak{p}^n)$. Similarly,

$$T^s_{1,0}v = q^4\int\limits_K \pi(k\begin{bmatrix}\varpi^2&&&\\&\varpi&&\\&&\varpi&\\&&&1\end{bmatrix})v\,dk. \tag{3.72}$$

By (3.72) and (3.69) we now have

$$\begin{aligned}
T^s_{1,0}(T^s_{0,1}v) &= q^4\int\limits_K \pi(k\begin{bmatrix}\varpi^2&&&\\&\varpi&&\\&&\varpi&\\&&&1\end{bmatrix})(T^s_{0,1}v)\,dk\\
&= \sum_{y,z\in\mathfrak{o}/\mathfrak{p}} q^4\int\limits_K \pi(k\begin{bmatrix}\varpi^2&&&\\&\varpi&&\\&&\varpi&\\&&&1\end{bmatrix}\begin{bmatrix}1&y&&z\varpi^{-n+1}\\&1&&\\&&1&-y\\&&&1\end{bmatrix}\begin{bmatrix}\varpi&&&\\&1&&\\&&\varpi&\\&&&1\end{bmatrix})v\,dk\\
&\quad+\sum_{c,y,z\in\mathfrak{o}/\mathfrak{p}} q^4\int\limits_K \pi(k\begin{bmatrix}\varpi^2&&&\\&\varpi&&\\&&\varpi&\\&&&1\end{bmatrix}\begin{bmatrix}1&&y&z\varpi^{-n+1}\\&1&c&y\\&&1&\\&&&1\end{bmatrix}\begin{bmatrix}\varpi&&&\\&\varpi&&\\&&1&\\&&&1\end{bmatrix})v\,dk\\
&= \sum_{y,z\in\mathfrak{o}/\mathfrak{p}} q^4\int\limits_K \pi(k\begin{bmatrix}1&y\varpi&&z\varpi^{-n+3}\\&1&&\\&&1&-y\varpi\\&&&1\end{bmatrix}\begin{bmatrix}\varpi^3&&&\\&\varpi&&\\&&\varpi^2&\\&&&1\end{bmatrix})v\,dk\\
&\quad+\sum_{c,y,z\in\mathfrak{o}/\mathfrak{p}} q^4\int\limits_K \pi(k\begin{bmatrix}1&&y\varpi&z\varpi^{-n+3}\\&1&c&y\varpi\\&&1&\\&&&1\end{bmatrix}\begin{bmatrix}\varpi^3&&&\\&\varpi^2&&\\&&\varpi&\\&&&1\end{bmatrix})v\,dk\\
&= q^6\int\limits_K \pi(k\begin{bmatrix}\varpi^3&&&\\&\varpi&&\\&&\varpi^2&\\&&&1\end{bmatrix})v\,dk + q^7\int\limits_K \pi(k\begin{bmatrix}\varpi^3&&&\\&\varpi^2&&\\&&\varpi&\\&&&1\end{bmatrix})v\,dk\\
&= (q^6+q^7)\int\limits_K \pi(k\begin{bmatrix}\varpi^3&&&\\&\varpi^2&&\\&&\varpi&\\&&&1\end{bmatrix})v\,dk.
\end{aligned}$$

Similarly, by (3.71) and (3.70),

$$\begin{aligned}
T^s_{0,1}(T^s_{1,0}v) &= (q^2+q^3)\int\limits_K \pi(k\begin{bmatrix}\varpi&&&\\&\varpi&&\\&&1&\\&&&1\end{bmatrix})T^s_{1,0}v\,dk\\
&= \sum_{\substack{x,y\in\mathfrak{o}/\mathfrak{p}\\ z\in\mathfrak{o}/\mathfrak{p}^2}} (q^2+q^3)
\end{aligned}$$

$$\times \int_K \pi(k \begin{bmatrix} \varpi & & & \\ & \varpi & & \\ & & 1 & \\ & & & 1 \end{bmatrix} \begin{bmatrix} 1 & x & y & z\varpi^{-n+1} \\ & 1 & & y \\ & & 1 & -x \\ & & & 1 \end{bmatrix} \begin{bmatrix} \varpi^2 & & & \\ & \varpi & & \\ & & \varpi & \\ & & & 1 \end{bmatrix}) v \, dk$$

$$= \sum_{\substack{x,y \in \mathfrak{o}/\mathfrak{p} \\ z \in \mathfrak{o}/\mathfrak{p}^2}} (q^2 + q^3) \int_K \pi(k \begin{bmatrix} 1 & x & y\varpi & z\varpi^{-n+2} \\ & 1 & & y\varpi \\ & & 1 & -x \\ & & & 1 \end{bmatrix} \begin{bmatrix} \varpi^3 & & & \\ & \varpi^2 & & \\ & & \varpi & \\ & & & 1 \end{bmatrix}) v \, dk$$

$$= (q^6 + q^7) \int_K \pi(k \begin{bmatrix} \varpi^3 & & & \\ & \varpi^2 & & \\ & & \varpi & \\ & & & 1 \end{bmatrix}) v \, dk.$$

It now follows that $T^s_{1,0}(T^s_{0,1}v) = T^s_{0,1}(T^s_{1,0}v)$. □

Lemma 3.8.3 *Let (π, V) be a smooth representation of* GSp(4, F) *for which the center acts trivially, and let n be an integer such that $n \geq 1$. Let $v \in V(n)$. Then*

$$(1 + q^{-1}) p_n T^s_{0,1} v = T_{0,1} v \tag{3.73}$$

and

$$(1 + q^{-1}) p_n T^s_{1,0} v = T_{1,0} v. \tag{3.74}$$

Proof By (3.35), (3.69), and the assumption that $v \in V(n)$,

$$(q+1) p_n T^s_{0,1} v = q\tau_n T^s_{0,1} v + t_n T^s_{0,1} v$$

$$= q\tau_n \sum_{y,z \in \mathfrak{o}/\mathfrak{p}} \pi(\begin{bmatrix} 1 & y & & z\varpi^{-n+1} \\ & 1 & & \\ & & 1 & -y \\ & & & 1 \end{bmatrix} \begin{bmatrix} \varpi & & & \\ & 1 & & \\ & & \varpi & \\ & & & 1 \end{bmatrix}) v$$

$$+ q\tau_n \sum_{c,y,z \in \mathfrak{o}/\mathfrak{p}} \pi(\begin{bmatrix} 1 & & y & z\varpi^{-n+1} \\ & 1 & c & y \\ & & 1 & \\ & & & 1 \end{bmatrix} \begin{bmatrix} \varpi & & & \\ & \varpi & & \\ & & 1 & \\ & & & 1 \end{bmatrix}) v$$

$$+ t_n \sum_{y,z \in \mathfrak{o}/\mathfrak{p}} \pi(\begin{bmatrix} 1 & y & & z\varpi^{-n+1} \\ & 1 & & \\ & & 1 & -y \\ & & & 1 \end{bmatrix} \begin{bmatrix} \varpi & & & \\ & 1 & & \\ & & \varpi & \\ & & & 1 \end{bmatrix}) v$$

$$+ t_n \sum_{c,y,z \in \mathfrak{o}/\mathfrak{p}} \pi(\begin{bmatrix} 1 & & y & z\varpi^{-n+1} \\ & 1 & c & y \\ & & 1 & \\ & & & 1 \end{bmatrix} \begin{bmatrix} \varpi & & & \\ & \varpi & & \\ & & 1 & \\ & & & 1 \end{bmatrix}) v$$

$$= q \sum_{y,z \in \mathfrak{o}/\mathfrak{p}} \pi(\begin{bmatrix} 1 & y & & z\varpi^{-n} \\ & 1 & & \\ & & 1 & -y \\ & & & 1 \end{bmatrix} \begin{bmatrix} \varpi & & & \\ & 1 & & \\ & & \varpi & \\ & & & 1 \end{bmatrix}) v$$

$$
\begin{aligned}
&+q \sum_{c,y,z\in\mathfrak{o}/\mathfrak{p}} \pi(\begin{bmatrix} 1 & & y & z\varpi^{-n} \\ & 1 & c & y \\ & & 1 & \\ & & & 1 \end{bmatrix}\begin{bmatrix} \varpi & & & \\ & \varpi & & \\ & & 1 & \\ & & & 1 \end{bmatrix})v \\
&+q \sum_{y\in\mathfrak{o}/\mathfrak{p}} \pi(t_n \begin{bmatrix} 1 & y & & \\ & 1 & & \\ & & 1 & -y \\ & & & 1 \end{bmatrix}\begin{bmatrix} \varpi & & & \\ & 1 & & \\ & & \varpi & \\ & & & 1 \end{bmatrix})v \\
&+q \sum_{c,y\in\mathfrak{o}/\mathfrak{p}} \pi(t_n \begin{bmatrix} 1 & & y & \\ & 1 & c & y \\ & & 1 & \\ & & & 1 \end{bmatrix}\begin{bmatrix} \varpi & & & \\ & \varpi & & \\ & & 1 & \\ & & & 1 \end{bmatrix})v \\
&= qT_{0,1}v.
\end{aligned}
$$

The last equality follows from Lemma 6.1.2 of [105]. This proves (3.73). Next, by (3.35), (3.70), and the assumption that $v \in V(n)$,

$$
\begin{aligned}
(q+1)p_n T^s_{1,0}v &= q\tau_n T^s_{1,0}v + t_n T^s_{1,0}v \\
&= q \sum_{\substack{x,y\in\mathfrak{o}/\mathfrak{p}\\ z\in\mathfrak{o}/\mathfrak{p}^2}} \pi(\begin{bmatrix} 1 & x & y & z\varpi^{-n} \\ & 1 & & y \\ & & 1 & -x \\ & & & 1 \end{bmatrix}\begin{bmatrix} \varpi^2 & & & \\ & \varpi & & \\ & & \varpi & \\ & & & 1 \end{bmatrix})v \\
&\quad + q \sum_{x,y,z\in\mathfrak{o}/\mathfrak{p}} \pi(t_n \begin{bmatrix} 1 & x & y & z\varpi^{-n+1} \\ & 1 & & y \\ & & 1 & -x \\ & & & 1 \end{bmatrix}\begin{bmatrix} \varpi^2 & & & \\ & \varpi & & \\ & & \varpi & \\ & & & 1 \end{bmatrix})v \\
&= qT_{1,0}v.
\end{aligned}
$$

The last equality again follows from Lemma 6.1.2 of [105]. □

Lemma 3.8.4 *Let (π, V) be a smooth representation of* GSp(4, F) *for which the center acts trivially. Let n be an integer such that $n \geq 1$. Let $v \in V_s(n)$. Then*

$$T^s_{0,1}v = q^2\sigma_n\theta v \quad \text{and} \quad T^s_{1,0}v = q^4\sigma_n\tau_n v. \tag{3.75}$$

Assume that $n \geq 2$. Then we also have

$$T^s_{1,0}v = q^4\tau_{n-1}\sigma_{n-1}v, \tag{3.76}$$

and $T^s_{1,0}v = 0$ if and only if $\sigma_{n-1}v = 0$.

Proof Let $K = \mathrm{K}_s(\mathfrak{p}^n)$. Let dk be the Haar measure on K that assigns K volume 1. Let $v \in V_s(n)$. Then

$$
\begin{aligned}
&q^2\sigma_n\theta v\\
&= q^2 \int\limits_K \pi(k)\pi(\begin{bmatrix} \varpi &&& \\ & 1 && \\ && 1 & \\ &&& \varpi^{-1}\end{bmatrix})\theta v\, dk \qquad \text{(by (3.62))}\\
&= q^2 \int\limits_K \pi(k)\pi(\begin{bmatrix} \varpi &&& \\ & 1 && \\ && 1 & \\ &&& \varpi^{-1}\end{bmatrix}\begin{bmatrix} 1 &&& \\ & 1 && \\ && \varpi & \\ &&& \varpi\end{bmatrix})v\, dk\\
&\quad + q^3 \int\limits_K \int\limits_{\mathfrak{o}} \pi(k)\pi(\begin{bmatrix} \varpi &&& \\ & 1 && \\ && 1 & \\ &&& \varpi^{-1}\end{bmatrix}\begin{bmatrix} 1 &&& \\ & 1 & c & \\ && 1 & \\ &&& 1\end{bmatrix}\begin{bmatrix} 1 &&& \\ & \varpi && \\ && 1 & \\ &&& \varpi\end{bmatrix})v\, dc\, dk \qquad \text{(by (3.42))}\\
&= q^2 \int\limits_K \pi(k)\pi(\begin{bmatrix} \varpi &&& \\ & 1 && \\ && \varpi & \\ &&& 1\end{bmatrix})v\, dk + q^3 \int\limits_K \pi(k)\pi(\begin{bmatrix} \varpi &&& \\ & \varpi && \\ && 1 & \\ &&& 1\end{bmatrix})v\, dk\\
&= (q^2+q^3) \int\limits_K \pi(k)\pi(\begin{bmatrix} \varpi &&& \\ & \varpi && \\ && 1 & \\ &&& 1\end{bmatrix})v\, dk\\
&= T^s_{0,1}v \qquad \text{(by (3.71))}.
\end{aligned}
$$

A similar proof shows that $T^s_{1,0}v = q^4\sigma_n\tau_n v$. Assume that $n \geq 2$. Then $\sigma_n\tau_n v = \tau_{n-1}\sigma_{n-1}v$ by (6) of Lemma 3.7.3. Finally, we have $T^s_{1,0}v = 0$ if and only if $\sigma_{n-1}v = 0$ by $T^s_{1,0}v = q^4\sigma_n\tau_n v$ and Lemma 3.7.5. □

3.9 Commutation Relations

In this section we derive commutation relations for the operators we defined in the previous sections. These relations will play an important role in some of the subsequent chapters.

Lemma 3.9.1 *Let (π, V) be a smooth representation of* GSp(4, F) *for which the center acts trivially. Let n be an integer such that $n \geq 0$, and $v \in V_s(n)$. Then*

$$\theta\tau_n v = \tau_{n+1}\theta v, \tag{3.77}$$

$$\theta t_n v = t_{n+1}\theta v, \tag{3.78}$$

$$\theta\rho'_n v = \rho'_{n+1}\theta v, \tag{3.79}$$

$$\theta p_n v = p_{n+1}\theta v. \tag{3.80}$$

Proof It is easy to verify these statements using the defining formulas. □

Let (π, V) be a smooth representation of $\mathrm{GSp}(4, F)$ for which the center acts trivially. For $v \in V_s(1)$ define

$$e(v) = \int\limits_{\mathfrak{o}}\int\limits_{\mathfrak{o}}\int\limits_{\mathfrak{o}} \pi(\begin{bmatrix} 1 & x\varpi^{-1} & y\varpi^{-1} & z\varpi^{-1} \\ & 1 & & y\varpi^{-1} \\ & & 1 & -x\varpi^{-1} \\ & & & 1 \end{bmatrix})v\,dx\,dy\,dz. \tag{3.81}$$

It is easy to verify that $e(v) \in V_s(2)$ for $v \in V_s(1)$, so that this defines a linear map $e : V_s(1) \to V_s(2)$.

Lemma 3.9.2 *Let (π, V) be a smooth representation for which the center acts trivially. Assume that $v \in V_s(1)$ is such that the paramodularization $p_1(T^s_{1,0}v)$ is zero. Then*

$$e(v) = q^{-4}\eta T^s_{1,0}v - q^{-2}\tau_2 T^s_{1,0}v + q^{-2}\tau_1 T^s_{1,0}v. \tag{3.82}$$

Proof By (3.51),

$$\rho'_1 T^s_{1,0}v = q^{-1}\eta T^s_{1,0}v - q\tau_2 T^s_{1,0}v$$

is in $V(2)$, and in particular in $V_s(2)$. Hence, both sides of (3.82) are in $V_s(2)$. By Lemma 3.5.4, to prove (3.82), it is enough to prove that

$$\tau_2 e(v) = \tau_2\big(q^{-4}\eta T^s_{1,0}v - q^{-2}\tau_2 T^s_{1,0}v + q^{-2}\tau_1 T^s_{1,0}v\big). \tag{3.83}$$

This is equivalent to

$$\tau_2 e(v) = q^{-4}\tau_2\eta T^s_{1,0}v. \tag{3.84}$$

It follows from the definitions (3.81) and (3.70) that (3.84) holds. □

Lemma 3.9.3 *Let (π, V) be a smooth representation of* $\mathrm{GSp}(4, F)$ *for which the center acts trivially, and let n be an integer. Let $v \in V_s(n)$. Then*

$$T^s_{0,1}\tau_n v = \tau_n T^s_{0,1}v \qquad \text{if } n \geq 1, \tag{3.85}$$

$$T^s_{1,0}\tau_n v = \tau_n T^s_{1,0}v \qquad \text{if } n \geq 1, \tag{3.86}$$

$$T^s_{1,0}\theta v = q^2 T^s_{0,1}\tau_n v \qquad \text{if } n \geq 0, \tag{3.87}$$

$$T^s_{0,1}\theta v = \theta T^s_{0,1}v + q^3\tau_n v - q^3\eta\sigma_{n-1}v \qquad \text{if } n \geq 2, \tag{3.88}$$

$$T^s_{0,1}\theta v = \theta T^s_{0,1}v + q^3\tau_1 v - q^3 e(v) \qquad \text{if } n = 1, \tag{3.89}$$

$$T^s_{1,0}\theta' v = q^4 v + qT^s_{1,0}\tau_n v, \qquad \text{if } n \geq 1 \text{ and } v \in V(n), \tag{3.90}$$

$$T^s_{0,1}\rho'_n v = q\theta v - q\tau_n T^s_{0,1}v \qquad \text{if } n \geq 1 \text{ and } p_n(v) = 0, \tag{3.91}$$

$$T^s_{0,1}\rho'_n v = q\theta v + \tau_n T^s_{0,1} v \qquad \text{if } n \geq 1 \text{ and } v \in V(n), \tag{3.92}$$

$$T^s_{1,0}\rho'_n v = q^3\tau_n v - q\tau_n T^s_{1,0} v \qquad \text{if } n \geq 1 \text{ and } p_n(v) = 0, \tag{3.93}$$

$$T^s_{1,0}\rho'_n v = q^3\tau_n v + \tau_n T^s_{1,0} v \qquad \text{if } n \geq 1 \text{ and } v \in V(n). \tag{3.94}$$

In these equations, if $T^s_{0,1}$ or $T^s_{1,0}$ appears on the left-hand side, then this operator is an endomorphism of $V_s(n+1)$, and if $T^s_{0,1}$ or $T^s_{1,0}$ appears on the right-hand side, then this operator is an endomorphism of $V_s(n)$.

Proof Equations (3.85)–(3.87) follow from straightforward calculations. To prove (3.88) and (3.89), we begin by proving a preliminary identity. Let $K = \mathrm{K}_s(\mathfrak{p}^{n+1})$, and let dk be the Haar measure on K that assigns K volume 1. Let $v \in V_s(n)$; as before, n is an integer such that $n \geq 1$. Then:

$$\int\limits_K \int\limits_{\mathfrak{o}} \pi(k \begin{bmatrix} 1 & y\varpi^{-1} & & \\ & 1 & & \\ & & 1 & -y\varpi^{-1} \\ & & & 1 \end{bmatrix}) v \, dy \, dk$$

$$= \int\limits_K \int\limits_{\mathfrak{o}} \int\limits_{\mathfrak{o}} \pi(k \begin{bmatrix} 1 & & & \\ & 1 & b & \\ & & 1 & \\ & & & 1 \end{bmatrix} \begin{bmatrix} 1 & y\varpi^{-1} & & \\ & 1 & & \\ & & 1 & -y\varpi^{-1} \\ & & & 1 \end{bmatrix}) v \, dy \, db \, dk$$

$$= \int\limits_K \int\limits_{\mathfrak{o}} \int\limits_{\mathfrak{o}} \pi(k \begin{bmatrix} 1 & y\varpi^{-1} & & \\ & 1 & & \\ & & 1 & -y\varpi^{-1} \\ & & & 1 \end{bmatrix} \begin{bmatrix} 1 & & -yb\varpi^{-1} & y^2b\varpi^{-2} \\ & 1 & b & -yb\varpi^{-1} \\ & & 1 & \\ & & & 1 \end{bmatrix}) v \, dy \, db \, dk$$

$$= \int\limits_K \int\limits_{\mathfrak{o}} \int\limits_{\mathfrak{o}^\times} \pi(k \begin{bmatrix} 1 & y\varpi^{-1} & & \\ & 1 & & \\ & & 1 & -y\varpi^{-1} \\ & & & 1 \end{bmatrix} \begin{bmatrix} 1 & & -yb\varpi^{-1} & y^2b\varpi^{-2} \\ & 1 & & -yb\varpi^{-1} \\ & & 1 & \\ & & & 1 \end{bmatrix}) v \, dy \, db \, dk$$

$$+ \int\limits_K \int\limits_{\mathfrak{o}} \int\limits_{\mathfrak{p}} \pi(k \begin{bmatrix} 1 & y\varpi^{-1} & & \\ & 1 & & \\ & & 1 & -y\varpi^{-1} \\ & & & 1 \end{bmatrix} \begin{bmatrix} 1 & & -yb\varpi^{-1} & y^2b\varpi^{-2} \\ & 1 & & -yb\varpi^{-1} \\ & & 1 & \\ & & & 1 \end{bmatrix}) v \, dy \, db \, dk$$

$$= \int\limits_K \int\limits_{\mathfrak{o}} \int\limits_{\mathfrak{o}^\times} \pi(k \begin{bmatrix} 1 & y\varpi^{-1} & & \\ & 1 & & \\ & & 1 & -y\varpi^{-1} \\ & & & 1 \end{bmatrix} \begin{bmatrix} 1 & & -b\varpi^{-1} & yb\varpi^{-2} \\ & 1 & & -b\varpi^{-1} \\ & & 1 & \\ & & & 1 \end{bmatrix}) v \, dy \, db \, dk$$

$$+ q^{-1} \int\limits_K \pi(k) v \, dk$$

$$= \int\limits_K \int\limits_{\mathfrak{o}} \int\limits_{\mathfrak{o}} \pi(k \begin{bmatrix} 1 & y\varpi^{-1} & & \\ & 1 & & \\ & & 1 & -y\varpi^{-1} \\ & & & 1 \end{bmatrix} \begin{bmatrix} 1 & & -b\varpi^{-1} & yb\varpi^{-2} \\ & 1 & & -b\varpi^{-1} \\ & & 1 & \\ & & & 1 \end{bmatrix}) v \, dy \, db \, dk$$

$$- \int\limits_K \int\limits_{\mathfrak{o}} \int\limits_{\mathfrak{p}} \pi(k \begin{bmatrix} 1 & y\varpi^{-1} & & \\ & 1 & & \\ & & 1 & -y\varpi^{-1} \\ & & & 1 \end{bmatrix} \begin{bmatrix} 1 & & -b\varpi^{-1} & yb\varpi^{-2} \\ & 1 & & -b\varpi^{-1} \\ & & 1 & \\ & & & 1 \end{bmatrix}) v \, dy \, db \, dk$$

$$+q^{-1}\int\limits_K \pi(k)v\,dk$$

$$=\int\limits_K\int\limits_{\mathfrak{o}}\int\limits_{\mathfrak{o}} \pi(k\begin{bmatrix}1 & y\varpi^{-1} & -b\varpi^{-1} & \\ & 1 & & -b\varpi^{-1} \\ & & 1 & -y\varpi^{-1} \\ & & & 1\end{bmatrix})v\,dy\,db\,dk$$

$$-q^{-1}\int\limits_K\int\limits_{\mathfrak{o}} \pi(k\begin{bmatrix}1 & & -b\varpi^{-1} & \\ & 1 & & -b\varpi^{-1} \\ & & 1 & \\ & & & 1\end{bmatrix})v\,db\,dk+q^{-1}\int\limits_K \pi(k)v\,dk$$

$$=\int\limits_K\int\limits_{\mathfrak{o}}\int\limits_{\mathfrak{o}}\int\limits_{\mathfrak{o}} \pi(k\begin{bmatrix}1 & y\varpi^{-1} & b\varpi^{-1} & z\varpi^{-n} \\ & 1 & & b\varpi^{-1} \\ & & 1 & -y\varpi^{-1} \\ & & & 1\end{bmatrix})v\,dy\,db\,dk$$

$$-q^{-1}\int\limits_K\int\limits_{\mathfrak{o}} \pi(k\begin{bmatrix}1 & b\varpi^{-1} & & \\ & 1 & & \\ & & 1 & -b\varpi^{-1} \\ & & & 1\end{bmatrix})v\,db\,dk+q^{-1}\int\limits_K \pi(k)v\,dk.$$

We have proven that

$$(q+1)\int\limits_K\int\limits_{\mathfrak{o}} \pi(k\begin{bmatrix}1 & y\varpi^{-1} & & \\ & 1 & & \\ & & 1 & -y\varpi^{-1} \\ & & & 1\end{bmatrix})v\,dy\,dk$$

$$=\int\limits_K \pi(k)v\,dk+q\int\limits_K\int\limits_{\mathfrak{o}}\int\limits_{\mathfrak{o}}\int\limits_{\mathfrak{o}} \pi(k\begin{bmatrix}1 & y\varpi^{-1} & b\varpi^{-1} & z\varpi^{-n} \\ & 1 & & b\varpi^{-1} \\ & & 1 & -y\varpi^{-1} \\ & & & 1\end{bmatrix})v\,dy\,db\,dk. \tag{3.95}$$

Next, as in the proof of Lemma 3.8.2, we have

$$T_{0,1}^s\theta v=(q^2+q^3)\int\limits_K \pi(k\begin{bmatrix}\varpi & & & \\ & \varpi & & \\ & & 1 & \\ & & & 1\end{bmatrix})\theta v\,dk.$$

Hence, by (3.42),

$$T_{0,1}^s\theta v=(q^2+q^3)\int\limits_K \pi(k\begin{bmatrix}\varpi & & & \\ & \varpi & & \\ & & 1 & \\ & & & 1\end{bmatrix}\begin{bmatrix}1 & & & \\ & 1 & & \\ & & \varpi & \\ & & & \varpi\end{bmatrix})v\,dk$$

$$+q(q^2+q^3)\int\limits_K\int\limits_{\mathfrak{o}} \pi(k\begin{bmatrix}\varpi & & & \\ & \varpi & & \\ & & 1 & \\ & & & 1\end{bmatrix}\begin{bmatrix}1 & & & \\ & 1 & c & \\ & & 1 & \\ & & & 1\end{bmatrix}\begin{bmatrix}1 & & & \\ & \varpi & & \\ & & 1 & \\ & & & \varpi\end{bmatrix})v\,dc\,dk$$

$$=(q^2+q^3)\int\limits_K \pi(k)v\,dk+(q^3+q^4)\int\limits_K \pi(k\begin{bmatrix}\varpi & & & \\ & \varpi^2 & & \\ & & 1 & \\ & & & \varpi\end{bmatrix})v\,dk.$$

We also have, by (3.45), and then (3.69),

$$\theta T^s_{0,1}v = (q+1)\int\limits_K \pi(k\begin{bmatrix}1&&&\\&\varpi&&\\&&1&\\&&&\varpi\end{bmatrix})T^s_{0,1}v\,dk$$

$$= \sum_{y,z\in\mathfrak{o}/\mathfrak{p}}(q+1)\int\limits_K \pi(k\begin{bmatrix}1&&&\\&\varpi&&\\&&1&\\&&&\varpi\end{bmatrix}\begin{bmatrix}1&y&&z\varpi^{-n+1}\\&1&&\\&&1&-y\\&&&1\end{bmatrix}\begin{bmatrix}\varpi&&&\\&1&&\\&&\varpi&\\&&&1\end{bmatrix})v\,dk$$

$$+\sum_{c,y,z\in\mathfrak{o}/\mathfrak{p}}(q+1)\int\limits_K \pi(k\begin{bmatrix}1&&&\\&\varpi&&\\&&1&\\&&&\varpi\end{bmatrix}\begin{bmatrix}1&&y&z\varpi^{-n+1}\\&1&c&y\\&&1&\\&&&1\end{bmatrix}\begin{bmatrix}\varpi&&&\\&\varpi&&\\&&1&\\&&&1\end{bmatrix})v\,dk$$

$$= (q^3+q^2)\int\limits_K\int\limits_{\mathfrak{o}} \pi(k\begin{bmatrix}1&y\varpi^{-1}&&\\&1&&\\&&1&-y\varpi^{-1}\\&&&1\end{bmatrix})v\,dy\,dk$$

$$+(q^4+q^3)\int\limits_K \pi(k\begin{bmatrix}\varpi&&&\\&\varpi^2&&\\&&1&\\&&&\varpi\end{bmatrix})v\,dk.$$

Hence, by these formulas for $T^s_{0,1}\theta v$, $\theta T^s_{0,1}v$, and (3.95),

$$T^s_{0,1}\theta v - \theta T^s_{0,1}v$$

$$= (q^2+q^3)\int\limits_K \pi(k)v\,dk - (q^3+q^2)\int\limits_K\int\limits_{\mathfrak{o}} \pi(k\begin{bmatrix}1&y\varpi^{-1}&&\\&1&&\\&&1&-y\varpi^{-1}\\&&&1\end{bmatrix})v\,dy\,dk$$

$$= q^3\int\limits_K \pi(k)v\,dk - q^3\int\limits_K\int\limits_{\mathfrak{o}}\int\limits_{\mathfrak{o}}\int\limits_{\mathfrak{o}} \pi(k\begin{bmatrix}1&y\varpi^{-1}&b\varpi^{-1}&z\varpi^{-n}\\&1&&b\varpi^{-1}\\&&1&-y\varpi^{-1}\\&&&1\end{bmatrix})v\,dy\,db\,dk.$$

The first term on the right-hand side equals $q^3\tau_n v$ by (3.39). This proves (3.88) and (3.89) since

$$\int\limits_K\int\limits_{\mathfrak{o}}\int\limits_{\mathfrak{o}}\int\limits_{\mathfrak{o}} \pi(k\begin{bmatrix}1&y\varpi^{-1}&b\varpi^{-1}&z\varpi^{-n}\\&1&&b\varpi^{-1}\\&&1&-y\varpi^{-1}\\&&&1\end{bmatrix})v\,dy\,db\,dk = \begin{cases}\eta\sigma_{n-1}v & \text{if } n\geq 2,\\ e(v) & \text{if } n=1,\end{cases}$$

by (3.60) and (3.81).

Finally, to complete the proof, we note that: (3.90) follows from (3.70) and (2.29); (3.91) follows from (3.51); (3.92) follows from (3.49); (3.93) follows from (3.51); and (3.94) follows from (3.49). □

3.10 A Result About Eigenvalues

Let (π, V) be a smooth representation of $\mathrm{GSp}(4, F)$ for which the center acts trivially. In the following lemma we prove results that show that the spectrum of the endomorphism $T^s_{1,0}$ of $V_s(n)$ for $n \geq 1$ is contained in the union of $\{0\}$ and the spectrum of the endomorphism $T^s_{1,0}$ of $V_s(1)$.

Lemma 3.10.1 *Let (π, V) be a smooth representation of* $\mathrm{GSp}(4, F)$ *for which the center acts trivially. Let n be an integer such that $n \geq 1$.*

(1) *Assume that $V_s(1) = 0$. Then the operator $(T^s_{1,0})^{n-1}$ is zero on $V_s(n)$.*
(2) *Assume that $T^s_{1,0}$ acts on $V_s(1)$ by multiplication by $c \in \mathbb{C}$. Then the operator $(T^s_{1,0} - c \cdot \mathrm{id})(T^s_{1,0})^{n-1}$ is zero on $V_s(n)$. Moreover, if $n \geq 2$, then*

$$\mathrm{im}((T^s_{1,0})^k) = \tau_{n-1} V_s(n-k) \tag{3.96}$$

and $\dim(\mathrm{im}((T^s_{1,0})^k)) = \dim(V_s(n-k))$ *for $1 \leq k \leq n-1$.*
(3) *Let $\{\mu_1, \ldots, \mu_r\}$ be the set of non-zero eigenvalues of $T^s_{1,0}$ on $V_s(1)$, and let $\{\mu'_1, \ldots, \mu'_{r'}\}$ be the set of non-zero eigenvalues of $T^s_{1,0}$ on $V_s(n)$. Then*

$$\{\mu_1, \ldots, \mu_r\} = \{\mu'_1, \ldots, \mu'_{r'}\}. \tag{3.97}$$

Proof (1). We may assume that $n \geq 2$. Let $v \in V_s(n)$. It follows from (3.60) that

$$\sigma_{n-k}\sigma_{n-k+1} \cdots \sigma_{n-1} v = \int\limits_{\mathfrak{o}} \int\limits_{\mathfrak{o}} \int\limits_{\mathfrak{o}} \pi(\begin{bmatrix} 1 & x & y & z\varpi^{-n+k+1} \\ & 1 & & y \\ & & 1 & -x \\ & & & 1 \end{bmatrix} \eta^{-k}) v \, dx \, dy \, dz \tag{3.98}$$

for $1 \leq k \leq n-1$. Also, by (3.70),

$$(T^s_{1,0})^k v = q^{4k} \int\limits_{\mathfrak{o}} \int\limits_{\mathfrak{o}} \int\limits_{\mathfrak{o}} \pi(\begin{bmatrix} 1 & x & y & z\varpi^{-n+1} \\ & 1 & & \\ & & 1 & \\ & & & 1 \end{bmatrix} \eta^{-k}) v \, dx \, dy \, dz. \tag{3.99}$$

It follows that

$$q^{4k} \tau_{n-1} \sigma_{n-k}\sigma_{n-k+1} \cdots \sigma_{n-1} v = (T^s_{1,0})^k v \tag{3.100}$$

for $1 \leq k \leq n-1$. For $k = n-1$,

$$q^{4n-4} \tau_{n-1} \sigma_1 \sigma_2 \ldots \sigma_{n-1} v = (T^s_{1,0})^{n-1} v. \tag{3.101}$$

Since $\sigma_1\sigma_2 \ldots \sigma_{n-1} v \in V_s(1) = 0$, we conclude that $(T^s_{1,0})^{n-1} v = 0$.

(2). We may assume that $n \geq 2$. Let $v \in V_s(n)$. By applying $T^s_{1,0} - c \cdot \mathrm{id}$ to (3.101) we obtain

$$q^{4n-4}(T^s_{1,0} - c \cdot \mathrm{id})\tau_{n-1}\sigma_1\sigma_2 \dots \sigma_{n-1} v = (T^s_{1,0} - c \cdot \mathrm{id})(T^s_{1,0})^{n-1} v. \qquad (3.102)$$

By (3.86),

$$q^{4n-4}\tau_{n-1}(T^s_{1,0} - c \cdot \mathrm{id})\sigma_1\sigma_2 \dots \sigma_{n-1} v = (T^s_{1,0} - c \cdot \mathrm{id})(T^s_{1,0})^{n-1} v. \qquad (3.103)$$

By hypothesis $T^s_{1,0} - c \cdot \mathrm{id}$ acts by zero on $V_s(1)$. This proves that $(T^s_{1,0} - c \cdot \mathrm{id})(T^s_{1,0})^{n-1}$ is zero on $V_s(n)$. Equation (3.96) follows from (3.100) and Proposition 3.7.4. Lemma 3.5.4 (1) implies the last assertion.

(3). We may assume that $n \geq 2$. Let $\mu \in \{\mu'_1, \dots, \mu'_{r'}\}$. Let $v \in V_s(n)$ be a non-zero vector such that $T^s_{1,0} v = \mu v$. By (3.101), the vector

$$v_1 := \sigma_1\sigma_2 \dots \sigma_{n-1} v \in V_s(1)$$

is non-zero. We have

$$\begin{aligned}
\tau_{n-1}(T^s_{1,0} - \mu \cdot \mathrm{id})v_1 &= (T^s_{1,0} - \mu \cdot \mathrm{id})\tau_{n-1}\tau_{n-2} \cdots \tau_1 v_1 \qquad \text{(by (3.86))}\\
&= (T^s_{1,0} - \mu \cdot \mathrm{id})\tau_{n-1}\tau_{n-2} \cdots \tau_1\sigma_1\sigma_2 \dots \sigma_{n-1} v\\
&= q^{4-4n}(T^s_{1,0} - \mu \cdot \mathrm{id})(T^s_{1,0})^{n-1} v \qquad \text{(by (3.101))}\\
&= q^{4-4n}(T^s_{1,0})^{n-1}(T^s_{1,0} - \mu \cdot \mathrm{id}) v\\
&= 0.
\end{aligned}$$

Lemma 3.5.4 (1) now implies that $(T^s_{1,0} - \mu \cdot \mathrm{id})v_1 = 0$. Hence, μ is a non-zero eigenvalue for $T^s_{1,0}$ on $V_s(1)$, so that $\mu \in \{\mu_1, \dots, \mu_r\}$.

Assume now that $\mu \in \{\mu_1, \dots, \mu_r\}$. Let $v \in V_s(1)$ be a non-zero vector such that $T^s_{0,1} v = \mu v$. Then by (3.86),

$$T^s_{1,0}\tau_{n-1} v = \tau_{n-1} T^s_{1,0} v = \mu\tau_{n-1} v.$$

The vector $\tau_{n-1} v \in V_s(n)$ is non-zero by (1) of Lemma 3.5.4. Hence, μ is a non-zero eigenvalue for $T^s_{1,0}$ on $V_s(n)$, so that $\mu \in \{\mu'_1, \dots, \mu'_{r'}\}$. □

Chapter 4
Some Induced Representations

In this chapter we prove a number of important results concerning stable Klingen vectors in representations of GSp(4, F) induced from the Siegel parabolic subgroup P of GSp(4, F). These results are key ingredients for the dimension formulas proved in Chap. 5.

4.1 Double Coset Representatives

We begin by determining a system of representatives for $P\backslash \mathrm{GSp}(4, F)/\mathrm{K}_s(\mathfrak{p}^n)$ for an integer n such that $n \geq 1$. For an integer i, let

$$M_i = \begin{bmatrix} 1 & & & \\ & 1 & & \\ \varpi^i & & 1 & \\ & \varpi^i & & 1 \end{bmatrix}. \tag{4.1}$$

Let t_n be the element of GSp(4, F) defined in (3.8). The following lemma is the main result of this section.

Lemma 4.1.1 *Let n be an integer such that $n \geq 1$. Then*

$$\begin{aligned} \mathrm{GSp}(4, F) = P\mathrm{K}_s(\mathfrak{p}^n) \sqcup \bigsqcup_{0<i\leq n/2} PM_i\mathrm{K}_s(\mathfrak{p}^n) \\ \sqcup Pt_n\mathrm{K}_s(\mathfrak{p}^n) \sqcup \bigsqcup_{0<i<n/2} PM_it_n\mathrm{K}_s(\mathfrak{p}^n). \end{aligned} \tag{4.2}$$

Proof By Proposition 5.1.2 of [105] we have

$$\mathrm{GSp}(4, F) = P\mathrm{K}(\mathfrak{p}^n) \sqcup \bigsqcup_{0<i\leq n/2} PM_i\mathrm{K}(\mathfrak{p}^n). \tag{4.3}$$

J. Johnson-Leung et al., *Stable Klingen Vectors and Paramodular Newforms*,
Lecture Notes in Mathematics 2342, https://doi.org/10.1007/978-3-031-45177-5_4

Using (3.10), it follows that

$$\begin{aligned}
\mathrm{GSp}(4,F) &= \bigcup_{v\in\mathfrak{o}/\mathfrak{p}} P\begin{bmatrix}1&&v\varpi^{-n}&\\&1&&\\&&1&\\&&&1\end{bmatrix}\mathrm{K}_s(\mathfrak{p}^n)\ \cup\ Pt_n\mathrm{K}_s(\mathfrak{p}^n)\\
&\quad\cup\bigcup_{0<i\le n/2}\ \bigcup_{v\in\mathfrak{o}/\mathfrak{p}} PM_i\begin{bmatrix}1&&v\varpi^{-n}&\\&1&&\\&&1&\\&&&1\end{bmatrix}\mathrm{K}_s(\mathfrak{p}^n)\\
&\quad\cup\bigcup_{0<i\le n/2} PM_it_n\mathrm{K}_s(\mathfrak{p}^n)\\
&= P\mathrm{K}_s(\mathfrak{p}^n)\ \cup\ Pt_n\mathrm{K}_s(\mathfrak{p}^n)\\
&\quad\cup\bigcup_{0<i\le n/2}\ \bigcup_{v\in\mathfrak{o}/\mathfrak{p}} PM_i\begin{bmatrix}1&&v\varpi^{-n}&\\&1&&\\&&1&\\&&&1\end{bmatrix}\mathrm{K}_s(\mathfrak{p}^n)\\
&\quad\cup\bigcup_{0<i\le n/2} PM_it_n\mathrm{K}_s(\mathfrak{p}^n).
\end{aligned}$$

Since $\begin{bmatrix}v&&&\\&1&&\\&&v&\\&&&1\end{bmatrix}$ for $v\in\mathfrak{o}^\times$ commutes with M_i, we have

$$\begin{aligned}
\mathrm{GSp}(4,F) = &\bigcup_{0<i\le n/2} PM_i\begin{bmatrix}1&&\varpi^{-n}&\\&1&&\\&&1&\\&&&1\end{bmatrix}\mathrm{K}_s(\mathfrak{p}^n)\\
&\cup\Big(P\mathrm{K}_s(\mathfrak{p}^n)\ \cup\bigcup_{0<i\le n/2} PM_i\mathrm{K}_s(\mathfrak{p}^n)\Big)\\
&\cup\Big(Pt_n\mathrm{K}_s(\mathfrak{p}^n)\ \cup\bigcup_{0<i\le n/2} PM_it_n\mathrm{K}_s(\mathfrak{p}^n)\Big).
\end{aligned}$$

From the disjointness of (4.3) it follows that

$$\begin{aligned}
\mathrm{GSp}(4,F) = &\bigsqcup_{0<i\le n/2} PM_i\begin{bmatrix}1&&\varpi^{-n}&\\&1&&\\&&1&\\&&&1\end{bmatrix}\mathrm{K}_s(\mathfrak{p}^n)\\
&\cup\Big(P\mathrm{K}_s(\mathfrak{p}^n)\ \sqcup\bigsqcup_{0<i\le n/2} PM_i\mathrm{K}_s(\mathfrak{p}^n)\Big)\\
&\cup\Big(Pt_n\mathrm{K}_s(\mathfrak{p}^n)\ \sqcup\bigsqcup_{0<i\le n/2} PM_it_n\mathrm{K}_s(\mathfrak{p}^n)\Big).
\end{aligned}\tag{4.4}$$

If n is even, then there are additional identities

$$\begin{aligned} &M_{n/2}\begin{bmatrix} 1 & & & \varpi^{-n} \\ & 1 & & \\ & & 1 & \\ & & & 1 \end{bmatrix}\begin{bmatrix} -1 & & & \\ & 1 & & \\ & & -1 & \\ & & & 1 \end{bmatrix}\begin{bmatrix} 1 & & & \\ \varpi^{n} & 1 & 1+\varpi^{n/2} & \\ -\varpi^{n} & -1 & -\varpi^{n/2} & \\ \varpi^{n}(1-\varpi^{n/2}) & & -\varpi^{n} & 1 \end{bmatrix} \\ &= \begin{bmatrix} -1 & & & \\ & 1 & & \\ & & -1 & \\ & & & 1 \end{bmatrix}\begin{bmatrix} & \varpi^{-n/2} & & \\ -\varpi^{n/2} & 1 & & \\ & & & -\varpi^{-n/2} \\ & & \varpi^{n/2} & 1 \end{bmatrix}\begin{bmatrix} 1 & & -\varpi^{-n/2} & -\varpi^{-n} \\ & 1 & \varpi^{n/2} & -\varpi^{-n/2} \\ & & 1 & \\ & & & 1 \end{bmatrix} M_{n/2} \end{aligned} \tag{4.5}$$

and

$$\begin{aligned} &\begin{bmatrix} & -\varpi^{-n/2} & & \\ \varpi^{n/2} & & & \\ & & & \varpi^{-n/2} \\ & & -\varpi^{n/2} & \end{bmatrix}\begin{bmatrix} 1 & & -\varpi^{-n/2} & \\ & 1 & & -\varpi^{-n/2} \\ & & 1 & \\ & & & 1 \end{bmatrix} M_{n/2}t_n \\ &= M_{n/2}\begin{bmatrix} 1 & & & \\ & & -1 & \\ & 1 & & \\ & & & 1 \end{bmatrix}. \end{aligned} \tag{4.6}$$

They show that if n is even, then

$$PM_{n/2}\begin{bmatrix} 1 & & & \varpi^{-n} \\ & 1 & & \\ & & 1 & \\ & & & 1 \end{bmatrix}\mathrm{Kl}(\mathfrak{p}^n) = PM_{n/2}\mathrm{Kl}(\mathfrak{p}^n) = PM_{n/2}t_n\mathrm{Kl}(\mathfrak{p}^n). \tag{4.7}$$

Hence, if n is even, then we get from (4.4) that

$$\begin{aligned} \mathrm{GSp}(4, F) = &\bigsqcup_{0<i<n/2} PM_i\begin{bmatrix} 1 & & & \varpi^{-n} \\ & 1 & & \\ & & 1 & \\ & & & 1 \end{bmatrix}\mathrm{K}_s(\mathfrak{p}^n) \\ &\cup\Big(P\mathrm{K}_s(\mathfrak{p}^n) \sqcup \bigsqcup_{0<i\le n/2} PM_i\mathrm{K}_s(\mathfrak{p}^n)\Big) \\ &\cup\Big(Pt_n\mathrm{K}_s(\mathfrak{p}^n) \sqcup \bigsqcup_{0<i<n/2} PM_it_n\mathrm{K}_s(\mathfrak{p}^n)\Big). \end{aligned} \tag{4.8}$$

If n is odd, then we can never have $i = n/2$, so that (4.8) holds for both n even and n odd. The identity

$$\begin{aligned} &M_i\begin{bmatrix} 1 & & & \varpi^{-n} \\ & 1 & & \\ & & 1 & \\ & & & 1 \end{bmatrix}\begin{bmatrix} -1 & & & \\ & 1 & & \\ & & -1 & \\ & & & 1 \end{bmatrix}\begin{bmatrix} 1 & & & \\ & 1+\varpi^{n-2i} & -\varpi^{n-2i} & \\ & -1 & 1 & \\ \varpi^{n}+\varpi^{2n-2i} & & & 1 \end{bmatrix} \\ &= \begin{bmatrix} -1 & & & \\ & 1 & & \\ & & -1 & \\ & & & 1 \end{bmatrix}\begin{bmatrix} 1 & \varpi^{-i} & & \\ \varpi^{n-i} & 1+\varpi^{n-2i} & & \\ & & 1 & -\varpi^{-i} \\ & & -\varpi^{n-i} & 1+\varpi^{n-2i} \end{bmatrix}\begin{bmatrix} 1 & & \varpi^{n-3i} & -\varpi^{n-4i}-\varpi^{-2i} \\ & 1 & -\varpi^{n-2i} & \varpi^{n-3i} \\ & & 1 & \\ & & & 1 \end{bmatrix} M_it_n \end{aligned} \tag{4.9}$$

shows that

$$PM_i \begin{bmatrix} 1 & & \varpi^{-n} & \\ & 1 & & \\ & & 1 & \\ & & & 1 \end{bmatrix} \mathrm{Kl}(\mathfrak{p}^n) = PM_i t_n \mathrm{Kl}(\mathfrak{p}^n) \tag{4.10}$$

for $0 < i \le n/2$. Hence, we can omit the first line on the right-hand side of (4.8), and get

$$\begin{aligned} \mathrm{GSp}(4, F) = &\Big(P\mathrm{K}_s(\mathfrak{p}^n) \sqcup \bigsqcup_{0<i\le n/2} PM_i\mathrm{K}_s(\mathfrak{p}^n) \Big) \\ &\cup \Big(Pt_n\mathrm{K}_s(\mathfrak{p}^n) \sqcup \bigsqcup_{0<i<n/2} PM_i t_n\mathrm{K}_s(\mathfrak{p}^n) \Big). \end{aligned} \tag{4.11}$$

It follows from (4.3) and (4.11) that the only possible equalities between the involved double cosets are

$$PM_i\mathrm{K}_s(\mathfrak{p}^n) = PM_i t_n\mathrm{K}_s(\mathfrak{p}^n) \tag{4.12}$$

for some i with $0 < i < n/2$, or

$$P\mathrm{K}_s(\mathfrak{p}^n) = Pt_n\mathrm{K}_s(\mathfrak{p}^n). \tag{4.13}$$

Assume that $0 < i < n/2$, and that

$$M_i k = pM_i t_n. \tag{4.14}$$

for some $k \in \mathrm{K}_s(\mathfrak{p}^n)$ and $p \in P$. Then

$$M_i^{-1} pM_i t_n = k \in \mathrm{K}_s(\mathfrak{p}^n). \tag{4.15}$$

Write

$$p = \begin{bmatrix} A & \\ & uA' \end{bmatrix} \begin{bmatrix} 1 & B \\ & 1 \end{bmatrix} \quad \text{with} \quad A = \begin{bmatrix} a_1 & a_2 \\ a_3 & a_4 \end{bmatrix} \quad \text{and} \quad B = \begin{bmatrix} b_1 & b_2 \\ b_3 & b_1 \end{bmatrix},$$

where $a_1, a_2, a_3, a_4, b_1, b_2, b_3 \in F$ and $u \in F^\times$. Let k_{ij}, $1 \le i, j \le 4$, be the entries of the matrix k in (4.15). Calculations show that

$$k_{14} + \varpi^{i-n} k_{13} = -a_1\varpi^{-n}. \tag{4.16}$$

Since k is in $\mathrm{K}_s(\mathfrak{p}^n)$, we obtain

$$-a_1\varpi^{-n} \in \mathfrak{p}^{-n+1} + \mathfrak{p}^{i-n} = \mathfrak{p}^{-n+1}, \tag{4.17}$$

which implies that

$$a_1 \in \mathfrak{p}. \tag{4.18}$$

It follows from (4.15) that

$$p = \left[\begin{smallmatrix} A & \\ & uA' \end{smallmatrix}\right]\left[\begin{smallmatrix} 1 & B \\ & 1 \end{smallmatrix}\right] \in P \cap M_i \mathrm{K}(\mathfrak{p}^n) M_i^{-1}. \tag{4.19}$$

By Lemma 5.2.1 of [105], we have

$$A = \left[\begin{smallmatrix} a_1 & a_2 \\ a_3 & a_4 \end{smallmatrix}\right] \in \begin{bmatrix} \mathfrak{o} & \mathfrak{p}^{-i} \\ \mathfrak{p}^{n-i} & \mathfrak{o} \end{bmatrix} \quad \text{and} \quad \det(A) \in \mathfrak{o}^\times. \tag{4.20}$$

Since $\mathfrak{p}^{n-2i} \subset \mathfrak{p}$ by assumption, it follows that $a_1, a_4 \in \mathfrak{o}^\times$. Now we have a contradiction with (4.18), proving that (4.12) is impossible.

Now assume that (4.13) holds. Then $pt_n \in \mathrm{K}_s(\mathfrak{p}^n)$ for some $p \in P$. Hence, also $pt_n s_2 \in \mathrm{K}_s(\mathfrak{p}^n)$. Calculations show that $pt_n s_2$ is of the form $\left[\begin{smallmatrix} A & B \\ C & 0 \end{smallmatrix}\right]$ for some 2×2 matrices A, B, and C with entries from F. We claim that $\mathrm{K}_s(\mathfrak{p}^n)$ does not contain any matrices of this kind. If $\left[\begin{smallmatrix} A & B \\ C & 0 \end{smallmatrix}\right] \in \mathrm{K}_s(\mathfrak{p}^n)$, then $\det(B)\det(C) \in \mathfrak{o}^\times$. On the other hand, $\det(B) \in \mathfrak{p}^{-n+1}$ and $\det(C) \in \mathfrak{p}^n$ by (3.3). Hence, $\det(B)\det(C) \in \mathfrak{p}$, a contradiction. This proves our claim, and concludes the proof. □

Remark 4.1.2 It follows from (3.13) that

$$P\mathrm{K}(\mathfrak{p}^n) = P\mathrm{K}_s(\mathfrak{p}^n) \sqcup Pt_n\mathrm{K}_s(\mathfrak{p}^n) \tag{4.21}$$

(the disjointness follows from Lemma 4.1.1). It follows from (3.13) and (4.10) that

$$PM_i\mathrm{K}(\mathfrak{p}^n) = PM_i\mathrm{K}_s(\mathfrak{p}^n) \sqcup PM_it_n\mathrm{K}_s(\mathfrak{p}^n) \quad \text{for } 0 < i < n/2. \tag{4.22}$$

It follows from (3.13) and (4.7) that if n is even, then

$$PM_{n/2}\mathrm{K}(\mathfrak{p}^n) = PM_{n/2}\mathrm{K}_s(\mathfrak{p}^n). \tag{4.23}$$

4.2 Stable Klingen Vectors in Siegel Induced Representations

Let (π, W) be an admissible representation of $\mathrm{GL}(2, F)$ admitting a central character ω_π, and let σ be a character of $F^\times$. Assume that $\omega_\pi\sigma^2 = 1$, so that the induced representation $\pi \rtimes \sigma$ has trivial central character. Let V be the standard model of $\pi \rtimes \sigma$. In this section we will determine the stable Klingen level of $\pi \rtimes \sigma$, and, in the case π is irreducible, the dimensions of the vector spaces $V_s(n)$ of stable

Klingen vectors for all integers $n \geq 0$. Explicitly, V consists of the locally constant functions $f : \mathrm{GSp}(4, F) \to W$ such that

$$f(\begin{bmatrix} A & * \\ & uA' \end{bmatrix} g) = |u^{-1} \det(A)|^{3/2} \sigma(u) \pi(A) f(g) \tag{4.24}$$

for $\begin{bmatrix} A & * \\ & uA' \end{bmatrix} \in P$ and $g \in \mathrm{GSp}(4, F)$ (see Sect. 2.2). Consequently, if n is an integer such that $n \geq 1$, and $f \in V_s(n)$, then by Lemma 4.1.1 and (4.24), f is determined by its values on the set

$$X = \{1, t_n\} \cup \{M_i \mid 0 < i \leq n/2\} \cup \{M_i t_n \mid 0 < i < n/2\}. \tag{4.25}$$

The main result of this section is Theorem 4.2.7. We begin with some lemmas that will be useful in determining the elements of $V_s(n)$ supported on a particular element of X.

Lemma 4.2.1 *Let n be an integer such that $n \geq 0$. Let $g \in \mathrm{GSp}(4, F)$, $u \in F^\times$, and $A \in \mathrm{GL}(2, F)$. If there exists $X \in \mathrm{M}(2, F)$ of the form $X = \begin{bmatrix} x & y \\ z & x \end{bmatrix}$ such that $\begin{bmatrix} A & \\ & uA' \end{bmatrix} \begin{bmatrix} 1 & X \\ & 1 \end{bmatrix} \in P \cap g\mathrm{K}_s(\mathfrak{p}^n) g^{-1}$, then $u \in \mathfrak{o}^\times$ and $\det(A) \in \mathfrak{o}^\times$.*

Proof Consider the homomorphism

$$P \cap g\mathrm{K}_s(\mathfrak{p}^n) g^{-1} \longrightarrow F^\times \times F^\times \quad \text{defined by} \quad \begin{bmatrix} B & * \\ & vB' \end{bmatrix} \longmapsto (v, \det(B)).$$

Its image is a compact subgroup of $F^\times \times F^\times$, and is therefore contained in $\mathfrak{o}^\times \times \mathfrak{o}^\times$. This implies that $u \in \mathfrak{o}^\times$ and $\det(A) \in \mathfrak{o}^\times$. □

Lemma 4.2.2 *Let n be an integer such that $n \geq 0$. Let $u \in F^\times$ and $A \in \mathrm{GL}(2, F)$. The following statements are equivalent*

(1) *There exists an $X \in \mathrm{M}(2, F)$ of the form $X = \begin{bmatrix} x & y \\ z & x \end{bmatrix}$ such that*

$$\begin{bmatrix} A & \\ & uA' \end{bmatrix} \begin{bmatrix} 1 & X \\ & 1 \end{bmatrix} \in P \cap \mathrm{K}_s(\mathfrak{p}^n). \tag{4.26}$$

(2) *The following conditions are satisfied:*

- $u \in \mathfrak{o}^\times$,
- $\det(A) \in \mathfrak{o}^\times$,
- $A \in \begin{bmatrix} \mathfrak{o} & \mathfrak{o} \\ \mathfrak{p}^n & \mathfrak{o} \end{bmatrix}$.

Proof Assume that (1) holds. Then $u \in \mathfrak{o}^\times$ and $\det(A) \in \mathfrak{o}^\times$ by Lemma 4.2.1. The assumption $\begin{bmatrix} A & \\ & uA' \end{bmatrix} \begin{bmatrix} 1 & X \\ & 1 \end{bmatrix} \in P \cap \mathrm{K}_s(\mathfrak{p}^n)$, along with (3.3), implies that $A \in \begin{bmatrix} \mathfrak{o} & \mathfrak{o} \\ \mathfrak{p}^n & \mathfrak{o} \end{bmatrix}$. Thus, (2) holds. If (2) holds, then (1) holds with $X = 0$. □

Lemma 4.2.3 *Let n be an integer such that $n \geq 0$. Let $u \in F^\times$ and $A \in \mathrm{GL}(2, F)$. The following statements are equivalent:*

(1) *There exists an $X \in \mathrm{M}(2, F)$ of the form $X = \left[\begin{smallmatrix} x & y \\ z & x \end{smallmatrix}\right]$ such that*

$$\left[\begin{smallmatrix} A & \\ & uA' \end{smallmatrix}\right]\left[\begin{smallmatrix} 1 & X \\ & 1 \end{smallmatrix}\right] \in P \cap t_n \mathrm{K}_s(\mathfrak{p}^n) t_n^{-1}. \tag{4.27}$$

(2) *The following conditions are satisfied:*

- $u \in \mathfrak{o}^\times$,
- $\det(A) \in \mathfrak{o}^\times$,
- $A \in \left[\begin{smallmatrix} \mathfrak{o} & \mathfrak{o} \\ \mathfrak{p}^n & \mathfrak{o} \end{smallmatrix}\right]$.

Proof The proof is similar to the proof of Lemma 4.2.2. □

Lemma 4.2.4 *Let n be an integer such that $n \geq 1$. Let $u \in F^\times$ and $A = \left[\begin{smallmatrix} a & b \\ c & d \end{smallmatrix}\right] \in \mathrm{GL}(2, F)$ be given. For $1 \leq i < n/2$ the following statements are equivalent:*

(1) *There exists an $X \in \mathrm{M}(2, F)$ of the form $X = \left[\begin{smallmatrix} x & y \\ z & x \end{smallmatrix}\right]$ such that*

$$\left[\begin{smallmatrix} A & \\ & uA' \end{smallmatrix}\right]\left[\begin{smallmatrix} 1 & X \\ & 1 \end{smallmatrix}\right] \in P \cap M_i \mathrm{K}_s(\mathfrak{p}^n) M_i^{-1}. \tag{4.28}$$

Here M_i is as in (4.1).

(2) *The following conditions are satisfied:*

- $u \in \mathfrak{o}^\times$,
- $\det(A) \in \mathfrak{o}^\times$,
- $u \det(A)^{-1} \in 1 + \mathfrak{p}^i$,
- $A \in \left[\begin{smallmatrix} \mathfrak{o} & \mathfrak{p}^{-i} \\ \mathfrak{p}^{n-i} & \mathfrak{o} \end{smallmatrix}\right]$.

Proof We begin with a preliminary observation. Let $X = \left[\begin{smallmatrix} x & y \\ z & x \end{smallmatrix}\right] \in \mathrm{M}(2, F)$. Then using (3.3), a calculation shows that $\left[\begin{smallmatrix} A & \\ & uA' \end{smallmatrix}\right]\left[\begin{smallmatrix} 1 & X \\ & 1 \end{smallmatrix}\right] \in P \cap M_i \mathrm{K}_s(\mathfrak{p}^n) M_i^{-1}$, or equivalently, $M_i^{-1}\left[\begin{smallmatrix} A & \\ & uA' \end{smallmatrix}\right]\left[\begin{smallmatrix} 1 & X \\ & 1 \end{smallmatrix}\right] M_i \in \mathrm{K}_s(\mathfrak{p}^n)$, if and only if $u \in \mathfrak{o}^\times$ and the following (4.29)–(4.32) are fulfilled:

$$A(1 + \varpi^i X) \in \left[\begin{smallmatrix} \mathfrak{o} & \mathfrak{o} \\ \mathfrak{p}^n & \mathfrak{o} \end{smallmatrix}\right], \tag{4.29}$$

$$AX \in \left[\begin{smallmatrix} \mathfrak{o} & \mathfrak{p}^{-n+1} \\ \mathfrak{o} & \mathfrak{o} \end{smallmatrix}\right], \tag{4.30}$$

$$uA' - A - \varpi^i AX \in \left[\begin{smallmatrix} \mathfrak{p}^{n-i} & \mathfrak{p}^{-i} \\ \mathfrak{p}^{n-i} & \mathfrak{p}^{n-i} \end{smallmatrix}\right], \tag{4.31}$$

$$uA' - \varpi^i AX \in \left[\begin{smallmatrix} \mathfrak{o} & \mathfrak{o} \\ \mathfrak{p}^n & \mathfrak{o} \end{smallmatrix}\right]. \tag{4.32}$$

Now assume that (1) holds; we will prove (2). By the previous paragraph, (4.29)–(4.32) hold. By Lemma 4.2.1 we have $u \in \mathfrak{o}^\times$ and $\det(A) \in \mathfrak{o}^\times$. By (4.31) and (4.32) we get

$$A = \left[\begin{smallmatrix} a & b \\ c & d \end{smallmatrix}\right] \in \begin{bmatrix} \mathfrak{p}^{n-i} & \mathfrak{p}^{-i} \\ \mathfrak{p}^{n-i} & \mathfrak{p}^{n-i} \end{bmatrix} + \begin{bmatrix} \mathfrak{o} & \mathfrak{o} \\ \mathfrak{p}^{n} & \mathfrak{o} \end{bmatrix} = \begin{bmatrix} \mathfrak{o} & \mathfrak{p}^{-i} \\ \mathfrak{p}^{n-i} & \mathfrak{o} \end{bmatrix},$$

hence $a, d \in \mathfrak{o}$, $b \in \mathfrak{p}^{-i}$ and $c \in \mathfrak{p}^{n-i}$. Since $i < n/2$, we have $bc \in \mathfrak{p}^{n-2i} \subset \mathfrak{p}$, and it follows from $ad - bc \in \mathfrak{o}^\times$ that $a, d \in \mathfrak{o}^\times$. By (4.30) and (4.31) we get

$$uA' - A \in \begin{bmatrix} \mathfrak{p}^{n-i} & \mathfrak{p}^{-i} \\ \mathfrak{p}^{n-i} & \mathfrak{p}^{n-i} \end{bmatrix} + \begin{bmatrix} \mathfrak{p}^{i} & \mathfrak{p}^{i-n+1} \\ \mathfrak{p}^{i} & \mathfrak{p}^{i} \end{bmatrix} = \begin{bmatrix} \mathfrak{p}^{i} & \mathfrak{p}^{i-n+1} \\ \mathfrak{p}^{i} & \mathfrak{p}^{i} \end{bmatrix}.$$

Since $A' = \frac{1}{ad-bc}\left[\begin{smallmatrix} a & -b \\ -c & d \end{smallmatrix}\right]$, it follows that $(\frac{u}{ad-bc} - 1)a \in \mathfrak{p}^i$, hence $\frac{u}{ad-bc} \in 1 + \mathfrak{p}^i$. Therefore all the conditions in (2) are fulfilled.

Assume that (2) holds. We define

$$X := \varpi^{-i}\Big(\frac{u}{ad-bc} A^{-1} \begin{bmatrix} a & \varpi^i b \\ \varpi^i c & d \end{bmatrix} - 1\Big).$$

It is then easy to verify that X has the required form and that (4.29)–(4.32) are fulfilled. By the first paragraph, (1) holds. □

Lemma 4.2.5 *Let n be an integer such that $n \geq 1$. Let $u \in F^\times$ and $A = \left[\begin{smallmatrix} a & b \\ c & d \end{smallmatrix}\right] \in \mathrm{GL}(2, F)$ be given. For $1 \leq i < n/2$ the following statements are equivalent:*

(1) *There exists an $X \in \mathrm{M}(2, F)$ of the form $X = \left[\begin{smallmatrix} x & y \\ z & x \end{smallmatrix}\right]$ such that*

$$\left[\begin{smallmatrix} A & \\ & uA' \end{smallmatrix}\right]\left[\begin{smallmatrix} 1 & X \\ & 1 \end{smallmatrix}\right] \in P \cap M_i t_n \mathrm{K}_s(\mathfrak{p}^n) t_n^{-1} M_i^{-1}. \tag{4.33}$$

Here M_i is as in (4.1).

(2) *The following conditions are satisfied:*

- $u \in \mathfrak{o}^\times$,
- $\det(A) \in \mathfrak{o}^\times$,
- $u \det(A)^{-1} \in 1 + \mathfrak{p}^i$,
- $A \in \begin{bmatrix} \mathfrak{o} & \mathfrak{p}^{-i} \\ \mathfrak{p}^{n-i+1} & \mathfrak{o} \end{bmatrix}$.

Proof We begin with a preliminary observation. Let $X = \left[\begin{smallmatrix} x & y \\ z & x \end{smallmatrix}\right] \in \mathrm{M}(2, F)$. Then using (3.3), a calculation shows that $\left[\begin{smallmatrix} A & \\ & uA' \end{smallmatrix}\right]\left[\begin{smallmatrix} 1 & X \\ & 1 \end{smallmatrix}\right] \in P \cap M_i t_n \mathrm{K}_s(\mathfrak{p}^n) t_n^{-1} M_i^{-1}$, or equivalently, $M_i^{-1}\left[\begin{smallmatrix} A & \\ & uA' \end{smallmatrix}\right]\left[\begin{smallmatrix} 1 & X \\ & 1 \end{smallmatrix}\right] M_i \in t_n \mathrm{K}_s(\mathfrak{p}^n) t_n^{-1}$, if and only if $u \in \mathfrak{o}^\times$ and the following (4.34)–(4.37) are fulfilled:

$$A(1 + \varpi^i X) \in \begin{bmatrix} \mathfrak{o} & \mathfrak{o} \\ \mathfrak{p}^n & \mathfrak{o} \end{bmatrix}, \tag{4.34}$$

$$AX \in \begin{bmatrix} \mathfrak{o} & \mathfrak{p}^{-n} \\ \mathfrak{o} & \mathfrak{o} \end{bmatrix}, \tag{4.35}$$

$$uA' - A - \varpi^i AX \in \begin{bmatrix} \mathfrak{p}^{n-i} & \mathfrak{p}^{-i} \\ \mathfrak{p}^{n-i+1} & \mathfrak{p}^{n-i} \end{bmatrix}, \tag{4.36}$$

$$uA' - \varpi^i AX \in \begin{bmatrix} \mathfrak{o} & \mathfrak{o} \\ \mathfrak{p}^n & \mathfrak{o} \end{bmatrix}. \tag{4.37}$$

Now assume that (1) holds; we will prove (2). By the previous paragraph, (4.34)–(4.37) hold. By Lemma 4.2.1 we have $u \in \mathfrak{o}^\times$ and $\det(A) \in \mathfrak{o}^\times$. By (4.36) and (4.37) we get

$$A = \begin{bmatrix} a & b \\ c & d \end{bmatrix} \in \begin{bmatrix} \mathfrak{p}^{n-i} & \mathfrak{p}^{-i} \\ \mathfrak{p}^{n-i+1} & \mathfrak{p}^{n-i} \end{bmatrix} + \begin{bmatrix} \mathfrak{o} & \mathfrak{o} \\ \mathfrak{p}^n & \mathfrak{o} \end{bmatrix} = \begin{bmatrix} \mathfrak{o} & \mathfrak{p}^{-i} \\ \mathfrak{p}^{n-i+1} & \mathfrak{o} \end{bmatrix},$$

hence $a, d \in \mathfrak{o}$, $b \in \mathfrak{p}^{-i}$, and $c \in \mathfrak{p}^{n-i+1}$. Since $i < n/2$, we have $bc \in \mathfrak{p}^{n-2i+1} \subset \mathfrak{p}$, and it follows from $ad - bc \in \mathfrak{o}^\times$ that $a, d \in \mathfrak{o}^\times$. By (4.35) and (4.36) we get

$$uA' - A \in \begin{bmatrix} \mathfrak{p}^{n-i} & \mathfrak{p}^{-i} \\ \mathfrak{p}^{n-i+1} & \mathfrak{p}^{n-i} \end{bmatrix} + \begin{bmatrix} \mathfrak{p}^i & \mathfrak{p}^{i-n} \\ \mathfrak{p}^i & \mathfrak{p}^i \end{bmatrix} = \begin{bmatrix} \mathfrak{p}^i & \mathfrak{p}^{i-n} \\ \mathfrak{p}^i & \mathfrak{p}^i \end{bmatrix}.$$

Since $A' = \frac{1}{ad-bc}\begin{bmatrix} a & -b \\ -c & d \end{bmatrix}$, it follows that $(\frac{u}{ad-bc} - 1)a \in \mathfrak{p}^i$, hence $\frac{u}{ad-bc} \in 1 + \mathfrak{p}^i$. Therefore all the conditions in (2) are fulfilled.

Assume that (2) holds. We define

$$X := \varpi^{-i}\Big(\frac{u}{ad - bc} A^{-1} \begin{bmatrix} a & \varpi^i b \\ \varpi^i c & d \end{bmatrix} - 1\Big).$$

It is then easy to verify that X has the required form and that (4.34)–(4.37) are fulfilled. By the first paragraph, (1) holds. □

Lemma 4.2.6 *Let n be an even integer such that $n \geq 2$. Let $u \in F^\times$ and $A = \begin{bmatrix} a & b \\ c & d \end{bmatrix} \in \mathrm{GL}(2, F)$ be given. The following statements are equivalent:*

(1) *There exists an $X \in \mathrm{M}(2, F)$ of the form $X = \begin{bmatrix} x & y \\ z & x \end{bmatrix}$ such that*

$$\begin{bmatrix} A & \\ & uA' \end{bmatrix}\begin{bmatrix} 1 & X \\ & 1 \end{bmatrix} \in P \cap M_{n/2}\mathrm{K}_s(\mathfrak{p}^n)M_{n/2}^{-1}. \tag{4.38}$$

(2) *The following conditions are satisfied:*

- $u \in \mathfrak{o}^\times$,
- $\det(A) \in \mathfrak{o}^\times$,
- $u\det(A)^{-1} \in 1 + \mathfrak{p}^{n/2}$,
- $A \in \begin{bmatrix} \mathfrak{o} & \mathfrak{p}^{-n/2+1} \\ \mathfrak{p}^{n/2} & \mathfrak{o} \end{bmatrix}$.

Proof We begin with a preliminary observation. Let $X = \begin{bmatrix} x & y \\ z & x \end{bmatrix} \in \mathrm{M}(2, F)$. Then using (3.3), a calculation shows that

$$\begin{bmatrix} A & \\ & uA' \end{bmatrix}\begin{bmatrix} 1 & X \\ & 1 \end{bmatrix} \in P \cap M_{n/2}\mathrm{K}_s(\mathfrak{p}^n)M_{n/2}^{-1},$$

or equivalently, $M_{n/2}^{-1}\left[\begin{smallmatrix} A & \\ & uA' \end{smallmatrix}\right]\left[\begin{smallmatrix} 1 & X \\ & 1 \end{smallmatrix}\right] M_{n/2} \in \mathrm{K}_s(\mathfrak{p}^n)$, if and only if $u \in \mathfrak{o}^\times$ and the following (4.39)–(4.42) are fulfilled:

$$A(1 + \varpi^{n/2} X) \in \left[\begin{smallmatrix} \mathfrak{o} & \mathfrak{o} \\ \mathfrak{p}^n & \mathfrak{o} \end{smallmatrix}\right], \tag{4.39}$$

$$AX \in \left[\begin{smallmatrix} \mathfrak{o} & \mathfrak{p}^{-n+1} \\ \mathfrak{o} & \mathfrak{o} \end{smallmatrix}\right], \tag{4.40}$$

$$uA' - A - \varpi^{n/2} AX \in \left[\begin{smallmatrix} \mathfrak{p}^{n/2} & \mathfrak{p}^{-n/2} \\ \mathfrak{p}^{n/2} & \mathfrak{p}^{n/2} \end{smallmatrix}\right], \tag{4.41}$$

$$uA' - \varpi^{n/2} AX \in \left[\begin{smallmatrix} \mathfrak{o} & \mathfrak{o} \\ \mathfrak{p}^n & \mathfrak{o} \end{smallmatrix}\right]. \tag{4.42}$$

Assume that (1) holds; we will prove (2). By the previous paragraph, (4.39)–(4.42) hold. By Lemma 4.2.1 we have $u \in \mathfrak{o}^\times$ and $\det(A) \in \mathfrak{o}^\times$. From (4.40) and (4.42) we get

$$uA' = \frac{u}{ad-bc}\left[\begin{smallmatrix} a & -b \\ -c & d \end{smallmatrix}\right] \in \left[\begin{smallmatrix} \mathfrak{p}^{n/2} & \mathfrak{p}^{-n/2+1} \\ \mathfrak{p}^{n/2} & \mathfrak{p}^{n/2} \end{smallmatrix}\right] + \left[\begin{smallmatrix} \mathfrak{o} & \mathfrak{o} \\ \mathfrak{p}^n & \mathfrak{o} \end{smallmatrix}\right] = \left[\begin{smallmatrix} \mathfrak{o} & \mathfrak{p}^{-n/2+1} \\ \mathfrak{p}^{n/2} & \mathfrak{o} \end{smallmatrix}\right],$$

hence $a, d \in \mathfrak{o}$, $b \in \mathfrak{p}^{-n/2+1}$, and $c \in \mathfrak{p}^{n/2}$. Since $bc \in \mathfrak{p}$ and $ad - bc \in \mathfrak{o}^\times$, we have $a, d \in \mathfrak{o}^\times$. By (4.40) and (4.41) we get

$$uA' - A \in \left[\begin{smallmatrix} \mathfrak{p}^{n/2} & \mathfrak{p}^{-n/2} \\ \mathfrak{p}^{n/2} & \mathfrak{p}^{n/2} \end{smallmatrix}\right] + \left[\begin{smallmatrix} \mathfrak{p}^{n/2} & \mathfrak{p}^{-n/2+1} \\ \mathfrak{p}^{n/2} & \mathfrak{p}^{n/2} \end{smallmatrix}\right] = \left[\begin{smallmatrix} \mathfrak{p}^{n/2} & \mathfrak{p}^{-n/2} \\ \mathfrak{p}^{n/2} & \mathfrak{p}^{n/2} \end{smallmatrix}\right].$$

Since $A' = \frac{1}{ad-bc}\left[\begin{smallmatrix} a & -b \\ -c & d \end{smallmatrix}\right]$, it follows that $(\frac{u}{ad-bc} - 1)a \in \mathfrak{p}^{n/2}$, and hence $\frac{u}{ad-bc} \in 1 + \mathfrak{p}^{n/2}$. Therefore, all the conditions in (2) are satisfied.

Assume that (2) holds. We define

$$X := \varpi^{-n/2}\left(\frac{u}{ad-bc} A^{-1}\left[\begin{smallmatrix} a & \varpi^{n/2} b \\ \varpi^{n/2} c & d \end{smallmatrix}\right] - 1\right).$$

It is then easy to verify that X has the required form and that (4.39)–(4.42) are fulfilled. By the first paragraph, (1) holds. □

Let n be an integer such that $n \geq 0$. If $n = 0$, then we define $\Gamma_0(\mathfrak{p}^n) = \mathrm{GL}(2, \mathfrak{o})$. If $n > 0$, then we define

$$\Gamma_0(\mathfrak{p}^n) := \left\{\left[\begin{smallmatrix} a & b \\ c & d \end{smallmatrix}\right] \in \mathrm{GL}(2, \mathfrak{o}) \mid c \in \mathfrak{p}^n\right\}. \tag{4.43}$$

Evidently, $\Gamma_0(\mathfrak{p}^n)$ is a subgroup of $\mathrm{GL}(2, \mathfrak{o})$. Let (τ, W) be an admissible representation of $\mathrm{GL}(2, F)$. Assume that τ admits a central character, and that this character is trivial. We define $\tau^{\Gamma_0(\mathfrak{p}^n)} = \{w \in W \mid \tau(k)w = w \text{ for } k \in \Gamma_0(\mathfrak{p}^n)\}$. If $\tau^{\Gamma_0(\mathfrak{p}^m)} \neq 0$ for some m, then we let N_τ be the smallest such m; for convenience, if $\tau^{\Gamma_0(\mathfrak{p}^m)} = 0$ for all $m \geq 0$, then we also define $N_\tau = \infty$. We call N_τ the *level* of τ.

Theorem 4.2.7 *Let (π, W) be an admissible representation of* $\mathrm{GL}(2, F)$ *admitting a central character ω_π, and let σ be a character of $F^\times$. We assume that $\omega_\pi \sigma^2 = 1$, so that the induced representation $\pi \rtimes \sigma$ has trivial central character. Let V be the space of $\pi \rtimes \sigma$.*

(1) *The minimal stable Klingen level of $\pi \rtimes \sigma$ is*

$$N_{\pi \rtimes \sigma, s} = \begin{cases} N_{\sigma\pi} & \text{if } a(\sigma) = 0, \\ 2a(\sigma) & \text{if } a(\sigma) > 0 \text{ and } N_{\sigma\pi} = 0, \\ N_{\sigma\pi} + 2a(\sigma) - 1 & \text{if } a(\sigma) > 0 \text{ and } N_{\sigma\pi} > 0. \end{cases} \tag{4.44}$$

If this number is finite, then

$$\dim V_s(N_{\pi \rtimes \sigma, s}) = \begin{cases} \dim \left(V_{\sigma\pi}^{\mathrm{GL}(2,\mathfrak{o})}\right) & \text{if } a(\sigma) = 0 \text{ and } N_{\sigma\pi} = 0, \\ 2 \dim \left(V_{\sigma\pi}^{\Gamma_0(\mathfrak{p}^{N_{\sigma\pi}})}\right) & \text{if } a(\sigma) = 0 \text{ and } N_{\sigma\pi} > 0, \\ \dim \left(V_{\sigma\pi}^{\Gamma_0(\mathfrak{p})}\right) & \text{if } a(\sigma) > 0 \text{ and } N_{\sigma\pi} = 0, \\ \dim \left(V_{\sigma\pi}^{\Gamma_0(\mathfrak{p}^{N_{\sigma\pi}})}\right) & \text{if } a(\sigma) > 0 \text{ and } N_{\sigma\pi} > 0. \end{cases}$$

(2) *Assume that π is irreducible and infinite-dimensional. If $a(\sigma) = 0$ and $N_{\sigma\pi} = 0$, then*

$$\dim V_s(n) = \frac{n^2 + 5n + 2}{2} \quad \text{for all integers } n \geq 0.$$

If $a(\sigma) = 0$ and $N_{\sigma\pi} > 0$, then for all integers $n \geq 0$

$$\dim V_s(n) = \begin{cases} \dfrac{(n - N_{\pi \rtimes \sigma, s} + 1)(n - N_{\pi \rtimes \sigma, s} + 4)}{2} & \text{if } n \geq N_{\pi \rtimes \sigma, s}, \\ 0 & \text{if } n < N_{\pi \rtimes \sigma, s}. \end{cases}$$

If $a(\sigma) > 0$ and $N_{\sigma\pi} = 0$, then for all integers $n \geq 0$

$$\dim V_s(n) = \begin{cases} \dfrac{(n - N_{\pi \rtimes \sigma, s} + 1)(n - N_{\pi \rtimes \sigma, s} + 4)}{2} & \text{if } n \geq N_{\pi \rtimes \sigma, s}, \\ 0 & \text{if } n < N_{\pi \rtimes \sigma, s}. \end{cases}$$

If $a(\sigma) > 0$ and $N_{\sigma\pi} > 0$, then for all integers $n \geq 0$

$$\dim V_s(n) = \begin{cases} \dfrac{(n - N_{\pi \rtimes \sigma, s} + 1)(n - N_{\pi \rtimes \sigma, s} + 2)}{2} & \text{if } n \geq N_{\pi \rtimes \sigma, s}, \\ 0 & \text{if } n < N_{\pi \rtimes \sigma, s}. \end{cases}$$

(3) *Assume that* $\pi = \chi 1_{\mathrm{GL}(2)}$ *for a character* χ *of* $F^\times$ *and that* $\sigma\chi$ *is unramified. Then*

$$\dim V_s(n) = n - N_{\pi \rtimes \sigma, s} + 1 \qquad \text{for all } n \geq N_{\pi \rtimes \sigma, s}.$$

(4) *If* $\pi = \chi 1_{\mathrm{GL}(2)}$ *for a character* χ *of* $F^\times$ *and* $\sigma\chi$ *is ramified, then* $V_s(n) = 0$ *for all integers* $n \geq 0$.

Proof Let W be the space of π. We note that since $\omega_\pi \sigma^2 = 1$, the representation $\sigma\pi$, which also has space W, has trivial central character. Thus, the remarks preceding the statement of the theorem apply to $\sigma\pi$. Let n be an integer such that $n \geq 1$. Our first task is to find a formula for $\dim V_s(n)$ in terms of σ and π. By Lemma 4.1.1 and (4.24), the elements of $V_s(n)$ are determined by their values on the set X from (4.25). It follows that

$$\dim V_s(n) = \sum_{g \in X} \dim V_s(n)_g,$$

where $V_s(n)_g = \{f \in V_s(n) \mid f \text{ is supported on } P g \mathrm{K}_s(\mathfrak{p}^n)\}$ for $g \in X$. Let $g \in X$. Define $C_g : V_s(n)_g \to W$ by $C_g(f) = f(g)$ for $f \in V_s(n)_g$. This linear map is injective; therefore, $\dim V_s(n)_g = \dim \mathrm{im}(C_g)$. It is straightforward to verify that $\mathrm{im}(C_g) = W_g$, where

$$W_g = \{w \in W \mid \sigma(u)\pi(A)w = w \text{ for } \left[\begin{smallmatrix} A & * \\ & uA' \end{smallmatrix}\right] \in P \cap g\mathrm{K}_s(\mathfrak{p}^n)g^{-1}\}. \tag{4.45}$$

For this, note that if $\left[\begin{smallmatrix} A & * \\ & uA' \end{smallmatrix}\right] \in P \cap g\mathrm{K}_s(\mathfrak{p}^n)g^{-1}$, then $u, \det(A) \in \mathfrak{o}^\times$ by Lemma 4.2.1. Hence,

$$\dim V_s(n) = \sum_{g \in X} \dim W_g. \tag{4.46}$$

Next, we calculate the dimensions of the spaces W_g for $g \in X$.

Assume that $g = 1$. By (4.45) and Lemma 4.2.2, we have

$$W_1 = \{w \in W \mid \sigma(u)\pi(A)w = w \text{ for } u \in \mathfrak{o}^\times, A \in \Gamma_0(\mathfrak{p}^n)\}.$$

It follows that if σ is ramified, then $W_1 = 0$, and if σ is unramified, then

$$W_1 = (\sigma\pi)^{\Gamma_0(\mathfrak{p}^n)}. \tag{4.47}$$

Thus,

$$\dim W_1 = \begin{cases} 0 & \text{if } \sigma \text{ is ramified,} \\ \dim(\sigma\pi)^{\Gamma_0(\mathfrak{p}^n)} & \text{if } \sigma \text{ is unramified.} \end{cases} \tag{4.48}$$

A similar argument using (4.45) and Lemma 4.2.3 proves that

$$\dim W_{t_n} = \begin{cases} 0 & \text{if } \sigma \text{ is ramified,} \\ \dim(\sigma\pi)^{\Gamma_0(\mathfrak{p}^n)} & \text{if } \sigma \text{ is unramified.} \end{cases} \tag{4.49}$$

Assume that $g = M_i$ for some $0 < i < n/2$. By (4.45) and Lemma 4.2.4 we have

$$W_{M_i} = \left\{ w \in W \mid \sigma(u)\pi(A)w = w \quad \text{for} \quad \begin{matrix} u \in \mathfrak{o}^\times, \\ A \in \begin{bmatrix} \mathfrak{o} & \mathfrak{p}^{-i} \\ \mathfrak{p}^{n-i} & \mathfrak{o} \end{bmatrix}, \\ \det(A) \in \mathfrak{o}^\times, \\ u\det(A)^{-1} \in 1 + \mathfrak{p}^i \end{matrix} \right\}.$$

We see that if $W_{M_i} \neq 0$, then $\sigma(1 + \mathfrak{p}^i) = 1$, i.e., $i \geq a(\sigma)$. Assume that $i \geq a(\sigma)$. Then

$$\begin{aligned} W_{M_i} &= \{w \in W \mid (\sigma\pi)(A)w = w \text{ for } A \in \begin{bmatrix} \mathfrak{o} & \mathfrak{p}^{-i} \\ \mathfrak{p}^{n-i} & \mathfrak{o} \end{bmatrix} \text{ with } \det(A) \in \mathfrak{o}^\times\} \\ &= (\sigma\pi)(\begin{bmatrix} 1 & \\ & \varpi^i \end{bmatrix}) \left((\sigma\pi)^{\Gamma_0(\mathfrak{p}^{n-2i})}\right). \end{aligned} \tag{4.50}$$

Hence, for $0 < i < n/2$,

$$\dim W_{M_i} = \begin{cases} 0 & \text{if } i < a(\sigma), \\ \dim(\sigma\pi)^{\Gamma_0(\mathfrak{p}^{n-2i})} & \text{if } i \geq a(\sigma). \end{cases} \tag{4.51}$$

A similar argument using Lemma 4.2.5 proves that for $0 < i < n/2$,

$$\dim W_{M_i t_n} = \begin{cases} 0 & \text{if } i < a(\sigma), \\ \dim(\sigma\pi)^{\Gamma_0(\mathfrak{p}^{n-2i+1})} & \text{if } i \geq a(\sigma). \end{cases} \tag{4.52}$$

Finally, assume that n is even. By (4.45) and Lemma 4.2.6 we have

$$W_{M_{n/2}} = \left\{ w \in W \mid \sigma(u)\pi(A)w = w \quad \text{for} \quad \begin{matrix} u \in \mathfrak{o}^\times, \\ A \in \begin{bmatrix} \mathfrak{o} & \mathfrak{p}^{-n/2+1} \\ \mathfrak{p}^{n/2} & \mathfrak{o} \end{bmatrix}, \\ \det(A) \in \mathfrak{o}^\times, \\ u\det(A)^{-1} \in 1 + \mathfrak{p}^{n/2} \end{matrix} \right\}.$$

Evidently, if $W_{M_{n/2}} \neq 0$, then $\sigma(1 + \mathfrak{p}^{n/2}) = 1$, i.e., $n/2 \geq a(\sigma)$. Assume that $n/2 \geq a(\sigma)$. Then

$$\begin{aligned} W_{M_{n/2}} &= \{w \in W \mid (\sigma\pi)(A)w = w \text{ for } A \in \begin{bmatrix} \mathfrak{o} & \mathfrak{p}^{-n/2+1} \\ \mathfrak{p}^{n/2} & \mathfrak{o} \end{bmatrix} \text{ with } \det(A) \in \mathfrak{o}^\times\} \\ &= (\sigma\pi)(\begin{bmatrix} 1 & \\ & \varpi^{n/2-1} \end{bmatrix}) \left((\sigma\pi)^{\Gamma_0(\mathfrak{p})}\right). \end{aligned} \tag{4.53}$$

It follows that if n is even, then

$$\dim W_{M_{n/2}} = \begin{cases} 0 & \text{if } n/2 < a(\sigma), \\ \dim(\sigma\pi)^{\Gamma_0(\mathfrak{p})} & \text{if } n/2 \geq a(\sigma). \end{cases} \tag{4.54}$$

By (4.46), (4.48), (4.49), (4.51), (4.52), and (4.54) we now see that if $a(\sigma) = 0$, then

$$\dim V_s(n) = 2\dim(\sigma\pi)^{\Gamma_0(\mathfrak{p}^n)} + \sum_{0<i<\frac{n}{2}} \Big((\sigma\pi)^{\Gamma_0(\mathfrak{p}^{n-2i})} + \dim(\sigma\pi)^{\Gamma_0(\mathfrak{p}^{n-2i+1})}\Big)$$
$$+ \begin{Bmatrix} \dim(\sigma\pi)^{\Gamma_0(\mathfrak{p})} & \text{if } n \text{ even} \\ 0 & \text{if } n \text{ odd} \end{Bmatrix}, \tag{4.55}$$

and if $a(\sigma) > 0$, then

$$\dim V_s(n) = \sum_{a(\sigma)\leq i<\frac{n}{2}} \Big(\dim(\sigma\pi)^{\Gamma_0(\mathfrak{p}^{n-2i})} + \dim(\sigma\pi)^{\Gamma_0(\mathfrak{p}^{n-2i+1})}\Big)$$
$$+ \begin{Bmatrix} \dim(\sigma\pi)^{\Gamma_0(\mathfrak{p})} & \text{if } n \text{ even and } a(\sigma) \leq \frac{n}{2} \\ 0 & \text{if } n \text{ odd or } a(\sigma) > \frac{n}{2} \end{Bmatrix}. \tag{4.56}$$

The formulas (4.55) and (4.56) hold for all $n \geq 1$.

(1). The assertion is easy to prove if $a(\sigma) = 0$ and $N_{\sigma\pi} = 0$. Assume therefore that either $a(\sigma) > 0$ or $N_{\sigma\pi} > 0$. Then $V_s(0) = 0$. It follows from (4.55) and (4.56) that the minimal stable Klingen level is

$$N_{\pi\rtimes\sigma,s} = \begin{cases} N_{\sigma\pi} & \text{if } a(\sigma) = 0, \\ N_{\sigma\pi} + 2a(\sigma) - 1 & \text{if } a(\sigma) > 0 \text{ and } N_{\sigma\pi} \geq 1, \\ 2a(\sigma) & \text{if } a(\sigma) > 0 \text{ and } N_{\sigma\pi} = 0. \end{cases}$$

The dimension of $V_s(N_{\pi\rtimes\sigma,s})$ follows by substituting $N_{\pi\rtimes\sigma,s}$ into (4.55) resp. (4.56).

(2) follows from evaluating the formulas (4.55) and (4.56), keeping in mind that

$$\dim V_{\sigma\pi}^{\Gamma_0(\mathfrak{p}^m)} = \begin{cases} m - N_{\sigma\pi} + 1 & \text{if } m \geq N_{\sigma\pi}, \\ 0 & \text{if } m < N_{\sigma\pi}. \end{cases} \tag{4.57}$$

(3) and (4) follow similarly from (4.55) and (4.56). □

For future use, we point out some facts that are evident from the proof of Theorem 4.2.7. If the notation is as in Theorem 4.2.7, n is an integer such that $n \geq 1$, and $f \in V_s(n)$, then

$$f(1), f(t_n) \in \begin{cases} \{0\} & \text{if } \sigma \text{ is ramified}, \\ (\sigma\pi)^{\Gamma_0(\mathfrak{p}^n)} & \text{if } \sigma \text{ is unramified}, \end{cases} \tag{4.58}$$

$$(\sigma\pi)(\begin{bmatrix} 1 & \\ & \varpi^{-i} \end{bmatrix}) f(M_i) \in \begin{cases} \{0\} & \text{if } 0 < i < a(\sigma), \\ (\sigma\pi)^{\Gamma_0(\mathfrak{p}^{n-2i})} & \text{if } a(\sigma) \leq i < n/2, \end{cases} \tag{4.59}$$

$$(\sigma\pi)(\begin{bmatrix} 1 & \\ & \varpi^{-i} \end{bmatrix}) f(M_i t_n) \in \begin{cases} \{0\} & \text{if } 0 < i < a(\sigma), \\ (\sigma\pi)^{\Gamma_0(\mathfrak{p}^{n-2i+1})} & \text{if } a(\sigma) \leq i < n/2, \end{cases} \tag{4.60}$$

$$(\sigma\pi)(\begin{bmatrix} 1 & \\ & \varpi^{-(n/2-1)} \end{bmatrix}) f(M_{n/2}) \in (\sigma\pi)^{\Gamma_0(\mathfrak{p})} \quad \text{if } n \text{ is even.} \tag{4.61}$$

4.3 Non-Existence of Certain Vectors

The following result will be used in the proof of Proposition 5.1.3.

Lemma 4.3.1 *Let π be an admissible representation of* GL(2, F) *admitting a central character ω_π. Assume that either*

$$\pi \text{ does not contain any non-zero vectors fixed by } \mathrm{SL}(2, F), \tag{4.62}$$

or

$$\pi = \chi 1_{\mathrm{GL}(2)} \text{ for } \chi \text{ a character of } F^\times \text{ with } \chi \neq \nu^{1/2}. \tag{4.63}$$

Let σ be a character of $F^\times$. We assume that $\omega_\pi \sigma^2 = 1$, so that the induced representation $\pi \rtimes \sigma$ of GSp(4, F) *has trivial central character. Let V be the space of $\pi \rtimes \sigma$, and let n be an integer such that $n \geq 0$. Suppose that $f \in V_s(n)$ satisfies*

$$\tau_{n+1} f = q^{-2} \eta f. \tag{4.64}$$

Then $f = 0$.

Proof We assume that V is the standard model of $\pi \rtimes \sigma$ (see Sect. 2.2). We note that Lemma 3.6.2 implies that

$$-q\tau_n f = t_n f \tag{4.65}$$

since (4.64) holds by assumption. If $n = 0$, then (4.65) is $-f = f$, so that $f = 0$. Assume that $n \geq 1$. By Lemma 4.1.1,

$$\begin{aligned}\mathrm{GSp}(4, F) = P\mathrm{K}_s(\mathfrak{p}^n) &\sqcup \bigsqcup_{0<i\leq n/2} PM_i\mathrm{K}_s(\mathfrak{p}^n)\\ &\sqcup Pt_n\mathrm{K}_s(\mathfrak{p}^n) \sqcup \bigsqcup_{0<i<n/2} PM_it_n\mathrm{K}_s(\mathfrak{p}^n).\end{aligned} \tag{4.66}$$

To prove that $f = 0$ it will suffice to prove that f vanishes on 1, t_n, M_i and M_it_n for $0 < i \leq n/2$. Evaluating (4.64) at 1, and using (4.24), gives

$$f(1) = q^{-1/2}\pi(\begin{bmatrix} \varpi^{-1} & \\ & 1 \end{bmatrix})f(1). \tag{4.67}$$

Since $f \in V_s(n)$, we also deduce from (4.24) that

$$\pi(\begin{bmatrix} 1 & x \\ & 1 \end{bmatrix})f(1) = f(1) \quad \text{for } x \in \mathfrak{o}, \tag{4.68}$$

$$\pi(\begin{bmatrix} 1 & \\ y & 1 \end{bmatrix})f(1) = f(1) \quad \text{for } y \in \mathfrak{p}^n, \tag{4.69}$$

$$\pi(\begin{bmatrix} v & \\ & v^{-1} \end{bmatrix})f(1) = f(1) \quad \text{for } v \in \mathfrak{o}^\times. \tag{4.70}$$

By (4.67) and (4.69) we have

$$\pi(\begin{bmatrix} 1 & \\ y & 1 \end{bmatrix})f(1) = f(1) \quad \text{for } y \in F. \tag{4.71}$$

The identity

$$\begin{bmatrix} & 1 \\ -1 & \end{bmatrix} = \begin{bmatrix} -1 & \\ & -1 \end{bmatrix}\begin{bmatrix} 1 & -1 \\ & 1 \end{bmatrix}\begin{bmatrix} 1 & \\ 1 & 1 \end{bmatrix}\begin{bmatrix} 1 & -1 \\ & 1 \end{bmatrix}$$

now implies that $\pi(\begin{bmatrix} & 1 \\ -1 & \end{bmatrix})f(1) = f(1)$. This, along with (4.71), proves that $f(1)$ is fixed by $\mathrm{SL}(2, F)$ (see Satz A 5.4 of [42]). If (4.62) is satisfied, then it follows that $f(1) = 0$. Assume that (4.63) is satisfied. Then (4.67) becomes

$$f(1) = q^{-1/2}\chi(\varpi)^{-1}f(1). \tag{4.72}$$

By (4.24) and the assumption $f \in V_s(n)$,

$$f(1) = f(\begin{bmatrix} u & & & \\ & 1 & & \\ & & 1 & \\ & & & u^{-1} \end{bmatrix}) = \chi(u)f(1) \quad \text{for } u \in \mathfrak{o}^\times. \tag{4.73}$$

If $f(1) \neq 0$, then (4.72) and (4.73) imply that $\chi = \nu^{1/2}$, contradiction. Hence, $f(1) = 0$. Evaluating (4.65) at 1 yields $f(t_n) = -qf(1) = 0$.

To complete the proof we will prove that $f(M_i) = f(M_i t_n) = 0$ for $0 \le i \le n/2$ by induction on i. There are matrix identities

$$M_0 = \begin{bmatrix} & 1 & \varpi^{-n} & \\ \varpi^{n} & 1 & & \\ & & \varpi^{-n} & \\ & 1 & & \end{bmatrix} t_n \begin{bmatrix} 1 & & 1 & \\ & & -1 & \\ & 1 & & 1 \\ & & & 1 \end{bmatrix}, \tag{4.74}$$

$$M_0 t_n = \begin{bmatrix} & 1 & \varpi^{-n} & \\ \varpi^{n} & 1 & & \\ & & \varpi^{-n} & \\ & 1 & & \end{bmatrix} \begin{bmatrix} -1 & & & \\ & & -1 & \\ \varpi^{n} & 1 & & \\ & & \varpi^{n} & -1 \end{bmatrix}. \tag{4.75}$$

These identities imply that

$$P M_0 \mathrm{K}_s(\mathfrak{p}^n) = P t_n \mathrm{K}_s(\mathfrak{p}^n) \quad \text{and} \quad P M_0 t_n \mathrm{K}_s(\mathfrak{p}^n) = P \mathrm{K}_s(\mathfrak{p}^n).$$

Since $f(t_n) = 0$ and $f(1) = 0$, we conclude that $f(M_0) = 0$ and $f(M_0 t_n) = 0$. This proves our assertion for $i = 0$. Assume that $i > 0$, and that the assertion has been proven for $i - 1$. We have the following identity for $z \in \mathfrak{o}^\times$:

$$\begin{bmatrix} 1 & z\varpi^{-i} & & \\ z^{-1}\varpi^{n-i} & 1+\varpi^{n-2i} & & \\ & & 1 & -z\varpi^{-i} \\ & & -z^{-1}\varpi^{n-i} & 1+\varpi^{n-2i} \end{bmatrix} \begin{bmatrix} 1 & \varpi^{n-3i} & -z\varpi^{n-4i} - z\varpi^{-2i} & \\ & 1 - z^{-1}\varpi^{n-2i} & \varpi^{n-3i} & \\ & 1 & & \\ & & 1 & \end{bmatrix} M_i t_n$$
$$= M_i \begin{bmatrix} 1 & & -z\varpi^{-n} & \\ & 1 & & \\ & & 1 & \\ & & & 1 \end{bmatrix} \begin{bmatrix} & z & & \\ & 1+\varpi^{n-2i} & -z^{-1}\varpi^{n-2i} & \\ & -z & 1 & \\ \varpi^{n}+\varpi^{2n-2i} & & & z^{-1} \end{bmatrix}. \tag{4.76}$$

Now

$$(\tau_n f)(M_i) = \int\limits_{\mathfrak{o}} f(M_i \begin{bmatrix} 1 & & z\varpi^{-n} & \\ & 1 & & \\ & & 1 & \\ & & & 1 \end{bmatrix})\, dz$$
$$= \int\limits_{\mathfrak{p}} f(M_i \begin{bmatrix} 1 & & z\varpi^{-n} & \\ & 1 & & \\ & & 1 & \\ & & & 1 \end{bmatrix})\, dz + \int\limits_{\mathfrak{o}^\times} f(M_i \begin{bmatrix} 1 & & z\varpi^{-n} & \\ & 1 & & \\ & & 1 & \\ & & & 1 \end{bmatrix})\, dz$$
$$= q^{-1} f(M_i) + \int\limits_{\mathfrak{o}^\times} f(M_i \begin{bmatrix} 1 & & -z\varpi^{-n} & \\ & 1 & & \\ & & 1 & \\ & & & 1 \end{bmatrix})\, dz$$
$$= q^{-1} f(M_i) + \int\limits_{\mathfrak{o}^\times} \pi(\begin{bmatrix} 1 & z\varpi^{-i} \\ z^{-1}\varpi^{n-i} & 1+\varpi^{n-2i} \end{bmatrix}) f(M_i t_n)\, dz. \tag{4.77}$$

For the last equality we used (4.76). Also,

$$
\begin{aligned}
(\tau_{n+1}f)(M_i) &= \int\limits_{\mathfrak{o}} f(M_i\begin{bmatrix}1&&&z\varpi^{-n-1}\\&1&&\\&&1&\\&&&1\end{bmatrix})\,dz\\
&= \int\limits_{\mathfrak{p}} f(M_i\begin{bmatrix}1&&&z\varpi^{-n-1}\\&1&&\\&&1&\\&&&1\end{bmatrix})\,dz + \int\limits_{\mathfrak{o}^\times} f(M_i\begin{bmatrix}1&&&z\varpi^{-n-1}\\&1&&\\&&1&\\&&&1\end{bmatrix})\,dz\\
&= q^{-1}(\tau_n f)(M_i) + \int\limits_{\mathfrak{o}^\times} f(M_i\begin{bmatrix}1&&&-z\varpi^{-n-1}\\&1&&\\&&1&\\&&&1\end{bmatrix})\,dz\\
&= q^{-1}(\tau_n f)(M_i) + \int\limits_{\mathfrak{o}^\times} \pi(\begin{bmatrix}1&z\varpi^{-i}\\z^{-1}\varpi^{n+1-i}&1+\varpi^{n+1-2i}\end{bmatrix})f(M_it_{n+1})\,dz\\
&= q^{-1}(\tau_n f)(M_i) + \int\limits_{\mathfrak{o}^\times} \pi(\begin{bmatrix}1&z\varpi^{-i}\\z^{-1}\varpi^{n+1-i}&1+\varpi^{n+1-2i}\end{bmatrix})f(M_i\eta t_n)\,dz\\
&= q^{-1}(\tau_n f)(M_i) + \int\limits_{\mathfrak{o}^\times} \pi(\begin{bmatrix}1&z\varpi^{-i}\\z^{-1}\varpi^{n+1-i}&1+\varpi^{n+1-2i}\end{bmatrix})f(\eta M_{i-1}t_n)\,dz\\
&= q^{-1}(\tau_n f)(M_i) \qquad (4.78)\\
&\quad + q^{3/2}\int\limits_{\mathfrak{o}^\times} \pi(\begin{bmatrix}1&z\varpi^{-i}\\z^{-1}\varpi^{n+1-i}&1+\varpi^{n+1-2i}\end{bmatrix}\begin{bmatrix}\varpi^{-1}&\\&1\end{bmatrix})f(M_{i-1}t_n)\,dz.
\end{aligned}
$$

Here, we again used (4.76) in the fourth equality. Evaluating (4.65) and (4.64) at M_i gives

$$
-q(\tau_n f)(M_i) = f(M_it_n) \tag{4.79}
$$

and

$$
(\tau_{n+1}f)(M_i) = q^{-2}f(M_i\eta) = q^{-1/2}\pi(\begin{bmatrix}\varpi^{-1}&\\&1\end{bmatrix})f(M_{i-1}) = 0; \tag{4.80}
$$

note that $f(M_{i-1}) = 0$ by the induction hypothesis. It follows from (4.78), (4.80), and the induction hypothesis that $(\tau_n f)(M_i) = 0$. From (4.79) we have $f(M_it_n) = 0$. Since $f(M_it_n) = 0$ and $(\tau_n f)(M_i) = 0$, we also deduce from (4.77) that $f(M_i) = 0$. This concludes the proof. □

4.4 Characterization of Paramodular Vectors

To end this chapter, we prove a result which characterizes the paramodular vectors among the stable Klingen vectors in a Siegel induced representation.

Lemma 4.4.1 *Let (π, W) be an infinite-dimensional, irreducible, admissible representation of* $\mathrm{GL}(2, F)$ *with central character ω_π, and let σ be a character of $F^\times$. We assume that $\omega_\pi\sigma^2 = 1$, so that the induced representation $\pi \rtimes \sigma$ has trivial central character. Let V be the standard model of $\pi \rtimes \sigma$ (see Sect. 2.2).*

(1) *Let n be an integer such that $n \geq N_{\pi\rtimes\sigma,s}$ and let $f \in V_s(n)$. Then f lies in $V(n)$ if and only if the following conditions are satisfied.*

$$f(t_n) = f(1), \tag{4.81}$$

$$f(M_i t_n) = f(M_i) \quad \textit{for } 0 < i < n/2, \tag{4.82}$$

$$f(M_{n/2}) = \pi(\begin{bmatrix} & -\varpi^{-n/2} \\ \varpi^{n/2} & \end{bmatrix}) f(M_{n/2}) \quad \textit{if } n \textit{ is even.} \tag{4.83}$$

(2) *Let n be an integer such that $n \geq N_{\pi\rtimes\sigma,s}$ and $n \geq 1$, and let $f \in V_s(n)$. Then f lies in $V(n-1)$ if and only if the following conditions are satisfied.*

$$f(t_n) = q^{3/2}\pi(\begin{bmatrix} \varpi^{-1} & \\ & 1 \end{bmatrix}) f(1), \tag{4.84}$$

$$f(M_i t_n) = q^{3/2}\pi(\begin{bmatrix} \varpi^{-1} & \\ & 1 \end{bmatrix}) f(M_{i-1}) \quad \textit{for } 1 < i < n/2, \tag{4.85}$$

$$f(M_1 t_n) = q^{3-(3/2)n}\pi(\begin{bmatrix} & \varpi^{-1} \\ \varpi^{n-1} & \end{bmatrix}) f(1), \tag{4.86}$$

$$f(M_{n/2}) = q^{3/2}\pi(\begin{bmatrix} & -\varpi^{-n/2} \\ \varpi^{n/2-1} & \end{bmatrix}) f(M_{n/2-1}) \quad \textit{if } n \textit{ is even,} \tag{4.87}$$

$$f(M_{(n-1)/2}) = \pi(\begin{bmatrix} & -\varpi^{-(n-1)/2} \\ \varpi^{(n-1)/2} & \end{bmatrix}) f(M_{(n-1)/2}) \quad \textit{if } n \textit{ is odd.} \tag{4.88}$$

Proof We begin by noting that by (1) of Theorem 4.2.7,

$$N_{\pi\rtimes\sigma,s} = \begin{cases} N_{\sigma\pi} + 2a(\sigma) & \text{if } a(\sigma) = 0 \text{ or } N_{\sigma\pi} = 0, \\ N_{\sigma\pi} + 2a(\sigma) - 1 & \text{if } a(\sigma) > 0 \text{ and } N_{\sigma\pi} > 0. \end{cases} \tag{4.89}$$

We see that if $n \geq N_{\pi\rtimes\sigma,s}$, then $a(\sigma) \leq n/2$. We also note that the elements of V satisfy (4.24); this will be used repeatedly in the following proof.

(1). Let n be an integer such that $n \geq N_{\pi\rtimes\sigma,s}$. Let $A(n)$ be the subspace of all $f \in V_s(n)$ satisfying (4.81)–(4.83). We need to prove that $V(n) = A(n)$. Let $f \in V(n)$. Then $f(gt_n) = f(g)$ for all $g \in \mathrm{GSp}(4, F)$. Substituting $g = 1$ and $g = M_i$, we obtain (4.81)–(4.83); for (4.83), observe the matrix identity (4.6). Thus, $V(n) \subset A(n)$. To prove $V(n) = A(n)$, we will show that $\dim V(n) =$

$\dim A(n)$. The dimension of $V(n)$ is known: by Theorem 5.2.2 of [105], the minimal paramodular level of $\pi \rtimes \sigma$ is $N_{\pi \rtimes \sigma} = N_{\sigma\pi} + 2a(\sigma)$, and

$$\dim V(n) = \begin{cases} \left\lfloor \dfrac{(n - N_{\pi \rtimes \sigma} + 2)^2}{4} \right\rfloor & \text{if } n \geq N_{\pi \rtimes \sigma}, \\ 0 & \text{if } n < N_{\pi \rtimes \sigma}. \end{cases} \tag{4.90}$$

To estimate the dimension of $A(n)$ we need a preliminary result. For each integer i such that $a(\sigma) \leq i \leq n/2$ define $g_i \in \mathrm{GSp}(4, F)$ by

$$g_i = \begin{cases} 1 & \text{if } i = 0, \\ M_i & \text{if } 1 < i \leq n/2. \end{cases}$$

Let $f \in A(n)$. We claim that

$$(\sigma\pi)(\left[\begin{smallmatrix} 1 & \\ & \varpi^{-i} \end{smallmatrix}\right]) f(g_i) \in (\sigma\pi)^{\Gamma_0(\mathfrak{p}^{n-2i})} \qquad \text{for } a(\sigma) \leq i \leq n/2. \tag{4.91}$$

Let i be an integer such that $a(\sigma) \leq i \leq n/2$. If $i = 0$, so that necessarily $a(\sigma) = 0$, then (4.91) follows from (4.58). If $0 < i < n/2$, then (4.91) follows from (4.59). Assume that $i = n/2$, so that n is necessarily even. By (4.61) we have

$$(\sigma\pi)(\left[\begin{smallmatrix} 1 & \\ & \varpi^{-(n/2-1)} \end{smallmatrix}\right]) f(M_{n/2}) \in (\sigma\pi)^{\Gamma_0(\mathfrak{p})}. \tag{4.92}$$

Define $w = (\sigma\pi)(\left[\begin{smallmatrix} 1 & \\ & \varpi^{-n/2} \end{smallmatrix}\right]) f(M_{n/2})$. By (4.92), we have $(\sigma\pi)(\left[\begin{smallmatrix} 1 & \\ & \varpi \end{smallmatrix}\right]) w \in (\sigma\pi)^{\Gamma_0(\mathfrak{p})}$. This implies that

$$(\sigma\pi)(\left[\begin{smallmatrix} 1 & \\ y & 1 \end{smallmatrix}\right]) w = w \quad \text{for} \quad y \in \mathfrak{o}, \tag{4.93}$$

$$(\sigma\pi)(\left[\begin{smallmatrix} 1 & \\ & u \end{smallmatrix}\right]) w = w \quad \text{for} \quad u \in \mathfrak{o}^\times. \tag{4.94}$$

Since $f \in A(n)$, (4.83) holds; therefore,

$$(\sigma\pi)(\left[\begin{smallmatrix} & -1 \\ 1 & \end{smallmatrix}\right]) w = w. \tag{4.95}$$

By (4.93) and (4.95) we have $(\sigma\pi)(k)w = w$ for $k \in \mathrm{SL}(2, \mathfrak{o})$ (see Satz A 5.4 of [42]); by (4.94), recalling the definition of w, we conclude that

$$(\sigma\pi)(\left[\begin{smallmatrix} 1 & \\ & \varpi^{-n/2} \end{smallmatrix}\right]) f(M_{n/2}) \in (\sigma\pi)^{\mathrm{GL}(2,\mathfrak{o})}. \tag{4.96}$$

This completes the argument for (4.91). Next, let

$$B(n) = \bigoplus_{a(\sigma) \leq i \leq n/2} (\sigma\pi)^{\Gamma_0(\mathfrak{p}^{n-2i})}.$$

Define $\varphi : A(n) \to B(n)$ by

$$\varphi(f) = \bigoplus_{a(\sigma) \le i \le n/2} (\sigma\pi)(\begin{bmatrix} 1 & \\ & \varpi^{-i} \end{bmatrix}) f(g_i).$$

Then φ is a well-defined linear map by (4.91). We claim that φ is injective. Suppose that $f \in A(n)$ is such that $\varphi(f) = 0$. By the definition of φ, we have $f(g_i) = 0$ for all integers i such that $a(\sigma) \le i \le n/2$. Since (4.58) and (4.59) hold, and since f satisfies (4.81) and (4.82), it follows f vanishes on the set X from (4.25), and hence on $\mathrm{GSp}(4, F)$ by Lemma 4.1.1 and (4.24). Thus, φ is injective. It follows that $\dim A(n) \le \dim B(n)$. We will now prove that $\dim B(n) = \dim V(n)$. Evidently,

$$\dim B(n) = \sum_{a(\sigma) \le i \le n/2} \dim(\sigma\pi)^{\Gamma_0(\mathfrak{p}^{n-2i})}.$$

Hence, using (4.57),

$$\dim B(n) = \sum_{a(\sigma) \le i \le (n - N_{\sigma\pi})/2} n - N_{\sigma\pi} - 2i + 1.$$

Assume first that this sum is empty. Then $\dim B(n) = 0$ and $n < N_{\sigma\pi} + 2a(\sigma)$. Since $\dim V(n) = 0$ by (4.90), it follows that $\dim B(n) = \dim V(n) = 0$ in this case. Now assume that the sum is not empty. A calculation shows that the sum is

$$\left\lfloor \frac{(n - N_{\sigma\pi} - 2a(\sigma) + 2)^2}{4} \right\rfloor.$$

By (4.90) this is $\dim V(n)$, so that $\dim B(n) = \dim V(n)$, in all cases. Since

$$\dim V(n) \le \dim A(n) \le \dim B(n) = \dim V(n),$$

we get $\dim V(n) = \dim A(n)$; hence, $V(n) = A(n)$, completing the proof of (1).

(2). Let n be an integer such that $n \ge N_{\pi \rtimes \sigma, s}$ and $n \ge 1$. Let $C(n)$ be the subspace of all $f \in V_s(n)$ satisfying (4.84)–(4.88). We need to prove that $V(n-1) = C(n)$. Let $f \in V(n-1)$; we will prove that f satisfies (4.84)–(4.88). Since $f \in V(n-1)$ we have that $f(gt_{n-1}) = f(g)$ for all $g \in \mathrm{GSp}(4, F)$. Equation (4.84) follows by substituting $g = 1$ and using

$$t_{n-1} = \begin{bmatrix} & & \varpi & \\ & 1 & & \\ & & 1 & \\ & & & \varpi^{-1} \end{bmatrix} t_n. \tag{4.97}$$

Equation (4.85) follows by substituting $g = M_{i-1}$ and using (4.97) again. Equation (4.86) follows by substituting $g = M_0$ and using (4.97) and (4.74).

Equation (4.87) follows by evaluating f at both sides of the identity (4.6). Equation (4.88) follows by evaluating f at both sides of the identity

$$M_{(n-1)/2} = \begin{bmatrix} & -\varpi^{(1-n)/2} & & \varpi^{1-n} \\ \varpi^{(n-1)/2} & & -1 & \\ & & & \varpi^{-(n-1)/2} \\ & & -\varpi^{(n-1)/2} & \end{bmatrix} M_{(n-1)/2} t_{n-1} s_2.$$

Since f satisfies (4.84)–(4.88) it follows that $f \in C(n)$. Thus, $V(n-1) \subset C(n)$. To prove that $V(n-1) = C(n)$ we will prove that $\dim V(n-1) = \dim C(n)$. The dimension of $V(n-1)$ is known: by Theorem 5.2.2 of [105], the minimal paramodular level of $\pi \rtimes \sigma$ is $N_{\pi \rtimes \sigma} = N_{\sigma\pi} + 2a(\sigma)$, and

$$\dim V(n-1) = \begin{cases} \left\lfloor \dfrac{(n - N_{\pi \rtimes \sigma} + 1)^2}{4} \right\rfloor & \text{if } n \geq N_{\pi \rtimes \sigma} + 1, \\ 0 & \text{if } n < N_{\pi \rtimes \sigma} + 1. \end{cases} \tag{4.98}$$

We estimate the dimension of $C(n)$ as follows. For each integer i such that $a(\sigma) \leq i < n/2$ define $h_i \in \mathrm{GSp}(4, F)$ by

$$h_i = \begin{cases} 1 & \text{if } i = 0, \\ M_i & \text{if } 1 < i < n/2. \end{cases}$$

Let $f \in C(n)$. An argument as in the proof of (1) shows that

$$(\sigma\pi)(\begin{bmatrix} 1 & \\ & \varpi^{-i} \end{bmatrix}) f(h_i) \in (\sigma\pi)^{\Gamma_0(\mathfrak{p}^{n-1-2i})} \quad \text{for } a(\sigma) \leq i < n/2. \tag{4.99}$$

For this, note since $f \in V(n-1)$ we have $f \in V_s(n-1)$. Now let

$$D(n) = \bigoplus_{a(\sigma) \leq i < n/2} (\sigma\pi)^{\Gamma_0(\mathfrak{p}^{n-1-2i})}.$$

Define $\rho : C(n) \to D(n)$ by

$$\rho(f) = \bigoplus_{a(\sigma) \leq i < n/2} (\sigma\pi)(\begin{bmatrix} 1 & \\ & \varpi^{-i} \end{bmatrix}) f(h_i).$$

Then ρ is a well-defined linear map by (4.99). As in the proof of (1), ρ is injective. It follows that $\dim C(n) \leq \dim D(n)$. We will prove that $\dim D(n) = \dim V(n-1)$. As in the proof of (1), we find that

$$\dim D(n) = \sum_{a(\sigma) \leq i \leq (n - N_{\sigma\pi})/2 - 1/2} n - N_{\sigma\pi} - 2i.$$

Assume first that this sum is empty. Then $\dim D(n) = 0$, and $n-1 < N_{\sigma\pi} + 2a(\sigma)$. By (4.98) we have $\dim V(n-1) = 0$, so that $\dim D(n) = \dim V(n-1) = 0$ in this case. Now assume that the sum is not empty. A calculation shows that the sum is

$$\left\lfloor \frac{(n - N_{\sigma\pi} - 2a(\sigma) + 1)^2}{4} \right\rfloor.$$

By (4.98) this is $\dim V(n-1)$, so that $\dim D(n) = \dim V(n-1)$, in all cases. Since

$$\dim V(n-1) \leq \dim C(n) \leq \dim D(n) = \dim V(n-1),$$

we get $\dim V(n-1) = \dim C(n)$; hence, $V(n-1) = C(n)$, completing the proof of (2). □

Chapter 5
Dimensions

Let (π, V) be an irreducible, admissible representation of $\mathrm{GSp}(4, F)$ with trivial central character. In this chapter we will compute the dimensions of $V_s(n)$ and $\bar{V}_s(n)$ for all non-negative integers n; the result appears in Theorem 5.6.1. To prove Theorem 5.6.1 we will use three tools. The first tool is the theory about induced representations from Chap. 4. The second tool is an important inequality. For non-negative integers n we will prove that

$$\dim V_s(n) \leq \dim V(n) + \dim V(n+1). \tag{5.1}$$

As a corollary of this upper bound on $\dim V_s(n)$, we will show that π admits non-zero stable Klingen vectors if and only if π is paramodular, and if π is paramodular, then $N_{\pi,s}$ is either $N_\pi - 1$ or N_π. Thus, paramodular representations fall into one of two classes. If $N_{\pi,s} = N_\pi - 1$, then we will say that π is a category 1 representation; if $N_{\pi,s} = N_\pi$, then we will say that π is a category 2 representation. Our third and final tool for proving Theorem 5.6.1 is a certain stable Klingen vector which we refer to as the shadow of a newform.

Theorem 5.6.1 has some significant immediate consequences. For example, in Corollary 5.7.1 we will prove that π is a category 1 paramodular representation if and only if the decomposition of the L-parameter of π into indecomposable representations contains no unramified one-dimensional factors. In this chapter we will also deduce results indicating that if π is a paramodular representation and $N_\pi \geq 2$, then the stable Klingen vectors in π exhibit a simple structure.

5.1 The Upper Bound

In this section we prove the upper bound (5.1). Our approach uses the P_3-quotient recalled in Sect. 2.2 as well as the non-existence result from Sect. 4.3.

J. Johnson-Leung et al., *Stable Klingen Vectors and Paramodular Newforms*,
Lecture Notes in Mathematics 2342, https://doi.org/10.1007/978-3-031-45177-5_5

Lemma 5.1.1 *Let (π, V) be an admissible representation of* GSp(4, F) *for which the center acts trivially. Let n be an integer such that $n \geq 0$. Let $p : V \to V_{Z^J}$ be the projection map. If $v \in V_s(n)$ is such that $p(v) = 0$, then $v = 0$.*

Proof Assume that $v \in V_s(n)$ and $p(v) = 0$. Then there exists a positive integer $M \geq n$ such that

$$\int\limits_{\mathfrak{p}^{-M}} \pi(\begin{bmatrix} 1 & & x & \\ & 1 & & \\ & & 1 & \\ & & & 1 \end{bmatrix})v\, dx = 0.$$

This implies that $(\tau_M \tau_{M-1} \ldots \tau_n)(v) = 0$. Applying (1) of Lemma 3.5.4 repeatedly proves that $v = 0$. □

Lemma 5.1.2 *Let (π, V) be an irreducible, admissible representation of the group* GSp(4, F) *with trivial central character. Assume that π is not of one of the following types:*

- *$\chi_1 \times \chi_2 \rtimes \sigma$ of type I with χ_1, χ_2, σ unramified and ($\chi_1 = 1$ or $\chi_2 = 1$);*
- *Type VIc with unramified σ;*
- *Type VId with unramified σ.*

Let n be an integer such that $n \geq 0$. Let $v \in V_s(n)$. If $\tau_{n+1} v = q^{-2}\eta v$, then $v = 0$.

Proof Assume that $\tau_{n+1} v = q^{-2}\eta v$. Assume that $v \neq 0$; we will obtain a contradiction. Applying the projection $p : V \to V_{Z^J}$, we get

$$p(v) = q^{-2} \begin{bmatrix} \varpi^{-1} & & \\ & \varpi^{-1} & \\ & & 1 \end{bmatrix} p(v). \tag{5.2}$$

The vector $p(v)$ is also non-zero by Lemma 5.1.1. Hence, $p(v)$ defines a non-zero vector w in an irreducible subquotient τ of the P_3-filtration of V_{Z^J} (see Sect. 2.2). This vector has the property that

$$\tau(P_3(\mathfrak{o}))w = w \tag{5.3}$$

and

$$w = q^{-2}\tau(\begin{bmatrix} \varpi^{-1} & & \\ & \varpi^{-1} & \\ & & 1 \end{bmatrix})w. \tag{5.4}$$

Assume that $\tau = \tau_{\mathrm{GL}(0)}^{P_3}(1)$ or $\tau = \tau_{\mathrm{GL}(1)}^{P_3}(\chi)$ for some character χ of $F^\times$; we will obtain a contradiction. By the definition of $\tau_{\mathrm{GL}(0)}^{P_3}(1)$ or $\tau_{\mathrm{GL}(1)}^{P_3}(\chi)$ there exists a compact set $X \subset P_3$ such that if $y \in P_3$ is such that $w(y) \neq 0$, then $y \in \begin{bmatrix} * & * & * \\ & 1 & * \\ & & 1 \end{bmatrix} X$. Fix $y \in P_3$ such that $w(y) \neq 0$. By (5.4), we also have $w(y \begin{bmatrix} \varpi^{-k} & & \\ & \varpi^{-k} & \\ & & 1 \end{bmatrix}) \neq 0$ for all $k \in \mathbb{Z}$. It follows that $y\begin{bmatrix} \varpi^{-k} & & \\ & \varpi^{-k} & \\ & & 1 \end{bmatrix} \in \begin{bmatrix} * & * & * \\ & 1 & * \\ & & 1 \end{bmatrix} X$ for all $k \in \mathbb{Z}$. In particular, we see that there exists a compact subset K of F such that the $(2, 1)$ and $(2, 2)$ entries of $y\begin{bmatrix} \varpi^{-k} & & \\ & \varpi^{-k} & \\ & & 1 \end{bmatrix}$ lie in K for all $k \in \mathbb{Z}$. This implies that $y_{21} = y_{22} = 0$, a contradiction. It follows that $\tau = \tau_{\mathrm{GL}(2)}^{P_3}(\rho)$ for some irreducible, admissible representation ρ of $\mathrm{GL}(2, F)$. By (5.3) the representation ρ is unramified. Condition (5.4) translates to $\omega_\rho = \nu^2$. Using Table A.5 of [105], we see that the only representations that admit $\tau_{\mathrm{GL}(2)}^{P_3}(\rho)$ with such ρ in their P_3-filtration are $\chi_1 \times \chi_2 \rtimes \sigma$ of type I with χ_1, χ_2, σ unramified and ($\chi_1 = 1$ or $\chi_2 = 1$), or VIc or VId with unramified σ (note that there is typo in Table A.5 of [105]: The entry for Vd in the "s.s.(V_0/V_1)" column should be $\tau_{\mathrm{GL}(2)}^{P_3}(\nu(\nu^{-\frac{1}{2}}\sigma \times \nu^{-\frac{1}{2}}\xi\sigma)))$; by assumption, π is not one of these representations. This is a contradiction. □

Proposition 5.1.3 *Let (π, V) be an irreducible, admissible representation of the group* $\mathrm{GSp}(4, F)$ *with trivial central character. Let n be an integer such that $n \geq 0$. Assume that $v \in V_s(n)$ satisfies*

$$\tau_{n+1}v = q^{-2}\eta v. \tag{5.5}$$

Then $v = 0$.

Proof By Lemma 5.1.2, we may assume that π is one of the following types:

- $\chi_1 \times \chi_2 \rtimes \sigma$ of type I with χ_1, χ_2, σ unramified and ($\chi_1 = 1$ or $\chi_2 = 1$);
- Type VIc with unramified σ;
- Type VId with unramified σ.

We claim that π is a subrepresentation of $\pi' \rtimes \sigma'$ for some admissible representation π' of $\mathrm{GL}(2, F)$ satisfying the hypotheses of Lemma 4.3.1 and character σ' of $F^\times$ such that $\omega_{\pi'}\sigma'^2 = 1$. This is clear in the first case because $\chi_1 \times \chi_2$ is irreducible. Assume that π is as in the second case, so that $\pi = L(\nu^{\frac{1}{2}}\mathrm{St}_{\mathrm{GL}(2)}, \nu^{-\frac{1}{2}}\sigma)$ for some unramified character σ of $F^\times$ with $\sigma^2 = 1$. By (2.20), π is a quotient of $\nu^{\frac{1}{2}}\mathrm{St}_{\mathrm{GL}(2)} \rtimes \nu^{-\frac{1}{2}}\sigma$. Taking contragredients and using $\pi^\vee \cong \pi$, we see that π is a subrepresentation of $\nu^{-\frac{1}{2}}\mathrm{St}_{\mathrm{GL}(2)} \rtimes \nu^{\frac{1}{2}}\sigma^{-1}$. Assume that π is as in the third case, so that $\pi \cong L(\nu, 1_{F^\times} \rtimes \nu^{-\frac{1}{2}}\sigma)$ for some unramified character of $F^\times$ such that $\sigma^2 = 1$. Then π is a quotient of $\nu^{\frac{1}{2}}1_{\mathrm{GL}(2)} \rtimes \nu^{-\frac{1}{2}}\sigma$ by (2.20). Taking contragredients, we see that π is a subrepresentation of $\nu^{-\frac{1}{2}}1_{\mathrm{GL}(2)} \rtimes \nu^{\frac{1}{2}}\sigma^{-1}$. This proves our claim in all cases.

We now apply Lemma 4.3.1 to conclude that $v = 0$. □

For the following result recall the paramodularization operator $p_n : V_s(n) \to V(n)$ defined in (3.18).

Lemma 5.1.4 *Let (π, V) be a smooth representation of* $\mathrm{GSp}(4, F)$ *for which the center of* $\mathrm{GSp}(4, F)$ *acts trivially, and assume that the subspace of vectors of V*

fixed by Sp(4, *F*) *is trivial. Let n be an integer such that* $n \geq 0$. *Define*

$$c_n : V_s(n) \longrightarrow V(n) \oplus V(n+1)$$

by

$$c_n(v) = (p_n v, p_{n+1}\tau_n v)$$

for $v \in V_s(n)$. *Let* $v \in V_s(n)$. *If* $c_n(v) = 0$, *then*

$$\tau_{n+1}v = q^{-2}\eta v. \tag{5.6}$$

Proof Assume that $c_n(v) = 0$. Then $p_n v = 0$ and $p_{n+1}\tau_n v = 0$. The result follows now from Lemma 3.6.2. □

Theorem 5.1.5 *Let n be an integer such that* $n \geq 0$. *Let* (π, V) *be an irreducible, admissible representation of* GSp(4, *F*) *with trivial central character. Then*

$$\dim V_s(n) \leq \dim V(n) + \dim V(n+1). \tag{5.7}$$

Proof It suffices to show that the map c_n defined in Lemma 5.1.4 is injective. Assume first that π is one-dimensional, so that $\pi = \chi 1_{\mathrm{GSp}(4,F)}$ for some character χ of $F^\times$. If χ is ramified, then $V_s(n) = V(n) = V(n+1) = 0$ so that c_n is injective. If χ is unramified, then $c_n(v) = (v, v)$ for $v \in V_s(n)$, so that c_n is injective. Assume that π is not one-dimensional, so that π is infinite-dimensional. Then the subspace of vectors in V fixed by Sp(4, *F*) is trivial. Let $v \in V_s(n)$ be such that $c_n(v) = 0$. By Lemma 5.1.4, $\tau_{n+1}v = q^{-2}\eta v$. Hence, $v = 0$ by Proposition 5.1.3. □

Corollary 5.1.6 *Let* (π, V) *be an irreducible, admissible representation of* GSp(4, *F*) *with trivial central character. Then* π *admits non-zero paramodular vectors if and only if* π *admits non-zero stable Klingen vectors. If* π *is paramodular, then* $N_{\pi,s} = N_\pi - 1$ *or* $N_{\pi,s} = N_\pi$.

Proof If π is paramodular, then π admits non-zero stable Klingen vectors because $V(n) \subset V_s(n)$ for all integers $n \geq 0$. Assume that π admits non-zero stable Klingen vectors. Then there exists an integer n such that $n \geq 0$ and $V_s(n) \neq 0$. By (5.7), $0 < \dim V_s(n) \leq V(n) + \dim V(n+1)$, so that π is paramodular. For the final assertion, assume that π is paramodular. Then

$$0 < \dim V_s(N_{\pi,s}) \leq \dim V(N_{\pi,s}) + \dim V(N_{\pi,s}+1)$$

by (5.7). It follows that $N_\pi \leq N_{\pi,s}$ or $N_\pi \leq N_{\pi,s} + 1$. Hence, $N_\pi - 1 \leq N_{\pi,s}$. We also have

$$0 < \dim V(N_\pi) \leq \dim V_s(N_\pi)$$

because $V(N_\pi) \subset V_s(N_\pi)$. This implies that $N_{\pi,s} \le N_\pi$. We now have $N_\pi - 1 \le N_{\pi,s} \le N_\pi$ so that $N_{\pi,s} = N_\pi - 1$ or $N_{\pi,s} = N_\pi$. □

Let (π, V) be an irreducible, admissible representation of $\mathrm{GSp}(4, F)$ with trivial central character. Assume that π is paramodular. By Corollary 5.1.6 we have $N_{\pi,s} = N_\pi - 1$ or $N_{\pi,s} = N_\pi$. If $N_{\pi,s} = N_\pi - 1$, then we will say that π is a *category 1 representation*; if $N_{\pi,s} = N_\pi$, then we will say that π is a *category 2 representation*. We note that if π is a category 1 representation, then $\bar{N}_{\pi,s}$ is also equal to $N_\pi - 1$. Also, it is evident that if π is a category 1 representation, then $N_\pi \ge 2$. Thus, we have the following diagram:

$$\begin{array}{ll} N_\pi & \\ \left.\begin{array}{c} 0 \\ 1 \end{array}\right\} & \text{only category 2 representations} \\ \left.\begin{array}{c} 2 \\ 3 \\ \vdots \end{array}\right\} & \text{category 1 or category 2 representations} \end{array}$$

5.2 The Shadow of a Newform

The following lemma defines the shadow of a newform and proves some basic properties. As we shall see, the shadow of a newform is a key example of a stable Klingen vector. This vector was originally introduced in Sect. 7.4 of [105]. In the following lemma, in (5.10) and (5.11), the vector v_{new} is regarded as an element of $V_s(N_\pi + 1)$, while in (5.12), v_{new} is regarded as an element of $V_s(N_\pi)$. We recall the level lowering operators $s_n, \sigma_n : V_s(n+1) \to V_s(n)$ defined in (3.57) and (3.60).

Lemma 5.2.1 *Let (π, V) be an irreducible, admissible representation of the group* $\mathrm{GSp}(4, F)$ *with trivial central character. Assume that π is paramodular and that $N_\pi \ge 2$. Let v_{new} be a newform for π, i.e., a non-zero element of the one-dimensional space $V(N_\pi)$. Define*

$$v_s = q^3 s_{N_\pi - 1} v_{\mathrm{new}}, \tag{5.8}$$

so that

$$v_s = q^3 \int\limits_{\mathfrak{o}} \int\limits_{\mathfrak{o}} \int\limits_{\mathfrak{o}} \pi(\begin{bmatrix} 1 &&& \\ x\varpi^{N_\pi - 1} & 1 && \\ y\varpi^{N_\pi - 1} && 1 & \\ z\varpi^{N_\pi - 1} & y\varpi^{N_\pi - 1} & -x\varpi^{N_\pi - 1} & 1 \end{bmatrix}) v_{\mathrm{new}}\, dx\, dy\, dz. \tag{5.9}$$

We refer to v_s *as the* shadow *of the newform* $v_{\rm new}$. *The vector* v_s *is contained in the space* $V_s(N_\pi - 1)$. *Moreover,*

$$t_{N_\pi-1} v_s = q^3 \sigma_{N_\pi} v_{\rm new}, \tag{5.10}$$

$$\tau_{N_\pi-1} v_s = -q^2 \sigma_{N_\pi} v_{\rm new}, \tag{5.11}$$

$$v_s = -q^2 \sigma_{N_\pi-1} v_{\rm new}, \tag{5.12}$$

$$p_{N_\pi} \tau_{N_\pi-1} v_s = -(q+q^2)^{-1} \mu_\pi v_{\rm new}, \tag{5.13}$$

$$\rho'_{N_\pi-1} v_s = q^{-1} \mu_\pi v_{\rm new}, \tag{5.14}$$

$$-q^2 \tau_{N_\pi-1} v_s = T^s_{1,0} v_{\rm new}, \tag{5.15}$$

$$q^{-1} \mu_\pi v_{\rm new} = q^3 \tau_{N_\pi} \tau_{N_\pi-1} \sigma_{N_\pi-1} v_{\rm new} - q\eta\sigma_{N_\pi-1} v_{\rm new}. \tag{5.16}$$

Proof The vector v_s is contained in $V_s(N_\pi - 1)$ by the definition of $s_{N_\pi-1}$. To prove (5.10) we calculate:

$$\begin{aligned} t_{N_\pi-1} v_s &= q^3 t_{N_\pi-1} s_{N_\pi-1} v_{\rm new} \\ &= q^3 \sigma_{N_\pi} t_{N_\pi} v_{\rm new} \qquad \text{(by (5) of Lemma 3.7.3)} \\ &= q^3 \sigma_{N_\pi} v_{\rm new} \qquad \text{(since } v_{\rm new} \in V(N_\pi)\text{)}. \end{aligned}$$

Equation (5.11) follows from (5.10) because $-q\tau_{N_\pi-1} v_s = t_{N_\pi-1} v_s$ (by (3.35), since $p_{N_\pi-1} v_s = 0$). For (5.12), we have:

$$\begin{aligned} \tau_{N_\pi-1} v_s &= -q^2 \sigma_{N_\pi} v_{\rm new} \qquad \text{(by (5.11))} \\ &= -q^2 \sigma_{N_\pi} \tau_{N_\pi} v_{\rm new} \qquad \text{(since } v_{\rm new} \in V(N_\pi)\text{)} \\ &= -q^2 \tau_{N_\pi-1} \sigma_{N_\pi-1} v_{\rm new} \qquad \text{(by (6) of Lemma 3.7.3)}. \end{aligned}$$

Hence, $\tau_{N_\pi-1}(q^2 \sigma_{N_\pi-1} v_{\rm new} + v_s) = 0$. Since $\tau_{N_\pi-1} : V_s(N_\pi - 1) \to V(N_\pi)$ is injective by Lemma 3.5.4, we get $q^2 \sigma_{N_\pi-1} v_{\rm new} + v_s = 0$; this is (5.12). To prove (5.13), let dk be the Haar measure on $\mathrm{GSp}(4, F)$ that assigns $\mathrm{K}(\mathfrak{p}^n)$ volume 1. We have

$$\begin{aligned} p_{N_\pi} \tau_{N_\pi-1} v_s &= -q^{-1} p_{N_\pi} t_{N_\pi-1} v_s \qquad \text{(by (3.35))} \\ &= -q^2 p_{N_\pi} \sigma_{N_\pi} v_{\rm new} \qquad \text{(by (5.10))} \\ &= -q^2 \int\limits_{\mathrm{K}(\mathfrak{p}^n)} \int\limits_{\mathfrak{o}} \int\limits_{\mathfrak{o}} \end{aligned}$$

$$\times \int\limits_{\mathfrak{o}} \pi(k \begin{bmatrix} 1 & x & y & z\varpi^{-N_\pi+1} \\ & 1 & & y \\ & & 1 & -x \\ & & & 1 \end{bmatrix} \begin{bmatrix} \varpi & & & \\ & 1 & & \\ & & 1 & \\ & & & \varpi^{-1} \end{bmatrix}) v_{\mathrm{new}}\, dx\, dy\, dz\, dk$$

$$= -q^2 \int\limits_{\mathrm{K}(\mathfrak{p}^n)} \pi(k \begin{bmatrix} \varpi^2 & & & \\ & \varpi & & \\ & & \varpi & \\ & & & 1 \end{bmatrix}) v_{\mathrm{new}}\, dk$$

$$= -q^2(q^3+q^4)^{-1} T_{1,0} v_{\mathrm{new}} \qquad \text{(by Lemma 6.1.2 of [105])}$$

$$= -(q+q^2)^{-1} \mu_\pi v_{\mathrm{new}}.$$

This is (5.13). And

$$\rho'_{N_\pi-1} v_s = (1+q^{-1}) p_{N_\pi} t_{N_\pi-1} v_s \qquad \text{(by (3.47))}$$

$$= -q(1+q^{-1}) p_{N_\pi} \tau_{N_\pi-1} v_s \qquad \text{(by (3.35), since } p_{N_\pi-1} v_s = 0)$$

$$= q^{-1} \mu_\pi v_{\mathrm{new}} \qquad \text{(by (5.13))}.$$

This proves (5.14). Next,

$$-q^2 \tau_{N_\pi-1} v_s = -q^2 \tau_{N_\pi-1}(-q^2 \sigma_{N_\pi-1} v_{\mathrm{new}}) \qquad \text{(by (5.12))}$$

$$= q^4 \tau_{N_\pi-1} \sigma_{N_\pi-1} v_{\mathrm{new}}$$

$$= T^s_{1,0} v_{\mathrm{new}} \qquad \text{(by (3.76))}.$$

This is (5.15). Finally,

$$q^{-1} \mu_\pi v_{\mathrm{new}} = \rho'_{N_\pi-1} v_s \qquad \text{(by (5.14))}$$

$$= q^{-1} \eta v_s + \tau_{N_\pi} t_{N_\pi-1} v_s \qquad \text{(by (3.48))}$$

$$= -q\eta \sigma_{N_\pi-1} v_{\mathrm{new}} + q^3 \tau_{N_\pi} \sigma_{N_\pi} v_{\mathrm{new}} \qquad \text{(by (5.12) and (5.10))}$$

$$= -q\eta \sigma_{N_\pi-1} v_{\mathrm{new}} + q^3 \tau_{N_\pi} \sigma_{N_\pi} \tau_{N_\pi} v_{\mathrm{new}} \qquad (v_{\mathrm{new}} \in V(N_\pi))$$

$$= -q\eta \sigma_{N_\pi-1} v_{\mathrm{new}} + q^3 \tau_{N_\pi} \tau_{N_\pi-1} \sigma_{N_\pi-1} v_{\mathrm{new}} \qquad \text{((6) of Lemma 3.7.3)}.$$

This is (5.16). □

With the notation and assumptions as in Lemma 5.2.1, a basic observation is that the shadow of a newform provides an example of a vector in $V_s(N_\pi - 1)$. By Corollary 5.1.6 the stable Klingen level $N_{\pi,s}$ of π is either $N_\pi - 1$ or N_π. Thus, if the shadow of a newform is non-zero, then $N_{\pi,s} = N_\pi - 1$.

5.3 Zeta Integrals and Diagonal Evaluation

We will sometimes use zeta integrals to investigate stable Klingen vectors in generic representations. Let (π, V) be a generic, irreducible, admissible representation of $\mathrm{GSp}(4, F)$ with trivial central character. Let $c_1, c_2 \in \mathfrak{o}^\times$, and let $V = \mathcal{W}(\pi, \psi_{c_1,c_2})$ be the Whittaker model of π with respect to ψ_{c_1,c_2} (see Sect. 2.2). Let n be an integer such that $n \geq 0$, and let $W \in V_s(n)$. Since W is invariant under the elements

$$\begin{bmatrix} 1 & x & & \\ & 1 & & \\ & & 1 & -x \\ & & & 1 \end{bmatrix} \quad \text{and} \quad \begin{bmatrix} 1 & & & \\ & 1 & y & \\ & & 1 & \\ & & & 1 \end{bmatrix}$$

for $x, y \in \mathfrak{o}$, it follows that for $a, b, c \in F^\times$,

$$W(\begin{bmatrix} a & & & \\ & b & & \\ & & cb^{-1} & \\ & & & ca^{-1} \end{bmatrix}) = 0 \qquad \text{if } v(a) < v(b) \text{ or } 2v(b) < v(c). \tag{5.17}$$

Let i and j be integers, and recall the element $\Delta_{i,j}$ defined in (2.6). By (5.17) we see that

$$W(\Delta_{i,j}) = 0 \qquad \text{if } i < 0 \text{ or } j < 0. \tag{5.18}$$

We recall from Sect. 2.2 that the zeta integral of W is

$$Z(s, W) = \int\limits_{F^\times}\int\limits_{F} W(\begin{bmatrix} a & & & \\ & a & & \\ & x & 1 & \\ & & & 1 \end{bmatrix})|a|^{s-\frac{3}{2}}\, dx\, d^\times a. \tag{5.19}$$

By Lemma 4.1.1 of [105], the zeta integral of W is given by the simplified formula

$$Z(s, W) = \int\limits_{F^\times} W(\begin{bmatrix} a & & & \\ & a & & \\ & & 1 & \\ & & & 1 \end{bmatrix})|a|^{s-\frac{3}{2}}\, d^\times a. \tag{5.20}$$

Moreover, by (5.17), we even have that

$$Z(s, W) = \int\limits_{\substack{F^\times \\ v(a) \geq 0}} W(\begin{bmatrix} a & & & \\ & a & & \\ & & 1 & \\ & & & 1 \end{bmatrix})|a|^{s-\frac{3}{2}}\, d^\times a. \tag{5.21}$$

The following lemma describes the zeta integrals of stable Klingen vectors obtained by applying the operators of Sect. 3.5.

Lemma 5.3.1 *Let (π, V) be a generic, irreducible, admissible representation of* $\mathrm{GSp}(4, F)$ *with trivial central character. Let $c_1, c_2 \in \mathfrak{o}^\times$, and let $V = \mathcal{W}(\pi, \psi_{c_1,c_2})$*

be the Whittaker model of π with respect to ψ_{c_1,c_2}. Let n be an integer such that $n \geq 0$ and let $W \in V_s(n)$. Then

$$Z(s, \tau_n W) = Z(s, W), \tag{5.22}$$

$$Z(s, \theta W) = q^{-s+3/2} Z(s, W), \tag{5.23}$$

$$Z(s, \eta W) = 0, \tag{5.24}$$

$$Z(s, \rho_n' W) = \begin{cases} Z(s, W) & \text{if } W \in V(n), \\ -q Z(s, W) & \text{if } p_n(W) = 0. \end{cases} \tag{5.25}$$

Proof It is straightforward to verify (5.22). For (5.23), we have by (3.42),

$$Z(s, \theta W) = \int\limits_{F^\times} W(\begin{bmatrix} a & & & \\ & a & & \\ & & 1 & \\ & & & 1 \end{bmatrix}\begin{bmatrix} 1 & & & \\ & 1 & & \\ & & \varpi & \\ & & & \varpi \end{bmatrix})|a|^{s-\frac{3}{2}}\, d^\times a$$

$$+ q \int\limits_{\mathfrak{o}} \int\limits_{F^\times} W(\begin{bmatrix} a & & & \\ & a & & \\ & & 1 & \\ & & & 1 \end{bmatrix}\begin{bmatrix} 1 & & & \\ & 1 & c & \\ & & 1 & \\ & & & 1 \end{bmatrix}\begin{bmatrix} 1 & & & \\ & \varpi & & \\ & & 1 & \\ & & & \varpi \end{bmatrix})|a|^{s-\frac{3}{2}}\, d^\times a$$

$$= q^{-s+\frac{3}{2}} Z(s, W) + q \int\limits_{\mathfrak{o}} \int\limits_{F^\times} \psi(c_2 ac) W(\begin{bmatrix} a & & & \\ & a\varpi & & \\ & & 1 & \\ & & & \varpi \end{bmatrix})|a|^{s-\frac{3}{2}}\, d^\times a.$$

It follows from (5.17) that

$$W(\begin{bmatrix} a & & & \\ & a\varpi & & \\ & & 1 & \\ & & & \varpi \end{bmatrix}) = 0$$

for all $a \in F^\times$. This proves (5.23). The assertion (5.24) has a similar proof. Finally, for (5.25), observe (3.49) and (3.51). □

To close this section we make some basic observations about stable Hecke operators for generic representations. Let (π, V) be a generic, irreducible, admissible representation of the group $\mathrm{GSp}(4, F)$ with trivial central character. Let $V = \mathcal{W}(\pi, \psi_{c_1,c_2})$ be the Whittaker model of π; as usual, $c_1, c_2 \in \mathfrak{o}^\times$. Let $i, j \in \mathbb{Z}$. At various points in this work we will evaluate elements W of $V_s(n)$ at $\Delta_{i,j}$ when n is an integer such that $n \geq 0$. As an initial observation we note that, by (5.18), if $i < 0$ or $j < 0$, then $W(\Delta_{i,j}) = 0$. As concerns the stable Hecke operators we have the following lemma.

Lemma 5.3.2 *Let (π, V) be a generic, irreducible, admissible representation of the group* $\mathrm{GSp}(4, F)$ *with trivial central character. Let $V = \mathcal{W}(\pi, \psi_{c_1,c_2})$ be the Whittaker model of π; as usual, $c_1, c_2 \in \mathfrak{o}^\times$. Let n be an integer such that $n \geq 0$,* *and let* $W \in V_s(n)$*. Let* $i, j \in \mathbb{Z}$*. If* $i, j \geq 0$*, then*

$$(T^s_{0,1} W)(\Delta_{i,j}) = q^2 W(\Delta_{i+1,j-1}) + q^3 W(\Delta_{i,j+1}), \tag{5.26}$$

$$(T^s_{1,0} W)(\Delta_{i,j}) = q^4 W(\Delta_{i+j,j}). \tag{5.27}$$

Proof These formulas follow from (3.69) and (3.70). □

5.4 Dimensions for Some Generic Representations

Let (π, V) be a generic, irreducible, admissible representation of $\mathrm{GSp}(4, F)$ with trivial central character, and assume that $N_\pi \geq 2$. Under the assumption that $\mu_\pi \neq 0$, we can prove the following result about the structure of $V_s(n)$ when $n \geq N_{\pi,s} = N_\pi - 1$. This theorem will be used in the proof of the main result of this chapter.

To prove this theorem we will need a result from [105] about the shadow of a newform in π which we now recall. Let V be the Whittaker model $\mathcal{W}(\pi, \psi_{c_1,c_2})$ of π, let W_{new} be a newform for V, and let W_s be the shadow of W_{new}; we have $T_{1,0} W_{\mathrm{new}} = \mu_\pi W_{\mathrm{new}}$. In Proposition 7.4.8 of [105] it is proven that

$$Z(s, W_s) = -q^{-2} \mu_\pi Z(s, W_{\mathrm{new}}). \tag{5.28}$$

Since $Z(s, W_{\mathrm{new}})$ is a non-zero multiple of $L(s, \pi)$ by Theorem 7.5.4 of [105], it follows that if $\mu_\pi \neq 0$, then $W_s \neq 0$, so that $N_{\pi,s} = N_\pi - 1$ by Corollary 5.1.6, and π is a category 1 representation. In (5.29) of the following theorem, to simplify notation, we write τ for the level raising operator $\tau_k : V_s(k) \to V_s(k+1)$ for each integer $k \geq 0$; we also note that by (3.77) the operators τ and θ commute.

Theorem 5.4.1 *Let* π *be a generic, irreducible, admissible representation of the group* $\mathrm{GSp}(4, F)$ *with trivial central character. Assume that* $N_\pi \geq 2$*. Let* W_{new} *be a newform for* π*, and let* W_s *be the shadow of* W_{new}*. Assume that* $\mu_\pi \neq 0$ *so that* $W_s \neq 0$*, and* π *is a category 1 representation. Then*

$$V_s(n) = V(n-1) \oplus V(n) \oplus \bigoplus_{\substack{i,j \geq 0 \\ i+j = n - N_\pi + 1}} \mathbb{C} \tau^i \theta^j W_s \tag{5.29}$$

and

$$\dim V_s(n) = \frac{(n - N_\pi + 2)(n - N_\pi + 3)}{2} \tag{5.30}$$

and

$$\dim \bar{V}_s(n) = n - N_\pi + 2 \tag{5.31}$$

for all $n \geq N_{\pi,s} = \bar{N}_{\pi,s} = N_\pi - 1$.

Proof First we will prove that the sum on the right-hand side of (5.29) is direct by induction on n. The case $n = N_{\pi,s}$ is clear. Assume that $n \geq N_{\pi,s} + 1$, and that the sum on the right-hand side of (5.29) is direct for $n - 1$. Suppose that

$$W_1 + W_2 + \sum_{\substack{i,j \geq 0 \\ i+j=n-N_\pi+1}} c_i \tau^i \theta^j W_s = 0 \tag{5.32}$$

for some $W_1 \in V(n-1)$, $W_2 \in V(n)$ and complex numbers $c_0, \ldots, c_{n-N_\pi+1}$. Taking zeta integrals, and using Lemma 5.3.1 and (5.28), we get

$$Z(s, W_1) + Z(s, W_2) = q^{-2}\mu_\pi \sum_{\substack{i,j \geq 0 \\ i+j=n-N_\pi+1}} (q^{-s+3/2})^j c_i Z(s, W_{\mathrm{new}}). \tag{5.33}$$

By Theorem 7.5.6 of [105] we have

$$Z(s, W_1) \in (\mathbb{C} + \mathbb{C}q^{-s} + \cdots + \mathbb{C}(q^{-s})^{n-N_\pi-1}) Z(s, W_{\mathrm{new}})$$

and

$$Z(s, W_2) \in (\mathbb{C} + \mathbb{C}q^{-s} + \cdots + \mathbb{C}(q^{-s})^{n-N_\pi}) Z(s, W_{\mathrm{new}})$$

It therefore follows from (5.33) that $c_0 = 0$. From (5.32) we thus obtain

$$W_1 + W_2 + \tau_{n-1} \sum_{\substack{i,j \geq 0 \\ i+j=n-N_\pi}} c_{i+1} \tau^i \theta^j W_s = 0. \tag{5.34}$$

The map $\tau_{n-1} : \bar{V}_s(n-1) \to \bar{V}_s(n)$ is injective by Lemma 3.5.4. It follows that

$$\sum_{\substack{i,j \geq 0 \\ i+j=n-N_\pi}} c_{i+1} \tau^i \theta^j W_s \in V(n-2) \oplus V(n-1).$$

By the induction hypothesis, $c_1 = \ldots = c_{n-N_\pi+1} = 0$. Hence, $W_1 + W_2 = 0$, which implies $W_1 = W_2 = 0$ by Theorem 2.3.1. We have proven that the sum on the right-hand side of (5.29) is direct.

Next, since the sum is direct, since $\dim V(m) = \lfloor (m - N_\pi + 2)^2/4 \rfloor$ for integers m such that $m \geq N_\pi$ by Theorem 7.5.6 of [105], and since $\lfloor k^2/4 \rfloor + \lfloor (k+1)^2/4 \rfloor =$

$k(k+1)/2$ for any integer k,

$$\begin{aligned}\dim V_s(n) &\geq \dim V(n-1) + \dim V(n) + n - N_\pi + 2 \\ &= \left\lfloor \frac{(n-N_\pi+1)^2}{4} \right\rfloor + \left\lfloor \frac{(n-N_\pi+2)^2}{4} \right\rfloor + n - N_\pi + 2 \\ &= \frac{(n-N_\pi+2)(n-N_\pi+3)}{2}.\end{aligned}$$

On the other hand, by Theorem 5.1.5,

$$\begin{aligned}\dim V_s(n) &\leq \dim V(n) + \dim V(n+1) \\ &= \left\lfloor \frac{(n-N_\pi+2)^2}{4} \right\rfloor + \left\lfloor \frac{(n-N_\pi+3)^2}{4} \right\rfloor \\ &= \frac{(n-N_\pi+2)(n-N_\pi+3)}{2}.\end{aligned}$$

The equality (5.30) follows. It is also now clear that we have equality in (5.29) and that (5.31) holds. □

A similar result for arbitrary generic representations will be proven in Theorem 7.2.2.

5.5 Dimensions for Some Non-Generic Representations

In this section we will prove a number of lemmas about the dimensions of the spaces of stable Klingen vectors in various non-generic representations. These results will be used in the proof of Theorem 5.6.1, the main result of this chapter.

Lemma 5.5.1 *Let (π, V) be a representation of type IIIb, so that there exist characters χ and σ of $F^\times$ with $\chi \notin \{1, \nu^{\pm 2}\}$ such that $\pi = \chi \rtimes \sigma 1_{\mathrm{GSp}(2)}$. Assume that π has trivial central character, i.e., $\chi\sigma^2 = 1$. Then, for all integers n such that $n \geq 0$,*

$$\dim V_s(n) = \begin{cases} 2n+1 & \text{if } \sigma \text{ is unramified,} \\ 0 & \text{if } \sigma \text{ is ramified.} \end{cases}$$

Proof If σ is ramified, then π is not paramodular by Theorem 3.4.3 of [105]; hence $V_s(n) = 0$ for all integers $n \geq 0$ by Corollary 5.1.6. Assume that σ is unramified. Then χ is also unramified. By Sect. 2.2 there is an exact sequence

$$0 \longrightarrow \chi \rtimes \sigma \mathrm{St}_{\mathrm{GSp}(2)} \longrightarrow \chi \times \nu \rtimes \nu^{-\frac{1}{2}}\sigma \longrightarrow \pi \longrightarrow 0.$$

By Theorem 4.2.7, $\chi \times \nu \rtimes \nu^{-\frac{1}{2}}\sigma$ has stable Klingen level 0, and the dimension of the space of stable Klingen vectors of level $\mathfrak{p}^n$ in this representation is $(n^2+5n+2)/2$ for integers $n \geq 0$. By Theorem 5.4.1 and Table A.14 of [105], the IIIa representation $\chi \rtimes \sigma \mathrm{St}_{\mathrm{GSp}(2)}$ has stable Klingen level 1, and the dimension of the space of stable Klingen vectors of level $\mathfrak{p}^n$ in this representation is $n(n+1)/2$ for integers $n \geq 1$. The statement of the lemma now follows from the exact sequence. □

Lemma 5.5.2 *Let (π, V) be a representation of type IVb, so that there exists a character σ of $F^\times$ such that $\pi = L(\nu^2, \nu^{-1}\sigma \mathrm{St}_{\mathrm{GSp}(2)})$. Assume that π has trivial central character, i.e., $\sigma^2 = 1$. Then, for all integers $n \geq 0$,*

$$\dim V_s(n) = \begin{cases} n & \text{if } \sigma \text{ is unramified,} \\ 0 & \text{if } \sigma \text{ is ramified.} \end{cases}$$

Proof If σ is ramified, then π is not paramodular by Theorem 3.4.3 of [105]; hence $V_s(n) = 0$ for all integers $n \geq 0$ by Corollary 5.1.6. Assume that σ is unramified. By (2.14) there is an exact sequence

$$0 \longrightarrow \pi \longrightarrow \nu^{\frac{3}{2}} 1_{\mathrm{GL}(2)} \rtimes \nu^{-\frac{3}{2}}\sigma \longrightarrow \sigma 1_{\mathrm{GSp}(4)} \longrightarrow 0.$$

By Theorem 4.2.7, $\nu^{\frac{3}{2}} 1_{\mathrm{GL}(2)} \rtimes \nu^{-\frac{3}{2}}\sigma$ has stable Klingen level 0, and the dimension of the space of stable Klingen vectors of level $\mathfrak{p}^n$ in this representation is $n+1$ for integers $n \geq 0$. It is clear that $\sigma 1_{\mathrm{GSp}(4)}$ has stable Klingen level 0, and the dimension of the space of stable Klingen vectors of level $\mathfrak{p}^n$ in this representation is 1 for integers $n \geq 0$. The statement of the lemma now follows from the exact sequence. □

Lemma 5.5.3 *Let (π, V) be a representation of type IVc, so that there exists a character σ of $F^\times$ such that $\pi = L(\nu^{\frac{3}{2}} \mathrm{St}_{\mathrm{GL}(2)}, \nu^{-\frac{3}{2}}\sigma)$. Assume that π has trivial central character, i.e., $\sigma^2 = 1$. Then, for all integers n such that $n \geq 0$,*

$$\dim V_s(n) = \begin{cases} 2n & \text{if } \sigma \text{ is unramified,} \\ 0 & \text{if } \sigma \text{ is ramified.} \end{cases}$$

Proof If σ is ramified, then π is not paramodular by Theorem 3.4.3 of [105]; hence $V_s(n) = 0$ for all integers $n \geq 0$ by Corollary 5.1.6. Assume that σ is unramified. By (2.14) there is an exact sequence

$$0 \longrightarrow \sigma \mathrm{St}_{\mathrm{GSp}(4)} \longrightarrow \nu^{\frac{3}{2}} \mathrm{St}_{\mathrm{GL}(2)} \rtimes \nu^{-\frac{3}{2}}\sigma \longrightarrow \pi \longrightarrow 0.$$

By Theorem 4.2.7, $\nu^{\frac{3}{2}} \mathrm{St}_{\mathrm{GL}(2)} \rtimes \nu^{-\frac{3}{2}}\sigma$ has stable Klingen level 1, and the dimension of the space of stable Klingen vectors of level $\mathfrak{p}^n$ in this representation is $n(n+3)/2$ for integers $n \geq 1$. By Theorem 5.4.1 and Table A.14 of [105], the

IVa representation $\sigma\mathrm{St}_{\mathrm{GSp}(4)}$ has stable Klingen level 2, and the dimension of the space of stable Klingen vectors of level $\mathfrak{p}^n$ in this representation is $(n-1)n/2$ for integers $n \geq 2$. The statement of the lemma now follows from the exact sequence. □

Lemma 5.5.4 *Let (π, V) be a representation of type Vd, so that there exist a non-trivial quadratic character ξ of $F^\times$ and a character σ of $F^\times$ such that $\pi = L(\nu\xi, \xi \rtimes \nu^{-\frac{1}{2}}\sigma)$. Assume that π has trivial central character, i.e., $\sigma^2 = 1$. Then, for all integers n such that $n \geq 0$,*

$$\dim V_s(n) = \begin{cases} 1 & \textit{if } \xi \textit{ and } \sigma \textit{ are unramified,} \\ 0 & \textit{if } \xi \textit{ or } \sigma \textit{ is ramified.} \end{cases}$$

Proof If ξ or σ is ramified, then π is not paramodular by Theorem 3.4.3 of [105]; hence $V_s(n) = 0$ for all integers $n \geq 0$ by Corollary 5.1.6. Assume that ξ and σ are unramified. By Table A.12 of [105]

$$\dim V(n) = \begin{cases} 1 & \text{if } n \text{ is even,} \\ 0 & \text{if } n \text{ is odd,} \end{cases}$$

for all integers $n \geq 0$. Since $V_s(0) = V(0)$ and $V_s(n)$ contains $V(n-1) \oplus V(n)$ for integers $n \geq 1$, it follows that $\dim V_s(n) \geq 1$ for all integers $n \geq 0$. By Theorem 5.1.5, we also have $\dim V_s(n) \leq 1$ for all integers $n \geq 0$. This completes the proof. □

Lemma 5.5.5 *Let (π, V) be a representation of type Vb, so that there exist a non-trivial quadratic character ξ of $F^\times$ and a character σ of $F^\times$ such that $\pi = L(\nu^{\frac{1}{2}}\xi\mathrm{St}_{\mathrm{GL}(2)}, \nu^{-\frac{1}{2}}\sigma)$. Assume that π has trivial central character, i.e., $\sigma^2 = 1$. Then, for all integers n such that $n \geq 0$,*

$$\dim V_s(n) = \begin{cases} n & \textit{if } \sigma \textit{ and } \xi \textit{ are unramified,} \\ n - 2a(\xi) + 1 & \textit{if } n \geq 2a(\xi), \sigma \textit{ is unramified, and } \xi \textit{ is ramified,} \\ 0 & \textit{if } n < 2a(\xi), \sigma \textit{ is unramified, and } \xi \textit{ is ramified,} \\ 0 & \textit{if } \sigma \textit{ is ramified.} \end{cases}$$

Proof By (2.17) there is an exact sequence

$$0 \longrightarrow \pi \longrightarrow \nu^{\frac{1}{2}}\xi 1_{\mathrm{GL}(2)} \rtimes \xi\nu^{-\frac{1}{2}}\sigma \longrightarrow L(\nu\xi, \xi \rtimes \nu^{-\frac{1}{2}}\sigma) \longrightarrow 0.$$

The dimensions of the spaces of stable Klingen vectors of level $\mathfrak{p}^n$ for integers $n \geq 0$ in $L(\nu\xi, \xi \rtimes \nu^{-\frac{1}{2}}\sigma)$ and in $\nu^{\frac{1}{2}}\xi 1_{\mathrm{GL}(2)} \rtimes \nu^{-\frac{1}{2}}\xi\sigma$ are given by Lemma 5.5.4 and Theorem 4.2.7, respectively. The statement of the lemma now follows from the exact sequence. □

Lemma 5.5.6 *Let (π, V) be a representation of type Vc, so that there exist a non-trivial quadratic character ξ of $F^\times$ and a character σ of $F^\times$ such that $\pi = L(\nu^{\frac{1}{2}}\xi \mathrm{St}_{\mathrm{GL}(2)}, \xi\nu^{-\frac{1}{2}}\sigma)$. Assume that π has trivial central character, i.e., $\sigma^2 = 1$. Then, for all integers n such that $n \geq 0$,*

$$\dim V_s(n) = \begin{cases} n & \text{if } \sigma \text{ and } \xi \text{ are unramified,} \\ n - 2a(\xi) + 1 & \text{if } n \geq 2a(\xi), \xi\sigma \text{ is unramified, and } \xi \text{ is ramified,} \\ 0 & \text{if } n < 2a(\xi), \xi\sigma \text{ is unramified, and } \xi \text{ is ramified,} \\ 0 & \text{if } \xi\sigma \text{ is ramified.} \end{cases}$$

Proof This follows from Lemma 5.5.5. □

Lemma 5.5.7 *Let (π, V) be a representation of type VIc, so that there exists a character σ of $F^\times$ such that $\pi = L(\nu^{\frac{1}{2}}\mathrm{St}_{\mathrm{GL}(2)}, \nu^{-\frac{1}{2}}\sigma)$. Assume that π has trivial central character, i.e., $\sigma^2 = 1$. Then, for all integers n such that $n \geq 0$,*

$$\dim V_s(n) = \begin{cases} n & \text{if } \sigma \text{ is unramified,} \\ 0 & \text{if } \sigma \text{ is ramified.} \end{cases}$$

Proof If σ is ramified, then π is not paramodular by Theorem 3.4.3 of [105]; hence $V_s(n) = 0$ for all integers $n \geq 0$ by Corollary 5.1.6. Assume that σ is unramified. By (2.20) there is an exact sequence

$$0 \longrightarrow \tau(S, \nu^{-\frac{1}{2}}\sigma) \longrightarrow \nu^{\frac{1}{2}}\mathrm{St}_{\mathrm{GL}(2)} \rtimes \nu^{-\frac{1}{2}}\sigma \longrightarrow \pi \longrightarrow 0.$$

By Theorem 4.2.7, $\nu^{\frac{1}{2}}\mathrm{St}_{\mathrm{GL}(2)} \rtimes \nu^{-\frac{1}{2}}\sigma$ has stable Klingen level 1, and the dimension of the space of stable Klingen vectors of level $\mathfrak{p}^n$ in this representation is $n(n+3)/2$ for integers $n \geq 1$. By Theorem 5.4.1 and Table A.14 of [105], the VIa representation $\tau(S, \nu^{-\frac{1}{2}}\sigma)$ has stable Klingen level 1, and the dimension of the space of stable Klingen vectors of level $\mathfrak{p}^n$ in this representation is $n(n+1)/2$ for integers $n \geq 1$. The statement of the lemma follows now from the exact sequence. □

Lemma 5.5.8 *Let (π, V) be a representation of type VId, so that there exists a character σ of $F^\times$ such that $\pi = L(\nu, 1_{F^\times} \rtimes \nu^{-\frac{1}{2}}\sigma)$. Assume that π has trivial central character, i.e., $\sigma^2 = 1$. Then, for all integers n such that $n \geq 0$,*

$$\dim V_s(n) = \begin{cases} n+1 & \text{if } \sigma \text{ is unramified,} \\ 0 & \text{if } \sigma \text{ is ramified.} \end{cases}$$

Proof If σ is ramified, then π is not paramodular by Theorem 3.4.3 of [105]; hence $V_s(n) = 0$ for all integers $n \geq 0$ by Corollary 5.1.6. Assume that σ is unramified.

By (2.20) there is an exact sequence

$$0 \longrightarrow \tau(T, \nu^{-\frac{1}{2}}\sigma) \longrightarrow \nu^{\frac{1}{2}}1_{\mathrm{GL}(2)} \rtimes \nu^{-\frac{1}{2}}\sigma \longrightarrow \pi \longrightarrow 0.$$

By Theorem 3.4.3 of [105] the IVb representation is not paramodular, and hence has no non-zero stable Klingen vectors of any level. It follows from the exact sequence that, for integers $n \geq 0$, the dimension of $V_s(n)$ is the same as the dimension of the space of stable Klingen vectors of level $\mathfrak{p}^n$ in $\nu^{\frac{1}{2}}1_{\mathrm{GL}(2)} \rtimes \nu^{-\frac{1}{2}}\sigma$; by Theorem 4.2.7 this dimension is $n + 1$. □

Lemma 5.5.9 *Let (π, V) be a representation of type XIb, so that there exists an irreducible, admissible, supercuspidal representation τ of $\mathrm{GL}(2, F)$ with trivial central character and a character σ of $F^\times$ such that $\pi = L(\nu^{\frac{1}{2}}\tau, \nu^{-\frac{1}{2}}\sigma)$. Assume that π has trivial central character, i.e., $\sigma^2 = 1$. Then, for all integers n such that $n \geq 0$,*

$$\dim V_s(n) = \begin{cases} n - N_\tau + 1 & \text{if } n \geq N_\tau \text{ and } \sigma \text{ is unramified,} \\ 0 & \text{if } n < N_\tau \text{ or } \sigma \text{ is ramified.} \end{cases}$$

Proof If σ is ramified, then π is not paramodular by Theorem 3.4.3 of [105]; hence $V_s(n) = 0$ for all integers $n \geq 0$ by Corollary 5.1.6. Assume that σ is unramified. By Sect. 2.2 there is an exact sequence

$$0 \longrightarrow \delta(\nu^{\frac{1}{2}}\tau, \nu^{-\frac{1}{2}}\sigma) \longrightarrow \nu^{\frac{1}{2}}\tau \rtimes \nu^{-\frac{1}{2}}\sigma \longrightarrow \pi \longrightarrow 0.$$

By Theorem 4.2.7, $\nu^{\frac{1}{2}}\tau \rtimes \nu^{-\frac{1}{2}}\sigma$ has stable Klingen level N_τ, and the dimension of the space of stable Klingen vectors of level $\mathfrak{p}^n$ in this representation is $(n - N_\tau + 1)(n - N_\tau + 4)/2$ for integers $n \geq N_\tau$. By Theorem 5.4.1 and Table A.14 of [105], the XIa representation $\delta(\nu^{\frac{1}{2}}\tau, \nu^{-\frac{1}{2}}\sigma)$ has stable Klingen level N_τ, and the dimension of the space of stable Klingen vectors of level $\mathfrak{p}^n$ in this representation is $(n - N_\tau + 1)(n - N_\tau + 2)/2$ for integers $n \geq N_\tau$. The statement of the lemma follows now from the exact sequence. □

5.6 The Table of Dimensions

We will now prove the main result of this chapter.

Theorem 5.6.1 *For every irreducible, admissible representation (π, V) of the group $\mathrm{GSp}(4, F)$ with trivial central character, the stable Klingen level $N_{\pi,s}$, the dimensions of the spaces $V_s(n)$, the quotient stable Klingen level $\bar{N}_{\pi,s}$, and the dimensions of the spaces $\bar{V}_s(n)$ are given in Table A.3.*

Proof We begin by noting that Table A.12 of [105] indicates, for every π, whether or not π is paramodular, and lists N_π in the case that π is paramodular. If π is not paramodular, then this is indicated as such in Table A.3; if π is paramodular and $\bar{N}_{\pi,s}$ is not defined, then the entry for $\bar{N}_{\pi,s}$ is $-$. Also, in this proof we will sometimes use Theorem 5.4.1; the hypothesis $\mu_\pi \neq 0$ used in this theorem can be verified using Table A.14 of [105] (this information also appears in Table A.4 of this work).

The entries in the table are now verified as follows. For Groups I and II, the entries follow from Theorem 4.2.7. For Group IIIa, the entries follow from Theorem 5.4.1. For Group IIIb, the entries follow from Lemma 5.5.1. For Group IVa, the entries follow from Theorem 5.4.1. For Group IVb, the entries follow from Lemma 5.5.2. For Group IVc, the entries follow from Lemma 5.5.3. For Group IVd, the entries are easily verified as π is a twist of the trivial representation. For Group Va, the entries follow from Theorem 5.4.1. For Group Vb, the entries follow from Lemma 5.5.5. For Group Vc, the entries follow from Lemma 5.5.6. For Group Vd, the entries follow from Lemma 5.5.4. For Group VIa, the entries follow from Theorem 5.4.1. For Group VIb, we note that these representations are never paramodular. For Group VIc, the entries follow from Lemma 5.5.7. For Group VId, the entries follow from Lemma 5.5.8. For Groups VII and VIIIa, the entries follow from Theorem 5.4.1. For Group VIIIb, we note that these representations are never paramodular. For Group IXa, the entries follow from Theorem 5.4.1. For Group IXb, we note that these representations are never paramodular. For Group X, the entries follow from Theorem 4.2.7. For Group XIa, the entries follow from Theorem 5.4.1. For Group XIb, the entries follow from Lemma 5.5.9. If π is generic and supercuspidal, the entries follow from Theorem 5.4.1; the assertion that $a = N_\pi \geq 4$ will be proven in Theorem 8.4.7. Finally, if π is non-generic and supercuspidal, then π is never paramodular. □

Using Theorem 5.6.1 we can update the schematic diagram of paramodular representations from Fig. 2.1 on p. 87. The result is shown in Fig. 5.1 on p. 164. We note that every paramodular Saito-Kurokawa representation is a category 2 paramodular representation. Also, the only non-generic category 1 representations are the IVb representations with σ unramified. For the convenience of the reader we also include Table A.4 on p. 343. This table lists the irreducible, admissible representations π of GSp(4, F) with trivial central character; if π is paramodular, then the table lists the paramodular level N_π, the Atkin-Lehner eigenvalue ε_π, the paramodular Hecke eigenvalues λ_π and μ_π, whether π is category 1 or category 2, and any comments about π.

5.7 Some Consequences

To conclude this chapter we will derive some consequences from Theorem 5.6.1. Our first result provides a remarkable characterization of category 1 paramodular representations.

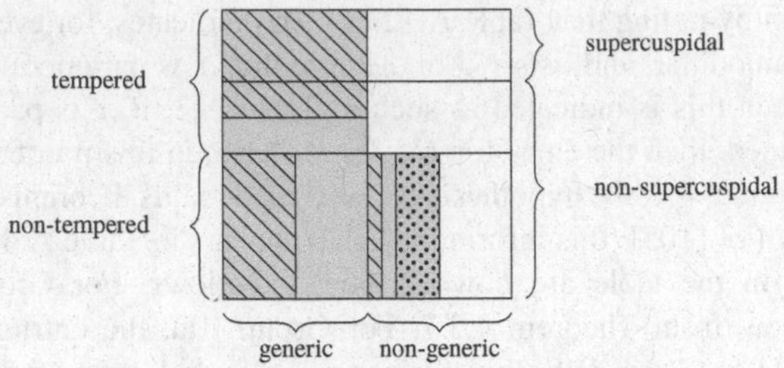

Fig. 5.1 Paramodular representations with categories. A schematic diagram of the paramodular representations among all irreducible, admissible representations of GSp(4, F) with trivial central character. Paramodular representations are in gray and category 1 paramodular representations are hatched. The paramodular Saito-Kurokawa representations are dotted; all of these are category 2 paramodular representations

Corollary 5.7.1 *Let* (π, V) *be an irreducible, admissible representation of* GSp(4, F) *with trivial central character. Assume that* π *is paramodular. Then the following are equivalent.*

(1) π *is a category 1 paramodular representation, i.e.,* $N_{\pi,s} = N_\pi - 1$.
(2) *The decomposition of the L-parameter of* π *into indecomposable representations contains no unramified one-dimensional factors.*

Proof For non-supercuspidal representations, the equivalence of (1) and (2) follows by inspecting Table A.3 from Theorem 5.6.1 and Table A.7 of [105]. Non-generic supercuspidals do not admit paramodular vectors by Theorem 3.4.3 of [105]. Assume that π is a generic and supercuspidal. Then $N_{\pi,s} = N_\pi - 1$ by Table A.3. The L-parameter of π is a discrete series parameter, and thus does not factor through the Levi of any proper parabolic of Sp(4, $\mathbb{C}$). Suppose that the decomposition of the L-parameter (φ, W) contains a one-dimensional factor, so that $W = V_1 \oplus V_2$ for some W'_F-subspaces V_1, V_2 of W with $\dim(V_1) = 1$; we will obtain a contradiction. Let $\langle\cdot,\cdot\rangle$ be the non-degenerate symplectic form on W preserved by φ. Let $x \in V_1$ be non-zero, so that $V_1 = \mathbb{C}x$. Choose a non-zero $y \in V_2$ such that $\langle y, v\rangle = 0$ for all $v \in V_2$; such a y exists because V_2 is odd-dimensional. Since W is non-degenerate, we may assume that $\langle x, y\rangle = 1$. Let $U = (\mathbb{C}x)^\perp \cap V_2$. We have

$$\begin{aligned}\dim(U) &= \dim((\mathbb{C}x)^\perp) + \dim(V_2) - \dim\left((\mathbb{C}x)^\perp + V_2\right)\\ &= 3 + 3 - 4 = 2.\end{aligned}$$

Clearly, $y \notin U$, so that $W = \mathbb{C}x \oplus U \oplus \mathbb{C}y$ with $(\mathbb{C}x + \mathbb{C}y) \perp U$. Let x', y' be a basis of U such $\langle x', y'\rangle = 1$. Then x, x', y', y is an ordered basis of W, with respect to which the symplectic form is J, the matrix defined in (2.3). We now have that

$$\varphi(W_F') \subset \begin{bmatrix} * & & & \\ & * & * & * \\ & * & * & * \\ & & & * \end{bmatrix} \cap \mathrm{Sp}(4, \mathbb{C}).$$

This implies that $\varphi(W_F')$ is contained in the Levi component of the Klingen parabolic, a contradiction. □

An examination of Table A.3 also establishes the following corollary.

Corollary 5.7.2 *Let (π, V) be an irreducible, admissible representation of* $\mathrm{GSp}(4, F)$ *with trivial central character. Assume that π is paramodular. Then* $\dim V_s(N_{\pi,s})$ *is 1 or 2. Further, assume that $N_\pi \geq 2$. If π is generic, then*

$$\dim V_s(N_{\pi,s}) = \begin{cases} 1 & \text{if } N_{\pi,s} = N_\pi - 1, \\ 2 & \text{if } N_{\pi,s} = N_\pi, \end{cases} \tag{5.35}$$

and if π is non-generic, then

$$\dim V_s(N_{\pi,s}) = 1. \tag{5.36}$$

The previous corollary suggests that the theory of stable Klingen vectors is regular if the minimal paramodular level is at least two, and this observation is reinforced by the next proposition. The following result provides practical criteria for identifying category 1 paramodular representations under the assumption that the minimal paramodular level is at least two. We will use this proposition in the next chapter.

Proposition 5.7.3 *Let (π, V) be an irreducible, admissible representation of the group* $\mathrm{GSp}(4, F)$ *with trivial central character. Assume that π is paramodular, and that $N_\pi \geq 2$. Let $v_{\mathrm{new}} \in V(N_\pi)$ be a newform, i.e., a non-zero element of $V(N_\pi)$. Let $v_s \in V_s(N_\pi - 1)$ be the shadow of v_{new} as defined in Lemma 5.2.1. Then the following are equivalent:*

(1) *π is a category 1 paramodular representation, i.e., $N_{\pi,s} = N_\pi - 1$.*
(2) *The $T_{1,0}$-eigenvalue μ_π on v_{new} is non-zero.*
(3) $\sigma_{N_\pi - 1} v_{\mathrm{new}} \neq 0$.
(4) $\sigma_{N_\pi} v_{\mathrm{new}} \neq 0$.
(5) $T_{1,0}^s v_{\mathrm{new}} \neq 0$.
(6) $v_s \neq 0$.

Proof (1) ⇔ (2) follows by inspecting Table A.3 from Theorem 5.6.1 and Table A.14 of [105].

(3) ⇔ (4). By (6) of Lemma 3.7.3 we have $\sigma_{N_\pi} \tau_{N_\pi} v_{\mathrm{new}} = \tau_{N_\pi - 1} \sigma_{N_\pi - 1} v_{\mathrm{new}}$. Since $v_{\mathrm{new}} \in V(N_\pi)$ we have $\tau_{N_\pi} v_{\mathrm{new}} = v_{\mathrm{new}}$, so that $\sigma_{N_\pi} v_{\mathrm{new}} = \tau_{N_\pi - 1} \sigma_{N_\pi - 1} v_{\mathrm{new}}$. The injectivity of $\tau_{N_\pi - 1} : V_s(N_\pi - 1) \to V_s(N_\pi)$ from (1) of Lemma 3.5.4 now implies that (3) and (4) are equivalent.

(3) $\Leftrightarrow$ (5) follows from Lemma 3.8.4.

(4) $\Leftrightarrow$ (6) follows from (5.10).

(6) $\Rightarrow$ (1) follows from Corollary 5.1.6.

(2) $\Rightarrow$ (5). By (3.74) we have $(1+q^{-1})p_{N_\pi}T^s_{1,0}v_{\text{new}} = T_{1,0}v_{\text{new}}$. Hence, if $T_{1,0}v_{\text{new}} \neq 0$, then $T^s_{1,0}v_{\text{new}} \neq 0$. □

Chapter 6
Hecke Eigenvalues and Minimal Levels

Let π be a paramodular, irreducible, admissible representation of $\mathrm{GSp}(4, F)$ with trivial central character such that $N_\pi \geq 2$. In this chapter we will consider the action of the stable Klingen operators $T^s_{0,1}$ and $T^s_{1,0}$ on the vector spaces $V_s(N_{\pi,s})$ and $V_s(N_\pi)$. We will show that the actions of $T^s_{0,1}$ and $T^s_{1,0}$ can be described in terms of the paramodular Hecke eigenvalues λ_π and μ_π, and in the final section we will explain how these results can be used to compute the paramodular Hecke eigenvalues λ_π and μ_π, and determine whether π is non-generic, using explicit upper block operators (see Sect. 3.4 for the definition). As outlined in the introduction of this work, this has useful applications to Siegel paramodular newforms. Before beginning, we mention that throughout this chapter we will often use the equivalences of Proposition 5.7.3.

6.1 At the Minimal Stable Klingen Level

Let π be a paramodular, irreducible, admissible representation of $\mathrm{GSp}(4, F)$ with trivial central character such that $N_\pi \geq 2$. In the following theorem we describe the action of the stable Klingen Hecke operators $T^s_{0,1}$ and $T^s_{1,0}$ on $V_s(N_{\pi,s})$ via the paramodular Hecke eigenvalues λ_π and μ_π. The proof of this theorem uses the theory developed in the preceding chapters as well as some results from [105].

Theorem 6.1.1 *Let (π, V) be an irreducible, admissible representation of the group $\mathrm{GSp}(4, F)$ with trivial central character. Assume that π is paramodular and that $N_\pi \geq 2$. Let $v_{\mathrm{new}} \in V(N_\pi)$ be a newform, and let v_s be the shadow of v_{new}.*

(1) *Assume that π is a category 1 representation, so that $N_{\pi,s} = N_\pi - 1$ and $\mu_\pi \neq 0$. Then $V_s(N_{\pi,s}) = V_s(N_\pi - 1)$ is one-dimensional and*

$$T^s_{0,1} v_s = \lambda_\pi v_s \quad \textit{and} \quad T^s_{1,0} v_s = (\mu_\pi + q^2) v_s. \tag{6.1}$$

J. Johnson-Leung et al., *Stable Klingen Vectors and Paramodular Newforms*, Lecture Notes in Mathematics 2342, https://doi.org/10.1007/978-3-031-45177-5_6

(2) *Assume that* π *is a category 2 representation, so that* $N_{\pi,s} = N_\pi$ *and* $\mu_\pi = 0$. *Let* $v_{\rm new} \in V(N_\pi)$ *be a newform. The vector space* $V_s(N_{\pi,s})$ *is spanned by the vectors* $v_{\rm new}$ *and* $T^s_{0,1} v_{\rm new}$. *If* $v_{\rm new}$ *is not an eigenvector for* $T^s_{0,1}$, *so that* $V_s(N_{\pi,s})$ *is two-dimensional, then* π *is generic, and the matrix of* $T^s_{0,1}$ *in the ordered basis* $v_{\rm new}, T^s_{0,1} v_{\rm new}$ *is*

$$\begin{bmatrix} 0 & -q^3 \\ 1 & \lambda_\pi \end{bmatrix}, \tag{6.2}$$

so that

$$q^3 v_{\rm new} + (T^s_{0,1})^2 v_{\rm new} = \lambda_\pi T^s_{0,1} v_{\rm new}. \tag{6.3}$$

If $v_{\rm new}$ *is an eigenvector for* $T^s_{0,1}$, *so that* $V_s(N_{\pi,s})$ *is one-dimensional, then* π *is non-generic, and*

$$T^s_{0,1} v_{\rm new} = (1 + q^{-1})^{-1} \lambda_\pi v_{\rm new}. \tag{6.4}$$

Proof (1). By Proposition 5.7.3 v_s is non-zero, and by Corollary 5.7.2 $V(N_{\pi,s})$ is one-dimensional. Let $\mu_{\pi,s}, \lambda_{\pi,s} \in \mathbb{C}$ be such that $T^s_{0,1} v_s = \mu_{\pi,s} v_s$ and $T^s_{1,0} v_s = \lambda_{\pi,s} v_s$. We need to prove that $\lambda_{\pi,s} = \lambda_\pi$ and $\mu_{\pi,s} = \mu_\pi + q^2$. We have

$$\begin{aligned} T^s_{0,1} \rho'_{N_\pi - 1} v_s &= q\theta v_s - q\tau_{N_\pi - 1} T^s_{0,1} v_s && \text{(by (3.91))} \\ q^{-1} \mu_\pi T^s_{0,1} v_{\rm new} &= q\theta v_s - q\lambda_{\pi,s} \tau_{N_\pi - 1} v_s && \text{(by (5.14))}. \end{aligned}$$

Applying p_{N_π} to this equation, we obtain:

$$\begin{aligned} q^{-1} \mu_\pi p_{N_\pi} T^s_{0,1} v_{\rm new} &= q p_{N_\pi} \theta v_s - q\lambda_{\pi,s} p_{N_\pi} \tau_{N_\pi - 1} v_s \\ (1+q)^{-1} \mu_\pi T_{0,1} v_{\rm new} &= q p_{N_\pi} \theta v_s - q\lambda_{\pi,s} p_{N_\pi} \tau_{N_\pi - 1} v_s && \text{(by (3.73))} \\ (1+q)^{-1} \mu_\pi \lambda_\pi v_{\rm new} &= q p_{N_\pi} \theta v_s + (1+q)^{-1} \lambda_{\pi,s} \mu_\pi v_{\rm new} && \text{(by (5.13))} \\ (1+q)^{-1} \mu_\pi \lambda_\pi v_{\rm new} &= q\theta p_{N_\pi - 1} v_s + (1+q)^{-1} \lambda_{\pi,s} \mu_\pi v_{\rm new} && \text{(by (3.80))} \\ (1+q)^{-1} \mu_\pi \lambda_\pi v_{\rm new} &= (1+q)^{-1} \lambda_{\pi,s} \mu_\pi v_{\rm new} && \text{(since } p_{N_\pi - 1} v_s = 0). \end{aligned}$$

Since $\mu_\pi \neq 0$ we obtain $\lambda_{\pi,s} = \lambda_\pi$, as desired.

Next,

$$\begin{aligned} T^s_{1,0} \rho'_{N_\pi - 1} v_s &= q^3 \tau_{N_\pi - 1} v_s - q\tau_{N_\pi - 1} T^s_{1,0} v_s && \text{(by (3.93))} \\ q^{-1} \mu_\pi T^s_{1,0} v_{\rm new} &= q^3 \tau_{N_\pi - 1} v_s - q\mu_{\pi,s} \tau_{N_\pi - 1} v_s && \text{(by (5.14))}. \end{aligned}$$

Applying p_{N_π} to this equation, we have:

$$q^{-1}\mu_\pi p_{N_\pi} T^s_{1,0} v_{\text{new}} = (q^3 - q\mu_{\pi,s}) p_{N_\pi} \tau_{N_\pi - 1} v_s$$

$$(1+q)^{-1}\mu_\pi T_{1,0} v_{\text{new}} = (q^3 - q\mu_{\pi,s}) p_{N_\pi} \tau_{N_\pi - 1} v_s \qquad \text{(by (3.74))}$$

$$(1+q)^{-1}\mu_\pi^2 v_{\text{new}} = -(q+q^2)^{-1}(q^3 - q\mu_{\pi,s})\mu_\pi v_{\text{new}} \qquad \text{(by (5.13))}.$$

Since $\mu_\pi \neq 0$ by Proposition 5.7.3 we obtain $\mu_{\pi,s} = \mu_\pi + q^2$. This completes the proof of (1).

(2). By Corollary 5.7.2, since $N_{\pi,s} = N_\pi$, the vector space $V_s(N_{\pi,s})$ is two-dimensional if π is generic, and $V_s(N_{\pi,s})$ is one-dimensional if π is non-generic. Thus, to prove that v_{new} and $T^s_{0,1} v_{\text{new}}$ span $V_s(N_{\pi,s})$ it will suffice to prove that v_{new} and $T^s_{0,1} v_{\text{new}}$ are linearly independent if π is generic. Assume that π is generic, and that v_{new} and $T^s_{0,1} v_{\text{new}}$ are linearly dependent; we will obtain a contradiction. Let $c \in \mathbb{C}$ be such that $T^s_{0,1} v_{\text{new}} = c v_{\text{new}}$. We have:

$$T^s_{0,1}\rho'_{N_\pi} v_{\text{new}} = q\theta v_{\text{new}} + \tau_{N_\pi} T^s_{0,1} v_{\text{new}} \qquad \text{(by (3.92))}$$

$$q^{-1} T^s_{0,1}\theta' v_{\text{new}} = q\theta v_{\text{new}} + c v_{\text{new}} \qquad \text{(by (3.49) and } v_{\text{new}} \in V(N_\pi)).$$

Applying $p_{N_\pi+1}$ to this equation, we obtain:

$$q^{-1} p_{N_\pi+1} T^s_{0,1}\theta' v_{\text{new}} = q p_{N_\pi+1}\theta v_{\text{new}} + c p_{N_\pi+1} v_{\text{new}}$$

$$T_{0,1}\theta' v_{\text{new}} = (q+q^2)\theta v_{\text{new}} + c\theta' v_{\text{new}} \qquad \text{(by (3.73), (3.35), and (2.29))}$$

$$\theta' T_{0,1} v_{\text{new}} + q^2\theta v_{\text{new}} = (q+q^2)\theta v_{\text{new}} + c\theta' v_{\text{new}} \qquad \text{(by (6.17) of [105])}$$

$$\lambda_\pi \theta' v_{\text{new}} + q^2\theta v_{\text{new}} = (q+q^2)\theta v_{\text{new}} + c\theta' v_{\text{new}}$$

$$(\lambda_\pi - c)\theta' v_{\text{new}} = q\theta v_{\text{new}}.$$

Since π is generic, the vectors $\theta' v_{\text{new}}$ and θv_{new} are linearly independent by Theorem 7.5.6 of [105]. It follows that $q = 0$, a contradiction. This completes the proof that v_{new} and $T^s_{0,1} v_{\text{new}}$ span $V_s(N_{\pi,s})$.

Next, assume that v_{new} is not an eigenvector for $T^s_{0,1}$, so that v_{new} and $T^s_{0,1} v_{\text{new}}$ are linearly independent. Then by Corollary 5.7.2 the vector space $V_s(N_{\pi,s})$ is two-dimensional and π is generic. We need to prove that the matrix of $T^s_{0,1}$ in the ordered basis $v_{\text{new}}, T^s_{0,1} v_{\text{new}}$ is as in (6.2). It is obvious that the first column of the matrix of $T^s_{0,1}$ is as in (6.2). To determine the second column of $T^s_{0,1}$, let $a, b \in \mathbb{C}$ be such that

$$(T^s_{0,1})^2 v_{\text{new}} = a v_{\text{new}} + b T^s_{0,1} v_{\text{new}}. \tag{6.5}$$

We need to prove that $a = -q^3$ and $b = \lambda_\pi$. To do this, we will assume that $V = \mathcal{W}(\pi, \psi_{c_1,c_2})$, the Whittaker model of π with respect to ψ_{c_1,c_2}; as usual, we assume that $c_1, c_2 \in \mathfrak{o}^\times$. As in (7.15) of [105], we define $c_{i,j} = v_{\text{new}}(i,j)$ for $i, j \in \mathbb{Z}$; here $\Delta_{i,j}$ is defined as in (2.6). Let $j \in \mathbb{Z}$ be such that $j \geq 0$. Evaluating the left-hand side of (6.5) at $\Delta_{0,j}$, we have by (5.26),

$$\begin{aligned}((T^s_{0,1})^2 v_{\text{new}})(\Delta_{0,j}) &= q^4 v_{\text{new}}(\Delta_{2,j-2}) + 2q^5 v_{\text{new}}(\Delta_{1,j}) + q^6 W(\Delta_{0,j+2}) \\ &= q^4 c_{2,j-2} + 2q^5 c_{1,j} + q^6 c_{0,j+2}. \end{aligned} \tag{6.6}$$

By Lemma 7.4.4 of [105],

$$c_{2,j-2} = q^{-4}(\mu_\pi + q^2) c_{1,j-2} = q^{-4}(\mu_\pi + q^2) q^{-4} \mu_\pi c_{0,j-2} = 0$$

and also $c_{1,j} = q^{-4} \mu_\pi c_{0,j} = 0$ because $\mu_\pi = 0$. Hence,

$$((T^s_{0,1})^2 v_{\text{new}}))(\Delta_{0,j}) = q^6 c_{0,j+2}. \tag{6.7}$$

Evaluating the right-hand side of (6.5) at $\Delta_{0,j}$, we have by (5.26),

$$\begin{aligned}(a v_{\text{new}} + b T^s_{0,1} v_{\text{new}})(\Delta_{0,j}) &= a v_{\text{new}}(\Delta_{0,j}) + b q^2 v_{\text{new}}(\Delta_{1,j-1}) + b q^3 v_{\text{new}}(\Delta_{0,j+1}) \\ &= a c_{0,j} + b q^2 c_{1,j-1} + b q^3 c_{0,j+1} \\ &= a c_{0,j} + b q^3 c_{0,j+1}, \end{aligned} \tag{6.8}$$

because again by Lemma 7.4.4 of [105], $c_{1,j-1} = q^{-4} \mu_\pi c_{0,j-1} = 0$ since $\mu_\pi = 0$. Since (6.7) and (6.8) are equal,

$$a c_{0,j} + b q^3 c_{0,j+1} = q^6 c_{0,j+2} \qquad \text{for all } j \geq 0.$$

By Lemma 7.4.4 of [105], $c_{0,j+2} = q^{-3} \lambda_\pi c_{0,j+1} - q^{-3} c_{0,j}$. Substituting, we find

$$a c_{0,j} + b q^3 c_{0,j+1} = q^3 \lambda_\pi c_{0,j+1} - q^3 c_{0,j} \qquad \text{for all } j \geq 0,$$

or

$$(a + q^3) c_{0,j} = q^3 (\lambda_\pi - b) c_{0,j+1} \qquad \text{for all } j \geq 0.$$

Now assume that $b \neq \lambda_\pi$. We then have

$$c_{0,j} = M^j c_{0,0} \qquad \text{for all } j \geq 0,$$

where

$$M = \frac{a + q^3}{q^3(\lambda_\pi - b)}.$$

A standard calculation using Lemma 4.1.1 and Lemma 4.1.2 of [105] shows that

$$Z(s, v_{\text{new}}) = (1 - q^{-1}) \sum_{j=0}^{\infty} c_{0,j} (q^{\frac{3}{2}} q^{-s})^j. \tag{6.9}$$

By (6.9) we thus obtain

$$Z(s, v_{\text{new}}) = \frac{(1 - q^{-1}) c_{0,0}}{1 - M q^{\frac{3}{2}} q^{-s}}. \tag{6.10}$$

On the other hand, by Proposition 7.4.5 of [105], since $\mu_\pi = 0$, we have

$$Z(s, v_{\text{new}}) = \frac{(1 - q^{-1}) c_{0,0}}{1 - q^{-\frac{3}{2}} \lambda_\pi q^{-s} + q^{-2s}}. \tag{6.11}$$

This contradicts (6.10). Hence, $b = \lambda_\pi$, which then implies $a = -q^3$, as desired.

Finally, assume that v_{new} is an eigenvector for $T^s_{0,1}$, so that $V_s(N_{\pi,s})$ is one-dimensional and spanned by v_{new}. Then, as we have already indicated above, π is non-generic by Corollary 5.7.2. Let $\lambda_{\pi,s} \in \mathbb{C}$ be such that $T^s_{0,1} v_{\text{new}} = \lambda_{\pi,s} v_{\text{new}}$. By (3.73),

$$\begin{aligned}
(1 + q^{-1}) \lambda_{\pi,s} v_{\text{new}} &= (1 + q^{-1}) p_{N_\pi} (\lambda_{\pi,s} v_{\text{new}}) \\
&= (1 + q^{-1}) p_{N_\pi} (T^s_{0,1} v_{\text{new}}) \\
&= T_{0,1} v_{\text{new}} \\
&= \lambda_\pi v_{\text{new}}.
\end{aligned}$$

Hence, $\lambda_{\pi,s} = (1 + q^{-1})^{-1} \lambda_\pi$. This completes the proof. □

As the next lemma shows, in case (1) of Theorem 6.1.1 the eigenvalue μ_π has a simple form, so that the action of $T^s_{1,0}$ on the shadow vector is also explicit.

Corollary 6.1.2 *Let π be an irreducible, admissible representation of the group* $\mathrm{GSp}(4, F)$ *with trivial central character. Assume that π is paramodular, $N_\pi \geq 2$, and π is a category 1 representation, so that $N_{\pi,s} = N_\pi - 1$ and $\mu_\pi \neq 0$. Let $v_{\text{new}} \in V(N_\pi)$ be a newform, and let $v_s \in V_s(N_\pi - 1)$ be the shadow of v_{new}. Then*

$$\mu_\pi = \begin{cases} -q^2 + \varepsilon_\pi q & \text{if } N_\pi = 2, \\ -q^2 & \text{if } N_\pi > 2, \end{cases} \tag{6.12}$$

so that

$$T^s_{1,0} v_s = \begin{cases} \varepsilon_\pi q v_s & \textit{if } N_\pi = 2, \\ 0 & \textit{if } N_\pi > 2. \end{cases} \tag{6.13}$$

Proof The assertion (6.12) follows from an inspection of Table A.4, and (6.13) now follows from (1) of Theorem 6.1.1. □

We mention that Corollary 6.1.2 is consistent with (3) of Lemma 3.10.1. To explain this, let the notation be as in Corollary 6.1.2, and assume that $N_\pi > 2$. Since $N_{\pi,s} = N_\pi - 1 > 1$, $V_s(1) = 0$, and so by (3) of Lemma 3.10.1 the only eigenvalue of $T^s_{1,0}$ on $V_s(N_\pi - 1)$ is 0. This is indeed the case by (6.13).

6.2 Non-Generic Paramodular Representations

Let (π, V) be a paramodular, irreducible, admissible representation of the group $\mathrm{GSp}(4, F)$ with trivial central character such that $N_\pi \geq 2$. In this section we determine under what conditions π is non-generic. This information will be used in the next two sections.

Lemma 6.2.1 *Let (π, V) be an irreducible, admissible representation of the group $\mathrm{GSp}(4, F)$ with trivial central character. Assume that π is paramodular and $N_\pi \geq 2$. Then π is a IVb representation if and only if $N_\pi = 2$, $\mu_\pi = -q^2 + q$, and $\lambda_\pi = \pm(1 + q^2)$.*

Proof If π is a IVb representation, then $N_\pi = 2$, $\mu_\pi = -q^2 + q$, and $\lambda_\pi = \pm(1 + q^2)$ by Table A.4 on p. 343. Conversely, assume that $N_\pi = 2$, $\mu_\pi = -q^2 + q$, and $\lambda_\pi = \pm(1 + q^2)$. Table A.4 along with $N_\pi = 2$ and $\mu_\pi = -q^2 + q$ then imply that π is either a IIIa representation of the form $\chi \rtimes \sigma \mathrm{St}_{\mathrm{GSp}(2)}$ for unramified characters χ and σ of $F^\times$ such that $\chi\sigma^2 = 1$ and $\chi \notin \{1, \nu^{\pm 2}\}$, or π is VIa representation of the form $\tau(S, \nu^{-1/2}\sigma)$ for an unramified character σ of $F^\times$ such that $\sigma^2 = 1$, or π is a IVb representation of the form $\pi = L(\nu^2, \nu^{-1}\sigma \mathrm{St}_{\mathrm{GSp}(2)})$ for an unramified character σ of $F^\times$ such that $\sigma^2 = 1$. Assume that the first possibility holds. Then (recalling the assumption $\lambda_\pi = \pm(1 + q^2)$) we have $q(\sigma(\varpi) + \sigma(\varpi)^{-1}) = \lambda_\pi = \pm(1 + q^2)$, so that $\sigma(\varpi) = \pm q^{\pm 1}$; this implies that $\chi = \nu^{\pm 2}$, a contradiction. The second possibility is similarly seen to lead to a contradiction. It follows that π is a IVb representation. □

Lemma 6.2.2 *Let (π, V) be an irreducible, admissible representation of the group $\mathrm{GSp}(4, F)$ with trivial central character. Assume that π is paramodular and $N_\pi \geq 2$. Let $v_{\mathrm{new}} \in V(N_\pi)$ be a newform. Then π is non-generic if and only if*

$$\mu_\pi = 0 \textit{ and } v_{\mathrm{new}} \in V(N_\pi) = V(N_{\pi,s}) \textit{ is an eigenvector for } T^s_{0,1} \tag{6.14}$$

or

$$N_\pi = 2,\ \mu_\pi = -q^2 + q,\ \text{and}\ \lambda_\pi = \pm(1 + q^2). \tag{6.15}$$

Proof Assume that π is non-generic. Then by inspection of Table A.4 on p. 343, either $\mu_\pi = 0$, so that (6.14) holds by (2) of Theorem 6.1.1, or π is a IVb representation of the form $\pi = L(\nu^2, \nu^{-1}\sigma \mathrm{St}_{\mathrm{GSp}(2)})$ for an unramified character σ of $F^\times$ such that $\sigma^2 = 1$, $N_\pi = 2$, $\lambda_\pi = \sigma(\varpi)(1 + q^2)$, and $\mu_\pi = -q^2 + q$, so that (6.15) holds. In the converse direction, if (6.14) holds, then π is non-generic by (2) of Theorem 6.1.1, and if (6.15) holds, then π is a IVb representation by Lemma 6.2.1, and is hence non-generic. □

Lemma 6.2.3 *Let (π, V) be an irreducible, admissible representation of the group* GSp(4, F) *with trivial central character. Assume that π is paramodular, $N_\pi \geq 2$, and π is a category 1 representation. Then the following are equivalent:*

(1) *π is non-generic;*
(2) *π is a IVb representation;*
(3) *$N_\pi = 2$, $\mu_\pi = -q^2 + q$, and $\lambda_\pi = \pm(1 + q^2)$.*

Proof (1) ⇔ (3). Assume (1). By Lemma 6.2.2 either (6.14) or (6.15) holds. Since $\mu_\pi \neq 0$ because π is a category 1 representation (see Proposition 5.7.3), (3) holds. Assume (3). Then π is non-generic by Lemma 6.2.2.

(2) ⇔ (3). This follows from Lemma 6.2.1. □

Lemma 6.2.4 *Let (π, V) be an irreducible, admissible representation of the group* GSp(4, F) *with trivial central character. Assume that π is paramodular, $N_\pi \geq 2$, and π is a category 2 representation. Then π is non-generic if and only if π is a Saito-Kurokawa representation.*

Proof This follows from an inspection of Table A.4. □

6.3 At the Minimal Paramodular Level

Let π be a paramodular, irreducible, admissible representation of GSp(4, F) with trivial central character such that $N_\pi \geq 2$. In this section we describe the action of the stable Klingen Hecke operators $T^s_{0,1}$ and $T^s_{1,0}$ on $V_s(N_\pi)$ in terms of the paramodular Hecke eigenvalues λ_π and μ_π. Since the case when π is a category 2 representation, so that $N_\pi = N_{\pi,s}$, was dealt with in (2) of Theorem 6.1.1, we need only consider category 1 representations, i.e., π such that $N_\pi = N_{\pi,s} + 1$. By Lemma 6.2.3, there are two cases to consider: π is generic, or π is a IVb representation. We first consider the case when π is generic.

Theorem 6.3.1 *Let (π, V) be a generic, irreducible, admissible representation of* GSp(4, F) *with trivial central character. Assume that $N_\pi \geq 2$ and that π is a category 1 representation, so that $N_{\pi,s} = N_\pi - 1$. Let $v_{\mathrm{new}} \in V(N_\pi)$ be a newform,* *and let v_s be the shadow of $v_{\rm new}$. The vector space $V_s(N_\pi)$ is three-dimensional and has ordered basis*

$$v_{\rm new}, \quad \tau_{N_\pi-1}v_s, \quad \theta v_s. \tag{6.16}$$

The matrix of the endomorphism $T^s_{0,1}$ of $V_s(N_\pi)$ with respect to (6.16) is

$$\begin{bmatrix} 0 & 0 & -\mu_\pi q^{-1}(\mu_\pi+q^2) \\ -q^2\mu_\pi^{-1}\lambda_\pi & \lambda_\pi & -q\mu_\pi \\ q^2\mu_\pi^{-1} & 0 & \lambda_\pi \end{bmatrix}, \tag{6.17}$$

and the matrix of the endomorphism $T^s_{1,0}$ of $V_s(N_\pi)$ with respect to (6.16) is

$$\begin{bmatrix} 0 & 0 & 0 \\ -q^2 & \mu_\pi+q^2 & q^2\lambda_\pi \\ 0 & 0 & 0 \end{bmatrix}. \tag{6.18}$$

The characteristic polynomial of $T^s_{0,1}$ on $V_s(N_\pi)$ is

$$p(T^s_{0,1}, V_s(N_\pi), X) = (X-\lambda_\pi)(X^2-\lambda_\pi X+q(\mu_\pi+q^2)), \tag{6.19}$$

and the characteristic polynomial of $T^s_{1,0}$ on $V_s(N_\pi)$ is

$$p(T^s_{1,0}, V_s(N_\pi), X) = (X-(\mu_\pi+q^2))X^2. \tag{6.20}$$

Proof It follows from Theorem 5.4.1 that $\dim V_s(N_\pi) = 3$, and that the vectors in (6.16) are linearly independent.

To prove that the matrix of $T^s_{0,1}$ of $V_s(N_\pi)$ with respect to (6.16) is as in (6.17) we proceed as follows. We have

$$\begin{aligned} T^s_{0,1}v_{\rm new} &= q\mu_\pi^{-1}T^s_{0,1}\rho'_{N_\pi-1}v_s && \text{(by (5.14))} \\ &= q\mu_\pi^{-1}(q\theta v_s - q\tau_{N_\pi-1}T^s_{0,1}v_s) && \text{(by (3.91))} \\ &= q^2\mu_\pi^{-1}\theta v_s - q^2\mu_\pi^{-1}\tau_{N_\pi-1}\lambda_\pi v_s && \text{(by (6.1)).} \end{aligned} \tag{6.21}$$

This verifies the first column of (6.17). For the second column we have:

$$\begin{aligned} T^s_{0,1}\tau_{N_\pi-1}v_s &= \tau_{N_\pi-1}T^s_{0,1}v_s && \text{(by (3.85))} \\ &= \lambda_\pi\tau_{N_\pi-1}v_s && \text{(by (6.1)).} \end{aligned}$$

This verifies the second column of (6.17). To obtain the third column we will first prove that

$$(T^s_{0,1})^2v_{\rm new} = -q(\mu_\pi+q^2)v_{\rm new} - (q^2\lambda_\pi^2\mu_\pi^{-1}+q^3)\tau_{N_\pi-1}v_s + q^2\lambda_\pi\mu_\pi^{-1}\theta v_s. \tag{6.22}$$

To prove (6.22), we begin with (6.21) and find that:

$$\begin{aligned}(T_{0,1}^s)^2 v_{\text{new}} &= T_{0,1}^s(T_{0,1}^s v_{\text{new}})\\ &= T_{0,1}^s\big(-q^2\mu_\pi^{-1}\lambda_\pi \tau_{N_\pi-1}v_s + q^2\mu_\pi^{-1}\theta v_s\big)\\ &= -q^2\mu_\pi^{-1}\lambda_\pi T_{0,1}^s\tau_{N_\pi-1}v_s + q^2\mu_\pi^{-1}T_{0,1}^s\theta v_s\\ &= -q^2\mu_\pi^{-1}\lambda_\pi \tau_{N_\pi-1}T_{0,1}^s W_s + q^2\mu_\pi^{-1}T_{0,1}^s\theta v_s \qquad \text{(by (3.85))}\\ &= -q^2\mu_\pi^{-1}\lambda_\pi^2 \tau_{N_\pi-1}v_s + q^2\mu_\pi^{-1}T_{0,1}^s\theta v_s \qquad \text{(by (6.1))}. \qquad (6.23)\end{aligned}$$

To proceed further we will need to look at cases. Assume first that $N_\pi \geq 3$. Then (6.23) becomes, by (3.88) with $n = N_\pi - 1$, and since $\sigma_{N_\pi-2}v_s = 0$ as $V_s(N_\pi - 2) = 0$,

$$\begin{aligned}(T_{0,1}^s)^2 v_{\text{new}} &= -q^2\mu_\pi^{-1}\lambda_\pi^2 \tau_{N_\pi-1}v_s + q^2\mu_\pi^{-1}\big(\theta T_{0,1}^s v_s + q^3\tau_{N_\pi-1}v_s - q^3\eta\sigma_{N_\pi-2}v_s\big)\\ &= -q^2\mu_\pi^{-1}\lambda_\pi^2 \tau_{N_\pi-1}v_s + q^2\mu_\pi^{-1}\big(\theta T_{0,1}^s v_s + q^3\tau_{N_\pi-1}v_s\big)\\ &= -q^2\mu_\pi^{-1}\lambda_\pi^2 \tau_{N_\pi-1}v_s + q^2\mu_\pi^{-1}\big(\lambda_\pi\theta v_s + q^3\tau_{N_\pi-1}v_s\big) \qquad \text{(by (6.1))}\\ &= (-q^2\mu_\pi^{-1}\lambda_\pi^2 + q^5\mu_\pi^{-1})\tau_{N_\pi-1}v_s + q^2\mu_\pi^{-1}\lambda_\pi\theta v_s.\end{aligned}$$

Since $N_\pi \geq 3$ and π is generic, we have $\mu_\pi = -q^2$ by Corollary 6.1.2. Hence,

$$(T_{0,1}^s)^2 v_{\text{new}} = (\lambda_\pi^2 - q^3)\tau_{N_\pi-1}v_s - \lambda_\pi\theta v_s.$$

This is (6.22). Now assume that $N_\pi = 2$. Then continuing from (6.23),

$$\begin{aligned}&(T_{0,1}^s)^2 v_{\text{new}}\\ &\quad= -q^2\mu_\pi^{-1}\lambda_\pi^2\tau_1 v_s + q^2\mu_\pi^{-1}\big(\theta T_{0,1}^s v_s + q^3\tau_1 v_s - q^3 e(v_s)\big) \qquad \text{(by (3.89))}\\ &\quad= -q^2\mu_\pi^{-1}\lambda_\pi^2\tau_1 v_s + q^2\mu_\pi^{-1}\big(\theta T_{0,1}^s v_s + q^3\tau_1 v_s\\ &\qquad - q^3\big(q^{-4}\eta T_{1,0}^s v_s - q^{-2}\tau_2 T_{1,0}^s v_s + q^{-2}\tau_1 T_{1,0}^s v_s\big)\big) \qquad \text{(by (3.82))}\\ &\quad= -q^2\mu_\pi^{-1}\lambda_\pi^2\tau_1 v_s + q^2\mu_\pi^{-1}\big(\lambda_\pi\theta v_s + q^3\tau_1 v_s\\ &\qquad - q^3\big(q^{-4}(\mu_\pi + q^2)\eta v_s - q^{-2}(\mu_\pi + q^2)\tau_2 v_s + q^{-2}(\mu_\pi + q^2)\tau_1 v_s\big)\big)\\ &\qquad \text{(by (6.1))}\\ &\quad= -q^2\mu_\pi^{-1}\lambda_\pi^2\tau_1 v_s + q^2\mu_\pi^{-1}\big(\lambda_\pi\theta v_s + q^3\tau_1 v_s\\ &\qquad - q^3(\mu_\pi + q^2)\big(q^{-3}(q^{-1}\eta v_s - q\tau_2 v_s) + q^{-2}\tau_1 v_s\big)\big)\\ &\quad= -q^2\mu_\pi^{-1}\lambda_\pi^2\tau_1 v_s + q^2\mu_\pi^{-1}\big(\lambda_\pi\theta v_s + q^3\tau_1 v_s\end{aligned}$$

$$
\begin{aligned}
&- q^3(\mu_\pi + q^2)\big(q^{-3}\rho_1' v_s + q^{-2}\tau_1 v_s)\big) \qquad \text{(by (3.51))} \\
&= -q^2\mu_\pi^{-1}\lambda_\pi^2\tau_1 v_s + q^2\mu_\pi^{-1}\big(\lambda_\pi\theta v_s + q^3\tau_1 v_s \\
&\quad - q^3(\mu_\pi + q^2)\big(q^{-4}\mu_\pi v_{\rm new} + q^{-2}\tau_1 v_s)\big) \qquad \text{(by (5.14))} \\
&= -q(\mu_\pi + q^2)v_{\rm new} - q^2(\mu_\pi^{-1}\lambda_\pi^2 + q)\tau_1 v_s + q^2\mu_\pi^{-1}\lambda_\pi\theta v_s.
\end{aligned}
$$

This is (6.22), and concludes the proof of (6.22) in all cases. We now return to the verification of the third column of (6.17). Equating the right-hand sides of (6.23) and (6.22) gives

$$
\begin{aligned}
&- q^2\mu_\pi^{-1}\lambda_\pi^2\tau_{N_\pi-1}v_s + q^2\mu_\pi^{-1}T_{0,1}^s(\theta v_s) \\
&\quad = -(q\mu_\pi + q^3)v_{\rm new} - (q^2\lambda_\pi^2\mu_\pi^{-1} + q^3)\tau_{N_\pi-1}v_s + q^2\lambda_\pi\mu_\pi^{-1}\theta v_s.
\end{aligned}
$$

Hence,

$$
\begin{aligned}
q^2\mu_\pi^{-1}T_{0,1}^s(\theta v_s) &= q^2\mu_\pi^{-1}\lambda_\pi^2\tau_{N_\pi-1}v_s \\
&\quad - (q\mu_\pi + q^3)v_{\rm new} - (q^2\lambda_\pi^2\mu_\pi^{-1} + q^3)\tau_{N_\pi-1}v_s + q^2\lambda_\pi\mu_\pi^{-1}\theta v_s \\
&= -(q\mu_\pi + q^3)v_{\rm new} - q^3\tau_{N_\pi-1}v_s + q^2\lambda_\pi\mu_\pi^{-1}\theta v_s.
\end{aligned}
$$

It follows that

$$
T_{0,1}^s(\theta v_s) = -\mu_\pi(q^{-1}\mu_\pi + q)v_{\rm new} - q\mu_\pi\tau_{N_\pi-1}v_s + \lambda_\pi\theta v_s.
$$

This verifies the third column of (6.17).

To prove that the matrix of $T_{1,0}^s$ of $V_s(N_\pi)$ with respect to (6.16) is as in (6.18), we note that by (5.15)

$$
T_{1,0}^s v_{\rm new} = -q^2\tau_{N_\pi-1}v_s.
$$

This verifies the first column of the matrix in (6.18). The second column in (6.18) follows from (3.86) and (6.1). The third column in (6.18) follows from:

$$
\begin{aligned}
T_{1,0}^s\theta v_s &= q^2 T_{0,1}^s\tau_{N_\pi-1}v_s \qquad \text{(by (3.87))} \\
&= q^2\tau_{N_\pi-1}T_{0,1}^s v_s \qquad \text{(by (3.85))} \\
&= q^2\tau_{N_\pi-1}\lambda_\pi v_s \qquad \text{(by (6.1))}.
\end{aligned}
$$

The final assertions about characteristic polynomials follow by standard calculations. □

Let π be a paramodular, irreducible, admissible representation of GSp(4, F) with trivial central character such that $N_\pi \geq 2$. The next proposition considers the actions of $T^s_{0,1}$ and $T^s_{1,0}$ on $V(N_\pi)$ when π is a IVb representation, as reviewed in (2.14). This proposition completes our analysis of the actions of $T^s_{0,1}$ and $T^s_{1,0}$ on $V(N_\pi)$ when π is a category 1 representation.

Proposition 6.3.2 *Let σ be an unramified character of $F^\times$ such that $\sigma^2 = 1$, and let π be the IVb representation $L(\nu^2, \nu^{-1}\sigma \mathrm{St}_{\mathrm{GSp}(2)})$. Then π has trivial central character, π is paramodular, $N_\pi = 2$, $\mu_\pi = -q^2 + q$, and $\lambda_\pi = \sigma(\varpi)(1+q^2)$, so that π is a category 1 representation. Let $v_{\mathrm{new}} \in V(2)$ be a newform, and let v_s be the shadow of v_{new}. The vector space $V_s(2)$ is two-dimensional and has ordered basis*

$$v_{\mathrm{new}}, \qquad \tau_1 v_s. \tag{6.24}$$

The matrix of the endomorphism $T^s_{0,1}$ on $V_s(2)$ with respect to (6.24) is

$$\begin{bmatrix} q^2(1+q^2)^{-1}\lambda_\pi & 0 \\ -q(1+q^2)^{-1}\lambda_\pi & \lambda_\pi \end{bmatrix} = \begin{bmatrix} \sigma(\varpi)q^2 & 0 \\ -\sigma(\varpi)q & \sigma(\varpi)(1+q^2) \end{bmatrix}, \tag{6.25}$$

and the matrix of the endomorphism $T^s_{1,0}$ on $V_s(2)$ with respect to (6.24) is

$$\begin{bmatrix} 0 & 0 \\ -q^2 & \mu_\pi+q^2 \end{bmatrix} = \begin{bmatrix} 0 & 0 \\ -q^2 & q \end{bmatrix}. \tag{6.26}$$

The characteristic polynomial of $T^s_{0,1}$ on $V_s(2)$ is

$$p(T^s_{0,1}, V_s(2), X) = (X - \lambda_\pi)(X - q^2(1+q^2)^{-1}\lambda_\pi), \tag{6.27}$$

and the characteristic polynomial of $T^s_{1,0}$ on $V_s(2)$ is

$$p(T^s_{1,0}, V_s(2), X) = (X - (\mu_\pi + q^2))X. \tag{6.28}$$

Proof That π is paramodular, $N_\pi = 2$, $\mu_\pi = -q^2 + q$, $\lambda_\pi = \sigma(\varpi)(1+q^2)$, and $V_s(2)$ is two-dimensional follow from Table A.3 on p. 339 and Table A.4 on p. 343. To prove that (6.24) is a basis for $V_s(2)$ it suffices to prove that v_{new} and $\tau_1 v_s$ are linearly independent; suppose otherwise. Then v_s is in the kernel of the map $\tau_1 : \bar{V}_s(1) \to \bar{V}_s(2)$; this contradicts the injectivity of this map from (2) of Lemma 3.5.4. Next, we prove that the matrix of $T^s_{1,0}$ on $V_s(2)$ with respect to (6.24) is as in (6.26). By (5.15) we have $T^s_{1,0} v_{\mathrm{new}} = -q^2 \tau_1 v_s$. This verifies the first column of (6.26). For the second column, we have

$$\begin{aligned} T^s_{1,0}\tau_1 v_s &= \tau_1 T^s_{1,0} v_s && \text{(by (3.86))} \\ &= (\mu_\pi + q^2)\tau_1 v_s && \text{(by (6.1))}. \end{aligned} \tag{6.29}$$

Next, we prove that the matrix of $T^s_{0,1}$ on $V_s(2)$ with respect to (6.24) is as in (6.25). We have

$$\begin{aligned} T^s_{0,1} v_{\text{new}} &= q\mu_\pi^{-1} T^s_{0,1}\rho'_1 v_s \qquad \text{(by (5.14))} \\ &= q\mu_\pi^{-1}(q\theta v_s - q\tau_{N_\pi-1} T^s_{0,1} v_s) \qquad \text{(by (3.91))} \\ &= q^2\mu_\pi^{-1}\theta v_s - q^2\lambda_\pi\mu_\pi^{-1}\tau_1 v_s \qquad \text{(by (6.1))}. \end{aligned} \tag{6.30}$$

To proceed further we need to express θv_s in the basis (6.24). Let $a, b \in \mathbb{C}$ be such that

$$\theta v_s = a v_{\text{new}} + b\tau_1 v_s. \tag{6.31}$$

Applying $T^s_{1,0}$ to this equation, we obtain

$$\begin{aligned} T^s_{1,0}\theta v_s &= a T^s_{1,0} v_{\text{new}} + b T^s_{1,0}\tau_1 v_s \\ &= -q^2 a\tau_1 v_s + (\mu_\pi + q^2) b\tau_1 v_s \qquad \text{(by (6.26))} \\ q^2 T^s_{0,1}\tau_1 v_s &= -q^2 a\tau_1 v_s + (\mu_\pi + q^2) b\tau_1 v_s \qquad \text{(by (3.87))} \\ q^2\tau_1 T^s_{0,1} v_s &= -q^2 a\tau_1 v_s + (\mu_\pi + q^2) b\tau_1 v_s \qquad \text{(by (3.85))} \\ q^2\lambda_\pi\tau_1 v_s &= (-q^2 a + (\mu_\pi + q^2) b)\tau_1 v_s \qquad \text{(by (6.1))}. \end{aligned}$$

It follows that

$$q^2\lambda_\pi = -q^2 a + (\mu_\pi + q^2) b. \tag{6.32}$$

Applying p_2 to (6.31), we have:

$$\begin{aligned} p_2\theta v_s &= a p_2 v_{\text{new}} + b p_2\tau_1 v_s \\ \theta p_1 v_s &= a v_{\text{new}} - (q + q^2)^{-1}\mu_\pi b v_{\text{new}} \qquad \text{(by (3.80) and (5.13))} \\ 0 &= (a - (q + q^2)^{-1}\mu_\pi b) v_{\text{new}} \qquad \text{(since } V_s(1) = 0\text{)}. \end{aligned}$$

Hence,

$$0 = a - (q + q^2)^{-1}\mu_\pi b. \tag{6.33}$$

Solving (6.32) and (6.33) for a and b, we obtain

$$a = \sigma(\varpi) q(1 - q), \qquad b = \sigma(\varpi) q(1 + q),$$

so that

$$\theta v_s = \sigma(\varpi)q(1-q)v_{\rm new} + \sigma(\varpi)q(1+q)\tau_1 v_s. \tag{6.34}$$

Returning to (6.30), and substituting for θv_s, we now have:

$$\begin{aligned} T^s_{0,1}v_{\rm new} &= q^2\mu_\pi^{-1}\sigma(\varpi)q(1-q)v_{\rm new} + (q^2\mu_\pi^{-1}\sigma(\varpi)q(1+q) - q^2\lambda_\pi\mu_\pi^{-1})\tau_1 v_s \\ &= q^2(1+q^2)^{-1}\lambda_\pi v_{\rm new} - q(1+q^2)^{-1}\lambda_\pi\tau_1 v_s. \end{aligned}$$

This verifies the first column of (6.25). The second column of (6.25) follows from (3.85) and (6.1). The final assertions about characteristic polynomials follow by standard computations. □

6.4 An Upper Block Algorithm

Let π be a paramodular, irreducible, admissible representation of $\mathrm{GSp}(4, F)$ with trivial central character such that $N_\pi \geq 2$, and let $v_{\rm new} \in V(N_\pi)$ be a newform. In this final section we describe how λ_π and μ_π can be computed from $v_{\rm new}$ using the upper block operators $\sigma_{N_\pi-1}$, $T^s_{0,1}$, and $T^s_{1,0}$; at the same time, this method also determines whether π is non-generic. Our algorithm is based on Proposition 5.7.3 and Theorem 6.1.1, and proceeds as follows. A flow chart version of this algorithm appears in Fig. 6.1.

(1) Calculate $v_s = -q^2\sigma_{N_\pi-1}v_{\rm new}$ (see (5.12)). If $v_s \neq 0$, then proceed to (2). If $v_s = 0$, then proceed to (3).

(2) Since $v_s \neq 0$, π is a category 1 representation so that $N_{\pi,s} = N_\pi - 1$ and $\mu_\pi \neq 0$ by Proposition 5.7.3. Calculate $T^s_{0,1}v_s$ and $T^s_{1,0}v_s$. By Theorem 6.1.1 we have that

$$\lambda_\pi v_s = T^s_{0,1}v_s \quad \text{and} \quad (\mu_\pi + q^2)v_s = T^s_{1,0}v_s.$$

Solve for λ_π and μ_π. By Lemma 6.2.3 π is non-generic if and only if $N_\pi = 2$, $\mu_\pi = -q^2 + q$, and $\lambda_\pi = \pm(1+q^2)$; in this case π is a IVb representation.

(3) Since $v_s = 0$, π is a category 2 representation so that $N_{\pi,s} = N_\pi$ and $\mu_\pi = 0$ by Proposition 5.7.3. Calculate $T^s_{0,1}v_{\rm new}$. If $T^s_{0,1}v_{\rm new}$ is a multiple of $v_{\rm new}$, then π is non-generic by Theorem 6.1.1 and is a Saito-Kurokawa representation by Lemma 6.2.4; proceed to (4). If $T^s_{0,1}v_{\rm new}$ is not a multiple of $v_{\rm new}$, then π is generic by Theorem 6.1.1; proceed to (5).

(4) By Theorem 6.1.1, we have that

$$\lambda_\pi v_{\rm new} = (1+q^{-1})T^s_{0,1}v_{\rm new}.$$

Solve for λ_π.

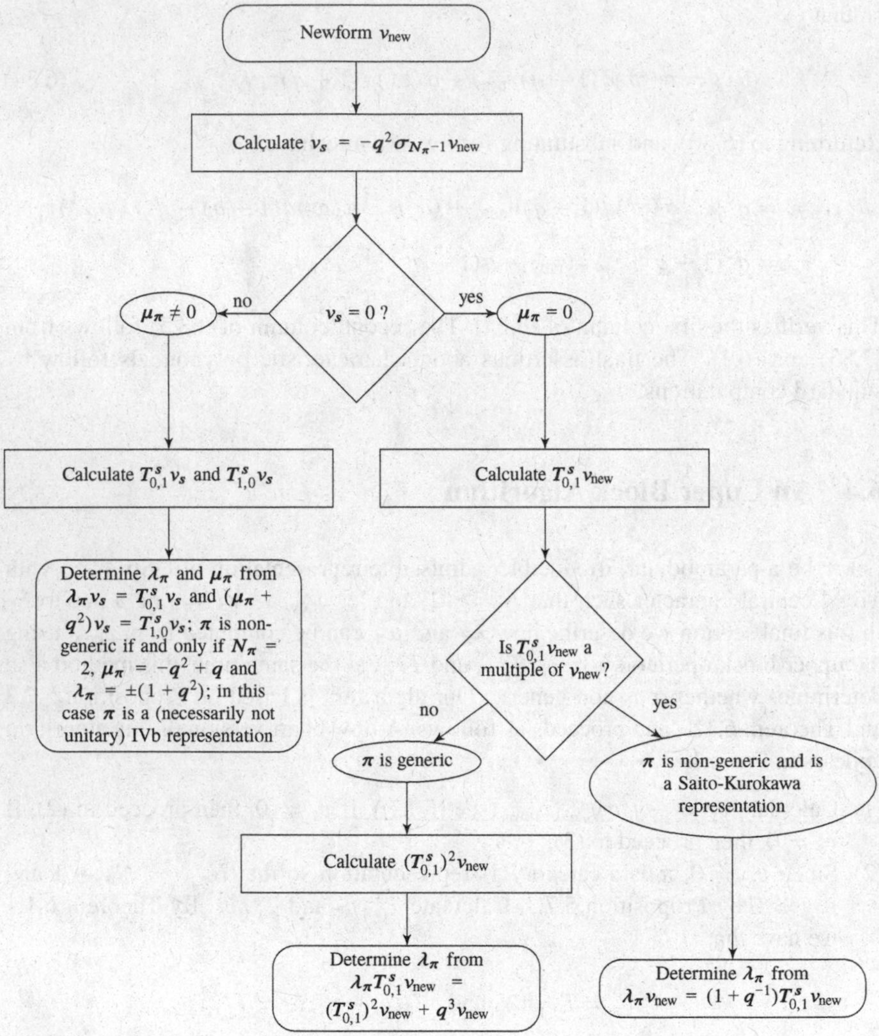

Fig. 6.1 Algorithm for computing paramodular Hecke eigenvalues

(5) Calculate $(T_{0,1})^2 v_{\mathrm{new}}$. By Theorem 6.1.1 we have that

$$\lambda_\pi T^s_{0,1} v_{\mathrm{new}} = (T_{0,1})^2 v_{\mathrm{new}} + q^3 v_{\mathrm{new}}.$$

Solve for λ_π.

Chapter 7
The Paramodular Subspace

Let (π, V) be an irreducible, admissible representation of $\mathrm{GSp}(4, F)$ with trivial central character, and let n be an integer such that $n \geq 0$. Assume that π is paramodular. In this chapter we investigate the relationship between $V_s(n)$ and its subspace $V(n-1) + V(n)$ of paramodular vectors. From Table A.3, it is already evident that for almost all non-generic π, including all Saito-Kurokawa representations, we have $V(n-1) + V(n) = V_s(n)$ for all $n \geq 0$; also, from Table A.3 we see that if π is generic, then $V(n-1) + V(n)$ is a proper subspace of $V_s(n)$ if $n \geq \max(N_\pi, 1)$. Thus, except for a few non-generic π, the investigation of the relation between $V_s(n)$ and its subspace $V(n-1) + V(n)$ is reduced to considering π that are generic. In Theorem 5.4.1 we proved that if π is generic and a category 1 representation, then

$$V_s(n) = V(n-1) \oplus V(n) \oplus \bigoplus_{\substack{i,j \geq 0 \\ i+j=n-N_\pi+1}} \mathbb{C}\tau^i \theta^j W_s$$

where W_s is the shadow of the newform W_{new} in $V(N_\pi)$. Most of this chapter is devoted to generalizing this result to all generic π, though we also give a complete account for all non-generic π.

7.1 Calculation of Certain Zeta Integrals

Let (π, V) be a generic, irreducible, admissible representation of $\mathrm{GSp}(4, F)$ with trivial central character, and let $V = \mathcal{W}(\pi, \psi_{c_1,c_2})$, the Whittaker model of π with respect to ψ_{c_1,c_2}; as usual, we assume that $c_1, c_2 \in \mathfrak{o}^\times$ (see Sect. 2.2). In this section we calculate certain zeta integrals for later use. As in earlier chapters, we let λ_π, μ_π, and ε_π be the eigenvalues of $T_{0,1}$, $T_{1,0}$, and u_{N_π}, respectively, on the one-dimensional space $V(N_\pi)$. Since π is infinite-dimensional, paramodular vectors

J. Johnson-Leung et al., *Stable Klingen Vectors and Paramodular Newforms*,
Lecture Notes in Mathematics 2342, https://doi.org/10.1007/978-3-031-45177-5_7

with distinct levels are linearly independent (see Theorem 2.3.1); thus, the subspace $V(n-1)+V(n)$ of $V_s(n)$ is a direct sum $V(n-1) \oplus V(n)$ for integers $n \geq 1$. Also, as in previous chapters, we will use the equivalences from Proposition 5.7.3.

We begin with some notation and a summary of some results from [105]. Let n be an integer such that $n \geq 0$, and let $W \in V(n)$. By Proposition 4.1.3 of [105] we have

$$Z(s, \theta' W) = q Z(s, W), \qquad Z(s, \theta W) = q^{-s+\frac{3}{2}} Z(s, W) \tag{7.1}$$

and

$$Z(s, \eta W) = 0. \tag{7.2}$$

For all integers i and j we recall that

$$\Delta_{i,j} = \begin{bmatrix} \varpi^{2i+j} & & & \\ & \varpi^{i+j} & & \\ & & \varpi^{i} & \\ & & & 1 \end{bmatrix}. \tag{7.3}$$

We have $\lambda(\Delta_{i,j}) = \varpi^{2i+j}$ for integers i and j. Let $W_{\text{new}} \in V(N_\pi)$ be a newform, i.e., a non-zero element of the one-dimensional space $V(N_\pi)$. For all integers i and j we set

$$c_{i,j} = W_{\text{new}}(\Delta_{i,j}). \tag{7.4}$$

By Corollary 4.3.8 of [105] the vector W_{new} is determined by the numbers $c_{i,j}$, and by (5.18) we have $c_{i,j} = 0$ if $i < 0$ or $j < 0$. As indicated in Theorem 2.3.4, $Z(s, W_{\text{new}})$ is a non-zero multiple of $L(s, \pi)$.

Lemma 7.1.1 *Let (π, V) be a generic, irreducible, admissible representation of* $\mathrm{GSp}(4, F)$ *with trivial central character, let $V = \mathcal{W}(\pi, \psi_{c_1,c_2})$, and let $W_{\text{new}} \in V(N_\pi)$ be a newform. Then*

$$c_{0,0} = \frac{Z(s, W_{\text{new}})}{1-q^{-1}} \begin{cases} (1 - q^{-\frac{3}{2}}\lambda_\pi q^{-s} + (\mu q^{-2} + 1 + q^{-2}) q^{-2s} \\ \quad -q^{-\frac{3}{2}}\lambda_\pi q^{-3s} + q^{-4s}) & \text{if } N_\pi = 0, \\ (1 - q^{-\frac{3}{2}}(\lambda_\pi + \varepsilon_\pi) q^{-s} + (\mu_\pi q^{-2} + 1) q^{-2s} \\ \quad +\varepsilon_\pi q^{-1/2} q^{-3s}) & \text{if } N_\pi = 1, \\ (1 - q^{-\frac{3}{2}}\lambda_\pi q^{-s} + (\mu_\pi q^{-2} + 1) q^{-2s}) & \text{if } N_\pi \geq 2, \end{cases} \tag{7.5}$$

and

$$\sum_{j=0}^{\infty} c_{1,j}(q^{-s+\frac{3}{2}})^j = \frac{Z(s, W_{\rm new})}{1-q^{-1}} \begin{cases} q^{-4}(\mu_\pi + 1 \\ \qquad -\lambda_\pi q^{\frac{1}{2}} q^{-s} + q^2 q^{-2s}) & \text{if } N_\pi = 0, \\ q^{-4}(\mu_\pi + \varepsilon_\pi q^{\frac{3}{2}} q^{-s}) & \text{if } N_\pi = 1, \\ q^{-4}\mu_\pi & \text{if } N_\pi \geq 2. \end{cases} \tag{7.6}$$

Proof Equation (7.5) follows from Proposition 7.1.4, Proposition 7.2.5 and Proposition 7.4.5 of [105]. Equation (7.6) follows from Lemma 7.1.3 of [105] (combined with (7.5)), the first display in the proof of Proposition 7.2.5 of [105], and Lemma 7.4.4 of [105]. □

Lemma 7.1.2 *Let* (π, V) *be a generic, irreducible, admissible representation of* $\mathrm{GSp}(4, F)$ *with trivial central character. Let* $W_{\rm new} \in V(N_\pi)$ *be a newform. We consider* $W_{\rm new}$ *as an element of* $V_s(n)$*, where* $n = 1$ *if* $N_\pi = 0$ *and* $n = N_\pi$ *if* $N_\pi \geq 1$*. Then*

$$Z(s, T^s_{0,1} W_{\rm new}) = \begin{cases} (\lambda_\pi - q^{-s+\frac{3}{2}}) Z(s, W_{\rm new}) & \text{if } N_\pi = 0, \\ (\lambda_\pi + \varepsilon_\pi - q^{-s+\frac{3}{2}}) Z(s, W_{\rm new}) & \text{if } N_\pi = 1, \\ (\lambda_\pi - q^{-s+\frac{3}{2}}) Z(s, W_{\rm new}) & \text{if } N_\pi \geq 2, \end{cases} \tag{7.7}$$

and

$$Z(s, T^s_{1,0} W_{\rm new}) = \begin{cases} (\mu_\pi + 1 - \lambda_\pi q^{\frac{1}{2}} q^{-s} + q^2 q^{-2s}) Z(s, W_{\rm new}) & \text{if } N_\pi = 0, \\ (\mu_\pi + \varepsilon_\pi q^{\frac{3}{2}} q^{-s}) Z(s, W_{\rm new}) & \text{if } N_\pi = 1, \\ \mu_\pi Z(s, W_{\rm new}) & \text{if } N_\pi \geq 2. \end{cases} \tag{7.8}$$

Proof By (5.21),

$$Z(s, T^s_{0,1} W_{\rm new}) = \int\limits_{\substack{F^\times \\ v(a) \geq 0}} (T^s_{0,1} W_{\rm new})(\begin{bmatrix} a &&& \\ & a && \\ && 1 & \\ &&& 1 \end{bmatrix}) |a|^{s-\frac{3}{2}} \, d^\times a.$$

Using (3.69), we have:

$$Z(s, T^s_{0,1} W_{\rm new})$$

$$= \sum_{y,z \in \mathfrak{o}/\mathfrak{p}} \int\limits_{\substack{F^\times \\ v(a) \geq 0}} W_{\rm new}(\begin{bmatrix} a &&& \\ & a && \\ && 1 & \\ &&& 1 \end{bmatrix} \begin{bmatrix} 1 & y && z\varpi^{-n+1} \\ & 1 && \\ && 1 & -y \\ &&& 1 \end{bmatrix} \begin{bmatrix} \varpi &&& \\ & 1 && \\ && \varpi & \\ &&& 1 \end{bmatrix}) |a|^{s-\frac{3}{2}} \, d^\times a$$

$$+\sum_{c,y,z\in\mathfrak{o}/\mathfrak{p}}\int\limits_{\substack{F^\times\\ v(a)\geq 0}} W_{\rm new}(\begin{bmatrix}a&&&\\&a&&\\&&1&\\&&&1\end{bmatrix}\begin{bmatrix}1&&y&z\varpi^{-n+1}\\&1&c&y\\&&1&\\&&&1\end{bmatrix}\begin{bmatrix}\varpi&&&\\&\varpi&&\\&&1&\\&&&1\end{bmatrix})|a|^{s-\frac{3}{2}}\,d^\times a$$

$$=q^2\int\limits_{\substack{F^\times\\ v(a)\geq 0}} W_{\rm new}(\begin{bmatrix}a&&&\\&a&&\\&&1&\\&&&1\end{bmatrix}\begin{bmatrix}\varpi&&&\\&1&&\\&&\varpi&\\&&&1\end{bmatrix})|a|^{s-\frac{3}{2}}\,d^\times a$$

$$+q^3\int\limits_{\substack{F^\times\\ v(a)\geq 0}} W_{\rm new}(\begin{bmatrix}a&&&\\&a&&\\&&1&\\&&&1\end{bmatrix}\begin{bmatrix}\varpi&&&\\&\varpi&&\\&&1&\\&&&1\end{bmatrix})|a|^{s-\frac{3}{2}}\,d^\times a$$

$$=q^2(1-q^{-1})\sum_{j=0}^{\infty}W_{\rm new}(\begin{bmatrix}\varpi^{j+1}&&&\\&\varpi^j&&\\&&\varpi&\\&&&1\end{bmatrix})q^{-j(s-\frac{3}{2})}$$

$$+q^3(1-q^{-1})\sum_{j=0}^{\infty}W_{\rm new}(\begin{bmatrix}\varpi^{j+1}&&&\\&\varpi^{j+1}&&\\&&1&\\&&&1\end{bmatrix})q^{-j(s-\frac{3}{2})}$$

$$=q^2(1-q^{-1})\sum_{j=0}^{\infty}c_{1,j-1}q^{-j(s-\frac{3}{2})}+q^3(1-q^{-1})\sum_{j=0}^{\infty}c_{0,j+1}q^{-j(s-\frac{3}{2})}$$

$$=q^2(1-q^{-1})\sum_{j=-1}^{\infty}c_{1,j}q^{-(j+1)(s-\frac{3}{2})}+q^3(1-q^{-1})\sum_{j=1}^{\infty}c_{0,j}q^{-(j-1)(s-\frac{3}{2})}$$

$$=q^{-s+\frac{7}{2}}(1-q^{-1})\sum_{j=-1}^{\infty}c_{1,j}q^{-j(s-\frac{3}{2})}+q^{s+\frac{3}{2}}(1-q^{-1})\sum_{j=1}^{\infty}c_{0,j}q^{-j(s-\frac{3}{2})}$$

$$=q^{-s+\frac{7}{2}}(1-q^{-1})\sum_{j=0}^{\infty}c_{1,j}q^{-j(s-\frac{3}{2})}$$

$$+q^{s+\frac{3}{2}}(1-q^{-1})\Big(\sum_{j=0}^{\infty}c_{0,j}q^{-j(s-\frac{3}{2})}-c_{0,0}\Big)$$

$$=q^{-s+\frac{7}{2}}(1-q^{-1})\sum_{j=0}^{\infty}c_{1,j}q^{-j(s-\frac{3}{2})}$$

$$+q^{s+\frac{3}{2}}Z(s,W_{\rm new})-q^{s+\frac{3}{2}}(1-q^{-1})c_{0,0}.$$

Substituting from (7.5) and (7.6) proves (7.7). Similarly, by (3.70),

$$Z(s, T^s_{1,0}W_{\rm new}) = \int\limits_{\substack{F^\times\\ v(a)\geq 0}} (T^s_{1,0}W_{\rm new})(\begin{bmatrix} a&&&\\&a&&\\&&1&\\&&&1\end{bmatrix})|a|^{s-\frac32}\,d^\times a$$

$$= \int\limits_{\substack{F^\times\\ v(a)\geq 0}} \sum_{\substack{x,y\in\mathfrak{o}/\mathfrak{p}\\ z\in\mathfrak{o}/\mathfrak{p}^2}} W_{\rm new}(\begin{bmatrix} a&&&\\&a&&\\&&1&\\&&&1\end{bmatrix}\begin{bmatrix}1&x&y&z\varpi^{-n+1}\\&1&&y\\&&1&-x\\&&&1\end{bmatrix}\begin{bmatrix}\varpi^2&&&\\&\varpi&&\\&&\varpi&\\&&&1\end{bmatrix})|a|^{s-\frac32}\,d^\times a$$

$$= q^4\int\limits_{\substack{F^\times\\ v(a)\geq 0}} W_{\rm new}(\begin{bmatrix} a&&&\\&a&&\\&&1&\\&&&1\end{bmatrix}\begin{bmatrix}\varpi^2&&&\\&\varpi&&\\&&\varpi&\\&&&1\end{bmatrix})|a|^{s-\frac32}\,d^\times a$$

$$= q^4(1-q^{-1})\sum_{j=0}^{\infty} W_{\rm new}(\begin{bmatrix}\varpi^{2+j}&&&\\&\varpi^{1+j}&&\\&&\varpi&\\&&&1\end{bmatrix})q^{-j(s-\frac32)}$$

$$= q^4(1-q^{-1})\sum_{j=0}^{\infty} c_{1,j}q^{-j(s-\frac32)}.$$

Hence, (7.8) follows from (7.6). □

7.2 Generic Representations

The main result of this section is Theorem 7.2.2, giving the structure of the spaces $V_s(n)$ for all generic, irreducible, admissible representations of GSp(4, F) with trivial central character. As a consequence, we will show that the linear dimension growth for the spaces $\bar{V}_s(n)$ proven in (5.31) for $N_\pi \geq 2$ and $\mu_\pi \neq 0$ holds in all cases. We also show that the maps $\theta : V_s(n) \to V_s(n+1)$ and $\theta : \bar{V}_s(n) \to \bar{V}_s(n+1)$ are injective.

Lemma 7.2.1 *Let π be a generic, irreducible, admissible representation of* GSp(4, F) *with trivial central character. Let $W_{\rm new} \in V(N_\pi)$ be a newform.*

(1) *Assume that $N_\pi = 0$. Then*

$$T^s_{0,1}W_{\rm new} \in \langle W_{\rm new}, \theta W_{\rm new}, \theta' W_{\rm new}\rangle, \qquad (7.9)$$

$$T^s_{1,0}W_{\rm new} \notin \langle W_{\rm new}, \theta W_{\rm new}, \theta' W_{\rm new}\rangle. \qquad (7.10)$$

Here, we view $W_{\rm new}$ as an element of $V_s(1)$ and apply the Hecke operators $T^s_{0,1}$, $T^s_{1,0}$ at this level.

(2) *Assume that* $N_\pi = 1$. *Then*

$$T^s_{0,1} W_{\rm new} \notin \mathbb{C} W_{\rm new} \quad \textit{and} \quad T^s_{1,0} W_{\rm new} \notin \mathbb{C} W_{\rm new}. \tag{7.11}$$

(3) *Assume that* $N_\pi \geq 2$. *Then*

$$T^s_{0,1} W_{\rm new} \notin \mathbb{C} W_{\rm new}. \tag{7.12}$$

Proof (1). By (7.8) we see that (7.10) holds. Next assume that $T^s_{0,1} W_{\rm new} \notin \langle W_{\rm new}, \theta W_{\rm new}, \theta' W_{\rm new} \rangle$. By (7.1) the vectors $\theta W_{\rm new}$ and $\theta' W_{\rm new}$ are linearly independent elements of $V(N_\pi + 1)$; since $W_{\rm new} \in V(N_\pi)$, the vectors $W_{\rm new}, \theta W_{\rm new}$, and $\theta' W_{\rm new}$ are linearly independent by Theorem 2.3.1. By our assumption, we now have that $W_{\rm new}, \theta W_{\rm new}, \theta' W_{\rm new}$ and $T^s_{0,1} W_{\rm new}$ span the four-dimensional space $V_s(1)$ (see Table A.3). By (7.7) we then have $Z(s, V_s(1)) = (\mathbb{C} + \mathbb{C} q^{-s}) Z(s, W_{\rm new})$, contradicting (7.8). This verifies (7.9).

(2) is immediate from (7.7) and (7.8).

(3) is immediate from (7.7). □

Let (π, V) be an irreducible, admissible representation of GSp(4, F) with trivial central character, and let n be an integer such that $n \geq 0$. We recall some definitions from Sect. 3.2. By definition, $\bar{V}_s(n) = V_s(n)/(V(n-1) \oplus V(n))$ if $n \geq 1$, and by definition $\bar{V}_s(0) = 0$. If $\bar{V}_s(n) \neq 0$ for some $n \geq 0$, then we let $\bar{N}_{\pi,s}$ be the smallest such integer n; if $\bar{V}_s(n) = 0$ for all $n \geq 0$, then we say that $\bar{N}_{\pi,s}$ is not defined. Clearly, $\bar{N}_{\pi,s} \geq 1$. The numbers $\bar{N}_{\pi,s}$ are listed in Table A.3 for all π. We note that if π is generic, then $\bar{N}_{\pi,s}$ is defined, and $\bar{N}_{\pi,s} = N_{\pi,s}$ except if π is an unramified type I representation; in the latter case, $N_{\pi,s} = 0$ and $\bar{N}_{\pi,s} = 1$. We can now prove the following generalization of Theorem 5.4.1. In (7.16) of the following theorem, to simplify notation, we write τ for the level raising operator $\tau_k : V_s(k) \to V_s(k+1)$ for each integer $k \geq 0$; we also note that by (3.77) the operators τ and θ commute.

Theorem 7.2.2 *Let* (π, V) *be a generic, irreducible, admissible representation of* GSp(4, F) *with trivial central character. Then* $\bar{V}_s(n) \neq 0$ *for some integer* $n \geq 0$. *Let* W_0 *be an element of* $V_s(\bar{N}_{\pi,s})$ *that is not in* $V(\bar{N}_{\pi,s} - 1) + V(\bar{N}_{\pi,s})$, *so that* W_0 *represents a non-zero element of* $\bar{V}_s(\bar{N}_{\pi,s})$. *Then*

$$Z(s, W_0) = P_0(q^{-s}) Z(s, W_{\rm new}), \tag{7.13}$$

where $P_0(X) \in \mathbb{C}[X]$ *is a non-zero polynomial of degree*

$$\deg P_0 = \begin{cases} 2 & \textit{if } N_\pi = 0, \\ 1 & \textit{if } N_\pi = 1, \\ 1 & \textit{if } N_\pi \geq 2 \textit{ and } \mu_\pi = 0, \\ 0 & \textit{if } N_\pi \geq 2 \textit{ and } \mu_\pi \neq 0. \end{cases} \tag{7.14}$$

We have

$$\bar{N}_{\pi,s} = N_\pi - 1 + \deg P_0. \tag{7.15}$$

Furthermore,

$$V_s(n) = V(n-1) \oplus V(n) \oplus \bigoplus_{\substack{i,j\geq 0\\ i+j=n-\bar{N}_{\pi,s}}} \mathbb{C}\tau^i\theta^j W_0 \tag{7.16}$$

and

$$\dim V_s(n) = \frac{(n-N_\pi+2)(n-N_\pi+3)}{2} - \deg(P_0) \tag{7.17}$$

for integers $n \geq N_{\pi;s}$. *In (7.16), if* $N_{\pi,s} = 0$ *and* $n = 0$*, then we take* $V(n-1) = V(-1)$ *to be the zero subspace, and the direct sum over* $i, j \geq 0$, $i+j = n-\bar{N}_{\pi,s} = -1$ *is empty. In addition,*

$$Z(s, V_s(n)) = (\mathbb{C} + \mathbb{C}q^{-s} + \ldots + \mathbb{C}(q^{-s})^{n-N_\pi+1})Z(s, W_{\rm new}) \tag{7.18}$$

for integers $n \geq \bar{N}_{\pi,s}$.

Proof It follows from Table A.3 that $\bar{V}_s(n) \neq 0$ for some n.

Assume that $N_\pi = 0$. Then $\bar{N}_{\pi,s} = 1$ and $\dim V_s(1) = 4$ by Table A.3. Lemma 7.2.1 implies that $V_s(1) = \langle W_{\rm new}, \theta W_{\rm new}, \theta' W_{\rm new}, T^s_{1,0} W_{\rm new}\rangle$. The vector W_0, when expressed as a linear combination of $W_{\rm new}, \theta W_{\rm new}, \theta' W_{\rm new}, T^s_{1,0} W_{\rm new}$, must have a non-zero $T^s_{1,0} W_{\rm new}$-component. Hence, (7.13) and (7.14) follow from (7.1) and (7.8).

Assume that $N_\pi = 1$, or that $N_\pi \geq 2$ and $\mu_\pi = 0$. Then $\bar{N}_{\pi,s} = N_\pi$ and $\dim V_s(N_\pi) = 2$ by Table A.3. Lemma 7.2.1 implies $V_s(N_\pi) = \langle W_{\rm new}, T^s_{0,1} W_{\rm new}\rangle$. The vector W_0, when expressed as a linear combination of $W_{\rm new}$ and $T^s_{0,1} W_{\rm new}$, must have a non-zero $T^s_{0,1} W_{\rm new}$-component. Hence, (7.13) and (7.14) follow from (7.1) and (7.7).

Assume that $N_\pi \geq 2$ and $\mu_\pi \neq 0$. Then $\bar{N}_{\pi,s} = N_\pi - 1$ by Table A.3. Let W_s be the shadow of $W_{\rm new}$ as defined in (5.8). By Theorem 5.4.1, W_s spans the one-dimensional space $V_s(N_\pi - 1)$. Hence, W_0 is a multiple of W_s. The assertions (7.13) and (7.14) follow from (5.28).

It is now easily verified that (7.15) holds in all cases.

Next we prove (7.16). Evidently, (7.16) holds for $N_\pi = 0$ and $n = 0$. We may therefore assume that $n > 0$. A calculation using Table A.3 shows that

$$\dim V_s(n) = \dim V(n-1) + \dim V(n) + (n - \bar{N}_{\pi,s} + 1).$$

For this, it is useful to note that $\dim V(m) = \lfloor (m-N_\pi+2)^2/4 \rfloor$ for integers m such that $m \geq N_\pi$ by Theorem 7.5.6 of [105] and that $\lfloor k^2/4 \rfloor + \lfloor (k+1)^2/4 \rfloor = k(k+1)/2$

for any integer k. It follows that we only need to prove that the sum on the right-hand side of (7.16) is direct. For this we use induction on n. By the choice of W_0, the sum is direct for $n = \max(\bar{N}_{\pi,s}, 1) = \max(N_{\pi,s}, 1)$. (The only case where $\bar{N}_{\pi,s}$ differs from $N_{\pi,s}$ is for $N_\pi = 0$, in which case $\bar{N}_{\pi,s} = 1$.) Assume that $n > \max(N_{\pi,s}, 1)$, and that the statement is true for $n-1$. Suppose that

$$W_1 + W_2 + \sum_{\substack{i,j\geq 0 \\ i+j=n-\bar{N}_{\pi,s}}} c_i \tau^i \theta^j W_0 = 0 \tag{7.19}$$

for some $W_1 \in V(n-1)$, $W_2 \in V(n)$ and complex numbers $c_0, \ldots, c_{n-\bar{N}_{\pi,s}}$. Taking zeta integrals and observing (7.13) and Lemma 5.3.1, we get

$$Z(s, W_1) + Z(s, W_2) = -P_0(q^{-s}) \sum_{\substack{i,j\geq 0 \\ i+j=n-\bar{N}_{\pi,s}}} (q^{-s+\frac{3}{2}})^j c_i Z(s, W_{\text{new}}). \tag{7.20}$$

By (7.1) and Theorem 7.5.6 of [105] we have

$$Z(s, V(n-1) + V(n)) = (\mathbb{C} + \mathbb{C}q^{-s} + \ldots + \mathbb{C}(q^{-s})^{n-N_\pi}) Z(s, W_{\text{new}}). \tag{7.21}$$

It therefore follows from (7.20) and (7.15) that $c_0 = 0$. From (7.19) we hence obtain

$$W_1 + W_2 + \tau_{n-1} \sum_{\substack{i,j\geq 0 \\ i+j=n-\bar{N}_{\pi,s}-1}} c_{i+1} \tau^i \theta^j W_s = 0. \tag{7.22}$$

The map $\tau_{n-1} : \bar{V}_s(n-1) \to \bar{V}_s(n)$ is injective by Lemma 3.5.4. It follows that

$$\sum_{\substack{i,j\geq 0 \\ i+j=n-\bar{N}_{\pi,s}-1}} c_{i+1} \tau^i \theta^j W_s \in V(n-2) \oplus V(n-1).$$

By the induction hypothesis, $c_1 = \ldots = c_{n-\bar{N}_{\pi,s}} = 0$. Hence, $W_1 + W_2 = 0$, which implies $W_1 = W_2 = 0$ by Theorem 2.3.1. This completes the proof of (7.16).

Equation (7.17) follows by taking dimensions on both sides of (7.16), observing (7.15) and using $\dim V(m) = \lfloor (m - N_\pi + 2)^2/4 \rfloor$ for integers m such that $m \geq N_\pi$ by Theorem 7.5.6 of [105], and $\lfloor k^2/4 \rfloor + \lfloor (k+1)^2/4 \rfloor = k(k+1)/2$ for any integer k. (Alternatively, (7.17) can be verified from Table A.3.)

Finally, (7.18) follows by taking zeta integrals of both sides of (7.16), observing Lemma 5.3.1, (7.15) and (7.21). □

We note that, by the proof of Theorem 7.2.2, possible choices for the vector W_0 are as follows:

$$W_0 = \begin{cases} T_{1,0}^s W_{\rm new} & \text{if } N_\pi = 0, \\ T_{0,1}^s W_{\rm new} \text{ or } T_{1,0}^s W_{\rm new} & \text{if } N_\pi = 1, \\ T_{0,1}^s W_{\rm new} & \text{if } N_\pi \geq 2 \text{ and } \mu_\pi = 0, \\ W_s & \text{if } N_\pi \geq 2 \text{ and } \mu_\pi \neq 0. \end{cases} \tag{7.23}$$

Here, W_s is the shadow of $W_{\rm new}$ as defined in (5.8).

The various cases of Theorem 7.2.2 are illustrated in Fig. 7.1.

Corollary 7.2.3 *Let (π, V) be a generic, irreducible, admissible representation of* GSp(4, *F*) *with trivial central character. Then*

$$\tilde{N}_{\pi,s} = \begin{cases} 1 & \textit{if } N_\pi = 0 \textit{ or } N_\pi = 1, \\ N_\pi - 1 & \textit{if } N_\pi \geq 2 \textit{ and } \mu_\pi \neq 0, \\ N_\pi & \textit{if } N_\pi \geq 2 \textit{ and } \mu_\pi = 0, \end{cases} \tag{7.24}$$

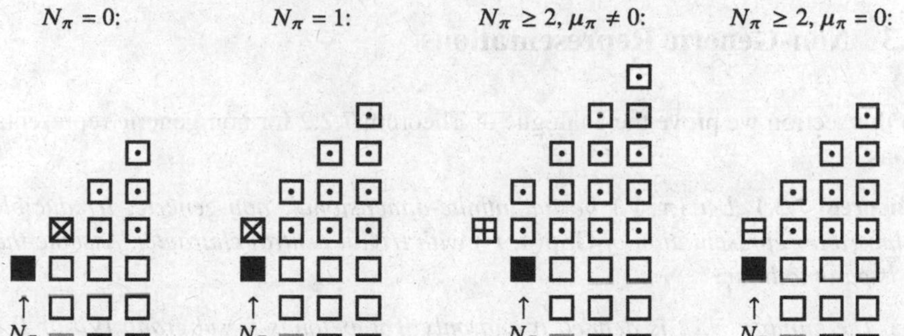

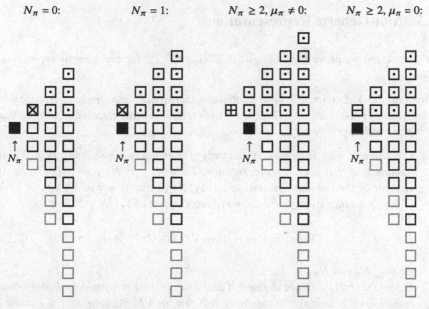

Fig. 7.1 Visualization of the generic cases. In each of the four diagrams, a vertical column of squares represents a basis for $V_s(n)$, beginning with $n = N_{\pi,s}$. The squares are as follows: ■ The paramodular newform; □ Descendants of ■ via θ, θ', η, making up the spaces $V(n)$ inside $V_s(n)$; □ The spaces $V(n-1)$ inside $V_s(n)$; ⊠ The vector $W_0 = T_{1,0}^s(■)$ in the cases $N_\pi = 0$ and $N_\pi = 1$; ⊞ The shadow vector; ⊟ The vector $W_0 = T_{0,1}^s(■)$; ⊡ Descendants of ⊠, ⊞, or ⊟ via θ and τ

and

$$\dim \bar{V}_s(n) = \begin{cases} n - \bar{N}_{\pi,s} + 1 & \text{for } n \geq \bar{N}_{\pi,s}, \\ 0 & \text{for } n < \bar{N}_{\pi,s}. \end{cases} \tag{7.25}$$

Let n be an integer such that $n \geq 0$. The operators $\tau, \theta : V_s(n) \to V_s(n+1)$, and also $\tau, \theta : \bar{V}_s(n) \to \bar{V}_s(n+1)$, are injective and satisfy $\theta\tau = \tau\theta$. Let $W_{s,\text{new}}$ be a non-zero element of the one-dimensional space $\bar{V}_s(\bar{N}_{\pi,s})$. Then the vectors

$$\tau^i \theta^j W_{s,\text{new}}, \qquad i, j \geq 0, \quad i + j = n - \bar{N}_{\pi,s}, \tag{7.26}$$

are a basis of $\bar{V}_s(n)$ for $n \geq \bar{N}_{\pi,s}$.

Proof Statements (7.24) and (7.25) follow from Table A.3. We already noticed in (3.77) that $\theta\tau = \tau\theta$. The injectivity of $\tau : V_s(n) \to V_s(n+1)$ and $\tau : \bar{V}_s(n) \to \bar{V}_s(n+1)$ was proven in Lemma 3.5.4. The injectivity of $\theta : V_s(n) \to V_s(n+1)$ and of $\theta : \bar{V}_s(n) \to \bar{V}_s(n+1)$ follows easily from (7.16). The last statement also follows from (7.16). □

7.3 Non-Generic Representations

In this section we prove the analogue of Theorem 7.2.2 for non-generic representations.

Theorem 7.3.1 *Let (π, V) be an infinite-dimensional, non-generic, irreducible, admissible representation of* GSp(4, F) *with trivial central character. Assume that π is paramodular.*

(1) *The number $\bar{N}_{\pi,s}$ is defined if and only if π belongs to subgroup IVb or IVc. Assume that π is such a representation. Then $N_{\pi,s} = \bar{N}_{\pi,s} = 1$. Let v_0 be an element of the one-dimensional space $V_s(\bar{N}_{\pi,s})$ that is not in $V(\bar{N}_{\pi,s} - 1) + V(\bar{N}_{\pi,s})$, so that v_0 represents a non-zero element of $\bar{V}_s(\bar{N}_{\pi,s})$. Then*

$$V_s(n) = V(n-1) \oplus V(n) \oplus \mathbb{C}\tau^{n-\bar{N}_{\pi,s}} v_0 \tag{7.27}$$

for $n \geq N_{\pi,s} = \bar{N}_{\pi,s} = 1$.

(2) *The number $\bar{N}_{\pi,s}$ is not defined if and only if π is a Saito-Kurokawa representation or π belongs to subgroup IIIb, Vd, or VId. Assume that π is such a representation. Then*

$$V_s(n) = V(n-1) \oplus V(n) \tag{7.28}$$

for $n \geq N_{\pi,s}$ In (7.28), if $N_{\pi,s} = 0$ and $n = 0$, then we take $V(n-1) = V(-1)$ to be the zero subspace.

Proof (1). All the assertions in (1) follow from an inspection of Table A.3, except (7.27). By Table A.3 dim $V_s(n) - \dim V(n-1) - \dim V(n) = 1$ for integers n such that $n \geq N_{\pi,s}$. Therefore, to verify (7.27), it suffices to prove that $\tau^{n-\bar{N}_{\pi,s}} v_0$ is not contained in $V(n-1) \oplus V(n)$. This follows from (2) of Lemma 3.5.4.

(2). The assertions in (2) follow from an inspection of Table A.3. □

7.4 Summary Statements

In this final section we provide statements that apply to all irreducible, admissible representations of GSp(4, F) with trivial central character. These results are corollaries of results from the previous sections.

Corollary 7.4.1 *Let* (π, V) *be an irreducible, admissible representation of* GSp(4, F) *with trivial central character. Assume that* π *is paramodular. The number* $\bar{N}_{\pi,s}$ *is not defined, so that* $V(n-1) + V(n) = V_s(n)$ *for all integers* $n \geq N_{\pi,s}$, *if and only if* π *is a Saito-Kurokawa representation or* π *belongs to subgroup IIIb, IVd, Vd, or VId.*

Proof Assume that $\bar{N}_{\pi,s}$ is not defined. By Sect. 7.2 we see that π is non-generic. Theorem 7.3.1 now implies that π is a Saito-Kurokawa representation or π belongs to subgroup IIIb, IVd, Vd, or VId (note that representations belonging to subgroup IVd are one-dimensional). The converse also follows from Sect. 7.2 and Theorem 7.3.1. □

Corollary 7.4.2 *Let* (π, V) *be an infinite-dimensional, irreducible, admissible representation of* GSp(4, F) *with trivial central character. Assume that* π *is paramodular. Define* $v_0 \in V$ *as follows. If* $\bar{N}_{\pi,s}$ *is not defined, set* $v_0 = 0$; *if* $\bar{N}_{\pi,s}$ *is defined, let* v_0 *be an element of* $V_s(\bar{N}_{\pi,s})$ *that is not contained in* $V(\bar{N}_{\pi,s}-1) + V(\bar{N}_{\pi,s})$. *For integers* n *such that* $n \geq N_{\pi,s}$ *let* $E(n)$ *be the subspace of* $V_s(n)$ *spanned by the vectors*

$$\tau^i \theta^j v_0, \qquad i, j \geq 0, \quad i + j = n - \bar{N}_{\pi,s}.$$

Then

$$V_s(n) = V(n-1) \oplus V(n) \oplus E(n)$$

for integers n *such that* $n \geq N_{\pi,s}$.

Proof This follows from Theorems 7.2.2 and 7.3.1. □

Corollary 7.4.3 *Let (π, V) be an irreducible, admissible representation of* $\mathrm{GSp}(4, F)$ *with trivial central character. Assume that π is paramodular. Assume that $\bar{N}_{\pi,s}$ is defined. Then*

$$\bar{N}_{\pi,s} = \begin{cases} 1 & \text{if } N_\pi = 0 \text{ or } N_\pi = 1, \\ N_\pi - 1 & \text{if } N_\pi \geq 2 \text{ and } \pi \text{ is a category 1 representation,} \\ N_\pi & \text{if } N_\pi \geq 2 \text{ and } \pi \text{ is a category 2 representation} \end{cases} \tag{7.29}$$

and

$$\dim \bar{V}_s(\bar{N}_{\pi,s}) = 1. \tag{7.30}$$

Let $v_{s,\mathrm{new}}$ be a non-zero element of the one-dimensional vector space $\bar{V}_s(\bar{N}_{\pi,s})$. Then the vectors

$$\tau^i \theta^j v_{s,\mathrm{new}}, \qquad i, j \geq 0, \quad i + j = n - \bar{N}_{\pi,s} \tag{7.31}$$

span $\bar{V}_s(n)$ for $n \geq \bar{N}_{\pi,s}$.

Proof By Corollary 7.4.1, the representation π is not a Saito-Kurokawa representation and π does not belong to subgroup IIIb, IVd, Vd, or VId; in particular, since π does not belong to subgroup IVd, π is infinite-dimensional. If π is generic, then (7.29) follows from (7.24) and Proposition 5.7.3. Assume that π is non-generic. By (1) of Theorem 7.3.1, π belongs to subgroup IVb or IVc. If π belongs to subgroup IVb, then $N_\pi = 2$, $N_{\pi,s} = 1$, and $\bar{N}_{\pi,s} = 1$ by Table A.3; this verifies (7.29). If π belongs to subgroup IVc, then $N_\pi = 1$, $N_{\pi,s} = 1$, and $\bar{N}_{\pi,s} = 1$ by Table A.3; this again verifies (7.29). Next, the assertion (7.30) follows from (7.24) if π is generic and Table A.3 if π is non-generic. Finally, the vectors in (7.31) span $\bar{V}_s(n)$ for $n \geq \bar{N}_{\pi,s}$ by Corollary 7.2.3 if π is generic, and by (1) of Theorem 7.3.1 if π is non-generic. □

Chapter 8
Further Results About Generic Representations

Let (π, V) be a generic, irreducible, admissible representation of $\mathrm{GSp}(4, F)$ with trivial central character. In this chapter we prove several additional results about stable Klingen vectors in π. Our first result generalizes a fundamental theorem from the paramodular theory. Assume that V is the Whittaker model of $\mathcal{W}(\pi, \psi_{c_1,c_2})$ of π, and let n be an integer such that $n \geq 0$. In the first section, we prove that if $W \in V_s(n)$, then $W \neq 0$ if and only if W does not vanish on the diagonal subgroup of $\mathrm{GSp}(4, F)$. To describe our second result, assume that $n \geq \max(N_{\pi,s}, 2)$, and recall the surjective level lowering operator $\sigma_{n-1} : V_s(n) \to V_s(n-1)$ from Sect. 3.7. We prove that

$$V_s(n) = (V(n-1) \oplus V(n)) + \ker(\sigma_{n-1})$$

for integers $n \geq \max(N_{\pi,s}, 2)$. Consequently, σ_{n-1} induces an isomorphism

$$(V(n-1) \oplus V(n))/\mathcal{K}_n \xrightarrow{\sim} V_s(n-1)$$

for integers $n \geq \max(N_{\pi,s}, 2)$, where $\mathcal{K}_n$ is the intersection of $V(n-1) \oplus V(n)$ with $\ker(\sigma_{n-1})$. We characterize $\mathcal{K}_n$ for $n \geq \max(N_{\pi,s}, 2)$; in particular, this subspace is at most two-dimensional. To prove these results we use the nonvanishing statement from the first section. Finally, in the last section, we consider π such that $L(s, \pi) = 1$; this includes all π that are supercuspidal. Assuming that $L(s, \pi) = 1$, we are able to define another, graphical, model for $V_s(n)$ for integers $n \geq N_{\pi,s}$. The existence of this alternative model also uses the result about the non-vanishing of stable Klingen vectors on the diagonal of $\mathrm{GSp}(4, F)$. In this model, our level changing operators have simple and visual interpretations. Finally, still under the hypothesis that $L(s, \pi) = 1$, we prove that $N_\pi \geq 4$.

Before beginning, we remind the reader that λ_π, μ_π, and ε_π are the eigenvalues of $T_{0,1}$, $T_{1,0}$, and u_{N_π}, respectively, on the one-dimensional space $V(N_\pi)$. Also, by

J. Johnson-Leung et al., *Stable Klingen Vectors and Paramodular Newforms*,
Lecture Notes in Mathematics 2342, https://doi.org/10.1007/978-3-031-45177-5_8

Theorem 2.3.1 the subspace $V(n-1)+V(n)$ of $V_s(n)$ is a direct sum $V(n-1)\oplus V(n)$ for $n \geq 1$. Finally, we will use the equivalences of Proposition 5.7.3.

8.1 Non-Vanishing on the Diagonal

Let (π, V) be a generic, irreducible, admissible representation of $\mathrm{GSp}(4, F)$ with trivial central character, and let n be an integer such that $n \geq 0$. Let $W \in V_s(n)$. In this section we prove that $W \neq 0$ if and only if W does not vanish on the diagonal subgroup of $\mathrm{GSp}(4, F)$. This result generalizes Corollary 4.3.8 of [105]. We begin by proving this statement for the elements of the subspace $V(n-1)\oplus V(n)$ of $V_s(n)$.

Proposition 8.1.1 *Let (π, V) be a generic, irreducible, admissible representation of* $\mathrm{GSp}(4, F)$ *with trivial central character, and let n be an integer such that $n \geq 0$. Let $V = \mathcal{W}(\pi, \psi_{c_1,c_2})$. Assume that $W \in V(n-1)\oplus V(n)$ is such that $W(\Delta_{i,j}) = 0$ for all $i, j \in \mathbb{Z}$; if $n = 0$, then we define $V(-1) = 0$. Then $W = 0$.*

Proof The statement is trivial if $n \leq N_\pi - 1$. For $n \geq N_\pi$ we use induction on n. The case $n = N_\pi$ follows from Corollary 4.3.8 of [105]. Assume that $n > N_\pi$, and that the statement holds for $n-1$. Let $W_1 \in V(n-1)$ and $W_2 \in V(n)$, and assume that

$$(W_1 + W_2)(\Delta_{i,j}) = 0 \tag{8.1}$$

for all $i, j \in \mathbb{Z}$; we need to prove that $W_1 = W_2 = 0$. Define $W' = q^{-1}\theta' W_1 + W_2$. Then $W' \in V(n)$ and:

$$\begin{aligned} Z(s, W') &= q^{-1}Z(s, \theta' W_1) + Z(s, W_2) \\ &= Z(s, W_1) + Z(s, W_2) \qquad \text{(by (7.1))} \\ &= Z(s, W_1 + W_2) \\ &= 0 \qquad \text{(by (8.1))}. \end{aligned}$$

Assume that $n \geq 2$. By the η-principle, Theorem 4.3.7 of [105], there exists $W_3 \in V(n-2)$ such that

$$W' = \eta W_3. \tag{8.2}$$

If $n = 1$, then we set $W_3 = 0$, and Eq. (8.2) still holds by Theorem 4.3.7 of [105]. Evaluating at $\Delta_{i,j}$, we get by (2.29)

$$q^{-1}W_1(\Delta_{i,j}\eta) + W_1(\Delta_{i,j}) + W_2(\Delta_{i,j}) = W_3(\Delta_{i,j}\eta).$$

By (8.1),

$$q^{-1}W_1(\Delta_{i,j}\eta) = W_3(\Delta_{i,j}\eta),$$

and hence

$$q^{-1}W_1(\Delta_{i,j}) = W_3(\Delta_{i,j})$$

for all $i, j \in \mathbb{Z}$. This implies that $(W_3 - q^{-1}W_1)(\Delta_{i,j}) = 0$ for all $i, j \in \mathbb{Z}$. Since $W_3 - q^{-1}W_1 \in V(n-2) + V(n-1)$ we have $W_1 = W_3 = 0$ by the induction hypothesis. Then also $W_2 = 0$. This concludes the proof. □

Proposition 8.1.2 *Let (π, V) be a generic, irreducible, admissible representation of* GSp(4, F) *with trivial central character, and let n be an integer such that $n \geq 0$. Let $V = \mathcal{W}(\pi, \psi_{c_1,c_2})$. Assume that $W \in V_s(n)$ is such that $W(\Delta_{i,j}) = 0$ for all $i, j \in \mathbb{Z}$. Then $W = 0$.*

Proof Write $W = W_1 + W_2$ with $W_1 \subset V(n)$ and $W_2 \in V_s(n)$ such that $p_n(W_2) = 0$ (see (3.21). By (3.51),

$$\rho_n' W_2 = q^{-1}\eta W_2 - q\tau_{n+1}W_2 \in V(n+1). \tag{8.3}$$

Hence,

$$\begin{aligned}(\rho_n' W_2)(\Delta_{i,j}) &= q^{-1}(\eta W_2)(\Delta_{i,j}) - q(\tau_{n+1}W_2)(\Delta_{i,j})\\ &= q^{-1}W_2(\Delta_{i,j}\eta) - qW_2(\Delta_{i,j})\\ &= -q^{-1}W_1(\Delta_{i,j}\eta) + qW_1(\Delta_{i,j}).\end{aligned} \tag{8.4}$$

It follows that

$$\begin{aligned}Z(s, \rho_n' W_2 - \theta' W_1) &= -q^{-1}Z(s, \eta W_1) + qZ(s, W_1) - Z(s, \theta' W_1)\\ &= 0 \qquad \text{(by (7.1) and (7.2))}.\end{aligned}$$

The vector $\rho_n' W_2 - \theta' W_1$ is in $V(n+1)$. If $n \geq 1$, then by the η-principle, Theorem 4.3.7 of [105], there exists $W_3 \in V(n-1)$ such that

$$\rho_n' W_2 - \theta' W_1 = \eta W_3. \tag{8.5}$$

If $n = 0$, then we set $W_3 = 0$, and (8.5) still holds by Theorem 4.3.7 of [105]. Hence, by (2.29),

$$\begin{aligned}(\rho_n' W_2)(\Delta_{i,j}) &= (\theta' W_1)(\Delta_{i,j}) + (\eta W_3)(\Delta_{i,j})\\ &= W_1(\Delta_{i,j}\eta) + qW_1(\Delta_{i,j}) + W_3(\Delta_{i,j}\eta)\end{aligned} \tag{8.6}$$

for all $i, j \in \mathbb{Z}$. Substituting into (8.4), we get

$$W_1(\Delta_{i,j}\eta) + qW_1(\Delta_{i,j}) + W_3(\Delta_{i,j}\eta) = -q^{-1}W_1(\Delta_{i,j}\eta) + qW_1(\Delta_{i,j}),$$
$$W_1(\Delta_{i,j}\eta) + W_3(\Delta_{i,j}\eta) = -q^{-1}W_1(\Delta_{i,j}\eta),$$
$$(1+q^{-1})W_1(\Delta_{i,j}\eta) + W_3(\Delta_{i,j}\eta) = 0$$

for all $i, j \in \mathbb{Z}$. It follows that

$$(1+q^{-1})W_1(\Delta_{i,j}) + W_3(\Delta_{i,j}) = 0$$

for all $i, j \in \mathbb{Z}$. By Proposition 8.1.1 and Theorem 2.3.1 we obtain $W_1 = W_3 = 0$. Hence, $W = W_2$, so that $p_n(W) = 0$. By (8.5), $\rho_n' W = 0$. By (3.51),

$$0 = \rho_n' W = q^{-1}\eta W - q\tau_{n+1}W. \tag{8.7}$$

Hence, $\tau_{n+1}W = q^{-2}\eta W$. Therefore $W = 0$ by Proposition 5.1.3. This completes the proof. □

8.2 The Kernel of a Level Lowering Operator

Let (π, V) be a generic, irreducible, admissible representation of $\mathrm{GSp}(4, F)$ with trivial central character, and let n be an integer such that $n \geq \max(N_{\pi,s}, 2)$. In Theorem 7.2.2 we proved the structural result (7.16) for the spaces $V_s(n)$. In this section we will prove that also $V_s(n) = (V(n-1) \oplus V(n)) + \ker(\sigma_{n-1})$, where $\sigma_{n-1} : V_s(n) \to V_s(n-1)$ is the surjective level lowering operator defined in Sect. 3.7. We will precisely determine the intersection $(V(n-1) \oplus V(n)) \cap \ker(\sigma_{n-1})$; it is at most two-dimensional.

Lemma 8.2.1 *Let (π, V) be a generic, irreducible, admissible representation of* $\mathrm{GSp}(4, F)$ *with trivial central character, and let n be an integer such that $n \geq 0$. Let $V = \mathcal{W}(\psi_{c_1,c_2})$. Assume that $i, j \in \mathbb{Z}$ with $i, j \geq 0$. Then, for $W \in V_s(n)$,*

$$(\sigma_{n+1}\theta W)(\Delta_{i,j}) = W(\Delta_{i+1,j-1}) + qW(\Delta_{i,j+1}),$$
$$(\sigma_{n+1}\eta W)(\Delta_{i,j}) = W(\Delta_{i,j}),$$
$$(\sigma_{n+1}\theta^2 W)(\Delta_{i,j}) = \begin{cases} W(\Delta_{i+1,j-2}) + 2qW(\Delta_{i,j}) + q^2W(\Delta_{i-1,j+2}) & \text{if } j \geq 1, \\ qW(\Delta_{i,0}) + q^2W(\Delta_{i-1,2}) & \text{if } j = 0. \end{cases}$$

Furthermore, for $W \in V(n)$,

$$\begin{aligned}(\sigma_{n+1}\theta' W)(\Delta_{i,j}) &= W(\Delta_{i,j}) + qW(\Delta_{i+1,j}),\\ (\sigma_{n+1}\theta\theta' W)(\Delta_{i,j}) &= W(\Delta_{i,j-1}) + qW(\Delta_{i-1,j+1})\\ &\quad + qW(\Delta_{i+1,j-1}) + q^2 W(\Delta_{i,j+1}),\\ (\sigma_{n+1}\theta'^2 W)(\Delta_{i,j}) &= W(\Delta_{i-1,j}) + 2qW(\Delta_{i,j}) + q^2 W(\Delta_{i+1,j}).\end{aligned}$$

Here, $\Delta_{i,j}$ is the diagonal matrix defined in (7.3).

Proof By (3.42), for $W \in V_s(n)$,

$$(\theta W)(\Delta_{i,j}) = \begin{cases} W(\Delta_{i,j-1}) + qW(\Delta_{i-1,j+1}) & \text{if } j \geq 0,\\ 0 & \text{if } j < 0,\end{cases} \tag{8.8}$$

and

$$(\theta^2 W)(\Delta_{i,j}) = \begin{cases} W(\Delta_{i,j-2}) + 2qW(\Delta_{i-1,j}) + q^2 W(\Delta_{i-2,j+2}) & \text{if } j \geq 1,\\ qW(\Delta_{i-1,0}) + q^2 W(\Delta_{i-2,2}) & \text{if } j = 0,\\ 0 & \text{if } j < 0.\end{cases} \tag{8.9}$$

By (2.29), for $W \in V(n)$,

$$(\theta' W)(\Delta_{i,j}) = W(\Delta_{i-1,j}) + qW(\Delta_{i,j}) \tag{8.10}$$

and

$$(\theta'^2 W)(\Delta_{i,j}) = W(\Delta_{i-2,j}) + 2qW(\Delta_{i-1,j}) + q^2 W(\Delta_{i,j}). \tag{8.11}$$

By (3.60), for $n \geq 1$ and $W \in V_s(n+1)$,

$$(\sigma_n W)(\Delta_{i,j}) = \begin{cases} W(\Delta_{i+1,j}) & \text{if } i \geq 0,\\ 0 & \text{if } i < 0.\end{cases} \tag{8.12}$$

Our asserted formulas follow easily from (8.8)–(8.12). □

Lemma 8.2.2 *Let (π, V) be a generic, irreducible, admissible representation of* GSp(4, *F*) *with trivial central character and $N_\pi = 0$. Let $W_{\text{new}} \in V(0)$ be non-zero. Then the subspace*

$$(V(1) + V(2)) \cap \ker(\sigma_1)$$

of $V_s(2)$ is two-dimensional, and is spanned by the vectors

$$W_1 = ((q^2 - q)\theta + \theta\theta' - \lambda_\pi \eta)W_{\text{new}}, \tag{8.13}$$

$$W_2 = (q(q^2 - 1)\theta' + q\theta^2 + \theta'^2 - (\mu_\pi + q^3 + q^2 + q + 1)\eta)W_{\text{new}}. \tag{8.14}$$

The map

$$\sigma_1 : V(1) + V(2) \longrightarrow V_s(1) \tag{8.15}$$

is surjective.

Proof We have

$$V(1) = \langle \theta W_{\text{new}}, \theta' W_{\text{new}} \rangle \quad \text{and} \quad V(2) = \langle \theta^2 W_{\text{new}}, \theta\theta' W_{\text{new}}, \theta'^2 W_{\text{new}}, \eta W_{\text{new}} \rangle$$

by Theorem 7.5.6 of [105]. Evidently, $W_1, W_2 \in V(1) + V(2) \subset V_s(2)$. We will show that $\sigma_1 W_1 = \sigma_1 W_2 = 0$. As in (7.4), let $c_{i,j} = W_{\text{new}}(\Delta_{i,j})$ for $i, j \in \mathbb{Z}$. By Lemma 8.2.1, for $i, j \in \mathbb{Z}$ with $i, j \geq 0$,

$$(\sigma_1 \theta W_{\text{new}})(\Delta_{i,j}) = c_{i+1,j-1} + q c_{i,j+1},$$
$$(\sigma_1 \theta' W_{\text{new}})(\Delta_{i,j}) = c_{i,j} + q c_{i+1,j},$$
$$(\sigma_1 \eta W_{\text{new}})(\Delta_{i,j}) = c_{i,j},$$
$$(\sigma_1 \theta^2 W_{\text{new}})(\Delta_{i,j}) = \begin{cases} c_{i+1,j-2} + 2q c_{i,j} + q^2 c_{i-1,j+2} & \text{if } j \geq 1, \\ q c_{i,0} + q^2 c_{i-1,2} & \text{if } j = 0, \end{cases}$$
$$(\sigma_1 \theta\theta' W_{\text{new}})(\Delta_{i,j}) = c_{i,j-1} + q c_{i-1,j+1} + q c_{i+1,j-1} + q^2 c_{i,j+1},$$
$$(\sigma_1 \theta'^2 W_{\text{new}})(\Delta_{i,j}) = c_{i-1,j} + 2q c_{i,j} + q^2 c_{i+1,j}.$$

Using Lemma 7.1.2 of [105], we get for $i, j \in \mathbb{Z}$ with $i, j \geq 0$,

$$\begin{aligned}
(\sigma_1 \theta W_{\text{new}})(\Delta_{i,j}) &= q^{-2}\lambda_\pi c_{i,j} - q^{-1} c_{i-1,j+1} - q^{-2} c_{i,j-1}, \\
(\sigma_1 \theta' W_{\text{new}})(\Delta_{i,j}) &= c_{i,j} + q c_{i+1,j}, \\
(\sigma_1 \eta W_{\text{new}})(\Delta_{i,j}) &= c_{i,j}, \\
(\sigma_1 \theta^2 W_{\text{new}})(\Delta_{i,j}) &= (q^{-1}\mu_\pi + q + q^{-1}) c_{i,j} - q^3 c_{i+1,j} - q^{-1} c_{i-1,j}, \\
(\sigma_1 \theta\theta' W_{\text{new}})(\Delta_{i,j}) &= q^{-1}\lambda_\pi c_{i,j} + (1 - q^{-1})(c_{i,j-1} + q c_{i-1,j+1}), \\
(\sigma_1 \theta'^2 W_{\text{new}})(\Delta_{i,j}) &= 2q c_{i,j} + q^2 c_{i+1,j} + c_{i-1,j}.
\end{aligned} \tag{8.16}$$

It follows that $(\sigma_1 W_1)(\Delta_{i,j}) = (\sigma_1 W_2)(\Delta_{i,j}) = 0$ for $i, j \in \mathbb{Z}$ with $i, j \geq 0$. By (5.18) and Proposition 8.1.2, $\sigma_1 W_1 = \sigma_1 W_2 = 0$. So far we have proven that

$\langle W_1, W_2\rangle \subset (V(1) + V(2)) \cap \ker(\sigma_1)$. From above, and by Table A.3 we have $\dim(V(1)+V(2)) = 6$ and $\dim V_s(1) = 4$; also, by (3.64) of Proposition 3.7.4, the three-dimensional subspace $V(0)+V(1)$ of $V_s(1)$ is contained in $\sigma_1(V(1)+V(2))$. Thus, to prove that $\sigma_1 : V(1) + V(2) \to V_s(1)$ is surjective and that $\langle W_1, W_2\rangle = (V(1)+V(2))\cap\ker(\sigma_1)$, it will suffice to prove that, say, $\sigma_1\theta' W_{\text{new}} \notin V(0)+V(1)$. We have

$$
\begin{aligned}
Z(s, \sigma_1\theta' W_{\text{new}}) &= (1-q^{-1})\sum_{j=0}^{\infty}(\sigma_1\theta' W_{\text{new}}) \\
&\quad \times (\begin{bmatrix} \varpi^j &&& \\ & \varpi^j && \\ && 1 & \\ &&& 1\end{bmatrix})(q^{-s+\frac{3}{2}})^j \quad \text{(by (5.21))} \\
&= (1-q^{-1})\sum_{j=0}^{\infty}(c_{0,j} + qc_{1,j})(q^{-s+\frac{3}{2}})^j \quad \text{(by (8.16))} \\
&= (1+q^{-3}\mu_\pi + q^{-3} - \lambda_\pi q^{-\frac{5}{2}}q^{-s} + q^{-1}q^{-2s}) \\
&\quad \times Z(s, W_{\text{new}}) \quad \text{(by (7.6))}.
\end{aligned}
$$

On the other hand, $Z(s, V(0) + V(1)) = (\mathbb{C} + \mathbb{C}q^{-s})Z(s, W_{\text{new}})$ since $V(0) + V(1)$ is spanned by $W_{\text{new}}, \theta W_{\text{new}}$, and $\theta' W_{\text{new}}$ and since (7.1) holds. It follows that $\sigma_1\theta' W_{\text{new}} \notin V(0) + V(1)$, as desired. □

Lemma 8.2.3 *Let (π, V) be a generic, irreducible, admissible representation of* GSp(4, F) *with trivial central character and $N_\pi = 1$. Let $W_{\text{new}} \in V(1)$ be non-zero. Then the subspace*

$$(V(1) + V(2)) \cap \ker(\sigma_1)$$

of $V_s(2)$ is one-dimensional, and is spanned by the vector

$$W_3 = (q(q^2-1) + \varepsilon_\pi q\theta + \theta')W_{\text{new}}. \tag{8.17}$$

The map

$$\sigma_1 : V(1) + V(2) \longrightarrow V_s(1) \tag{8.18}$$

is surjective.

Proof We have $V(1) = \langle W_{\text{new}}\rangle$ and $V(2) = \langle\theta W_{\text{new}}, \theta' W_{\text{new}}\rangle$ by Theorem 7.5.6 of [105]. Evidently, $W_3 \in V(1) + V(2)$. We will show that $\sigma_1 W_3 = 0$. As in (7.4), let $c_{i,j} = W_{\mathrm{new}}(\Delta_{i,j})$ for $i, j \in \mathbb{Z}$. By (8.8), (8.10) and (8.12), for $i, j \in \mathbb{Z}$ with $i, j \geq 0$,

$$\begin{aligned}
(\sigma_1 W_{\mathrm{new}})(\Delta_{i,j}) &= c_{i+1,j}, \\
(\sigma_1 \theta W_{\mathrm{new}})(\Delta_{i,j}) &= c_{i+1,j-1} + q c_{i,j+1}, \\
(\sigma_1 \theta' W_{\mathrm{new}})(\Delta_{i,j}) &= c_{i,j} + q c_{i+1,j}.
\end{aligned}$$

By Lemmas 7.2.1 and 7.2.2 of [105],

$$\begin{aligned}
(\sigma_1 W_{\mathrm{new}})(\Delta_{i,j}) &= c_{i+1,j}, \\
(\sigma_1 \theta W_{\mathrm{new}})(\Delta_{i,j}) &= -\varepsilon q^{-1}(c_{i,j} + q^3 c_{i+1,j}), \\
(\sigma_1 \theta' W_{\mathrm{new}})(\Delta_{i,j}) &= c_{i,j} + q c_{i+1,j}
\end{aligned} \tag{8.19}$$

for $i, j \in \mathbb{Z}$ with $i, j \geq 0$. It follows that $(\sigma_1 W_3)(\Delta_{i,j}) = 0$ for $i, j \in \mathbb{Z}$ with $i, j \geq 0$. By (5.18) and Proposition 8.1.2, $\sigma_1 W_3 = 0$. So far we have proven that $\langle W_3 \rangle \subset (V(1) + V(2)) \cap \ker(\sigma_1)$. From above and by Table A.3 we have $\dim(V(1) + V(2)) = 3$ and $\dim V_s(1) = 2$; also, by (3.64) of Proposition 3.7.4, the one-dimensional subspace $V(1)$ of $V_s(1)$ is contained in $\sigma_1(V(1) + V(2))$. Thus, to prove that $\sigma_1 : V(1) + V(2) \to V_s(1)$ is surjective and that $\langle W_3 \rangle = (V(1) + V(2)) \cap \ker(\sigma_1)$, it will suffice to prove that, say, $\sigma_1 \theta' W_{\mathrm{new}} \notin V(1)$. We have

$$\begin{aligned}
Z(s, \sigma_1 \theta' W_{\mathrm{new}}) &= (1 - q^{-1}) \sum_{j=0}^{\infty} (\sigma_1 \theta' W_{\mathrm{new}})\left(\begin{bmatrix} \varpi^j &&& \\ & \varpi^j && \\ && 1 & \\ &&& 1 \end{bmatrix}\right)(q^{-s+\frac{3}{2}})^j \quad \text{(by (5.21))} \\
&= (1 - q^{-1}) \sum_{j=0}^{\infty} (c_{0,j} + q c_{1,j})(q^{-s+\frac{3}{2}})^j \quad \text{(by (8.19))} \\
&= (1 + q^{-3}\mu_\pi + \varepsilon_\pi q^{-\frac{3}{2}} q^{-s}) Z(s, W_{\mathrm{new}}) \quad \text{(by (7.6))}.
\end{aligned}$$

On the other hand, $Z(s, V(1)) = \mathbb{C} Z(s, W_{\mathrm{new}})$ since $V(1)$ is spanned by W_{new}. It follows that $\sigma_1 \theta' W_{\mathrm{new}} \notin V(1)$, as desired. □

Lemma 8.2.4 *Let π be a generic, irreducible, admissible representation of the group* GSp(4, F) *with trivial central character, and let n be an integer such that $n \geq \max(N_{\pi,s}, 2) = \max(\bar{N}_{\pi,s}, 2)$. We consider the homomorphism*

$$\sigma_{n-1} : V_s(n) \to V_s(n-1).$$

(1) We have

$$\dim \ker(\sigma_{n-1}) = n - N_\pi + 2. \tag{8.20}$$

(2) The map

$$\ker(\sigma_{n-1}) \xrightarrow{\sim} Z(s, W_{\rm new})(\mathbb{C} + \mathbb{C}q^{-s} + \ldots + \mathbb{C}(q^{-s})^{n-N_\pi+1}) \tag{8.21}$$

defined by $W \mapsto Z(s, W)$ is an isomorphism of vector spaces.

Proof (1). By Proposition 3.7.4, the map $\sigma_{n-1} : V_s(n) \to V_s(n-1)$ is surjective. Hence, $\dim \ker(\sigma_{n-1}) = \dim V_s(n) - \dim V_s(n-1)$, Now (8.20) follows from (7.17).

(2). By (7.18), the map (8.21) is well-defined. By (8.20), it suffices to show that the map is injective. Hence, assume that $W \in \ker(\sigma_{n-1})$ and $Z(s, W) = 0$. The latter condition implies that

$$W(\Delta_{0,j}) = 0 \qquad \text{for all } j \in \mathbb{Z} \text{ with } j \geq 0. \tag{8.22}$$

It follows from (3.60) that

$$0 = (\sigma_{n-1}W)(\Delta_{i,j}) = W(\Delta_{i+1,j}) \qquad \text{for all } i, j \in \mathbb{Z} \text{ with } i, j \geq 0. \tag{8.23}$$

Hence, $W(\Delta_{i,j}) = 0$ for all $i, j \in \mathbb{Z}$ with $i, j \geq 0$, and then also for all $i, j \in \mathbb{Z}$ by (5.18). By Proposition 8.1.2 it follows that $W = 0$. This concludes the proof. □

Lemma 8.2.5 *Let π be a generic, irreducible, admissible representation of the group* GSp(4, F) *with trivial central character. Assume that $N_\pi \geq 2$. Let W_0 be as in Theorem 7.2.2. For any $n \geq \max(\bar{N}_{\pi,s}, 2) = \max(N_{\pi,s}, 2)$, and any $j \in \{0, \ldots, n - \bar{N}_{\pi,s}\}$, there exists a vector of the form*

$$W_1 + W_2 + \tau^i\theta^j W_0, \qquad W_1 \in V(n-1),\ W_2 \in V(n),\ i + j = n - \bar{N}_{\pi,s}, \tag{8.24}$$

which is in the kernel of $\sigma_{n-1} : V_s(n) \to V_s(n-1)$.

Proof We will use induction on n, distinguishing cases for the beginning of the induction.

Assume that $N_\pi = 0$. Then $\bar{N}_{\pi,s} = 1$ and the beginning of the induction is at $n = 2$. By Theorem 7.2.2,

$$V_s(2) = V(1) \oplus V(2) \oplus \mathbb{C}\tau W_0 \oplus \mathbb{C}\theta W_0.$$

By Lemma 8.2.2 and its proof, $\dim((V(1) \oplus V(2)) \cap \ker(\sigma_1)) = 2$, $\dim V_s(1) = 4$, and the map $\sigma_1 :\ V(1) \oplus V(2) \longrightarrow V_s(1)$ is surjective. It follows that the map

$$\sigma_1 :\ V(1) \oplus V(2) \oplus \mathbb{C}\tau W_0 \longrightarrow V_s(1)$$

is surjective and $\dim((V(1) \oplus V(2) \oplus \mathbb{C}\tau W_0) \cap \ker(\sigma_1)) = 3$. In particular, there exists an element in $(V(1) \oplus V(2) \oplus \mathbb{C}\tau W_0) \cap \ker(\sigma_1)$ with non-trivial τW_0 component. Similarly, we see that there exists an element in $(V(1) \oplus V(2) \oplus \mathbb{C}\theta W_0) \cap \ker(\sigma_1)$ with non-trivial θW_0 component. This proves the assertion for $N_\pi = 0$ and $n = 2$.

Assume that $N_\pi = 1$. Then $\bar{N}_{\pi,s} = 1$ and the beginning of the induction is at $n = 2$. By Theorem 7.2.2,

$$V_s(2) = V(1) \oplus V(2) \oplus \mathbb{C}\tau W_0 \oplus \mathbb{C}\theta W_0.$$

By Lemma 8.2.3 and its proof, $\dim((V(1) \oplus V(2)) \cap \ker(\sigma_1)) = 1$, $\dim V_s(1) = 2$, and the map $\sigma_1 : V(1) \oplus V(2) \longrightarrow V_s(1)$ is surjective. It follows that the map

$$\sigma_1 : V(1) \oplus V(2) \oplus \mathbb{C}\tau W_0 \longrightarrow V_s(1)$$

is surjective and $\dim((V(1) \oplus V(2) \oplus \mathbb{C}\tau W_0) \cap \ker(\sigma_1)) = 2$. In particular, there exists an element in $(V(1) \oplus V(2) \oplus \mathbb{C}\tau W_0) \cap \ker(\sigma_1)$ with non-trivial τW_0 component. Similarly, we see that there exists an element in $(V(1) \oplus V(2) \oplus \mathbb{C}\theta W_0) \cap \ker(\sigma_1)$ with non-trivial θW_0 component. This proves the assertion for $N_\pi = 1$ and $n = 2$.

Assume that $N_\pi = 2$ and $\mu_\pi \neq 0$. Then $\bar{N}_{\pi,s} = N_{\pi,s} = 1$ and the beginning of the induction is at $n = 2$. We have $\dim V_s(1) = 1$ and $\dim V_s(2) = 3$. By Theorem 7.2.2,

$$V_s(2) = V(2) \oplus \mathbb{C}\tau W_0 \oplus \mathbb{C}\theta W_0.$$

By Proposition 5.7.3, $\sigma_1 : V(2) \to V_s(1)$ is an isomorphism. It follows that $\dim((V(2) \oplus \mathbb{C}\tau W_0) \cap \ker(\sigma_1)) = \dim((V(2) \oplus \mathbb{C}\theta W_0) \cap \ker(\sigma_1)) = 1$. This proves the assertion for $N_\pi = 2$, $\mu_\pi \neq 0$ and $n = 2$.

Assume that $N_\pi \geq 3$ and $\mu_\pi \neq 0$. Then $\bar{N}_{\pi,s} = N_{\pi,s} = N_\pi - 1$ and the beginning of the induction is at $n = N_\pi - 1$. The assertion is true for $n = N_\pi - 1$, since $\sigma_{(N_\pi - 1) - 1}(W_0) = 0$.

Assume that $N_\pi \geq 2$ and $\mu_\pi = 0$. Then $\bar{N}_{\pi,s} = N_{\pi,s} = N_\pi$ and the beginning of the induction is at $n = N_\pi$. The assertion is true for $n = N_\pi$, since $\sigma_{N_\pi - 1}(W_0) = 0$.

We completed the beginning of the induction and proceed to the induction step. Assume that $n > \max(\bar{N}_{\pi,s}, 2)$, and that the assertion has been proven for $n - 1$. Hence, for any $j \in \{0, \ldots, n - 1 - \bar{N}_{\pi,s}\}$, there exists a vector of the form

$$W_1 + W_2 + \tau^i \theta^j W_0, \quad W_1 \in V(n-2),$$
$$W_2 \in V(n-1), \; i + j = n - 1 - \bar{N}_{\pi,s}, \tag{8.25}$$

which is in the kernel of $\sigma_{n-2} : V_s(n-1) \to V_s(n-2)$; then the vectors

$$\tau(W_1 + W_2 + \tau^i \theta^j W_0) = \tau(W_1 + W_2) + \tau^{i+1}\theta^j W_0, \qquad i + 1 + j = n - \bar{N}_{\pi,s}, \tag{8.26}$$

lie in the kernel of $\sigma_{n-1} : V_s(n) \to V_s(n-1)$ by (6) of Lemma 3.7.3, and also $\tau(W_1 + W_2) \in V(n-1) \oplus V(n)$ by the proof of Lemma 3.5.4. It remains to find a vector of the form

$$W_1' + W_2' + \theta^{n-\bar{N}_{\pi,s}} W_0, \qquad W_1' \in V(n-1),\ W_2' \in V(n), \tag{8.27}$$

which is in the kernel of $\sigma_{n-1} : V_s(n) \to V_s(n-1)$. Suppose that no such vector would exist. Then, by Theorem 7.2.2, every element of $\ker(\sigma_{n-1})$ would be of the form

$$W_1' + W_2' + \sum_{\substack{i+j=n-\bar{N}_{\pi,s} \\ j<n-\bar{N}_{\pi,s}}} c_i \tau^i \theta^j W_0, \qquad W_1' \in V(n-1),\ W_2' \in V(n),\ c_i \in \mathbb{C}. \tag{8.28}$$

By Theorem 7.2.2, in particular (7.15), it would follow that

$$Z(s, \ker(\sigma_{n-1})) \subset Z(s, W_{\text{new}})(\mathbb{C} + \mathbb{C}q^{-s} + \ldots + \mathbb{C}(q^{-s})^{n-N_\pi}). \tag{8.29}$$

This contradicts (8.21). □

Theorem 8.2.6 *Let π be a generic, irreducible, admissible representation of the group* GSp(4, F) *with trivial central character. Let n be an integer such that $n \geq \max(N_{\pi,s}, 2) = \max(\bar{N}_{\pi,s}, 2)$.*

(1) We have

$$V_s(n) = (V(n-1) \oplus V(n)) + \ker(\sigma_{n-1}). \tag{8.30}$$

(2) Let $\mathcal{K}_n = (V(n-1) \oplus V(n)) \cap \ker(\sigma_{n-1})$. Then

$$\dim(\mathcal{K}_n) = \begin{cases} 2 & \text{if } N_\pi = 0, \\ 1 & \text{if } N_\pi = 1, \\ 1 & \text{if } N_\pi \geq 2 \text{ and } \mu_\pi = 0, \\ 0 & \text{if } N_\pi \geq 2 \text{ and } \mu_\pi \neq 0. \end{cases} \tag{8.31}$$

If $N_\pi = 0$, then $\mathcal{K}_n = \langle \tau_{n-1} W_1, \tau_{n-1} W_2 \rangle$, where W_1, W_2 are the vectors defined in (8.13), (8.14)*; if $N_\pi = 1$, then $\mathcal{K}_n = \langle \tau_{n-1} W_3 \rangle$, where W_3 is the vector defined in* (8.17)*; if $N_\pi \geq 2$ and $\mu_\pi = 0$, then $\mathcal{K}_n = \langle \tau_{n-1} W_{\text{new}} \rangle$.*

(3) The operator σ_{n-1} induces an isomorphism

$$(V(n-1) \oplus V(n))/\mathcal{K}_n \xrightarrow{\sim} V_s(n-1). \tag{8.32}$$

Proof By (6) of Lemma 3.7.3, if $W \in V_s(m)$ lies in the kernel of σ_{m-1}, then $\tau_m W$ lies in the kernel of σ_m. It follows that

$$\mathcal{K}_n \supset \begin{cases} \langle \tau_{n-1} W_1, \tau_{n-1} W_2 \rangle & \text{if } N_\pi = 0, \\ \langle \tau_{n-1} W_3 \rangle & \text{if } N_\pi = 1, \\ \langle \tau_{n-1} W_{\text{new}} \rangle & \text{if } N_\pi \geq 2 \text{ and } \mu_\pi = 0. \end{cases} \tag{8.33}$$

Since τ_{n-1} is injective, it follows that the numbers on the right-hand side of (8.31) are lower bounds for the dimension of $\mathcal{K}_n$. Hence,

$$\dim(\mathcal{K}_n) \geq \deg(P_0), \tag{8.34}$$

where P_0 is the polynomial defined in Theorem 7.2.2.

Next, using Lemma 8.2.5, we see that there exist vectors $Y_0, \ldots, Y_{n-\bar{N}_{\pi,s}}$ in $V(n-1) \oplus V(n)$ such that

$$Y_i + \tau^i \theta^j W_0 \qquad 0 \leq i \leq n - \bar{N}_{\pi,s},\ i + j = n - \bar{N}_{\pi,s} \tag{8.35}$$

lie in $\ker(\sigma_{n-1})$. Now

$$\begin{aligned} n - N_\pi + 2 &= \dim(\ker(\sigma_{n-1})) \qquad \text{(by Lemma 8.2.4)} \\ &\geq \dim(\mathcal{K}_n) + n - \bar{N}_{\pi,s} + 1 \qquad \text{(by (7.16) and (8.35))} \\ &\geq \deg(P_0) + n - \bar{N}_{\pi,s} + 1 \qquad \text{(by (8.34))} \\ &= n - N_\pi + 2 \qquad \text{(by (7.15))}. \end{aligned}$$

It follows that $\dim(\mathcal{K}_n) = \deg(P_0)$. This proves (2).

To prove (1), we calculate as follows:

$$\begin{aligned} &\dim\big((V(n-1) \oplus V(n)) + \ker(\sigma_{n-1})\big) \\ &= \dim(V(n-1) \oplus V(n)) + \dim(\ker(\sigma_{n-1})) - \dim(\mathcal{K}_n) \\ &= \dim(V(n-1) \oplus V(n)) + n - N_\pi + 2 - \dim(\mathcal{K}_n) \qquad \text{(by (8.20))} \\ &= \left\lfloor \frac{(n - N_\pi + 1)^2}{4} \right\rfloor + \left\lfloor \frac{(n - N_\pi + 2)^2}{4} \right\rfloor \\ &\quad + n - N_\pi + 2 - \dim(\mathcal{K}_n) \qquad \text{(by Theorem 7.5.6 of [105])} \end{aligned}$$

$$= \frac{(n - N_\pi + 1)(n - N_\pi + 2)}{2} + n - N_\pi + 2 - \dim(\mathcal{K}_n)$$
$$= \frac{(n - N_\pi + 2)(n - N_\pi + 3)}{2} - \dim(\mathcal{K}_n)$$
$$= \dim V_s(n) \qquad \text{(by (7.17))}.$$

This proves (8.30).

Finally, (3) follows from (1) and Proposition 3.7.4. □

8.3 The Alternative Model

Let (π, V) be a generic, irreducible, admissible representation of $\mathrm{GSp}(4, F)$ with trivial central character. In this and the next section we will consider π with $L(s, \pi) = 1$; this is a substantial family of representations, and includes all π that are supercuspidal. We will prove two results about such π. First, in this section, for π with $L(s, \pi) = 1$, we will prove the existence of an alternative model for the spaces $V_s(n)$ of stable Klingen vectors in V for integers $n \geq N_{\pi,s}$. In this model, our level raising and lowering operators have simple visual interpretations. Second, in the next section, still for π with $L(s, \pi) = 1$, we will prove that $N_\pi \geq 4$ using zeta integrals.

We begin with some preliminary observations. Assume that $L(s, \pi) = 1$. Then Theorem 7.5.3 of [105] implies that $N_\pi \geq 2$, and that the Hecke eigenvalues λ_π and μ_π are given by $\lambda_\pi = 0$ and $\mu_\pi = -q^2$; consequently, we have $N_{\pi,s} = N_\pi - 1$ by Proposition 5.7.3, and π is a category 1 representation.

Let (π, V) be a generic, irreducible, admissible representation of $\mathrm{GSp}(4, F)$ with trivial central character. We will work in the Whittaker model $V = \mathcal{W}(\pi, \psi_{c_1,c_2})$ of π. For $W \in V$ and integers $i, j \in \mathbb{Z}$ with $i, j \geq 0$ we define

$$m(W)_{ij} = W(\Delta_{i,j})$$

and let $m(W)$ be the matrix

$$m(W) = (m(W)_{ij})_{0 \leq i,j < \infty} = \begin{bmatrix} W(\Delta_{0,0}) & W(\Delta_{0,1}) & W(\Delta_{0,2}) & \cdots \\ W(\Delta_{1,0}) & W(\Delta_{1,1}) & W(\Delta_{1,2}) & \cdots \\ W(\Delta_{2,0}) & W(\Delta_{2,1}) & W(\Delta_{2,2}) & \cdots \\ \vdots & \vdots & \vdots & \end{bmatrix}. \tag{8.36}$$

Evidently, $m(W)$ is an element of the complex vector space $\mathrm{M}_{\infty\times\infty}(\mathbb{C})$ consisting of all matrices $(m_{ij})_{0 \leq i,j < \infty}$ with $m_{ij} \in \mathbb{C}$ for $i, j \in \mathbb{Z}$ with $i, j \geq 0$. Now let n be an integer such that $n \geq 0$. By (5.18) and Proposition 8.1.2, the map $V_s(n) \to \mathrm{M}_{\infty\times\infty}(\mathbb{C})$ defined by $W \mapsto m(W)$ is injective. We denote by $\mathrm{M}_s(n)$ the $\mathbb{C}$ vector space of all $m(W)$ for $W \in V_s(n)$, so that the map

$$V_s(n) \xrightarrow{\sim} \mathrm{M}_s(n) \tag{8.37}$$

defined by $W \mapsto m(W)$ is an isomorphism. We refer to $\mathrm{M}_s(n)$ as the *alternative model* for the space $V_s(n)$ of stable Klingen vectors.

Our main result about the alternative model is an explicit description of the elements of $\mathrm{M}_s(n)$. Before we present and prove this result it will be convenient to realize our usual level changing operators in the alternative model. We will write the elements A of $\mathrm{M}_{\infty\times\infty}(\mathbb{C})$ as a column of rows,

$$A = \begin{bmatrix} r_0 \\ r_1 \\ r_2 \\ \vdots \end{bmatrix}.$$

We define two shift operations Left and Right on row vectors,

$$\mathrm{Left}[a_0, a_1, a_2, \ldots] = [a_1, a_2, a_3, \ldots],$$
$$\mathrm{Right}[a_0, a_1, a_2, \ldots] = [0, a_0, a_1, \ldots].$$

Using this notation we can describe the level changing operators θ, τ, η, and σ in the alternative model.

Proposition 8.3.1 *Let π be a generic, irreducible, admissible representation of the group* $\mathrm{GSp}(4, F)$ *with trivial central character, and let $V = \mathcal{W}(\pi, \psi_{c_1,c_2})$. Define*

$$\theta, \tau, \eta, \sigma : \mathrm{M}_{\infty\times\infty}(\mathbb{C}) \to \mathrm{M}_{\infty\times\infty}(\mathbb{C})$$

by

$$\theta(\begin{bmatrix} r_0 \\ r_1 \\ r_2 \\ \vdots \end{bmatrix}) = q \begin{bmatrix} 0 \\ \mathrm{Left}(r_0) \\ \mathrm{Left}(r_1) \\ \vdots \end{bmatrix} + \begin{bmatrix} \mathrm{Right}(r_0) \\ \mathrm{Right}(r_1) \\ \mathrm{Right}(r_2) \\ \vdots \end{bmatrix}, \qquad \tau(\begin{bmatrix} r_0 \\ r_1 \\ r_2 \\ \vdots \end{bmatrix}) = \begin{bmatrix} r_0 \\ r_1 \\ r_2 \\ \vdots \end{bmatrix},$$

and

$$\eta(\begin{bmatrix} r_0 \\ r_1 \\ r_2 \\ \vdots \end{bmatrix}) = \begin{bmatrix} 0 \\ r_0 \\ r_1 \\ \vdots \end{bmatrix}, \qquad \sigma(\begin{bmatrix} r_0 \\ r_1 \\ r_2 \\ \vdots \end{bmatrix}) = \begin{bmatrix} r_1 \\ r_2 \\ r_3 \\ \vdots \end{bmatrix}.$$

The diagrams

$$\begin{array}{ccc} V_s(n+1) & \xrightarrow{\sim} & \mathrm{M}_s(n+1) \\ {\scriptstyle\theta}\uparrow & & \uparrow{\scriptstyle\theta} \\ V_s(n) & \xrightarrow{\sim} & \mathrm{M}_s(n) \end{array}, \qquad \begin{array}{ccc} V_s(n+1) & \xrightarrow{\sim} & \mathrm{M}_s(n+1) \\ {\scriptstyle\tau}\uparrow & & \uparrow{\scriptstyle\tau} \\ V_s(n) & \xrightarrow{\sim} & \mathrm{M}_s(n) \end{array} \tag{8.38}$$

and

$$\begin{array}{ccc} V_s(n+2) & \xrightarrow{\sim} & \mathrm{M}_s(n+2) \\ \eta\uparrow & & \uparrow\eta \\ V_s(n) & \xrightarrow{\sim} & \mathrm{M}_s(n) \end{array}, \qquad \begin{array}{ccc} V_s(n+1) & \xrightarrow{\sim} & \mathrm{M}_s(n+1) \\ \sigma\downarrow & & \downarrow\sigma \\ V_s(n) & \xrightarrow{\sim} & \mathrm{M}_s(n) \end{array} \tag{8.39}$$

commute.

Proof This follows from (3.34), (3.42) and (3.60). □

We will also need the translation of the paramodular level raising operators to the $\mathrm{M}_{\infty\times\infty}(\mathbb{C})$ setting. Again let (π, V) be a generic, irreducible, admissible representation of $\mathrm{GSp}(4, F)$ with trivial central character, and assume that $V = \mathcal{W}(\pi, \psi_{c_1,c_2})$. Let n be an integer such that $n \geq 0$. We denote by $\mathrm{M}(n)$ the $\mathbb{C}$ vector space of all $m(W)$ for $W \in V(n)$. We have $\mathrm{M}(n) \subset \mathrm{M}_s(n)$, and the map $V(n) \to \mathrm{M}(n)$ defined by $W \mapsto m(W)$ for $W \in V(n)$ is an isomorphism of $\mathbb{C}$ vector spaces by Proposition 8.1.2. Define

$$\theta' : \mathrm{M}_{\infty\times\infty}(\mathbb{C}) \longrightarrow \mathrm{M}_{\infty\times\infty}(\mathbb{C})$$

by

$$\theta'(\begin{bmatrix} r_0 \\ r_1 \\ r_2 \\ \vdots \end{bmatrix}) = q \begin{bmatrix} r_0 \\ r_1 \\ r_2 \\ \vdots \end{bmatrix} + \begin{bmatrix} 0 \\ r_0 \\ r_1 \\ \vdots \end{bmatrix}.$$

Calculations using (2.27), (2.28) and (2.29) show that the diagrams

$$\begin{array}{ccc} V(n+1) & \xrightarrow{\sim} & \mathrm{M}(n+1) \\ \theta\uparrow & & \uparrow\theta \\ V(n) & \xrightarrow{\sim} & \mathrm{M}(n) \end{array}, \qquad \begin{array}{ccc} V(n+1) & \xrightarrow{\sim} & \mathrm{M}(n+1) \\ \theta'\uparrow & & \uparrow\theta' \\ V(n) & \xrightarrow{\sim} & \mathrm{M}(n) \end{array} \tag{8.40}$$

and

$$\begin{array}{ccc} V(n+2) & \xrightarrow{\sim} & \mathrm{M}(n+2) \\ \eta\uparrow & & \uparrow\eta \\ V(n) & \xrightarrow{\sim} & \mathrm{M}(n) \end{array} \tag{8.41}$$

commute.

We turn now to the main result of this subsection. Let n be an integer such that $n \geq 0$. We define

$$\mathrm{M}_\triangle(n) = \{A = (A_{ij})_{i,j\geq 0} \in \mathrm{M}_{\infty\times\infty}(\mathbb{C}) \mid A_{ij} = 0 \text{ for } i + j \geq n\}. \tag{8.42}$$

Hence, $\mathrm{M}_\triangle(0)$ is the zero matrix, and

$$\mathrm{M}_\triangle(1) = \begin{bmatrix} * & 0 & 0 & \cdots \\ 0 & 0 & 0 & \cdots \\ 0 & 0 & 0 & \cdots \\ \vdots & \vdots & \vdots & \end{bmatrix}, \qquad \mathrm{M}_\triangle(2) = \begin{bmatrix} * & * & 0 & \cdots \\ * & 0 & 0 & \cdots \\ 0 & 0 & 0 & \cdots \\ \vdots & \vdots & \vdots & \end{bmatrix}, \qquad \ldots$$

We have $\dim \mathrm{M}_\triangle(n) = \frac{n(n+1)}{2}$.

Proposition 8.3.2 *Let π be a generic, irreducible, admissible representation of the group* $\mathrm{GSp}(4, F)$ *with trivial central character, and let* $V = \mathcal{W}(\pi, \psi_{c_1,c_2})$. *Assume that* $L(s, \pi) = 1$. *Then*

$$\mathrm{M}_s(n) = \mathrm{M}_\triangle(n - (N_\pi - 2)) \tag{8.43}$$

for all integers n such that $n \geq N_{\pi,s} = N_\pi - 1$.

Proof By Theorem 5.4.1,

$$\dim V_s(n) = \begin{cases} \dfrac{(n - N_\pi + 2)(n - N_\pi + 3)}{2} & \text{for } n \geq N_\pi - 1, \\ 0 & \text{for } n < N_\pi - 1. \end{cases} \tag{8.44}$$

Hence, $\dim V_s(n) = \dim \mathrm{M}_\triangle(n - (N_\pi - 2))$ for $n \geq N_\pi - 1$. It is therefore enough to prove that $\mathrm{M}_s(n) \subset \mathrm{M}_\triangle(n - (N_\pi - 2))$ for $n \geq N_\pi - 1$. For this we will use induction on n. Let W_{new} be a non-zero element of the one-dimensional space $V(N_\pi)$, and let W_s be the shadow of W_{new}. By Theorem 5.4.1, the space $V_s(N_\pi - 1)$ is one-dimensional and spanned by W_s. Moreover, it follows from Lemma 7.4.1 and Corollary 7.4.6 of [105] that

$$m(W_s) = \begin{bmatrix} W_{\mathrm{new}}(\Delta_{0,0}) & 0 & 0 & \cdots \\ 0 & 0 & 0 & \cdots \\ 0 & 0 & 0 & \cdots \\ \vdots & \vdots & \vdots & \end{bmatrix}.$$

Hence, $m(W_s)$ is a non-zero element of $M_\triangle(1)$. From this it follows that $\mathrm{M}_s(n) \subset \mathrm{M}_\triangle(n - (N_\pi - 2))$ for $n = N_\pi - 1$. Now assume that $n > N_\pi - 1$, and that $M_s(k) \subset \mathrm{M}_\triangle(k - (N_\pi - 2))$ for $k = n - 1$. The second diagram in (8.38) shows that

$$\mathrm{M}_\triangle(n - 1 - (N_\pi - 2)) \subset \mathrm{M}_s(n). \tag{8.45}$$

Let $E_{ij} \in \mathrm{M}_{\infty\times\infty}(\mathbb{C})$ be the matrix with (i, j)-coefficient 1 and all other coefficients 0. For a positive integer k let $\mathrm{M}'_\Delta(k)$ be the subspace of $\mathrm{M}_\Delta(k)$ spanned by $E_{i,k-i-1}$ with $0 \le i \le k-2$. Hence,

$$\mathrm{M}'_\Delta(1) = \begin{bmatrix} 0 & 0 & \cdots \\ 0 & 0 & \cdots \\ \vdots & \vdots & \end{bmatrix}, \quad \mathrm{M}'_\Delta(2) = \begin{bmatrix} 0 & * & 0 & \cdots \\ 0 & 0 & 0 & \cdots \\ 0 & 0 & 0 & \cdots \\ \vdots & \vdots & \vdots & \end{bmatrix}, \quad \mathrm{M}'_\Delta(3) = \begin{bmatrix} 0 & 0 & * & 0 & \cdots \\ 0 & * & 0 & 0 & \cdots \\ 0 & 0 & 0 & 0 & \cdots \\ 0 & 0 & 0 & 0 & \cdots \\ \vdots & \vdots & \vdots & \vdots & \end{bmatrix}, \quad \cdots$$

The first diagram in (8.38) shows that

$$\mathrm{M}'_\Delta(n - (N_\pi - 2)) \subset \mathrm{M}_s(n). \tag{8.46}$$

The second diagram in (8.40) shows that

$$E_{n-N_\pi+1,0} \in (\theta')^{n-N_\pi}(W_{\mathrm{new}}) + \mathrm{M}_\Delta(n - 1 - (N_\pi - 2)) \subset m(V_s(n)). \tag{8.47}$$

The inclusions (8.45), (8.46) and (8.47) show that

$$\mathrm{M}_\Delta(n - (N_\pi - 2)) \subset \mathrm{M}_s(n). \tag{8.48}$$

This completes the proof. □

Using the alternative model we can provide an explicit description of the kernel of the level lowering operator σ_{n-1} when π is generic and $L(s, \pi) = 1$. For a positive integer k let $\mathrm{M}_0(k)$ be the subspace of $\mathrm{M}_\Delta(k)$ spanned by $E_{0,i}$ with $0 \le i \le k-1$. Hence,

$$\mathrm{M}_0(1) = \begin{bmatrix} * & 0 & \cdots \\ 0 & 0 & \cdots \\ \vdots & \vdots & \end{bmatrix}, \quad \mathrm{M}_0(2) = \begin{bmatrix} * & * & 0 & \cdots \\ 0 & 0 & 0 & \cdots \\ 0 & 0 & 0 & \cdots \\ \vdots & \vdots & \vdots & \end{bmatrix}, \quad \mathrm{M}_0(3) = \begin{bmatrix} * & * & * & 0 & \cdots \\ 0 & 0 & 0 & 0 & \cdots \\ 0 & 0 & 0 & 0 & \cdots \\ 0 & 0 & 0 & 0 & \cdots \\ \vdots & \vdots & \vdots & \vdots & \end{bmatrix}, \quad \cdots$$

Corollary 8.3.3 *Let π be a generic, irreducible, admissible representation of* GSp(4, F) *with trivial central character, and let $V = \mathcal{W}(\pi, \psi_{c_1,c_2})$. Assume that $L(s, \pi) = 1$. Under the isomorphism* (8.37), *we have*

$$\ker(\sigma_{n-1}) \xrightarrow{\sim} \mathrm{M}_0(n - (N_\pi - 2)) \tag{8.49}$$

for all $n \ge N_\pi$.

Proof This follows from Propositions 8.3.2 and 8.3.1. □

8.4 A Lower Bound on the Paramodular Level

Let (π, V) be a generic, irreducible, admissible representation of $\mathrm{GSp}(4, F)$ with trivial central character. In this final section we prove that if $L(s,\pi) = 1$, then $N_\pi \geq 4$. We begin with a result about representations of $\mathrm{GL}(2, F)$. In the following lemma, $\Gamma_0(\mathfrak{p})$ is the subgroup of $\left[\begin{smallmatrix} a & b \\ c & d \end{smallmatrix}\right]$ in $\mathrm{GL}(2, \mathfrak{o})$ such that $c \in \mathfrak{p}$.

Lemma 8.4.1 *Let (τ, W) be a smooth representation of $\mathrm{GL}(2, F)$. Let $w \in W$ be such that $\tau(k)w = w$ for $k \in \Gamma_0(\mathfrak{p})$. Assume that there exists an integer j such that $j \geq 1$ and*

$$\int\limits_{\mathfrak{p}^{-j}} \tau(\left[\begin{smallmatrix} 1 & y \\ & 1 \end{smallmatrix}\right])w\,dy = 0.$$

Then $w = 0$.

Proof Let X be the subset of $w \in W^{\Gamma_0(\mathfrak{p})}$ such that there exists an integer j such that $j \geq 1$ and

$$\int\limits_{\mathfrak{p}^{-j}} \tau(\left[\begin{smallmatrix} 1 & y \\ & 1 \end{smallmatrix}\right])w\,dy = 0. \tag{8.50}$$

The set X is a subspace of $W^{\Gamma_0(\mathfrak{p})}$; we need to prove that $X = 0$. Assume that $X \neq 0$; we will obtain a contradiction. Let j_0 be the smallest integer such that $j_0 \geq 1$ and there exists a non-zero element $w_0 \in X$ such that (8.50) holds with j and w replaced by j_0 and w_0, respectively. We will first prove that $j_0 = 1$. Suppose that $j_0 > 1$; we will obtain a contradiction.

We introduce an operator on W. Define $U : W \to W$ by

$$Uv = \tau(\left[\begin{smallmatrix} 1 & \\ & \varpi^{-1} \end{smallmatrix}\right]) \int\limits_{\mathfrak{p}^{-1}} \tau(\left[\begin{smallmatrix} 1 & x \\ & 1 \end{smallmatrix}\right])v\,dx$$

for $v \in W$. We claim that if $v \in W^{\Gamma_0(\mathfrak{p})}$, then $Uv \in W^{\Gamma_0(\mathfrak{p})}$. To see this, let $v \in W^{\Gamma_0(\mathfrak{p})}$. It is clear that $\tau(\left[\begin{smallmatrix} a & \\ & d \end{smallmatrix}\right])v = v$ and $\tau(\left[\begin{smallmatrix} 1 & b \\ & 1 \end{smallmatrix}\right])v = v$ for $a, d \in \mathfrak{o}^\times$ and $b \in \mathfrak{o}$. Let $c \in \mathfrak{p}$. Then

$$\begin{aligned} \tau(\left[\begin{smallmatrix} 1 & \\ c & 1 \end{smallmatrix}\right])Uv &= \tau(\left[\begin{smallmatrix} 1 & \\ & \varpi^{-1} \end{smallmatrix}\right]) \int\limits_{\mathfrak{p}^{-1}} \tau(\left[\begin{smallmatrix} (1+xc\varpi)^{-1} & x \\ & 1+xc\varpi \end{smallmatrix}\right]\left[\begin{smallmatrix} 1 & \\ (1+xc\varpi)^{-1}c\varpi & 1 \end{smallmatrix}\right])v\,dx \\ &= \tau(\left[\begin{smallmatrix} 1 & \\ & \varpi^{-1} \end{smallmatrix}\right]) \int\limits_{\mathfrak{p}^{-1}} \tau(\left[\begin{smallmatrix} 1 & (1+xc\varpi)^{-1}x \\ & 1 \end{smallmatrix}\right])v\,dx \\ &= Uv. \end{aligned}$$

It follows that $Uv \in W^{\Gamma_0(\mathfrak{p})}$.

By the last paragraph, $Uw_0 \in W^{\Gamma_0(\mathfrak{p})}$. Since $j_0 \geq 2$, and by the definition of U and the minimality of j_0, we must have $Uw_0 \neq 0$. Define $w_1 = Uw_0$. Then $w_1 \in W^{\Gamma_0(\mathfrak{p})}$. We have

$$\int\limits_{\mathfrak{p}^{-(j_0-1)}} \tau(\left[\begin{smallmatrix}1&y\\&1\end{smallmatrix}\right])w_1\,dy = \tau(\left[\begin{smallmatrix}1&\\&\varpi^{-1}\end{smallmatrix}\right])\int\limits_{\mathfrak{p}^{-j_0}} \tau(\left[\begin{smallmatrix}1&y\\&1\end{smallmatrix}\right])\,w_0\,dy = 0.$$

This implies that $w_1 \in X$, and contradicts the minimality of j_0; it follows that $j_0 = 1$.

We now have

$$\begin{aligned} 0 &= \int\limits_{\mathfrak{p}^{-1}} \tau(\left[\begin{smallmatrix}1&y\\&1\end{smallmatrix}\right])w_0\,dy \\ &= \int\limits_{\mathfrak{o}} \tau(\left[\begin{smallmatrix}1&y\\&1\end{smallmatrix}\right])w_0\,dy + \int\limits_{\mathfrak{o}^\times\varpi^{-1}} \tau(\left[\begin{smallmatrix}1&y\\&1\end{smallmatrix}\right])w_0\,dy \\ &= w_0 + q\int\limits_{\mathfrak{o}^\times} \tau(\left[\begin{smallmatrix}1&u\varpi^{-1}\\&1\end{smallmatrix}\right])w_0\,du. \end{aligned}$$

Hence, since $w_0 \in W^{\Gamma_0(\mathfrak{p})}$,

$$\begin{aligned} w_0 &= -q\int\limits_{\mathfrak{o}^\times} \tau(\left[\begin{smallmatrix}1&\\u^{-1}\varpi&1\end{smallmatrix}\right]\left[\begin{smallmatrix}u\varpi^{-1}&\\&u^{-1}\varpi\end{smallmatrix}\right]\left[\begin{smallmatrix}&1\\-1&\end{smallmatrix}\right]\left[\begin{smallmatrix}1&\\u^{-1}\varpi&1\end{smallmatrix}\right])w_0\,du \\ &= -q\int\limits_{\mathfrak{o}^\times} \tau(\left[\begin{smallmatrix}1&\\u^{-1}\varpi&1\end{smallmatrix}\right]\left[\begin{smallmatrix}\varpi^{-1}&\\&\varpi\end{smallmatrix}\right]\left[\begin{smallmatrix}&1\\-1&\end{smallmatrix}\right])w_0\,du \end{aligned}$$

$$w_0 = -q\tau(\left[\begin{smallmatrix}\varpi^{-1}&\\&\varpi\end{smallmatrix}\right]\left[\begin{smallmatrix}&1\\-1&\end{smallmatrix}\right])\int\limits_{\mathfrak{o}^\times} \tau(\left[\begin{smallmatrix}1&u^{-1}\varpi^{-1}\\&1\end{smallmatrix}\right])w_0\,du. \tag{8.51}$$

Using (8.51) we find that

$$\begin{aligned} w_0 &= q\int\limits_{\mathfrak{p}} \tau(\left[\begin{smallmatrix}1&\\c&1\end{smallmatrix}\right])w_0\,dc \\ &= -q^2\int\limits_{\mathfrak{p}} \tau(\left[\begin{smallmatrix}1&\\c&1\end{smallmatrix}\right])\tau(\left[\begin{smallmatrix}\varpi^{-1}&\\&\varpi\end{smallmatrix}\right]\left[\begin{smallmatrix}&1\\-1&\end{smallmatrix}\right])\int\limits_{\mathfrak{o}^\times} \tau(\left[\begin{smallmatrix}1&u^{-1}\varpi^{-1}\\&1\end{smallmatrix}\right])w_0\,du\,dc \\ &= -q^2\tau(\left[\begin{smallmatrix}\varpi^{-1}&\\&\varpi\end{smallmatrix}\right]\left[\begin{smallmatrix}&1\\-1&\end{smallmatrix}\right])\int\limits_{\mathfrak{o}^\times}\int\limits_{\mathfrak{p}} \tau(\left[\begin{smallmatrix}1&(-c+u^{-1}\varpi)\varpi^{-2}\\&1\end{smallmatrix}\right])w_0\,dc\,du \end{aligned}$$

$$= -q^2(1-q^{-1})\tau(\left[\begin{smallmatrix} \varpi^{-1} & \\ & \varpi \end{smallmatrix}\right]\left[\begin{smallmatrix} & 1 \\ -1 & \end{smallmatrix}\right]) \int\limits_{\mathfrak{p}} \tau(\left[\begin{smallmatrix} 1 & c\varpi^{-2} \\ & 1 \end{smallmatrix}\right])w_0\,dc$$

$$= -(1-q^{-1})\tau(\left[\begin{smallmatrix} \varpi^{-1} & \\ & \varpi \end{smallmatrix}\right]\left[\begin{smallmatrix} & 1 \\ -1 & \end{smallmatrix}\right]) \int\limits_{\mathfrak{p}^{-1}} \tau(\left[\begin{smallmatrix} 1 & c \\ & 1 \end{smallmatrix}\right])w_0\,dc$$

$$= 0,$$

where the last step follows because $j_0 = 1$. This contradicts $w_0 \neq 0$, and completes the proof. □

Let n be an integer such that $n \geq 2$. We define

$$\mathrm{Kl}_{s,1}(\mathfrak{p}^n) = \{g \in \mathrm{GSp}(4, F) \mid \lambda(g) \in \mathfrak{o}^\times\} \cap \begin{bmatrix} \mathfrak{o} & \mathfrak{o} & \mathfrak{o} & \mathfrak{p}^{-n+1} \\ \mathfrak{p}^{n-1} & \mathfrak{o} & \mathfrak{o} & \mathfrak{o} \\ \mathfrak{p}^n & \mathfrak{p} & \mathfrak{o} & \mathfrak{o} \\ \mathfrak{p}^n & \mathfrak{p}^n & \mathfrak{p}^{n-1} & \mathfrak{o} \end{bmatrix}.$$

Using (2.4) it is easy to verify that $\mathrm{Kl}_{s,1}(\mathfrak{p}^n)$ is a subgroup of $\mathrm{GSp}(4, F)$.

Lemma 8.4.2 *Let n be an integer such that $n \geq 2$. Then $\mathrm{Kl}_{s,1}(\mathfrak{p}^n)$ is equal to*

$$(\begin{bmatrix} 1 & \mathfrak{o} & \mathfrak{o} & \mathfrak{p}^{-n+1} \\ & 1 & & \mathfrak{o} \\ & & 1 & \mathfrak{o} \\ & & & 1 \end{bmatrix} \cap \mathrm{Kl}_{s,1}(\mathfrak{p}^n))(\begin{bmatrix} \mathfrak{o}^\times & & & \\ & \mathfrak{o} & \mathfrak{o} & \\ & \mathfrak{p} & \mathfrak{o} & \\ & & & \mathfrak{o}^\times \end{bmatrix} \cap \mathrm{Kl}_{s,1}(\mathfrak{p}^n))(\begin{bmatrix} 1 & & & \\ \mathfrak{p}^{n-1} & 1 & & \\ \mathfrak{p}^n & & 1 & \\ \mathfrak{p}^n & \mathfrak{p}^n & \mathfrak{p}^{n-1} & 1 \end{bmatrix} \cap \mathrm{Kl}_{s,1}(\mathfrak{p}^n)).$$

Proof Let $k \in \mathrm{Kl}_{s,1}(\mathfrak{p}^n)$; we need to prove that k is in the product. Let $k = (k_{ij})_{1 \leq i,j \leq 4}$. A calculation shows that $\det(k) \equiv k_{11}k_{22}k_{33}k_{44} \pmod{\mathfrak{p}}$; since $\det(k)^2 = \lambda(k)^4$ and $\lambda(k) \in \mathfrak{o}^\times$, we obtain $k_{11}, k_{22}, k_{33}, k_{44} \in \mathfrak{o}^\times$. It follows that

$$\begin{bmatrix} 1 & & & -k_{14}k_{44}^{-1} \\ & 1 & & \\ & & 1 & \\ & & & 1 \end{bmatrix} k \in \mathrm{Kl}(\mathfrak{p}^{n-1}).$$

Since $\mathrm{Kl}(\mathfrak{p}^{n-1})$ has an Iwahori decomposition, there exist $x, y, z \in \mathfrak{o}$, $t, \lambda \in \mathfrak{o}^\times$, $\left[\begin{smallmatrix} a & b \\ c & d \end{smallmatrix}\right] \in \mathrm{GL}(2, \mathfrak{o})$ with $ad - bc = \lambda$, and $x', y', z' \in \mathfrak{p}^{n-1}$ such that

$$\begin{bmatrix} 1 & & & -k_{14}k_{44}^{-1} \\ & 1 & & \\ & & 1 & \\ & & & 1 \end{bmatrix} k = \begin{bmatrix} 1 & x & y & z \\ & 1 & & y \\ & & 1 & -x \\ & & & 1 \end{bmatrix} \begin{bmatrix} t & & & \\ & a & b & \\ & c & d & \\ & & & \lambda t^{-1} \end{bmatrix} \begin{bmatrix} 1 & & & \\ x' & 1 & & \\ y' & & 1 & \\ z' & y' & -x' & 1 \end{bmatrix}.$$

From this equation we see that $y', z' \in \mathfrak{p}^n$, $x' \in \mathfrak{p}^{n-1}$, and $c \in \mathfrak{p}$. The lemma follows. □

For later use, we note that if n is an integer such that $n \geq 2$, then

$$u_n \mathrm{Kl}_{s,1}(\mathfrak{p}^n) u_n^{-1} = \mathrm{Kl}_{s,1}(\mathfrak{p}^n) \tag{8.52}$$

Here u_n is as in (2.31).

In the following we will use the Iwahori subgroup I defined in (9.1).

Lemma 8.4.3 *Let (π, V) be a smooth representation of* GSp(4, F) *for the which the center acts trivially. Assume that $V^I = 0$. Then*

$$\int\limits_{\mathfrak{o}}\int\limits_{\mathfrak{o}}\int\limits_{\mathfrak{o}} \pi(\begin{bmatrix} 1 & x\varpi^{-1} & y\varpi^{-1} & z\varpi^{-2} \\ & 1 & & y\varpi^{-1} \\ & & 1 & -x\varpi^{-1} \\ & & & 1 \end{bmatrix})w\,dx\,dy\,dz = 0 \tag{8.53}$$

for all $w \in V^{\mathrm{Kl}_s,1(\mathfrak{p}^2)}$.

Proof Let $w \in V^{\mathrm{Kl}_s,1(\mathfrak{p}^2)}$, and let u be the integral in (8.53); we need to prove that $u = 0$. Define $u_1 = \pi(\eta^{-1})u$. Given the assumption $V^I = 0$, to prove that $u = 0$ it will suffice to prove that $u_1 \in V^I$. Since $I = I_+T(\mathfrak{o})I_-$ (see (9.18)), it suffices to show that $\pi(k)u_1 = u_1$ for $k \in I_+$, $k \in T(\mathfrak{o})$, and $k \in I_-$; these statements can be verified by direct computations using $w \in V^{\mathrm{Kl}_s,1(\mathfrak{p}^2)}$. □

Let (π, V) be a smooth representation of GSp(4, F) for which the center acts trivially, and let n be an integer such that $n \geq 2$. As in Sect. 7.3 of [105], we define a linear map $R_{n-1} : V \to V$ by

$$R_{n-1}v = q \int\limits_{\mathfrak{o}} \pi(\begin{bmatrix} 1 & & & \\ x\varpi^{n-1} & 1 & & \\ & & 1 & \\ & & -x\varpi^{n-1} & 1 \end{bmatrix})v\,dx \tag{8.54}$$

for $v \in V$. Evidently, if $v \in V$ is invariant under the elements $\begin{bmatrix} 1 & & & \\ x & 1 & & \\ & & 1 & \\ & & -x & 1 \end{bmatrix}$ for $x \in \mathfrak{p}^n$, then

$$R_{n-1}v = \sum_{x \in \mathfrak{o}/\mathfrak{p}} \pi(\begin{bmatrix} 1 & & & \\ x\varpi^{n-1} & 1 & & \\ & & 1 & \\ & & -x\varpi^{n-1} & 1 \end{bmatrix})v. \tag{8.55}$$

We also define a linear map $S : V \to V$ by

$$Sv = \mathrm{vol}(\Gamma_0(\mathfrak{p}))^{-1} \int\limits_{\mathrm{GL}(2,\mathfrak{o})} \pi(\begin{bmatrix} 1 & & \\ & g & \\ & & \det(g) \end{bmatrix})v\,dg \tag{8.56}$$

for $v \in V$. Here, dg is a Haar measure on GL(2, F). We note that if $v \in V$ is invariant under the elements $\begin{bmatrix} 1 & & \\ & k & \\ & & \det(k) \end{bmatrix}$ for $k \in \Gamma_0(\mathfrak{p})$, then

$$Sv = \pi(s_2)v + q \int\limits_{\mathfrak{o}} \pi(\begin{bmatrix} 1 & & & \\ & 1 & & \\ & y & 1 & \\ & & & 1 \end{bmatrix})v\,dy. \tag{8.57}$$

The following result has some overlap with Lemma 3 of [130].

Lemma 8.4.4 *Let (π, V) be a smooth representation of* $\mathrm{GSp}(4, F)$ *for which the center acts trivially, and let n be an integer such that $n \geq 2$. If $v \in V_s(n)$, then $R_{n-1}v \in V^{\mathrm{Kl}_{s,1}(\mathfrak{p}^n)}$, and if $v \in V^{\mathrm{Kl}_{s,1}(\mathfrak{p}^n)}$, then $Sv \in V_s(n)$. The linear map $R_{n-1} : V_s(n) \to V^{\mathrm{Kl}_{s,1}(\mathfrak{p}^n)}$ is injective. If $V^I = 0$, then the map $S : V^{\mathrm{Kl}_{s,1}(\mathfrak{p}^2)} \to V_s(2)$ is injective.*

Proof If $v \in V_s(n)$, then calculations using the Iwahori decomposition from Lemma 8.4.2 show that $R_{n-1}v \in V^{\mathrm{Kl}_{s,1}(\mathfrak{p}^n)}$; if $v \in V^{\mathrm{Kl}_{s,1}(\mathfrak{p}^n)}$, then calculations using the Iwahori decomposition (3.14) show that $Sv \in V_s(n)$. Next, let $v \in V_s(n)$. Then by (8.55) and (8.57),

$$
\begin{aligned}
&S(R_{n-1}(v))\\
&= \sum_{y\in\mathfrak{o}/\mathfrak{p}} \pi(\begin{bmatrix}1&&&\\&1&&\\&y&1&\\&&&1\end{bmatrix})R_{n-1}(v) + \pi(s_2)R_{n-1}(v)\\
&= \sum_{x,y\in\mathfrak{o}/\mathfrak{p}} \pi(\begin{bmatrix}1&&&\\&1&&\\&y&1&\\&&&1\end{bmatrix}\begin{bmatrix}1&&&\\x\varpi^{n-1}&1&&\\&&1&\\&&-x\varpi^{n-1}&1\end{bmatrix})v\\
&\quad+ \sum_{x\in\mathfrak{o}/\mathfrak{p}} \pi(s_2\begin{bmatrix}1&&&\\x\varpi^{n-1}&1&&\\&&1&\\&&-x\varpi^{n-1}&1\end{bmatrix})v\\
&= \sum_{x,y\in\mathfrak{o}/\mathfrak{p}} \pi(\begin{bmatrix}1&&&\\x\varpi^{n-1}&1&&\\&&1&\\&&-x\varpi^{n-1}&1\end{bmatrix}\begin{bmatrix}1&&&\\&1&&\\yx\varpi^{n-1}&y&1&\\yx^2\varpi^{2n-2}&yx\varpi^{n-1}&&1\end{bmatrix})v\\
&\quad+ \sum_{x\in\mathfrak{o}/\mathfrak{p}} \pi(\begin{bmatrix}1&&&\\&1&&\\x\varpi^{n-1}&&1&\\&x\varpi^{n-1}&&1\end{bmatrix}s_2)v\\
&= qv + \sum_{x\in(\mathfrak{o}/\mathfrak{p})^\times}\sum_{y\in\mathfrak{o}/\mathfrak{p}} \pi(\begin{bmatrix}1&&&\\x\varpi^{n-1}&1&&\\&&1&\\&&-x\varpi^{n-1}&1\end{bmatrix}\begin{bmatrix}1&&&\\&1&&\\yx\varpi^{n-1}&&1&\\&yx\varpi^{n-1}&&1\end{bmatrix})v\\
&\quad+ \sum_{x\in\mathfrak{o}/\mathfrak{p}} \pi(\begin{bmatrix}1&&&\\&1&&\\x\varpi^{n-1}&&1&\\&x\varpi^{n-1}&&1\end{bmatrix})v\\
&= qv + \sum_{x\in(\mathfrak{o}/\mathfrak{p})^\times}\sum_{y\in\mathfrak{o}/\mathfrak{p}} \pi(\begin{bmatrix}1&&&\\x\varpi^{n-1}&1&&\\&&1&\\&&-x\varpi^{n-1}&1\end{bmatrix}\begin{bmatrix}1&&&\\&1&&\\y\varpi^{n-1}&&1&\\&y\varpi^{n-1}&&1\end{bmatrix})v\\
&\quad+ \sum_{y\in\mathfrak{o}/\mathfrak{p}} \pi(\begin{bmatrix}1&&&\\&1&&\\y\varpi^{n-1}&&1&\\&y\varpi^{n-1}&&1\end{bmatrix})v
\end{aligned}
$$

$$= qv + \sum_{x,y\in\mathfrak{o}/\mathfrak{p}} \pi(\begin{bmatrix} 1&&& \\ x\varpi^{n-1}&1&& \\ &&1& \\ &&-x\varpi^{n-1}&1 \end{bmatrix}\begin{bmatrix} 1&&& \\ &1&& \\ y\varpi^{n-1}&&1& \\ &y\varpi^{n-1}&&1 \end{bmatrix})v$$

$$= qv + \sum_{x,y\in\mathfrak{o}/\mathfrak{p}} \pi(\begin{bmatrix} 1&&& \\ &1&& \\ y\varpi^{n-1}&&1& \\ &y\varpi^{n-1}&&1 \end{bmatrix}\begin{bmatrix} 1&&& \\ x\varpi^{n-1}&1&& \\ &&1& \\ &&-x\varpi^{n-1}&1 \end{bmatrix})v$$

$$= qv + \sum_{y\in\mathfrak{o}/\mathfrak{p}} \pi(\begin{bmatrix} 1&&& \\ &1&& \\ y\varpi^{n-1}&&1& \\ &y\varpi^{n-1}&&1 \end{bmatrix})R_{n-1}(v).$$

By the last equation, if $R_{n-1}(v) = 0$, then $v = 0$, so that $R_{n-1} : V_s(n) \to V^{\mathrm{Kl}_{s,1}(\mathfrak{p}^n)}$ is injective.

Finally, assume that $V^I = 0$. Let $v \in V^{\mathrm{Kl}_{s,1}(\mathfrak{p}^2)}$, and assume that $Sv = 0$. Since $Sv = 0$, we have by (8.57),

$$\pi(s_2)v = -q\int\limits_{\mathfrak{o}} \pi(\begin{bmatrix} 1&&& \\ &1&& \\ &y&1& \\ &&&1 \end{bmatrix})v\,dy.$$

Applying u_2 to this equation, and using $\pi(u_2s_2u_2^{-1})\pi(u_2)v = \pi(t_2)\pi(u_2)v$, we obtain

$$\pi(t_2)v_1 = -q\int\limits_{\mathfrak{o}} \pi(\begin{bmatrix} 1&&&y\varpi^{-2} \\ &1&& \\ &&1& \\ &&&1 \end{bmatrix})v_1\,dy$$

where $v_1 = \pi(u_2)v$. Since u_2 normalizes $\mathrm{Kl}_{s,1}(\mathfrak{p}^2)$, it follows that $v_1 \in V^{\mathrm{Kl}_{s,1}(\mathfrak{p}^2)}$. If $x \in \mathfrak{o}$, then

$$t_2^{-1}\begin{bmatrix} 1&x\varpi^{-1}&& \\ &1&&x\varpi^{-1} \\ &&1& \\ &&&1 \end{bmatrix}t_2 = \begin{bmatrix} 1&&& \\ x\varpi&1&& \\ &&1& \\ &&-x\varpi&1 \end{bmatrix}.$$

It follows that $\pi(t_2)v_1$ is invariant under the elements

$$\begin{bmatrix} 1&x\varpi^{-1}&& \\ &1&&x\varpi^{-1} \\ &&1& \\ &&&1 \end{bmatrix}$$

for $x \in \mathfrak{o}$. Hence,

$$\pi(t_2)v_1 = -q\int\limits_{\mathfrak{o}}\int\limits_{\mathfrak{o}} \pi(\begin{bmatrix} 1&x\varpi^{-1}&&y\varpi^{-2} \\ &1&&x\varpi^{-1} \\ &&1& \\ &&&1 \end{bmatrix})v_1\,dy\,dx. \tag{8.58}$$

We now define a smooth representation (τ, W) of $\mathrm{GL}(2, F)$ by letting $W = V$ and setting $\tau(g)w = \pi(\begin{bmatrix} g & \\ & g' \end{bmatrix})w$ for $g \in \mathrm{GL}(2, F)$ and $w \in W$. Let $w = \pi(t_2)v_1$. Calculations using (8.58) and $v_1 \in V^{\mathrm{Kl}_{s,1}(\mathfrak{p}^2)}$ show that $\tau(k)w = w$ for $k \in \Gamma_0(\mathfrak{p})$. By Lemma 8.4.3 we have

$$\int\limits_{\mathfrak{p}^{-1}} \tau(\begin{bmatrix} 1 & y \\ & 1 \end{bmatrix})w \, dy = 0.$$

Lemma 8.4.1 now implies that $w = 0$, so that $v = 0$. □

Let (π, V) be a generic, irreducible, admissible representation with trivial central character. As in (7.14) of [105], if $W \in \mathcal{W}(\pi, \psi_{c_1,c_2})$, then we define

$$Z_N(s, W) = \int\limits_{F^\times} W(\begin{bmatrix} a & & & \\ & a & & \\ & & 1 & \\ & & & 1 \end{bmatrix})|a|^{s-\frac{3}{2}} \, d^\times a.$$

Lemma 8.4.5 *Let (π, V) be a generic, irreducible, admissible representation of the group* $\mathrm{GSp}(4, F)$ *with trivial central character. Let n be an integer such that $n \geq 2$. Let $W \in V_s(n)$. Then*

$$Z(s, \pi(s_2)R_{n-1}W) = \gamma(s, \pi)^{-1}q^{n/2-ns}Z_N(1 - s, \pi(u_n)W).$$

Proof The proof of this assertion is the same as the proof of Proposition 7.3.2 of [105], except that the last line of the last display in that proof is omitted. □

Lemma 8.4.6 *Let (π, V) be a generic, irreducible, admissible representation of the group* $\mathrm{GSp}(4, F)$ *with trivial central character. Assume that $L(s, \pi) = 1$, so that $N_\pi \geq 2$ and π is a category 1 representation. Let W_{new} be a non-zero vector in $V(N_\pi)$, and let $W_s \in V_s(N_\pi - 1)$ be the shadow of W_{new}. Then*

$$Z(s, \pi(s_2)R_{N_\pi-2}W_s) = 0.$$

Proof We will abbreviate $N = N_\pi$ in this proof. By Lemma 8.4.5 it suffices to prove that $Z_N(s, \pi(u_{N-1})W_s) = 0$. By (5.9),

$$W_s = q^3 \int\limits_{\mathfrak{o}^3} \pi(\begin{bmatrix} 1 & & & \\ x\varpi^{N-1} & 1 & & \\ y\varpi^{N-1} & & 1 & \\ z\varpi^{N-1} & y\varpi^{N-1} & -x\varpi^{N-1} & 1 \end{bmatrix})W_{\mathrm{new}} \, dx \, dy \, dz.$$

Hence,

$$\pi(u_{N-1})W_s = q^3 \int\limits_{\mathfrak{o}^3} \pi(\begin{bmatrix} 1 & & y & \\ x\varpi^{N-1} & 1 & z & y \\ & & 1 & \\ & & -x\varpi^{N-1} & 1 \end{bmatrix})\pi(u_{N-1})W_{\mathrm{new}} \, dx \, dy \, dz.$$

Therefore,

$$Z_N(s,\pi(u_{N-1})W_s)$$

$$= q^3 \int\limits_{F^\times}\int\limits_{\mathfrak{o}^3} W_{\mathrm{new}}(\begin{bmatrix} a&&&\\&a&&\\&&1&\\&&&1\end{bmatrix}\begin{bmatrix}1&&y&\\x\varpi^{N-1}&1&z&y\\&&1&\\&&-x\varpi^{N-1}&1\end{bmatrix}u_{N-1})|a|^{s-\frac{3}{2}}\,dx\,dy\,dz\,d^\times a$$

$$= q^3 \int\limits_{F^\times}\int\limits_{\mathfrak{o}^3} W_{\mathrm{new}}(\begin{bmatrix} a&&&\\&a&&\\&&1&\\&&&1\end{bmatrix}\begin{bmatrix}1&&y&\\x\varpi^{N-1}&1&z&y\\&&1&\\&&-x\varpi^{N-1}&1\end{bmatrix}$$

$$\times \begin{bmatrix}1&&&\\&1&&\\&&\varpi^{-1}&\\&&&\varpi^{-1}\end{bmatrix}u_N)|a|^{s-\frac{3}{2}}\,dx\,dy\,dz\,d^\times a$$

$$= q^3 \int\limits_{F^\times}\int\limits_{\mathfrak{o}^3} W_{\mathrm{new}}(\begin{bmatrix} a\varpi&&&\\&a\varpi&&\\&&1&\\&&&1\end{bmatrix}\begin{bmatrix}1&&y\varpi^{-1}&\\x\varpi^{N-1}&1&z\varpi^{-1}&y\varpi^{-1}\\&&1&\\&&-x\varpi^{N-1}&1\end{bmatrix}u_N)$$

$$\times |a|^{s-\frac{3}{2}}\,dx\,dy\,dz\,d^\times a$$

$$= q^{s+\frac{3}{2}}\varepsilon_\pi \int\limits_{F^\times}\int\limits_{\mathfrak{o}^3} W_{\mathrm{new}}(\begin{bmatrix} a&&&\\&a&&\\&&1&\\&&&1\end{bmatrix}\begin{bmatrix}1&&y\varpi^{-1}&\\&1&z\varpi^{-1}&y\varpi^{-1}\\&&1&\\&&&1\end{bmatrix}$$

$$\times \begin{bmatrix}1&&&\\x\varpi^{N-1}&1&&\\&&1&\\&&-x\varpi^{N-1}&1\end{bmatrix})|a|^{s-\frac{3}{2}}\,dx\,dy\,dz\,d^\times a$$

$$= q^{s+\frac{1}{2}}\varepsilon_\pi \int\limits_{F^\times}(\int\limits_{\mathfrak{o}} \psi(c_2az\varpi^{-1})\,dz)(R_{N-1}W_{\mathrm{new}})(\begin{bmatrix} a&&&\\&a&&\\&&1&\\&&&1\end{bmatrix})$$

$$\times |a|^{s-\frac{3}{2}}\,dx\,dy\,dz\,d^\times a$$

$$= q^{s+\frac{1}{2}}\varepsilon_\pi \int\limits_{\{a\in F^\times\,|\,v(a)\geq 1\}} (R_{N-1}W_{\mathrm{new}})(\begin{bmatrix} a&&&\\&a&&\\&&1&\\&&&1\end{bmatrix})|a|^{s-\frac{3}{2}}\,dx\,dy\,dz\,d^\times a.$$

By Lemma 2.4 of [103] we have $Z_N(s,R_{N-1}W_{\mathrm{new}}) = -q^3 Z_N(s,\pi(\eta^{-1})W_{\mathrm{new}})$. This implies that if $k\in\mathbb{Z}$, then

$$(R_{N-1}W_{\mathrm{new}})(\begin{bmatrix} \varpi^k&&&\\&\varpi^k&&\\&&1&\\&&&1\end{bmatrix}) = -q^3W_{\mathrm{new}}(\begin{bmatrix} \varpi^k&&&\\&\varpi^k&&\\&&1&\\&&&1\end{bmatrix}\eta^{-1}) = -q^3W_{\mathrm{new}}(\Delta_{1,k}).$$

By assumption, $L(s, \pi) = 1$. It follows from Theorem 7.5.1 and Corollary 7.4.6 of [105] that $W_{\text{new}}(\Delta_{i,j}) = 0$ if $i, j \in \mathbb{Z}$, $(i, j) \neq (0, 0)$ and $(i, j) \neq (1, 0)$. In particular, we have $W_{\text{new}}(\Delta_{1,k}) = 0$ if $k \in \mathbb{Z}$ and $k \neq 0$. It follows that

$$\int\limits_{\{a \in F^\times \mid v(a) \geq 1\}} (R_{N-1} W_{\text{new}})(\begin{bmatrix} a & & & \\ & a & & \\ & & 1 & \\ & & & 1 \end{bmatrix})|a|^{s-\frac{3}{2}}\, dx\, dy\, dz\, d^\times a = 0,$$

so that $Z_N(s, \pi(u_{N-1})W_s) = 0$. □

Theorem 8.4.7 *Let* (π, V) *be a generic, irreducible, admissible representation of* $\mathrm{GSp}(4, F)$ *with trivial central character. Assume that* $L(s, \pi) = 1$. *Then* $N_\pi \geq 4$.

Proof Since $L(s, \pi) = 1$ we have $N_\pi \geq 2$ and π is a category 1 representation. In particular, by iii) of Theorem 7.5.3 of [105] we must have $\lambda_\pi = 0$ and $\mu_\pi = -q^2$.

We first claim that $V^I = 0$. Suppose that $V^I \neq 0$; we will obtain a contradiction. Since $V^I \neq 0$, and since $N_\pi \geq 2$, by Table A.13 of [105] we see that π belongs to Group IIIa, IVa, Va, or VIa with unramified inducing data. This contradicts the values of λ_π and μ_π from Table A.14 of [105]; hence, $V^I = 0$.

Now assume that $N_\pi \leq 3$; we will obtain a contradiction. If $N_\pi = 2$, then since π is a category 1 representation we have $V_s(1) \neq 0$; this contradicts $V^I = 0$. Therefore, $N_\pi = 3$ and $N_{\pi,s} = 2$. By Theorem 5.4.1, the space $V_s(2)$ is one-dimensional and is spanned by the shadow W_s of a newform $W_{\text{new}} \in V(3)$. By Lemma 8.4.4 the maps $R_1 : V_s(2) \to V^{\mathrm{Kl}_{s,1}(\mathfrak{p}^2)}$ and $S : V^{\mathrm{Kl}_{s,1}(\mathfrak{p}^2)} \to V_s(2)$ are injective and hence isomorphisms. Let $c \in \mathbb{C}^\times$ be such that $SR_1 W_s = cW_s$. By the definition of S, this means that

$$cW_s = \pi(s_2)R_1 W_s + q \int\limits_{\mathfrak{o}} \pi(\begin{bmatrix} 1 & & & \\ & 1 & & \\ & z & 1 & \\ & & & 1 \end{bmatrix})R_1 W_s\, dy.$$

It follows that

$$cZ(s, W_s) = Z(s, \pi(s_2)R_1 W_s) + qZ(s, R_1 W_s).$$

By Lemma 8.4.6 we have $Z(s, \pi(s_2)R_1 W_s) = 0$. Also, by (5.28), $Z(s, W_s)$ is a non-zero constant. It follows that $Z(s, R_1 W_s) = C$ for some $C \in \mathbb{C}^\times$. Since $\dim V^{\mathrm{Kl}_{s,1}(\mathfrak{p}^2)} = 1$ and $\pi(u_2)V^{\mathrm{Kl}_{s,1}(\mathfrak{p}^2)} = V^{\mathrm{Kl}_{s,1}(\mathfrak{p}^2)}$, there exists $\varepsilon \in \{\pm 1\}$ such that $\pi(u_2)R_1 W_s = \varepsilon R_1 W_s$. Because $L(s, \pi) = 1$ we have $\gamma(s, \pi) = \varepsilon(s, \pi)$. By Corollary 7.5.5 of [105], $\varepsilon(s, \pi) = \varepsilon_\pi q^{-3(s-\frac{1}{2})}$. By (2.61) of [105] we have

$$Z(1 - s, \pi(u_2)R_1 W_s) = q^{2(s-\frac{1}{2})}\gamma(s, \pi)Z(s, R_1 W_s).$$

Substituting, we obtain

$$\varepsilon C = q^{2(s-\frac{1}{2})}\varepsilon_\pi q^{-3(s-\frac{1}{2})}C$$
$$\varepsilon = q^{-(s-\frac{1}{2})}\varepsilon_\pi.$$

This contradiction completes the proof. □

Chapter 9
Iwahori-Spherical Representations

Let (π, V) be an irreducible admissible representation of $\mathrm{GSp}(4, F)$ with trivial central character. We will say that π is *Iwahori-spherical* if the space V^I is non-zero, where $I \subset \mathrm{Kl}(\mathfrak{p})$ is the Iwahori subgroup of $\mathrm{GSp}(4, F)$ as defined below in (9.1). The goal of this chapter is to describe the actions of the stable Hecke operators $T^s_{0,1}$ and $T^s_{1,0}$ on $V_s(1)$ when π is Iwahori-spherical. Since $\mathrm{K}_s(\mathfrak{p}) = \mathrm{Kl}(\mathfrak{p})$, $V_s(1)$ is simply the space $V^{\mathrm{Kl}(\mathfrak{p})}$ of vectors fixed by the Klingen congruence subgroup $\mathrm{Kl}(\mathfrak{p})$ of level $\mathfrak{p}$. The dimensions of the spaces V^K for all standard parahoric subgroups K of $\mathrm{GSp}(4, F)$ and for all Iwahori-spherical π was calculated in [111]. We will use this information, along with the theory of Iwahori-spherical representations, to calculate the action of the stable Hecke operators.

9.1 Some Background

We begin by recalling some group theory about $\mathrm{GSp}(4, F)$ centered around the Iwahori subgroup. This will allow the introduction of the Iwahori-Hecke algebra, which will be used in the analysis of the stable Hecke operators $T^s_{0,1}$ and $T^s_{1,0}$ as mentioned above.

An Extended Tits System

As in (2.8), let T be the subgroup of $\mathrm{GSp}(4, F)$ consisting of diagonal matrices, and let $\mathrm{N}(T)$ be the normalizer of T in $\mathrm{GSp}(4, F)$. Also, let

$$I = \mathrm{GSp}(4, \mathfrak{o}) \cap \begin{bmatrix} \mathfrak{o} & \mathfrak{o} & \mathfrak{o} & \mathfrak{o} \\ \mathfrak{p} & \mathfrak{o} & \mathfrak{o} & \mathfrak{o} \\ \mathfrak{p} & \mathfrak{p} & \mathfrak{o} & \mathfrak{o} \\ \mathfrak{p} & \mathfrak{p} & \mathfrak{p} & \mathfrak{o} \end{bmatrix}. \tag{9.1}$$

J. Johnson-Leung et al., *Stable Klingen Vectors and Paramodular Newforms*, Lecture Notes in Mathematics 2342, https://doi.org/10.1007/978-3-031-45177-5_9

Then I is a compact, open subgroup of $\mathrm{GSp}(4, \mathfrak{o})$, and we refer to I as the *Iwahori subgroup* of $\mathrm{GSp}(4, F)$. The intersection $I \cap \mathrm{N}(T)$ is equal to $T(\mathfrak{o})$, where $T(\mathfrak{o}) = T \cap \mathrm{GSp}(4, \mathfrak{o})$ is the subgroup of T of elements with diagonal entries in $\mathfrak{o}^\times$, and $T(\mathfrak{o}) = I \cap \mathrm{N}(T)$ is a normal subgroup of $\mathrm{N}(T)$. We let

$$W^e = \mathrm{N}(T)/(I \cap \mathrm{N}(T)) = \mathrm{N}(T)/T(\mathfrak{o}), \tag{9.2}$$

and call this group the *Iwahori-Weyl group*, or the *extended affine Weyl group*. The group W^e is generated by the images of

$$s_0 = t_1 = \begin{bmatrix} &&&-\varpi^{-1}\\ &1&&\\ &&1&\\ \varpi&&& \end{bmatrix}, \quad s_1 = \begin{bmatrix} &1&&\\ 1&&&\\ &&&1\\ &&1& \end{bmatrix}, \quad s_2 = \begin{bmatrix} 1&&&\\ &&1&\\ &-1&&\\ &&&1 \end{bmatrix} \tag{9.3}$$

and

$$u_1 = \begin{bmatrix} &&1&\\ &&&-1\\ \varpi&&&\\ &-\varpi&& \end{bmatrix}. \tag{9.4}$$

We note that $s_0 = t_1$ and u_1 were already defined in (3.8) and (2.31), respectively. We let W^a be the subgroup of W^e generated by the images of s_0, s_1, and s_2; the subgroup W^a of W^e is called the *affine Weyl group*. The affine Weyl group W^a is a Coxeter group with Coxeter generators s_0, s_1 and s_2, and Coxeter graph as in Fig. 9.1 (see [16]). The group W^a is isomorphic to the $(2, 4, 4)$ triangle group; since $1/2 + 1/4 + 1/4 = 1$, W^a is infinite. The subgroup W of W^e generated by s_1 and s_2 is called the *Weyl group*. We have $W = (\mathrm{N}(T) \cap \mathrm{GSp}(4, \mathfrak{o}))/T(\mathfrak{o})$. The inclusion

$$W = (\mathrm{N}(T) \cap \mathrm{GSp}(4, \mathfrak{o}))/T(\mathfrak{o}) \xrightarrow{\sim} \mathrm{N}(T)/T$$

is surjective, so that W is naturally isomorphic to $\mathrm{N}(T)/T$. The Weyl group is a Coxeter group with Coxeter generators s_1 and s_2 and Coxeter graph as in Fig. 9.1. Finally, we let Ω be the subgroup of W^e generated by the image of u_1. The group Ω is an infinite cyclic group. It can be proven that the triple $(\mathrm{GSp}(4, F), I, \mathrm{N}(T))$ is a generalized Tits system (see [67]) with respect to W^a, the generators s_0, s_1 and s_2 of W^a, and Ω. This means that: W^e is the semi-direct product of Ω and the normal subgroup W^a; $\sigma I w \subset I\sigma w I \cup I \sigma I$ for $\sigma \in W^e$ and $w \in \{s_0, s_1, s_2\}$; $wIw^{-1} \neq I$ for $w \in \{s_0, s_1, s_2\}$; $u_1\{s_0, s_1, s_2\}u_1^{-1} = \{s_0, s_1, s_2\}$ in W^e; $u_1 I u_1^{-1} = I$; $I\rho \neq I$ for $\rho \in \Omega - \{1\}$; and $\mathrm{GSp}(4, F)$ is generated by I and $\mathrm{N}(T)$.

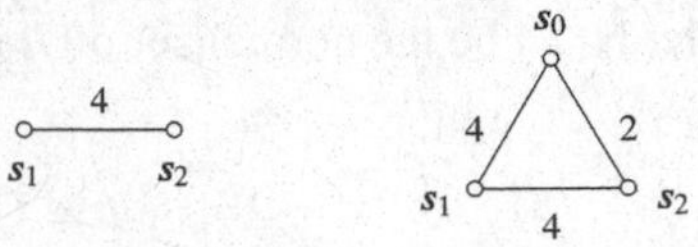

Fig. 9.1 The Coxeter graphs of the Weyl group W and affine Weyl group W^a, respectively

Parahoric Subgroups

A *standard parahoric subgroup* of $\mathrm{GSp}(4, F)$ is by definition a compact subgroup of $\mathrm{GSp}(4, F)$ containing the Iwahori subgroup I. Note that such a subgroup is necessarily open since it contains I. The standard parahoric subgroups of $\mathrm{GSp}(4, F)$ are in bijection with the proper subsets of $\{s_0, s_1, s_2\}$, the set of Coxeter generators of the affine Weyl group W^a (see Sect. 1 of [67]). More precisely, let J be a proper subset of $\{s_0, s_1, s_2\}$. Define W_J^a to be the subgroup of W^a generated by J. Since J is proper, W_J^a is finite. The standard parahoric subgroup corresponding to J is

$$K_J = \bigsqcup_{w \in W_J^a} IwI.$$

The union is disjoint because $\sqcup_{w \in W^e} IwI$ is disjoint (this is a property of generalized Tits systems). There are seven standard parahoric subgroups. The parahoric subgroup corresponding to the empty subset is $K_\emptyset = I$. The parahoric subgroup corresponding to $\{s_1\}$ is called the Siegel congruence subgroup of level $\mathfrak{p}$, and is often denoted by $\Gamma_0(\mathfrak{p})$. We have

$$K_{\{s_1\}} = \Gamma_0(\mathfrak{p}) = \mathrm{GSp}(4, \mathfrak{o}) \cap \begin{bmatrix} \mathfrak{o} & \mathfrak{o} & \mathfrak{o} & \mathfrak{o} \\ \mathfrak{o} & \mathfrak{o} & \mathfrak{o} & \mathfrak{o} \\ \mathfrak{p} & \mathfrak{p} & \mathfrak{o} & \mathfrak{o} \\ \mathfrak{p} & \mathfrak{p} & \mathfrak{o} & \mathfrak{o} \end{bmatrix}.$$

The parahoric subgroup corresponding to $\{s_2\}$ is the Klingen congruence subgroup $\mathrm{Kl}(\mathfrak{p})$ of level $\mathfrak{p}$ which was defined in (3.1). Thus,

$$K_{\{s_2\}} = \mathrm{Kl}(\mathfrak{p}) = \mathrm{GSp}(4, \mathfrak{o}) \cap \begin{bmatrix} \mathfrak{o} & \mathfrak{o} & \mathfrak{o} & \mathfrak{o} \\ \mathfrak{p} & \mathfrak{o} & \mathfrak{o} & \mathfrak{o} \\ \mathfrak{p} & \mathfrak{o} & \mathfrak{o} & \mathfrak{o} \\ \mathfrak{p} & \mathfrak{p} & \mathfrak{p} & \mathfrak{o} \end{bmatrix}.$$

For the parahoric subgroup corresponding to $\{s_0\}$ we have $W_{\{s_0\}}^a = \{1, s_0\}$ so that $K_{\{s_0\}} = I \sqcup Is_0I$. Since $u_1 s_2 u_1^{-1} = s_0$ as elements of W^a, there is an equality $K_{\{s_0\}} = u_1 K_{s_2} u_1^{-1}$. The parahoric subgroup for $\{s_1, s_2\}$ is $K_{\{s_1,s_2\}} = \mathrm{GSp}(4, \mathfrak{o})$. Note that $W_{\{s_1,s_2\}} = W$, the Weyl group. Since $u_1 s_1 u_1^{-1} = s_1$ in W^a, the parahoric subgroup corresponding to $\{s_0, s_1\}$ is $K_{\{s_0,s_1\}} = u_1 \mathrm{GSp}(4, \mathfrak{o}) u_1^{-1}$. Finally, the parahoric subgroup corresponding to $\{s_0, s_2\}$ is the paramodular group $\mathrm{K}(\mathfrak{p})$ of level $\mathfrak{p}$. The inclusion relations between the standard parahoric subgroups of $\mathrm{GSp}(4, F)$ are indicated in Fig. 9.2.

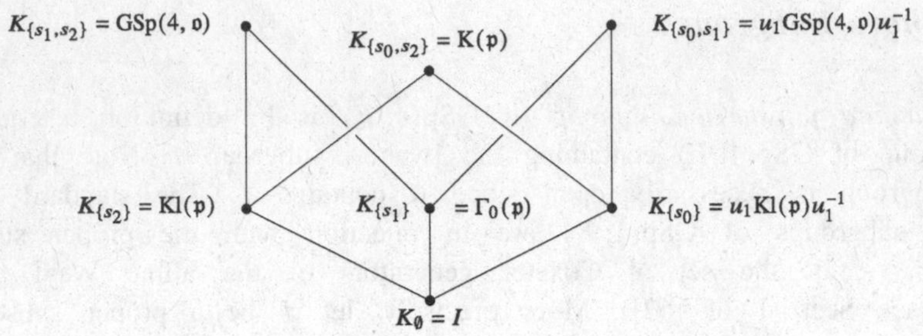

Fig. 9.2 Inclusion relations between the standard parahoric subgroups of GSp(4, F)

The Iwahori-Hecke Algebra

Next, we recall the Iwahori-Hecke algebra of GSp(4, F) associated to the Iwahori subgroup I. As a complex vector space, the *Iwahori-Hecke algebra* $\mathcal{H}(\mathrm{GSp}(4, F), I)$ is the set of all compactly supported left and right I-invariant complex-valued functions on GSp(4, F) with addition defined by pointwise addition of functions. If $T, T' \in \mathcal{H}(\mathrm{GSp}(4, F), I)$, then we define the product $T \cdot T'$ by

$$(T \cdot T')(x) = \int\limits_{\mathrm{GSp}(4,F)} T(xy^{-1})T'(y)\, d^I y$$

for $x \in \mathrm{GSp}(4, F)$. Here, $d^I y$ is the Haar measure on GSp(4, F) which gives I volume 1. If $T, T' \in \mathcal{H}(\mathrm{GSp}(4, F), I)$, then $T \cdot T' \in \mathcal{H}(\mathrm{GSp}(4, F), I)$, and with this definition $\mathcal{H}(\mathrm{GSp}(4, F), I)$ is a $\mathbb{C}$-algebra. For $g \in \mathrm{GSp}(4, F)$, we let $T_g \in \mathcal{H}(\mathrm{GSp}(4, F), I)$ be the characteristic function of IgI. The identity element of $\mathcal{H}(\mathrm{GSp}(4, F), I)$ is T_1, the characteristic function of $I \cdot 1 \cdot I = I$. We will write $e = T_1$. The element T_{u_1} is the characteristic function of $Iu_1I = u_1I = Iu_1$; recall that u_1 normalizes I. We will write $u_1 = T_{u_1}$. For $i = 0, 1$, and 2, let $e_i = T_{s_i}$, the characteristic function of Is_iI. By Sect. 5 of [67] (see also Ex. 25 c) of §2 of Chap. IV of [18]), the Iwahori-Hecke algebra $\mathcal{H}(\mathrm{GSp}(4, F), I)$ is generated by u_1, e_0, e_1, and e_2, and the following relations hold:

$$e_0u_1 = u_1e_2, \qquad e_1u_1 = u_1e_1, \qquad e_2u_1 = u_1e_0, \tag{9.5}$$

$$e_i^2 = (q-1)e_i + qe \qquad \text{for } i = 0, 1, 2, \tag{9.6}$$

$$e_0e_1e_0e_1 = e_1e_0e_1e_0, \qquad e_1e_2e_1e_2 = e_2e_1e_2e_1, \qquad e_0e_2 = e_2e_0. \tag{9.7}$$

Equation (9.6) implies that e_0, e_1, and e_2 are invertible. The element u_1 is also invertible, with inverse $T_{\varpi^{-1}u_1}$. For $w_1, w_2 \in W^a$ we have

$$T_{w_1w_2} = T_{w_1} \cdot T_{w_2} \qquad \text{if } \ell(w_1w_2) = \ell(w_1) + \ell(w_2). \tag{9.8}$$

Here, ℓ is the length function for the Coxeter group W^a. Also, if $x \in \mathrm{GSp}(4, F)$ normalizes I, and $g \in \mathrm{GSp}(4, F)$, then

$$T_x \cdot T_g = T_{xg} \quad \text{and} \quad T_g \cdot T_x = T_{gx}. \tag{9.9}$$

Finally, let $w \in W^a$, and let $IwI = \sqcup_{i \in X} x_i I$ be a disjoint decomposition; we define $q_w = \#X$. Since $\mathrm{GSp}(4, F)$ is unimodular, q_w is also the number $\#Y$, where $IwI = \sqcup_{j \in Y} I y_i$ is a disjoint decomposition. We have

$$q_{s_i} = q \qquad \text{for } i = 0, 1, 2; \tag{9.10}$$

for this, see Lemma 9.2.2. For $w_1, w_2 \in W^a$ one has

$$q_{w_1 w_2} = q_{w_1} q_{w_2} \qquad \text{if } \ell(w_1 w_2) = \ell(w_1) + \ell(w_2). \tag{9.11}$$

Representations

Now let (π, V) be a smooth representation of $\mathrm{GSp}(4, F)$. For $T \in \mathcal{H}(\mathrm{GSp}(4, F), I)$ and $v \in V$ we define

$$Tv = \int\limits_{\mathrm{GSp}(4,F)} T(g)\pi(g)v\, d^I g. \tag{9.12}$$

Calculations show that with this definition, V is an $\mathcal{H}(\mathrm{GSp}(4, F), I)$-module. Furthermore, if $v \in V^I$, and $T \in \mathcal{H}(\mathrm{GSp}(4, F), I)$, then $Tv \in V^I$. Here, V^I is the subspace of V of vectors fixed by I.

Volumes

Using (9.11) and the definitions of the parahoric subgroups, we have

$$\mathrm{vol}(\mathrm{GSp}(4, \mathfrak{o})) = \mathrm{vol}(K_{\{s_1,s_2\}}) = (1+q)^2(1+q^2), \tag{9.13}$$

$$\mathrm{vol}(\mathrm{K}(\mathfrak{p})) = \mathrm{vol}(K_{\{s_0,s_2\}}) = (1+q)^2, \tag{9.14}$$

$$\mathrm{vol}(\Gamma_0(\mathfrak{p})) = \mathrm{vol}(\mathrm{Kl}(\mathfrak{p})) = 1+q, \tag{9.15}$$

$$\mathrm{vol}(I) = 1. \tag{9.16}$$

Again, we use the Haar measure on $\mathrm{GSp}(4, F)$ that assigns I volume 1. Differences in volumes are reflected in vertical positioning in Fig. 9.2.

9.2 Action of the Iwahori-Hecke Algebra

Let (π, V) be an irreducible admissible representation of $\mathrm{GSp}(4, F)$ with trivial central character, and assume that π is Iwahori-spherical, i.e., $V^I \neq 0$. As mentioned at the beginning of this chapter, our goal is to calculate the action of the stable Hecke operators on $V_s(1)$. As a consequence of a theorem of Casselman (see Sect. 3.6 of [26] for a summary), π is an irreducible subquotient of $\chi_1 \times \chi_2 \rtimes \sigma$ for some unramified characters χ_1, χ_2, and σ of $F^\times$ such that $\chi_1\chi_2\sigma^2 = 1$. Consequently, as a first step toward our goal, in this section we will explicitly calculate the action of the Iwahori-Hecke algebra on $(\chi_1 \times \chi_2 \rtimes \sigma)^I$ for unramified characters χ_1, χ_2, and σ of $F^\times$.

For the remainder of this section, we fix unramified characters χ_1, χ_2, σ of $F^\times$. We define $\chi_1 \times \chi_2 \rtimes \sigma$, as in Sect. 2.2. The standard space of $\chi_1 \times \chi_2 \rtimes \sigma$ consists of the locally constant functions $f : \mathrm{GSp}(4, F) \to \mathbb{C}$ such that

$$f(\begin{bmatrix} a & * & * & * \\ & b & * & * \\ & & cb^{-1} & * \\ & & & ca^{-1} \end{bmatrix} g) = \chi_1(a)\chi_2(b)\sigma(c)|a^2b|\,|c|^{-3/2} f(g) \tag{9.17}$$

for all $g \in \mathrm{GSp}(4, F)$ and $a, b, c \in F^\times$. The action of $\mathrm{GSp}(4, F)$ on $\chi_1 \times \chi_2 \rtimes \sigma$ will be denoted by π. We let $(\chi_1 \times \chi_2 \rtimes \sigma)^I$ be the subspace of I-invariant vectors in $\chi_1 \times \chi_2 \rtimes \sigma$.

For the next lemma, we recall that the elements of the Weyl group W are the images of $1, s_1, s_2, s_2s_1, s_1s_2s_1, s_1s_2, s_1s_2s_1s_2$ and $s_2s_1s_2$.

Lemma 9.2.1 *The function* $(\chi_1 \times \chi_2 \rtimes \sigma)^I \to \mathbb{C}^8$ *defined by*

$$f \mapsto (f(1), f(s_1), f(s_2), f(s_2s_1), f(s_1s_2s_1), f(s_1s_2), f(s_1s_2s_1s_2), f(s_2s_1s_2))$$

is an isomorphism of $\mathbb{C}$*-vector spaces.*

Proof Let $f \in (\chi_1 \times \chi_2 \rtimes \sigma)^I$, and assume that f maps to zero. Let $g \in \mathrm{GSp}(4, F)$. The Iwasawa decomposition asserts that $\mathrm{GSp}(4, F) = B\mathrm{GSp}(4, \mathfrak{o})$, where B is the Borel subgroup defined in Sect. 2.2. By Sect. 9.1, we have $\mathrm{GSp}(4, \mathfrak{o}) = \sqcup_{w \in W} IwI$. The Iwahori subgroup I admits a decomposition $I = I_+T(\mathfrak{o})I_-$, where $T(\mathfrak{o})$ is as in Sect. 9.1,

$$I_+ = I \cap \begin{bmatrix} 1 & \mathfrak{o} & \mathfrak{o} & \mathfrak{o} \\ & 1 & \mathfrak{o} & \mathfrak{o} \\ & & 1 & \mathfrak{o} \\ & & & 1 \end{bmatrix} \quad \text{and} \quad I_- = I \cap \begin{bmatrix} 1 & & & \\ \mathfrak{p} & 1 & & \\ \mathfrak{p} & \mathfrak{p} & 1 & \\ \mathfrak{p} & \mathfrak{p} & \mathfrak{p} & 1 \end{bmatrix}. \tag{9.18}$$

We may thus write $g = butu'wk$ for some $b \in B$, $u \in I_+$, $t \in T(\mathfrak{o})$, $u' \in I_-$, $w \in W$, and $k \in I$. By (9.17), and the assumption that $f \in (\chi_1 \times \chi_2 \rtimes \sigma)^I$, there exists $c \in \mathbb{C}^\times$ such that $f(g) = f(butu'wk) = cf(u'w)$. Since $w^{-1}u'w \in I$, we have $f(g) = cf(u'w) = cf(ww^{-1}u'w) = cf(w)$. By assumption $f(w) = 0$; hence, $f(g) = 0$. It follows that our map is injective. To see that the map is surjective, let $w \in W$. Define $f_w : \mathrm{GSp}(4, F) \to \mathbb{C}$ in the following way. If $g \notin BIwI$, define $f_w(g) = 0$. If $g \in BIwI$, define

$$f_w(g) = \chi_1(a)\chi_2(b)\sigma(c)|a^2b|\,|c|^{-3/2}$$

where $g = pk$, with

$$p = \begin{bmatrix} a & * & * & * \\ & b & * & * \\ & & cb^{-1} & * \\ & & & ca^{-1} \end{bmatrix} \in B$$

for some $a, b, c \in F^\times$, and $k \in IwI$. Using that χ_1, χ_2, and σ are unramified, it is straightforward to verify that f_w is a well-defined element of $(\chi_1 \times \chi_2 \rtimes \sigma)^I$. Moreover, it is clear that f_w maps to the vector with entry 1 at the w-th position, and zeros elsewhere. It follows that our map is surjective. □

Lemma 9.2.2 *There are disjoint decompositions*

$$Is_0I = \bigsqcup_{x \in \mathfrak{o}/\mathfrak{p}} s_0 \begin{bmatrix} 1 & & & x\varpi^{-1} \\ & 1 & & \\ & & 1 & \\ & & & 1 \end{bmatrix} I, \tag{9.19}$$

$$Is_1I = \bigsqcup_{x \in \mathfrak{o}/\mathfrak{p}} s_1 \begin{bmatrix} 1 & & & \\ x & 1 & & \\ & & 1 & \\ & & -x & 1 \end{bmatrix} I, \tag{9.20}$$

$$Is_2I = \bigsqcup_{x \in \mathfrak{o}/\mathfrak{p}} s_2 \begin{bmatrix} 1 & & & \\ & 1 & & \\ & x & 1 & \\ & & & 1 \end{bmatrix} I. \tag{9.21}$$

Proof We first prove (9.20). It is easy to see that cosets on the right-hand side of (9.20) are disjoint. To see that the left-hand side of (9.20) is contained in the right-hand side, let $k \in I$; we need to prove that ks_1 is contained in the right-hand side of (9.20). We have

$$k = \begin{bmatrix} a_1 & a_2 & * & * \\ a_3 & a_4 & * & * \\ * & * & * & * \\ * & * & * & * \end{bmatrix}$$

for some $a_1, a_2, a_3, a_4 \in \mathfrak{o}$ with $a_1, a_4 \in \mathfrak{o}^\times$ and $a_3 \in \mathfrak{p}$. Hence,

$$s_1ks_1 = \begin{bmatrix} a_4 & a_3 & * & * \\ a_2 & a_1 & * & * \\ * & * & * & * \\ * & * & * & * \end{bmatrix} \in \begin{bmatrix} \mathfrak{o} & \mathfrak{p} & \mathfrak{o} & \mathfrak{o} \\ \mathfrak{o} & \mathfrak{o} & \mathfrak{o} & \mathfrak{o} \\ \mathfrak{p} & \mathfrak{p} & \mathfrak{o} & \mathfrak{p} \\ \mathfrak{p} & \mathfrak{p} & \mathfrak{o} & \mathfrak{o} \end{bmatrix}.$$

Since $a_4 \in \mathfrak{o}^\times$, there exists $x \in \mathfrak{o}$ such that $a_2 + xa_4 = 0$. It follows that

$$k' = \begin{bmatrix} 1 & & & \\ x & 1 & & \\ & & 1 & \\ & & -x & 1 \end{bmatrix} s_1ks_1 \in \begin{bmatrix} \mathfrak{o} & \mathfrak{p} & \mathfrak{o} & \mathfrak{o} \\ \mathfrak{p} & \mathfrak{o} & \mathfrak{o} & \mathfrak{o} \\ \mathfrak{p} & \mathfrak{p} & \mathfrak{o} & \mathfrak{p} \\ \mathfrak{p} & \mathfrak{p} & \mathfrak{o} & \mathfrak{o} \end{bmatrix}.$$

This inclusion, along with $k' \in \mathrm{GSp}(4, \mathfrak{o})$, also implies that $k'_{43} \in \mathfrak{p}$; hence, $k' \in I$. We now see that ks_1 is contained in the right-hand side of (9.20). The decomposition (9.21) is proven in a similar fashion. Finally, (9.19) follows from (9.21) by conjugation by u_1. □

By Lemma 9.2.1 and its proof, the vector space $(\chi_1 \times \chi_2 \rtimes \sigma)^I$ has as basis the functions f_w, $w \in W$, where f_w is the unique I-invariant function in $\chi_1 \times \chi_2 \rtimes \sigma$ with $f_w(w) = 1$ and $f_w(w') = 0$ for $w' \in W$, $w' \neq w$. It is convenient to order the basis as follows:

$$f_e, \quad f_1, \quad f_2, \quad f_{21}, \quad f_{121}, \quad f_{12}, \quad f_{1212}, \quad f_{212}, \tag{9.22}$$

where e is the identity element of W, and where we have abbreviated $f_1 = f_{s_1}$ and so on. Having fixed this basis, the operators e_0, e_1, e_2 and u on $(\chi_1 \times \chi_2 \rtimes \sigma)^I$ become 8×8 matrices. These are given in the following lemma. We use the notation

$$\alpha = \chi_1(\varpi), \qquad \beta = \chi_2(\varpi), \qquad \gamma = \sigma(\varpi) \tag{9.23}$$

for the Satake parameters.

Lemma 9.2.3 *Let χ_1, χ_2, and σ be unramified characters of $F^\times$. With respect to the basis (9.22) of $(\chi_1 \times \chi_2 \rtimes \sigma)^I$, the action of the elements e_1 and e_2 is given by the following matrices:*

$$e_1 = \begin{bmatrix} 0 & q & & & & & & \\ 1 & q-1 & & & & & & \\ & & 0 & q & & & & \\ & & 1 & q-1 & & & & \\ & & & & q-1 & 1 & & \\ & & & & q & 0 & & \\ & & & & & & q-1 & 1 \\ & & & & & & q & 0 \end{bmatrix}, \quad e_2 = \begin{bmatrix} 0 & 0 & q & 0 & 0 & 0 & 0 & 0 \\ 0 & 0 & 0 & 0 & 0 & q & 0 & 0 \\ 1 & 0 & q-1 & 0 & 0 & 0 & 0 & 0 \\ 0 & 0 & 0 & 0 & 0 & 0 & 0 & q \\ 0 & 0 & 0 & 0 & 0 & 0 & q & 0 \\ 0 & 1 & 0 & 0 & 0 & q-1 & 0 & 0 \\ 0 & 0 & 0 & 0 & 1 & 0 & q-1 & 0 \\ 0 & 0 & 0 & 1 & 0 & 0 & 0 & q-1 \end{bmatrix}$$

The action of u_1 is given by

$$u_1 = \begin{bmatrix} & & & & & & & \gamma q^{3/2} \\ & & & & & & \gamma q^{3/2} & \\ & & & & & \beta\gamma q^{1/2} & & \\ & & & & \beta\gamma q^{1/2} & & & \\ & & & \alpha\gamma q^{-1/2} & & & & \\ & & \alpha\gamma q^{-1/2} & & & & & \\ & \alpha\beta\gamma q^{-3/2} & & & & & & \\ \alpha\beta\gamma q^{-3/2} & & & & & & & \end{bmatrix}.$$

The matrix of e_0 is given by the matrix of $u_1 e_2 u_1^{-1}$.

Proof Let w, w' be in $\{1, s_1, s_2, s_2s_1, s_1s_2s_1, s_1s_2, s_1s_2s_1s_2, s_2s_1s_2\}$. By (9.12) and (9.20), we have

$$(e_1 f_w)(w') = \int\limits_{Is_1I} f_w(w'g)\, d^I g = \sum_{x\in\mathfrak{o}/\mathfrak{p}} f_w(w's_1 \begin{bmatrix} 1 & & & \\ x & 1 & & \\ & & 1 & \\ & & -x & 1 \end{bmatrix}).$$

Calculations now verify that the matrix of e_1 is as stated; for this, it is useful to employ (2.2). The formulas for e_2 and u_1 have similar proofs. The final claim follows from $e_0 = u_1 e_2 u_1^{-1}$ in the Iwahori-Hecke algebra. □

Projections and Bases

Let K be a parahoric subgroup of $\mathrm{GSp}(4, F)$ and let (π, V) be a smooth representation of $\mathrm{GSp}(4, F)$. We define

$$d_K : V \longrightarrow V^K$$

by

$$d_K v = \mathrm{vol}(K)^{-1} \int\limits_K \pi(g) v\, d^I g$$

for $v \in V$. Again we use the Haar measure on $\mathrm{GSp}(4, F)$ that gives I measure 1. Evidently, d_K is a projection onto the subspace V^K of V.

Lemma 9.2.4 *Let χ_1, χ_2, and σ be unramified characters of $F^\times$.*

(1) *The projection $d_{\Gamma_0(\mathfrak{p})} : \chi_1 \times \chi_2 \rtimes \sigma \to (\chi_1 \times \chi_2 \rtimes \sigma)^{\Gamma_0(\mathfrak{p})}$ is given by*

$$d_{\Gamma_0(\mathfrak{p})} = (1+q)^{-1}(e + e_1). \tag{9.24}$$

The subspace $(\chi_1 \times \chi_2 \rtimes \sigma)^{\Gamma_0(\mathfrak{p})}$ is four-dimensional and has basis

$$f_e + f_1, \qquad f_2 + f_{21}, \qquad f_{121} + f_{12}, \qquad f_{1212} + f_{212}. \tag{9.25}$$

(2) *The projection $d_{\mathrm{Kl}(\mathfrak{p})} : \chi_1 \times \chi_2 \rtimes \sigma \to (\chi_1 \times \chi_2 \rtimes \sigma)^{\mathrm{Kl}(\mathfrak{p})}$ is given by*

$$d_{\mathrm{Kl}(\mathfrak{p})} = (1+q)^{-1}(e + e_2). \tag{9.26}$$

The subspace $(\chi_1 \times \chi_2 \rtimes \sigma)^{\mathrm{Kl}(\mathfrak{p})}$ is four-dimensional and has basis

$$f_e + f_2, \qquad f_1 + f_{12}, \qquad f_{21} + f_{212}, \qquad f_{121} + f_{1212}. \tag{9.27}$$

(3) *The projection* $d_{\mathrm{K}(\mathfrak{p})} : \chi_1 \times \chi_2 \rtimes \sigma \to (\chi_1 \times \chi_2 \rtimes \sigma)^{\mathrm{K}(\mathfrak{p})}$ *is given by*

$$d_{\mathrm{K}(\mathfrak{p})} = (1+q)^{-2}(e + e_0 + e_2 + e_0 e_2). \tag{9.28}$$

The subspace $(\chi_1 \times \chi_2 \rtimes \sigma)^{\mathrm{K}(\mathfrak{p})}$ *is two-dimensional and has basis*

$$f_1^{\mathrm{para}} := f_e + f_2 + \alpha q^{-2}(f_{121} + f_{1212}), \tag{9.29}$$

$$f_2^{\mathrm{para}} := f_1 + f_{12} + \beta q^{-1}(f_{21} + f_{212}). \tag{9.30}$$

(4) *The projection* $d_{\mathrm{GSp}(4,\mathfrak{o})} : \chi_1 \times \chi_2 \rtimes \sigma \to (\chi_1 \times \chi_2 \rtimes \sigma)^{\mathrm{GSp}(4,\mathfrak{o})}$ *is given by*

$$d_{\mathrm{GSp}(4,\mathfrak{o})} = (1+q)^{-1}(1+q^2)^{-1}\big(e + e_1 + e_2 + e_1 e_2 + e_2 e_1 + e_1 e_2 e_1 + e_2 e_1 e_2 + e_1 e_2 e_1 e_2\big). \tag{9.31}$$

The subspace $(\chi_1 \times \chi_2 \rtimes \sigma)^{\mathrm{GSp}(4,\mathfrak{o})}$ *is one-dimensional and has basis*

$$f_0 = f_e + f_1 + f_2 + f_{21} + f_{121} + f_{12} + f_{1212} + f_{212}. \tag{9.32}$$

Proof Let K be one of $\Gamma_0(\mathfrak{p})$, $\mathrm{Kl}(\mathfrak{p})$, $\mathrm{K}(\mathfrak{p})$, or $\mathrm{GSp}(4,\mathfrak{o})$. The formula for d_K follows from the definition of d_K, (9.8), and the formula for $\mathrm{vol}(K)$ in Sect. 9.1. The subspace $(\chi_1 \times \chi_2 \rtimes \sigma)^K$ is exactly the 1-eigenspace of $d_K|_{(\chi_1 \times \chi_2 \rtimes \sigma)^I}$. It is straightforward to calculate this eigenspace using the formula for d_K, the matrices from Lemma 9.2.3, and a computer algebra program. □

Lemma 9.2.5 *Let* χ_1, χ_2, *and* σ *be unramified characters of* $F^\times$. *Let* f_0 *be the basis element of the one-dimensional space* $(\chi_1 \times \chi_2 \rtimes \sigma)^{\mathrm{GSp}(4,\mathfrak{o})}$ *defined in* (9.32). *Let* $\theta, \theta' : (\chi_1 \times \chi_2 \rtimes \sigma)^{\mathrm{GSp}(4,\mathfrak{o})} \to (\chi_1 \times \chi_2 \rtimes \sigma)^{\mathrm{K}(\mathfrak{p})}$ *be as in* (2.28) *and* (2.29). *We have*

$$\theta f_0 = \sigma(\varpi) q^{3/2}(1 + \chi_2(\varpi)) f_1^{\mathrm{para}} + \sigma(\varpi) q^{3/2}(1 + \chi_1(\varpi) q^{-1}) f_2^{\mathrm{para}},$$

$$\theta' f_0 = (\chi_1(\varpi)^{-1} q^2 + q) f_1^{\mathrm{para}} + (\chi_2(\varpi)^{-1} q + q) f_2^{\mathrm{para}}.$$

Assume further that $\chi_1 \chi_2 \sigma^2 = 1$, *so that* $\chi_1 \times \chi_2 \rtimes \sigma$ *has trivial central character. Then* θf_0 *and* $\theta' f_0$ *are linearly independent if and only if* $\chi_1 \chi_2 \neq \nu^{-1}$ *and* $\chi_1 \chi_2^{-1} \neq \nu^{-1}$.

Proof The vector θf_0 is contained in $(\chi_1 \times \chi_2 \rtimes \sigma)^{\mathrm{K}(\mathfrak{p})}$. Since f_1^{para} and f_2^{para} form a basis for $(\chi_1 \times \chi_2 \rtimes \sigma)^{\mathrm{K}(\mathfrak{p})}$ by Lemma 9.2.4, there exist $a, b \in \mathbb{C}$ such that $\theta f_0 = a f_1^{\mathrm{para}} + b f_2^{\mathrm{para}}$. Evaluating at $1, s_1 \in \mathrm{GSp}(4, F)$, we have $a = (\theta f_0)(1)$ and $b = (\theta f_0)(s_1)$. Calculations using the definition of θ, along with (9.17) show that a and b are as in the statement of the lemma. The expression for $\theta' f_0$ is similarly verified. The final assertion follows from the fact that $f_1^{\rm para}$ and $f_2^{\rm para}$ are always linearly independent. □

9.3 Stable Hecke Operators and the Iwahori-Hecke Algebra

Let (π, V) be a smooth representation of $\mathrm{GSp}(4, F)$ for which the center acts trivially. In this section we express the endomorphisms $T_{0,1}^s$ and $T_{1,0}^s$ of $V_s(1)$ defined in Sect. 3.8 in terms of certain elements of the Iwahori-Hecke algebra.

Lemma 9.3.1 *Let* (π, V) *be a smooth representation of* $\mathrm{GSp}(4, F)$. *If* $g \in \mathrm{GSp}(4, F)$ *and* $v \in V^I$, *then* $(e \circ \pi(g))(v) = q_g^{-1} T_g(v)$, *with* q_g *as in Sect. 9.1.*

Proof Let $g \in \mathrm{GSp}(4, F)$. We have a bijection

$$I/(I \cap gIg^{-1}) \xrightarrow{\sim} IgI/I \tag{9.33}$$

given by $x(I \cap gIg^{-1}) \mapsto xgI$. Let $v \in V^I$. Then

$$\begin{aligned}(e \circ \pi(g))(v) &= \int\limits_I \pi(h)\pi(g)v\, d^I h \\ &= \frac{1}{\#I/(I \cap gIg^{-1})} \sum_{x \in I/(I \cap gIg^{-1})} \pi(x)\pi(g)v \\ &= q_g^{-1} \sum_{x \in I/(I \cap gIg^{-1})} \pi(x)\pi(g)v.\end{aligned}$$

On the other hand,

$$T_g(v) = \int\limits_{IgI} \pi(h)v\, d^I h = \sum_{x \in IgI/I} \pi(x)v = \sum_{x \in I/(I \cap gIg^{-1})} \pi(xg)v. \tag{9.34}$$

The assertion follows. □

Lemma 9.3.2 *Let* (π, V) *be a smooth representation of* $\mathrm{GSp}(4, F)$ *for which the center acts trivially. Then, for* $v \in V_s(1)$,

$$T_{0,1}^s(v) = (1 + q^{-1}) d_{\mathrm{Kl}(\mathfrak{p})} e_2 e_1 e_2 u_1(v) \tag{9.35}$$

$$= (1 + q^{-1})(1 + q)^{-1}(e + e_2) e_2 e_1 e_2 u_1(v). \tag{9.36}$$

and

$$T_{1,0}^{s}(v) = d_{\mathrm{Kl}(\mathfrak{p})} e_1 e_2 e_1 e_0(v) \tag{9.37}$$

$$= (1+q)^{-1}(e+e_2)e_1 e_2 e_1 e_0(v). \tag{9.38}$$

Proof Let

$$h = \begin{bmatrix} \varpi & & & \\ & \varpi & & \\ & & 1 & \\ & & & 1 \end{bmatrix}.$$

Then

$$h = s_2 s_1 s_2 u_1 \begin{bmatrix} 1 & & & \\ & -1 & & \\ & & -1 & \\ & & & 1 \end{bmatrix}. \tag{9.39}$$

Hence,

$$\begin{aligned} T_h &= T_{s_2 s_1 s_2 u_1} \\ &= T_{s_2 s_1 s_2} T_{u_1} \qquad \text{(see (9.9))} \\ &= T_{s_2} T_{s_1} T_{s_2} T_{u_1} \qquad \text{(see (9.8))} \\ &= e_2 e_1 e_2 u_1. \end{aligned} \tag{9.40}$$

Let $v \in V_s(1) = V^{\mathrm{Kl}(\mathfrak{p})}$. By (3.71),

$$\begin{aligned} T_{0,1}^{s}(v) &= \frac{q^2+q^3}{\mathrm{vol}(\mathrm{Kl}(\mathfrak{p}))} \int\limits_{\mathrm{Kl}(\mathfrak{p})} \pi(k)\pi(h)v\, d^I k \\ &= (q^2+q^3) d_{\mathrm{Kl}(\mathfrak{p})}(\pi(h)v). \end{aligned} \tag{9.41}$$

Also, by Lemma 9.3.1,

$$e(\pi(h)v) = q_h^{-1} T_h v.$$

Applying $d_{\mathrm{Kl}(\mathfrak{p})}$ to this equation, we obtain

$$d_{\mathrm{Kl}(\mathfrak{p})}(\pi(h)v) = q_h^{-1} d_{\mathrm{Kl}(\mathfrak{p})}(T_h v).$$

Hence, by (9.41),

$$T_{0,1}^{s}(v) = (q^2+q^3) q_h^{-1} d_{\mathrm{Kl}(\mathfrak{p})}(T_h v).$$

By (9.40) this is

$$T^s_{0,1}(v) = (q^2 + q^3) q_h^{-1} d_{\mathrm{Kl}(\mathfrak{p})} e_2 e_1 e_2 u_1(v). \tag{9.42}$$

We have

$$\begin{aligned} q_h &= q_{s_2 s_1 s_2 u_1} \qquad \text{(by (9.39))} \\ &= q_{s_2 s_1 s_2} \\ &= q_{s_2} q_{s_1} q_{s_2} \qquad \text{(by (9.11))} \\ &= q^3. \qquad \text{(by (9.10))} \end{aligned}$$

Substituting into (9.42) now proves (9.35); (9.36) follows from (9.35) and (9.26).

Similarly, let

$$h' = \begin{bmatrix} \varpi & & & \\ & 1 & & \\ & & 1 & \\ & & & \varpi^{-1} \end{bmatrix} = s_1 s_2 s_1 s_0. \tag{9.43}$$

Then

$$\begin{aligned} T_{h'} &= T_{s_1 s_2 s_1 s_0} \\ &= T_{s_1} T_{s_2} T_{s_1} T_{s_0} \qquad \text{(see (9.8))} \\ &= e_1 e_2 e_1 e_0. \end{aligned} \tag{9.44}$$

Let $v \in V_s(1) = V^{P_2}$. By (3.72),

$$\begin{aligned} T^s_{1,0}(v) &= \frac{q^4}{\mathrm{vol}(\mathrm{Kl}(\mathfrak{p}))} \int\limits_{\mathrm{Kl}(\mathfrak{p})} \pi(k)\pi(h')v \, d^1 k \\ &= q^4 d_{\mathrm{Kl}(\mathfrak{p})}(\pi(h')v). \end{aligned} \tag{9.45}$$

By Lemma 9.3.1,

$$e(\pi(h')v) = q_{h'}^{-1} T_{h'} v.$$

Applying $d_{\mathrm{Kl}(\mathfrak{p})}$ to this equation yields

$$d_{\mathrm{Kl}(\mathfrak{p})}(\pi(h')v) = q_{h'}^{-1} d_{\mathrm{Kl}(\mathfrak{p})}(T_{h'} v). \tag{9.46}$$

Therefore, by (9.45),

$$T^s_{1,0}(v) = q^4 q_{h'}^{-1} d_{\mathrm{Kl}(\mathfrak{p})}(T_{h'} v).$$

By (9.44) we now have

$$T^s_{1,0}(v) = q^4 q_{h'}^{-1} d_{\mathrm{Kl}(\mathfrak{p})} e_1 e_2 e_1 e_0(v). \tag{9.47}$$

Now

$$\begin{aligned} q_{h'} &= q_{s_1 s_2 s_1 s_0} \qquad \text{(by (9.39))} \\ &= q_{s_1} q_{s_2} q_{s_1} q_{s_0} \qquad \text{(by (9.11))} \\ &= q^4. \qquad \text{(by (9.10))} \end{aligned}$$

Substituting into (9.47) proves (9.37); also, (9.38) follows from (9.37) and (9.26). □

9.4 Characteristic Polynomials

Let (π, V) be an irreducible, admissible representation of $\mathrm{GSp}(4, F)$ with trivial central character. In this section we calculate the characteristic polynomials of the stable Hecke operators $T^s_{0,1}$ and $T^s_{1,0}$ acting on the space $V_s(1)$ when π is Iwahori-spherical and $V_s(1)$ is non-zero. By a theorem of Casselman (see Sect. 3.6 of [26] for a summary), π is Iwahori-spherical if and only if π is an irreducible subquotient of $\chi_1 \times \chi_2 \rtimes \sigma$ for some unramified characters χ_1, χ_2, and σ of $F^\times$ such that $\chi_1\chi_2\sigma^2 = 1$. It follows that π is Iwahori-spherical if and only if π is a group I, II, III, IV, V, or VI representation with unramified inducing data. We begin this section by calculating the characteristic polynomials of $T^s_{0,1}$ and $T^s_{1,0}$ for three families of representations formed from unramified characters. These representations, which may be reducible, occur in the exact sequences involving group I-VI representations with unramified inducing data that are described in Sect. 2.2. With these lemmas in place, we then prove the main result of this chapter, Theorem 9.4.6.

In this section, if W is a finite-dimensional complex vector space, and $T : W \to W$ is a linear operator, then the characteristic polynomial of T is by definition

$$p(T, W, X) = \det(X \cdot 1_W - T).$$

Lemma 9.4.1 *Let*

$$\begin{array}{ccccccccc} 0 & \longrightarrow & W_1 & \longrightarrow & W_2 & \longrightarrow & W_3 & \longrightarrow & 0 \\ & & \downarrow{\scriptstyle T_1} & & \downarrow{\scriptstyle T_2} & & \downarrow{\scriptstyle T_3} & & \\ 0 & \longrightarrow & W_1 & \longrightarrow & W_2 & \longrightarrow & W_3 & \longrightarrow & 0 \end{array}$$

be a commutative diagram with exact rows, where W_1, W_2, and W_3 are finite-dimensional complex vector spaces, and all maps are linear. Then

$$p(T_2, W_2, X) = p(T_1, W_1, X)p(T_3, W_3, X).$$

Proof The proof is left to the reader. □

Lemma 9.4.2 *Let χ_1, χ_2, and σ be unramified characters of $F^\times$, and assume that $\chi_1\chi_2\sigma^2 = 1$, so that the center of $\mathrm{GSp}(4, F)$ acts trivially on $\chi_1 \times \chi_2 \rtimes \sigma$. Recall that $(\chi_1 \times \chi_2 \rtimes \sigma)^{\mathrm{Kl}(\mathfrak{p})}$ is four-dimensional with basis (9.27). We have*

$$\begin{aligned} p(T_{0,1}^s, (\chi_1 \times \chi_2 \rtimes \sigma)^{\mathrm{Kl}(\mathfrak{p})}, X) = &\Big(X - \chi_1(\varpi)\,(1 + \chi_2(\varpi))\sigma(\varpi)\,q^{\frac{3}{2}}\Big)\\ &\times \Big(X - \chi_2(\varpi)\,(1 + \chi_1(\varpi))\,\sigma(\varpi)q^{\frac{3}{2}}\Big)\\ &\times \Big(X - (1 + \chi_1(\varpi))\,\sigma(\varpi)q^{\frac{3}{2}}\Big)\\ &\times \Big(X - (1 + \chi_2(\varpi))\,\sigma(\varpi)q^{\frac{3}{2}}\Big) \end{aligned}$$

and

$$\begin{aligned} p(T_{1,0}^s, (\chi_1 \times \chi_2 \rtimes \sigma)^{\mathrm{Kl}(\mathfrak{p})}, X) = &\Big(X - \chi_1(\varpi)q^2\Big)\Big(X - \chi_2(\varpi)q^2\Big)\\ &\times \Big(X - \chi_2(\varpi)^{-1}q^2\Big)\Big(X - \chi_1(\varpi)^{-1}q^2\Big). \end{aligned}$$

Proof We work in $(\chi_1 \times \chi_2 \rtimes \sigma)^I$, using the basis (9.22) and the notation (9.23). By Lemma 9.2.4, the space $(\chi_1 \times \chi_2 \rtimes \sigma)^{\mathrm{Kl}(\mathfrak{p})}$ is a four-dimensional subspace of $(\chi_1 \times \chi_2 \rtimes \sigma)^I$, with basis (9.27). The elements $e_0, e_1, e_2, u_1 \in \mathcal{H}(\mathrm{GSp}(4, F), I)$ act on $(\chi_1 \times \chi_2 \rtimes \sigma)^I$, and have matrices as in Lemma 9.2.3 with respect to the basis (9.22). Using the formulas (9.36) and (9.38) for $T_{0,1}^s$ and $T_{1,0}^s$, respectively, in terms of e_0, e_1, e_2, and u_1, it is straightforward to verify that the matrix of $T_{0,1}^s$ on $(\chi_1 \times \chi_2 \rtimes \sigma)^{\mathrm{Kl}(\mathfrak{p})}$ with respect to the basis (9.27) is

$$\begin{bmatrix} \alpha(1{+}\beta)\gamma q^{3/2} & 0 & 0 & 0 \\ \alpha\gamma(q{-}1)q^{1/2} & \beta(1{+}\alpha)\gamma q^{3/2} & 0 & 0 \\ \alpha\beta\gamma(q{-}1)q^{-1/2} & (1{+}\alpha)\beta\gamma(q{-}1)q^{1/2} & (1{+}\alpha)\gamma q^{3/2} & 0 \\ \alpha(1{+}\beta)\gamma(q{-}1)q^{-1/2} & \alpha\beta\gamma(q{-}1)q^{1/2} & \alpha\gamma(q{-}1)q^{1/2} & (1{+}\beta)\gamma q^{3/2} \end{bmatrix},$$

and that the matrix of $T_{1,0}^s$ on $(\chi_1 \times \chi_2 \rtimes \sigma)^{\mathrm{Kl}(\mathfrak{p})}$ with respect to the basis (9.27) is

$$\begin{bmatrix} \alpha q^2 & 0 & 0 & 0 \\ \alpha(q{-}1)q & \beta q^2 & 0 & 0 \\ \alpha(q{-}1)q & (1{+}\beta)(q{-}1)q & \beta^{-1}q^2 & 0 \\ (q{-}1)(1{+}\alpha q) & (q{-}1)(q{-}1{+}\beta q) & (q{-}1)q\beta^{-1} & \alpha^{-1}q^2 \end{bmatrix}.$$

The formulas for the characteristic polynomials are now immediate. □

Lemma 9.4.3 *Let χ and σ be characters of $F^\times$.*

(1) *Assume that $\chi^2\sigma^2 = 1$. The representation $\chi 1_{\mathrm{GSp}(4,F)} \rtimes \sigma$ contains a non-zero vector fixed by $\mathrm{Sp}(4, F)$ if and only if $\chi = \nu^{-\frac{3}{2}}$.*
(2) *Assume that $\chi\sigma^2 = 1$. The representation $\chi \rtimes \sigma 1_{\mathrm{GSp}(2)}$ contains a non-zero vector fixed by $\mathrm{Sp}(4, F)$ if and only if $\chi = \nu^{-2}$.*

Proof (1) Assume that $\chi 1_{\mathrm{GL}(2)} \rtimes \sigma$ contains a non-zero vector f fixed by $\mathrm{Sp}(4, F)$. For $g \in \mathrm{GSp}(4, F)$, we define $g_1 = \begin{bmatrix} 1 & \\ & \lambda(g)^{-1} \end{bmatrix} g \in \mathrm{Sp}(4, F)$. If $g \in \mathrm{GSp}(4, F)$, then

$$\begin{aligned} f(g) &= f(\begin{bmatrix} 1 & \\ & \lambda(g) \end{bmatrix} g_1) \\ &= f(\begin{bmatrix} 1 & \\ & \lambda(g) \end{bmatrix}) \\ f(g) &= |\lambda(g)|^{-\frac{3}{2}}\sigma(\lambda(g))f(1). \end{aligned} \tag{9.48}$$

Since $f \neq 0$, it follows that $f(1) \neq 0$. Now let $a \in F^\times$. Then

$$\begin{aligned} f(\begin{bmatrix} a &&& \\ & 1 && \\ && 1 & \\ &&& a^{-1} \end{bmatrix}) &= |\det(\begin{bmatrix} a & \\ & 1 \end{bmatrix})|^{\frac{3}{2}}\chi(\det(\begin{bmatrix} a & \\ & 1 \end{bmatrix}))f(1) \\ f(1) &= |a|^{\frac{3}{2}}\chi(a)f(1). \end{aligned}$$

We conclude that $\chi = \nu^{-\frac{3}{2}}$. Now assume $\chi = \nu^{-\frac{3}{2}}$. Define $f : \mathrm{GSp}(4, F) \to \mathbb{C}$ by $f(g) = |\lambda(g)|^{-\frac{3}{2}}\sigma(\lambda(g))$. Calculations shows that f is a non-zero $\mathrm{Sp}(4, F)$ invariant element of $\chi 1_{\mathrm{GL}(2)} \rtimes \sigma$.

The proof of (2) is similar to the proof of (1). □

Lemma 9.4.4 *Let χ and σ be unramified characters of $F^\times$. Assume that $\chi^2\sigma^2 = 1$, $\chi^2 \neq \nu^{-1}$, and $\chi \neq \nu^{-3/2}$. The vector spaces $(\chi\mathrm{St}_{\mathrm{GL}(2)} \rtimes \sigma)^{\mathrm{Kl}(\mathfrak{p})}$ and $(\chi 1_{\mathrm{GL}(2)} \rtimes \sigma)^{\mathrm{Kl}(\mathfrak{p})}$ are both two-dimensional. We have*

$$\begin{aligned} p(T^s_{0,1}, (\chi\mathrm{St}_{\mathrm{GL}(2)} \rtimes \sigma)^{\mathrm{Kl}(\mathfrak{p})}, X) = X^2 - &\Big(2(\chi\sigma)(\varpi)q + \Big(\sigma(\varpi)+\sigma(\varpi)^{-1}\Big)q^{\frac{3}{2}}\Big)X \\ &+q^2+q^3+\Big(\chi(\varpi)+\chi(\varpi)^{-1}\Big)q^{\frac{5}{2}}, \end{aligned}$$

$$p(T^s_{1,0}, (\chi\mathrm{St}_{\mathrm{GL}(2)} \rtimes \sigma)^{\mathrm{Kl}(\mathfrak{p})}, X) = X^2 - \Big(\chi(\varpi)+\chi(\varpi)^{-1}\Big)q^{\frac{3}{2}}X + q^3,$$

$$p(T^s_{0,1}, (\chi 1_{\mathrm{GL}(2)} \rtimes \sigma)^{\mathrm{Kl}(\mathfrak{p})}, X) = X^2 - \Big(2(\chi\sigma)(\varpi)q^2 + \big(\sigma(\varpi)+\sigma(\varpi)^{-1}\big)q^{\frac{3}{2}}\Big)X$$
$$+q^3+q^4+\big(\chi(\varpi)+\chi(\varpi)^{-1}\big)q^{\frac{7}{2}},$$
$$p(T^s_{1,0}, (\chi 1_{\mathrm{GL}(2)} \rtimes \sigma)^{\mathrm{Kl}(\mathfrak{p})}, X) = X^2 - \big(\chi(\varpi)+\chi(\varpi)^{-1}\big)q^{5/2}X + q^5.$$

Proof Since there is an exact sequence

$$0 \longrightarrow \chi\mathrm{St}_{\mathrm{GL}(2)} \longrightarrow \nu^{1/2}\chi \times \nu^{-1/2}\chi \longrightarrow \chi 1_{\mathrm{GL}(2)} \longrightarrow 0$$

of $\mathrm{GL}(2, F)$ representations, there is an exact sequence

$$0 \longrightarrow \chi\mathrm{St}_{\mathrm{GL}(2)} \rtimes \sigma \longrightarrow (\nu^{1/2}\chi \times \nu^{-1/2}\chi) \rtimes \sigma \longrightarrow \chi 1_{\mathrm{GL}(2)} \rtimes \sigma \longrightarrow 0$$

of $\mathrm{GSp}(4, F)$ representations. There is an isomorphism $(\nu^{1/2}\chi \times \nu^{-1/2}\chi) \rtimes \sigma \xrightarrow{\sim} \nu^{1/2}\chi \times \nu^{-1/2}\chi \rtimes \sigma$ of $\mathrm{GSp}(4, F)$ representations given by $f \mapsto F_f$, where $F_f : \mathrm{GSp}(4, F) \to \mathbb{C}$ is defined by $F_f(g) = \big(f(g)\big)(\left[\begin{smallmatrix}1&\\&1\end{smallmatrix}\right])$ for f in the standard model of $(\nu^{1/2}\chi \times \nu^{-1/2}\chi) \rtimes \sigma$ and $g \in \mathrm{GSp}(4, F)$. We thus have an exact sequence

$$0 \longrightarrow \chi\mathrm{St}_{\mathrm{GL}(2)} \rtimes \sigma \longrightarrow \nu^{1/2}\chi \times \nu^{-1/2}\chi \rtimes \sigma \longrightarrow \chi 1_{\mathrm{GL}(2)} \rtimes \sigma \longrightarrow 0 \qquad (9.49)$$

of $\mathrm{GSp}(4, F)$ representations. Let v_0 be a newform in the space of $\chi\mathrm{St}_{\mathrm{GL}(2)} \subset \nu^{1/2}\chi \times \nu^{-1/2}\chi$ (see [110] for an account of newforms for representations of $\mathrm{GL}(2, F)$). By (26) of [110] we may assume that, as an element of $\nu^{1/2}\chi \times \nu^{-1/2}\chi$, $v_0(\left[\begin{smallmatrix}1&\\&1\end{smallmatrix}\right]) = q$ and $v_0(\left[\begin{smallmatrix}&-1\\1&\end{smallmatrix}\right]) = -1$. By the proof of Theorem 5.2.2 of [105], the function $F^a_{\mathrm{para}} : \mathrm{GSp}(4, F) \to \chi\mathrm{St}_{\mathrm{GL}(2)}$ defined by

$$F^a_{\mathrm{para}}(g) = |c^{-1}\det(A)|^{3/2}\sigma(c)(\chi\mathrm{St}_{\mathrm{GL}(2)})(A)v_0$$

for $g \in \mathrm{GSp}(4, F)$, $g = pk$, with $p = \left[\begin{smallmatrix}A&*\\&cA'\end{smallmatrix}\right] \in P$, $A \in \mathrm{GL}(2, F)$, $c \in F^\times$, and $k \in \mathrm{K}(\mathfrak{p})$, is a non-zero element of $(\chi\mathrm{St}_{\mathrm{GL}(2)} \rtimes \sigma)^{\mathrm{K}(\mathfrak{p})}$. We also write F^a_{para} for the image of F^a_{para} in $\nu^{1/2}\chi \times \nu^{-1/2}\chi \rtimes \sigma$. The element F^a_{para} is contained in $(\nu^{1/2}\chi \times \nu^{-1/2}\chi \rtimes \sigma)^{\mathrm{K}(\mathfrak{p})}$, which is two-dimensional and spanned by the vectors f_1^{para} and f_2^{para} from Lemma 9.2.4 (applied to $\nu^{1/2}\chi \times \nu^{-1/2}\chi \rtimes \sigma$). Calculations show that $F^a_{\mathrm{para}}(1) = q$ and $F^a_{\mathrm{para}}(s_1) = -1$. It follows that $F^a_{\mathrm{para}} = qf_1^{\mathrm{para}} - f_2^{\mathrm{para}}$. Further calculations using Lemma 9.2.3, Lemma 9.3.2, and the assumption $\chi^2 \neq \nu^{-1}$ show that F^a_{para} and $T^s_{0,1}F^a_{\mathrm{para}}$ are linearly independent. It follows that $(\chi\mathrm{St}_{\mathrm{GL}(2)} \rtimes \sigma)^{\mathrm{Kl}(\mathfrak{p})}$ is at least two-dimensional.

Next, there is also an exact sequence

$$0 \longrightarrow \chi 1_{\mathrm{GL}(2)} \rtimes \sigma \longrightarrow \nu^{-1/2}\chi \times \nu^{1/2}\chi \rtimes \sigma \longrightarrow \chi\mathrm{St}_{\mathrm{GL}(2)} \rtimes \sigma \longrightarrow 0. \qquad (9.50)$$

Let F_0^b be a non-zero element of $(\chi 1_{\mathrm{GL}(2)} \rtimes \sigma)^{\mathrm{GSp}(4,\mathfrak{o})}$. We regard F_0^b as an element of $(\nu^{-1/2}\chi \times \nu^{1/2}\chi \rtimes \sigma)^{\mathrm{GSp}(4,\mathfrak{o})}$ via (9.50). The vector θF_0^b is non-zero by Lemma 9.2.5 and the assumption $\chi^2 \neq \nu^{-1}$. Also, F_0^b and θF_0^b are linearly independent by Theorem 2.3.1; this uses (1) of Lemma 9.4.3 and the assumption $\chi \neq \nu^{-\frac{3}{2}}$. Since F_0^b and θF_0^b are elements of $(\chi 1_{\mathrm{GL}(2)} \rtimes \sigma)^{\mathrm{Kl}(\mathfrak{p})}$, it follows that $(\chi 1_{\mathrm{GL}(2)} \rtimes \sigma)^{\mathrm{Kl}(\mathfrak{p})}$ is at least two-dimensional. Since $(\chi \mathrm{St}_{\mathrm{GL}(2)} \rtimes \sigma)^{\mathrm{Kl}(\mathfrak{p})}$ and $(\chi 1_{\mathrm{GL}(2)} \rtimes \sigma)^{\mathrm{Kl}(\mathfrak{p})}$ are at least two-dimensional, and since $(\nu^{1/2}\chi \times \nu^{-1/2}\chi \rtimes \sigma)^{\mathrm{Kl}(\mathfrak{p})}$ is four-dimensional, we conclude that $(\chi \mathrm{St}_{\mathrm{GL}(2)} \rtimes \sigma)^{\mathrm{Kl}(\mathfrak{p})}$ and $(\chi 1_{\mathrm{GL}(2)} \rtimes \sigma)^{\mathrm{Kl}(\mathfrak{p})}$ are two-dimensional.

Using Lemmas 9.2.3 and 9.3.2, it now follows that the matrix of $T_{0,1}^s$ on $(\chi \mathrm{St}_{\mathrm{GL}(2)} \rtimes \sigma)^{\mathrm{Kl}(\mathfrak{p})}$ with respect to the basis $F_{\mathrm{para}}^a, T_{0,1}^s F_{\mathrm{para}}^a$ is

$$\begin{bmatrix} 0 & -(1+q^{1/2}(\chi(\varpi)+\chi(\varpi)^{-1})+q)q^2 \\ 1 & 2(\chi\sigma)(\varpi)q+(\sigma(\varpi)+\sigma(\varpi)^{-1})q^{3/2} \end{bmatrix}$$

and the matrix of $T_{1,0}^s$ on $(\chi \mathrm{St}_{\mathrm{GL}(2)} \rtimes \sigma)^{\mathrm{Kl}(\mathfrak{p})}$ with respect to the basis $F_{\mathrm{para}}^a, T_{0,1}^s F_{\mathrm{para}}^a$ is

$$\begin{bmatrix} -q & -q^2(q+1+q^{1/2}(\chi(\varpi)+\chi(\varpi)^{-1}))(\chi\sigma)(\varpi) \\ (\chi\sigma)(\varpi) & q(1+q^{1/2}(\chi(\varpi)+\chi(\varpi)^{-1})) \end{bmatrix}.$$

Similarly, the matrix of $T_{0,1}^s$ on $(\chi 1_{\mathrm{GL}(2)} \rtimes \sigma)^{\mathrm{Kl}(\mathfrak{p})}$ with respect to the basis $F_0^b, \theta F_0^b$ is

$$\begin{bmatrix} (\chi\sigma)(\varpi)(q+q^2)+(\sigma(\varpi)+\sigma(\varpi)^{-1})q^{3/2} & q^2+q^3+(\chi\sigma)(\varpi)(\sigma(\varpi)+\sigma(\varpi)^{-1})q^{5/2} \\ -1 & (\chi\sigma)(\varpi)(q^2-q) \end{bmatrix},$$

and the matrix of $T_{1,0}^s$ on $(\chi 1_{\mathrm{GL}(2)} \rtimes \sigma)^{\mathrm{Kl}(\mathfrak{p})}$ with respect to the basis $F_0^b, \theta F_0^b$ is

$$\begin{bmatrix} q^2+(\chi\sigma)(\varpi)(\sigma(\varpi)+\sigma(\varpi)^{-1})q^{5/2} & (\chi\sigma)(\varpi)(q^3+q^4)+(\sigma(\varpi)+\sigma(\varpi)^{-1})q^{7/2} \\ -(\chi\sigma)(\varpi)q & -q^2 \end{bmatrix}.$$

The assertions about characteristic polynomials follow by direct calculations. □

Lemma 9.4.5 *Let χ and σ be unramified characters of $F^\times$. Assume that $\chi\sigma^2 = 1$, $\chi \neq 1$, and $\chi \neq \nu^{-2}$. The vector space $(\chi \rtimes \sigma \mathrm{St}_{\mathrm{GSp}(2)})^{\mathrm{Kl}(\mathfrak{p})}$ is one-dimensional, and the vector space $(\chi \rtimes \sigma 1_{\mathrm{GSp}(2)})^{\mathrm{Kl}(\mathfrak{p})}$ is three-dimensional. We have*

$$p(T_{0,1}^s, (\chi \rtimes \sigma \mathrm{St}_{\mathrm{GSp}(2)})^{\mathrm{Kl}(\mathfrak{p})}, X) = X - q\left(\sigma(\varpi) + \sigma(\varpi)^{-1}\right),$$

$$p(T_{1,0}^s, (\chi \rtimes \sigma \mathrm{St}_{\mathrm{GSp}(2)})^{\mathrm{Kl}(\mathfrak{p})}, X) = X - q,$$

$$\begin{aligned} p(T_{0,1}^s, (\chi \rtimes \sigma 1_{\mathrm{GSp}(2)})^{\mathrm{Kl}(\mathfrak{p})}, X) &= X^3 - \left(\sigma(\varpi) + \sigma(\varpi)^{-1}\right) q(1+2q)X^2 \\ &\quad + \left(1 + 3q + (\chi(\varpi)+\chi(\varpi)^{-1})q\right) q^2(q+1)X \end{aligned}$$

$$- \left(\sigma(\varpi)+\sigma(\varpi)^{-1}\right) q^4(q+1)^2,$$

$$p(T^s_{1,0}, (\chi \rtimes \sigma 1_{\mathrm{GSp}(2)})^{\mathrm{Kl}(\mathfrak{p})}, X) = X^3 - \left(\left(\chi(\varpi)+\chi(\varpi)^{-1}\right) q^2 + q^3\right) X^2 + \left(\left(\chi(\varpi)+\chi(\varpi)^{-1}\right) q^5 + q^4\right) X - q^7.$$

Proof Since there is an exact sequence

$$0 \longrightarrow \chi \mathrm{St}_{\mathrm{GSp}(2)} \longrightarrow \nu \rtimes \nu^{-1/2}\chi \longrightarrow \chi 1_{\mathrm{GSp}(2)} \longrightarrow 0$$

of representations of $\mathrm{GSp}(2, F) = \mathrm{GL}(2, F)$, there is an exact sequence

$$0 \longrightarrow \chi \rtimes \sigma \mathrm{St}_{\mathrm{GSp}(2)} \longrightarrow \chi \rtimes (\nu \rtimes \nu^{-1/2}\sigma) \longrightarrow \chi \rtimes \sigma 1_{\mathrm{GSp}(2)} \longrightarrow 0$$

of representations of $\mathrm{GSp}(4, F)$. There is an isomorphism $\chi \rtimes (\nu \rtimes \nu^{-1/2}\sigma) \xrightarrow{\sim} \chi \times \nu \rtimes \nu^{-1/2}\sigma$ of $\mathrm{GSp}(4, F)$ representations given by $f \mapsto F_f$, where $F_f : \mathrm{GSp}(4, F) \to \mathbb{C}$ is defined by $F_f(g) = \big(f(g)\big)(\left[\begin{smallmatrix} 1 & \\ & 1 \end{smallmatrix}\right])$ for f in the standard model of $\chi \rtimes (\nu \rtimes \nu^{-1/2}\sigma)$ and $g \in \mathrm{GSp}(4, F)$. We thus have an exact sequence

$$0 \longrightarrow \chi \rtimes \sigma \mathrm{St}_{\mathrm{GSp}(2)} \longrightarrow \chi \times \nu \rtimes \nu^{-1/2}\sigma \longrightarrow \chi \rtimes \sigma 1_{\mathrm{GSp}(2)} \longrightarrow 0 \qquad (9.51)$$

of $\mathrm{GSp}(4, F)$ representations. Let v_0 be a newform in the space of $\sigma \mathrm{St}_{\mathrm{GSp}(2)} \subset \nu \rtimes \nu^{-1/2}\sigma$ (again, see [110] for an account of newforms for representations of $\mathrm{GL}(2, F)$). By (26) of [110] we may assume that, as an element of $\nu \rtimes \nu^{-1/2}\sigma = \nu^{1/2}\sigma \times \nu^{1/2}\sigma$, $v_0(\left[\begin{smallmatrix} 1 & \\ & 1 \end{smallmatrix}\right]) = q$ and $v_0(\left[\begin{smallmatrix} & -1 \\ 1 & \end{smallmatrix}\right]) = -1$. Let

$$L_1 = \begin{bmatrix} 1 & & & \\ \varpi & 1 & & \\ & & 1 & \\ & & -\varpi & 1 \end{bmatrix},$$

as on p. 153 of [105]. By the proof of Theorem 5.4.2 of [105], the function $F^a_{\mathrm{para}} : \mathrm{GSp}(4, F) \to \sigma \mathrm{St}_{\mathrm{GSp}(2)}$ defined by $F^a_{\mathrm{para}}(g) = 0$ for $g \notin PL_1\mathrm{K}(\mathfrak{p}^2)$ and by

$$F^a_{\mathrm{para}}(g) = |y^2 \det(A)^{-1}| \chi(y) (\sigma \mathrm{St}_{\mathrm{GSp}(2)})(A) v_0$$

for $g = pL_1k$, $p = \begin{bmatrix} y & * & * \\ & A & * \\ & & y^{-1}\det(A) \end{bmatrix}$, $y \in F^\times$, $A \in \mathrm{GL}(2, F)$, and $k \in \mathrm{K}(\mathfrak{p}^2)$, is a well-defined non-zero element in $(\chi \rtimes \sigma \mathrm{St}_{\mathrm{GSp}(2)})^{\mathrm{K}(\mathfrak{p}^2)}$. By definition, the vector $\sigma_1 F^a_{\mathrm{para}}$ is contained in $(\chi \rtimes \sigma \mathrm{St}_{\mathrm{GSp}(2)})^{\mathrm{Kl}(\mathfrak{p})}$. Moreover, a calculation using (3.60) and (2.2) shows that $(\sigma_1 F^a_{\mathrm{para}})(s_1) \neq 0$, so that $(\chi \rtimes \sigma \mathrm{St}_{\mathrm{GSp}(2)})^{\mathrm{Kl}(\mathfrak{p})}$ is at least one-dimensional.

Next, there is also an exact sequence

$$0 \longrightarrow \chi \rtimes \sigma 1_{\mathrm{GSp}(2)} \longrightarrow \chi \times \nu^{-1} \rtimes \nu^{1/2}\sigma \longrightarrow \chi \rtimes \sigma \mathrm{St}_{\mathrm{GSp}(2)} \longrightarrow 0 \qquad (9.52)$$

of $\mathrm{GSp}(4, F)$ representations. The representation $\chi \rtimes \sigma 1_{\mathrm{GSp}(2)}$ is unramified by Table A.12 of [105]; let F_0^b be a non-zero element of $(\chi \rtimes \sigma 1_{\mathrm{GSp}(2)})^{\mathrm{GSp}(4,\mathfrak{o})}$. We regard F_0^b as an element of $\chi \times \nu^{-1} \rtimes \nu^{1/2}\sigma$ via the inclusion from (9.52). By Lemma 9.2.5 and Theorem 2.3.1, the vectors F_0^b, θF_0^b, and $\theta' F_0^b$ are linearly independent; the application of Theorem 2.3.1 uses (2) of Lemma 9.4.3 and the assumption $\chi \neq \nu^{-2}$. It follows that $(\chi \rtimes \sigma 1_{\mathrm{GSp}(2)})^{\mathrm{Kl}(\mathfrak{p})}$ is at least three-dimensional. Since $(\chi \times \nu^{-1} \rtimes \nu^{1/2}\sigma)^{\mathrm{Kl}(\mathfrak{p})}$ is four-dimensional by Lemma 9.2.4, we see from (9.52) that $(\chi \rtimes \sigma \mathrm{St}_{\mathrm{GSp}(2)})^{\mathrm{Kl}(\mathfrak{p})}$ is one-dimensional and that $(\chi \rtimes \sigma \mathrm{St}_{\mathrm{GSp}(2)})^{\mathrm{Kl}(\mathfrak{p})}$ is three-dimensional. The vectors F_0^b, θF_0^b, and $\theta' F_0^b$ are thus a basis for $(\chi \rtimes \sigma \mathrm{St}_{\mathrm{GSp}(2)})^{\mathrm{Kl}(\mathfrak{p})}$; let W be the subspace of $\chi \times \nu^{-1} \rtimes \nu^{1/2}\sigma$ spanned by these vectors, regarded as elements of $\chi \times \nu^{-1} \rtimes \nu^{1/2}\sigma$. Using Lemmas 9.2.3 and 9.3.2, it now follows that the matrix of $T_{0,1}^s$ in the basis F_0^b, θF_0^b, and $\theta' F_0^b$ for $(\chi \rtimes \sigma \mathrm{St}_{\mathrm{GSp}(2)})^{\mathrm{Kl}(\mathfrak{p})}$ is

$$\begin{bmatrix} q(q+1)(\sigma(\varpi)+\sigma(\varpi)^{-1}) & 2q^2(q+1) & q^2(q+1)(\sigma(\varpi)+\sigma(\varpi)^{-1}) \\ -1 & q^2(\sigma(\varpi)+\sigma(\varpi)^{-1}) & (q-1)q \\ 0 & -q(q+1) & 0 \end{bmatrix},$$

and the matrix of $T_{1,0}^s$ in the same basis is

$$\begin{bmatrix} q^2(q+\chi(\varpi)+\chi(\varpi)^{-1}) & q^3(q+1)(\sigma(\varpi)+\sigma(\varpi)^{-1}) & q^3(2q+\chi(\varpi)+\chi(\varpi)^{-1}) \\ -q(\sigma(\varpi)+\sigma(\varpi)^{-1}) & -q^2 & -q^2(\sigma(\varpi)+\sigma(\varpi)^{-1}) \\ q & 0 & q^2 \end{bmatrix}.$$

The formulas for $p(T_{0,1}^s, (\chi 1_{\mathrm{GL}(2)} \rtimes \sigma)^{\mathrm{Kl}(\mathfrak{p})}, X)$ and $p(T_{1,0}^s, (\chi 1_{\mathrm{GL}(2)} \rtimes \sigma)^{\mathrm{Kl}(\mathfrak{p})}, X)$ are calculated using these matrices. In view of Lemma 9.4.1 and (9.52), the formulas for $p(T_{0,1}^s, (\mathrm{St}_{\mathrm{GL}(2)} \rtimes \sigma)^{\mathrm{Kl}(\mathfrak{p})}, X)$ and $p(T_{1,0}^s, (\mathrm{St}_{\mathrm{GL}(2)} \rtimes \sigma)^{\mathrm{Kl}(\mathfrak{p})}, X)$ are obtained by dividing

$$p(T_{0,1}^s, (\chi \times \nu^{-1} \rtimes \nu^{1/2}\sigma)^{\mathrm{Kl}(\mathfrak{p})}, X) \quad \text{and} \quad p(T_{1,0}^s, (\chi \times \nu^{-1} \rtimes \nu^{1/2}\sigma)^{\mathrm{Kl}(\mathfrak{p})}, X)$$

from Lemma 9.4.2 by

$$p(T_{0,1}^s, (\chi 1_{\mathrm{GL}(2)} \rtimes \sigma)^{\mathrm{Kl}(\mathfrak{p})}, X) \quad \text{and} \quad p(T_{1,0}^s, (\chi 1_{\mathrm{GL}(2)} \rtimes \sigma)^{\mathrm{Kl}(\mathfrak{p})}, X)$$

respectively. □

Theorem 9.4.6 *Let (π, V) be an irreducible, admissible representation of the group* $\mathrm{GSp}(4, F)$ *with trivial central character. The space $V_s(1)$ is non-zero if and only if π is one of the representations in Table A.5. If π is one of the representations in Table A.5, then π is paramodular, and the characteristic polynomials of $T_{0,1}^s$*

and $T^s_{1,0}$ acting on $V_s(1)$ are as in Tables A.5 and A.6; in the latter table, λ_π, μ_π, and ε_π are the eigenvalues of the paramodular Hecke operators $T_{0,1}$ and $T_{1,0}$, and the Atkin-Lehner operator $\pi(u_{N_\pi})$, respectively, on a paramodular newform of π.

Proof Assume that $V_s(1) = V^{\mathrm{Kl}(\mathfrak{p})}$ is non-zero. Then V^I is non-zero, i.e., V is Iwahori-spherical. By a theorem of Casselman (see Sect. 3.6 of [26] for a summary), π is an irreducible subquotient of $\chi_1 \times \chi_2 \rtimes \sigma$ for some unramified characters χ_1, χ_2, and σ of $F^\times$ such that $\chi_1\chi_2\sigma^2 = 1$. Therefore, π is an entry in Table 3 of [111]; examining this table, and using the assumption $V_s(1) \neq 0$, we now see that π is one of the representations in Table A.5.

Conversely, assume that π is one of the representations in Table A.5. Then π is an entry in Table 3 of [111], and by inspection of this table we see that $V_s(1) \neq 0$.

Assume that π is an entry in Table A.5. Then π is paramodular by Table A.12 of [105].

Now let π be one of representations in Table A.5; we will prove the characteristic polynomials $p(T^s_{0,1}, V_s(1), X)$ and $p(T^s_{1,0}, V_s(1), X)$ are as in Table A.5. If π belongs to Group I, II or III, then the formulas for the characteristic polynomials follow from Lemma 9.4.2, Lemma 9.4.4, and Lemma 9.4.5.

Assume that π belongs to group IVb, so that $\pi = L(\nu^2, \nu^{-1}\sigma \mathrm{St}_{\mathrm{GSp}(2)})$ for some unramified character σ of $F^\times$ with $\sigma^2 = 1$. By (2.14), there is an exact sequence

$$0 \to (\sigma \mathrm{St}_{\mathrm{GSp}(4)})^{\mathrm{Kl}(\mathfrak{p})} \to (\nu^2 \rtimes \nu^{-1}\sigma \mathrm{St}_{\mathrm{GSp}(2)})^{\mathrm{Kl}(\mathfrak{p})} \to L(\nu^2, \nu^{-1}\mathrm{St}_{\mathrm{GSp}(2)})^{\mathrm{Kl}(\mathfrak{p})} \to 0.$$

By Table 3 of [111], $(\sigma \mathrm{St}_{\mathrm{GSp}(4)})^{\mathrm{Kl}(\mathfrak{p})} = 0$; hence

$$p(T^s_{0,1}, V_s(1), X) = p(T^s_{0,1}, (\nu^2 \rtimes \nu^{-1}\sigma \mathrm{St}_{\mathrm{GSp}(2)})^{\mathrm{Kl}(\mathfrak{p})}, X),$$
$$p(T^s_{1,0}, V_s(1), X) = p(T^s_{1,0}, (\nu^2 \rtimes \nu^{-1}\sigma \mathrm{St}_{\mathrm{GSp}(2)})^{\mathrm{Kl}(\mathfrak{p})}, X).$$

The formulas for $p(T^s_{0,1}, V_s(1), X)$ and $p(T^s_{1,0}, V_s(1), X)$ follow from Lemma 9.4.5.

If π belongs to group IVc, then the argument is similar to the IVb case.

If π belongs to group IVd, so that $\pi = \sigma 1_{\mathrm{GSp}(4)}$ for some unramified character σ of $F^\times$ with $\sigma^2 = 1$, then the formulas for the characteristic polynomials follow by direct calculations from the involved definitions.

Assume that π belongs to group Vb, so that $\pi = L(\nu^{\frac{1}{2}}\xi \mathrm{St}_{\mathrm{GL}(2)}, \nu^{-\frac{1}{2}}\sigma)$ with σ and ξ unramified characters of $F^\times$, $\xi \neq 1$, $\xi^2 = 1$, and $\sigma^2 = 1$. By Table 3 of [111], $V(1)$ and $V_s(1)$ are both one-dimensional, and $V(0) = 0$. It follows that $V(1) = V_s(1)$ and that the paramodular level N_π of π is 1; let v be a non-zero element of $V(1) = V_s(1)$. Let $\lambda, \mu \in \mathbb{C}$ be such that $T_{0,1}v = \lambda v$ and $T_{1,0}v = \mu v$. By Table A.9 of [105] we have $\lambda = \sigma(\varpi)(q^2 - 1)$ and $\mu = -q(q+1)$. Let $c_{0,1}, c_{1,0} \in \mathbb{C}$ be such that $T^s_{0,1}v = c_{0,1}v$ and $T^s_{1,0}v = c_{1,0}v$. By Lemma 3.8.3 we obtain $c_{0,1} = \sigma(\varpi)(q^2 - q)$ and $c_{1,0} = -q^2$. The formulas for $p(T^s_{0,1}, V_s(1), X)$ and $p(T^s_{1,0}, V_s(1), X)$ now follow immediately.

If π belongs to group Vc, then the computation is similar to the Vb case.

If π belongs to group Va or Vd, then the computation is similar to the IVb case, using (2.17), Lemma 9.4.1, Lemma 9.4.4, and the Vb or Vc case.

If π belongs to group VId, then the computation is similar to the IVb case, using (2.20), Lemma 9.4.1, and Lemma 9.4.4 (note that $\tau(T, \nu^{-\frac{1}{2}}\sigma)^{\mathrm{Kl}(\mathfrak{p})} = 0$ by Table 3 of [111]).

If π belongs to group VIc, then the computation is similar to the Vb case (again, $V(1)$ and $V_s(1)$ are both one-dimensional by Table 3 of [111]).

If π belongs to group VIa, then the computation is similar to the IVb case, using (2.20), Lemma 9.4.1, Lemma 9.4.4, and the VIc case; this completes the verification of the characteristic polynomials in Table A.5.

Finally, using Table A.9 of [105], it is straightforward to verify that the characteristic polynomials of Table A.5 can be written in terms of the paramodular eigenvalues λ_π, μ_π, and ε_π as in Table A.6. □

Part II
Siegel Modular Forms

Chapter 10
Background on Siegel Modular Forms

The remainder of this text explores applications of the local theory developed in the first part of this work to Siegel modular forms of degree two. In this chapter, we recall some essential definitions. In the next chapter we translate the operators on stable Klingen vectors defined in previous chapters to modular forms.

10.1 Basic Definitions

The Symplectic Similitude Group

Let R be a commutative ring with identity 1. In this second part of this work we let

$$J = \begin{bmatrix} & & 1 & \\ & & & 1 \\ -1 & & & \\ & -1 & & \end{bmatrix}, \tag{10.1}$$

and we define $\mathrm{GSp}(4, R)$ and $\mathrm{Sp}(4, R)$ using J. Thus, $\mathrm{GSp}(4, R)$ is defined to be the set of g in $\mathrm{GL}(4, R)$ such that ${}^t g J g = \lambda J$ for some $\lambda \in R^\times$. If $g \in \mathrm{GSp}(4, R)$, then the unit $\lambda \in R^\times$ such that ${}^t g J g = \lambda J$ is unique, and will be denoted by $\lambda(g)$. The set $\mathrm{GSp}(4, R)$ is a subgroup of $\mathrm{GL}(4, R)$. If $\left[\begin{smallmatrix} A & B \\ C & D \end{smallmatrix}\right] \in \mathrm{GSp}(4, R)$, then

$$g^{-1} = \lambda(g)^{-1} \begin{bmatrix} {}^tD & -{}^tB \\ -{}^tC & {}^tA \end{bmatrix}. \tag{10.2}$$

J. Johnson-Leung et al., *Stable Klingen Vectors and Paramodular Newforms*,
Lecture Notes in Mathematics 2342, https://doi.org/10.1007/978-3-031-45177-5_10

We define Sp(4, R) to be the subgroup of $g \in$ GSp(4, R) such that $\lambda(g) = 1$, i.e., ${}^tgJg = J$. In Part I of this work we defined GSp(4) and Sp(4) with respect to

$$\begin{bmatrix} &&&1\\ &&1&\\ &-1&&\\ -1&&& \end{bmatrix}. \tag{10.3}$$

We will convert between Part I and Part II of this work by conjugating by the matrix

$$\begin{bmatrix} &1&&\\ 1&&&\\ &&1&\\ &&&1 \end{bmatrix}. \tag{10.4}$$

This matrix is its own inverse, and the conjugate of GSp(4, R) as defined with respect to (10.1) is GSp(4, R) as defined with respect to (10.3). When $R = \mathbb{R}$, we also define GSp(4, $\mathbb{R}$)$^+$ as the subgroup of $g \in$ GSp(4, $\mathbb{R}$) such that $\lambda(g) > 0$.

The Siegel Upper Half-Space

We define $\mathcal{H}_2$ to be the subset of M(2, $\mathbb{C}$) consisting of the matrices $Z = X + iY$ with $X, Y \in$ M(2, $\mathbb{R}$) such that ${}^tX = X$, ${}^tY = Y$, and Y is positive-definite. We refer to $\mathcal{H}_2$ as the *Siegel upper half-space of degree* 2. The set $\mathcal{H}_2$ is a simply connected, open subset of Sym(2, $\mathbb{C}$), the $\mathbb{C}$ vector space of 2×2 symmetric matrices with entries from $\mathbb{C}$. The group GSp(4, $\mathbb{R}$)$^+$ acts on $\mathcal{H}_2$ via the formula

$$g\langle Z\rangle = (AZ + B)(CZ + D)^{-1}$$

for $g = \left[\begin{smallmatrix} A & B \\ C & D \end{smallmatrix}\right] \in$ GSp(4, $\mathbb{R}$)$^+$ and $Z \in \mathcal{H}_2$; in particular, $CZ + D \in$ GL(2, $\mathbb{C}$). The action of GSp(4, $\mathbb{R}$)$^+$ on $\mathcal{H}_2$ is transitive. Define j : GSp(4, $\mathbb{R}$)$^+ \times \mathcal{H}_2 \to \mathbb{C}^\times$ by $j(g, Z) = \det(CZ + D)$ for $g = \left[\begin{smallmatrix} A & B \\ C & D \end{smallmatrix}\right] \in$ GSp(4, $\mathbb{R}$)$^+$ and $Z \in \mathcal{H}_2$. The function j satisfies the following cocycle condition

$$j(g_1g_2, Z) = j(g_1, g_2\langle Z\rangle)j(g_2, Z) \tag{10.5}$$

for $Z \in \mathcal{H}_2$ and $g_1, g_2 \in$ GSp(4, $\mathbb{R}$)$^+$. Next, let $F : \mathcal{H}_2 \to \mathbb{C}$ be a function, and let k be an integer such that $k > 0$. For $g \in$ GSp(4, $\mathbb{R}$)$^+$ we define $F\big|_k g : \mathcal{H}_2 \to \mathbb{C}$ by

$$(F\big|_k g)(Z) = \lambda(g)^k j(g, Z)^{-k} F(g\langle Z\rangle)$$

for $Z \in \mathcal{H}_2$. If $g_1, g_2 \in$ GSp(4, $\mathbb{R}$)$^+$, then $(F\big|_k g_1)\big|_k g_2 = F\big|_k g_1g_2$. We will abbreviate

$$I = \begin{bmatrix} i & \\ & i \end{bmatrix}. \tag{10.6}$$

Additional Notation

If A and B are square matrices of the same size, then we define $A[B] = {}^tBAB$, and we define $\mathrm{Tr}(A)$ to be the trace of A. Let N be an integer such that $N > 0$ and let p be a prime number. There exist unique integers n and M such that $n \geq 0$, $M > 0$, p and M are relatively prime, and $N = Mp^n$; we define $v_p(N) = n$. Given a prime number in $\mathbb{Z}$ denoted by a letter of the italic roman font, we will denote the prime ideal in the ring of integers of the corresponding local field by the same letter in the Fraktur font. For example, if p is a prime number in $\mathbb{Z}$, then the prime ideal $p\mathbb{Z}_p$ of $\mathbb{Z}_p$ will be denoted by $\mathfrak{p}$. Let p be a prime of $\mathbb{Z}$, let n be an integer such that $n \geq 0$, and let (π, V) be a smooth representation of $\mathrm{GSp}(4, \mathbb{Q}_p)$ for which the center acts trivially. In this second part, we will write

$$V(\mathfrak{p}^n) = \{v \in V \mid \pi(k)v = v \text{ for } k \in \mathrm{K}(\mathfrak{p}^n)\} \tag{10.7}$$

and

$$V_s(\mathfrak{p}^n) = \{v \in V \mid \pi(k)v = v \text{ for } k \in \mathrm{K}_s(\mathfrak{p}^n)\}. \tag{10.8}$$

Here, $\mathrm{K}(\mathfrak{p}^n)$ is the local paramodular group defined in (2.25), and $\mathrm{K}_s(\mathfrak{p}^n)$ is the local stable Klingen subgroup defined in (3.2) (with the change in the definition of $\mathrm{GSp}(4)$ mentioned at the beginning of this section). Previously, in Part I, the subspaces $V(\mathfrak{p}^n)$ and $V_s(\mathfrak{p}^n)$ were denoted by $V(n)$ and $V_s(n)$, respectively; in this second part we need the notation to also reflect the choice of prime. We will denote the adeles of $\mathbb{Q}$ by $\mathbb{A}$, and we denote the finite adeles of $\mathbb{Q}$ by $\mathbb{A}_{\mathrm{fin}}$. Let $g \in \mathrm{GSp}(4, \mathbb{Q})$ and let v be a place of $\mathbb{Q}$. Then g_v will denote the element of $\mathrm{GSp}(4, \mathbb{A})$ that is g at the place v and 1 at all other places.

10.2 Modular Forms

Congruence Subgroups

Let N be an integer such that $N > 0$. We define

$$\Gamma_0'(N) = \mathrm{Sp}(4, \mathbb{Q}) \cap \begin{bmatrix} \mathbb{Z} & N\mathbb{Z} & \mathbb{Z} & \mathbb{Z} \\ \mathbb{Z} & \mathbb{Z} & \mathbb{Z} & \mathbb{Z} \\ \mathbb{Z} & N\mathbb{Z} & \mathbb{Z} & \mathbb{Z} \\ N\mathbb{Z} & N\mathbb{Z} & N\mathbb{Z} & \mathbb{Z} \end{bmatrix}, \tag{10.9}$$

$$\mathrm{K}(N) = \mathrm{Sp}(4, \mathbb{Q}) \cap \begin{bmatrix} \mathbb{Z} & N\mathbb{Z} & \mathbb{Z} & \mathbb{Z} \\ \mathbb{Z} & \mathbb{Z} & \mathbb{Z} & N^{-1}\mathbb{Z} \\ \mathbb{Z} & N\mathbb{Z} & \mathbb{Z} & \mathbb{Z} \\ N\mathbb{Z} & N\mathbb{Z} & N\mathbb{Z} & \mathbb{Z} \end{bmatrix}, \tag{10.10}$$

and

$$\mathrm{K}_s(N) = \mathrm{Sp}(4, \mathbb{Q}) \cap \begin{bmatrix} \mathbb{Z} & N\mathbb{Z} & \mathbb{Z} & \mathbb{Z} \\ \mathbb{Z} & \mathbb{Z} & \mathbb{Z} & N_s^{-1}\mathbb{Z} \\ \mathbb{Z} & N\mathbb{Z} & \mathbb{Z} & \mathbb{Z} \\ N\mathbb{Z} & N\mathbb{Z} & N\mathbb{Z} & \mathbb{Z} \end{bmatrix}, \tag{10.11}$$

where

$$N_s = N \prod_{p|N} \frac{1}{p}. \tag{10.12}$$

In (10.12) $p \mid N$ means that p runs over the primes dividing N. Using (10.2) it is easy to verify that $\Gamma_0'(N)$, $\mathrm{K}(N)$, and $\mathrm{K}_s(N)$ are subgroups of $\mathrm{Sp}(4, \mathbb{Q})$. The group $\Gamma_0'(N)$ is the *Klingen congruence subgroup* of level N, the group $\mathrm{K}(N)$ is the *paramodular congruence subgroup* of level N, and the group $\mathrm{K}_s(N)$ is the *stable Klingen congruence subgroup* of level N. We note that if N is a square-free integer, then $\mathrm{K}_s(N) = \Gamma_0'(N)$. The groups $\Gamma_0'(N)$, $\mathrm{K}(N)$, and $\mathrm{K}_s(N)$ are commensurable with $\mathrm{Sp}(4, \mathbb{Z})$ (see 6.3 of Chap. II on p. 126 of [42] for the definition of commensurable).

Siegel Modular Forms

Let k be an integer such that $k > 0$, and let Γ be a subgroup of $\mathrm{Sp}(4, \mathbb{Q})$ commensurable with $\mathrm{Sp}(4, \mathbb{Z})$. A *Siegel modular form of weight k with respect to* Γ is a holomorphic function $F : \mathcal{H}_2 \to \mathbb{C}$ such that $F\big|_k \gamma = F$ for $\gamma \in \Gamma$. If $\Gamma = \mathrm{K}(N)$ for some positive integer N, then F is often referred to as a *paramodular form*; if $\Gamma = \mathrm{K}_s(N)$ for some positive integer N, then we call F a *stable Klingen form*. We denote by $M_k(\Gamma)$ the $\mathbb{C}$ vector space of all Siegel modular forms of weight k with respect to Γ. If $F \in M_k(\Gamma)$, then we say that F is a *cusp form* if F satisfies 6.9 of Chap. II on p. 129 of [42]; we let $S_k(\Gamma)$ be the subspace of $F \in M_k(\Gamma)$ such that F is a cusp form. We define two sets of positive semi-definite symmetric matrices,

$$A(N) = \{\left[\begin{smallmatrix} \alpha & \beta \\ \beta & \gamma \end{smallmatrix}\right] \mid \alpha, 2\beta, \gamma \in \mathbb{Z},\ N \mid \gamma,\ \alpha\gamma - \beta^2 \geq 0,\ \alpha \geq 0,\ \gamma \geq 0\}, \tag{10.13}$$

$$B(N) = \{\left[\begin{smallmatrix} \alpha & \beta \\ \beta & \gamma \end{smallmatrix}\right] \mid \alpha, 2\beta, \gamma \in \mathbb{Z},\ N_s \mid \gamma,\ \alpha\gamma - \beta^2 \geq 0,\ \alpha \geq 0,\ \gamma \geq 0\}. \tag{10.14}$$

We denote the subset of positive definite elements of $A(N)$ and $B(N)$ by $A(N)^+$ and $B(N)^+$, respectively. We note that $A(N) \subset B(N) \subset A(1)$. Define $\Gamma_0(N)_\pm$ to be the subgroup of $\left[\begin{smallmatrix} a & b \\ c & d \end{smallmatrix}\right] \in \mathrm{GL}(2, \mathbb{Z})$ such that $c \equiv 0 \pmod{N}$. The group $\Gamma_0(N)_\pm$ acts on $A(N)$ and $B(N)$ via the definition

$$g \cdot S = gS\,{}^{\mathrm{t}}g = S[{}^{\mathrm{t}}g] \tag{10.15}$$

for $g \in \Gamma_0(N)_\pm$ and $S \in A(N)$ or $S \in B(N)$. This action preserves the subsets $A(N)^+$ and $B(N)^+$. If $F \in M_k(\mathrm{K}(N))$, then F has a *Fourier expansion* of the form

$$F(Z) = \sum_{S \in A(N)} a(S) e^{2\pi i \mathrm{Tr}(SZ)}, \tag{10.16}$$

and if $F \in M_k(\mathrm{K}_s(N))$, then F has a Fourier expansion of the form

$$F(Z) = \sum_{S \in B(N)} a(S) e^{2\pi i \mathrm{Tr}(SZ)}. \tag{10.17}$$

If F is a cusp form, then the sum is over only $A(N)^+$ in (10.16) and over only $B(N)^+$ in (10.17). Let F be in $M_k(\mathrm{K}(N))$ or $M_k(\mathrm{K}_s(N))$ with Fourier expansion as in (10.16) or (10.17). Since $F\big|_k \left[\begin{smallmatrix} {}^{\mathrm{t}}g^{-1} & \\ & g \end{smallmatrix}\right] = F$ for $g \in \Gamma_0(N)_\pm$, it follows that if $F \in M_k(\mathrm{K}(N))$ (respectively, $F \in M_k(\mathrm{K}_s(N))$), then for all $S \in A(N)$ (respectively, $S \in B(N)$) we have

$$a(g \cdot S) = a(gS\,{}^{\mathrm{t}}g) = \det(g)^k a(S) \tag{10.18}$$

for all $g \in \Gamma_0(N)_\pm$.

Fourier-Jacobi Expansions

Let k be an integer such that $k > 0$. The elements of $M_k(\mathrm{K}(N))$ and $M_k(\mathrm{K}_s(N))$ have Fourier-Jacobi expansions. To explain this, let m be an integer such that $m \geq 0$, let $\mathcal{H}_1$ be the complex upper half-plane, and let $f : \mathcal{H}_1 \times \mathbb{C} \to \mathbb{C}$ be a function. For $\lambda, \mu, \kappa \in \mathbb{R}$ define

$$[\lambda, \mu, \kappa] = \begin{bmatrix} 1 & & & \mu \\ \lambda & 1 & \mu & \kappa \\ & & 1 & -\lambda \\ & & & 1 \end{bmatrix}.$$

The elements $[\lambda, \mu, \kappa]$ for $\lambda, \mu, \kappa \in \mathbb{R}$ form a group $H(\mathbb{R})$ under multiplication of matrices called the *Heisenberg group*. The group law is given by

$$[\lambda, \mu, \kappa] \cdot [\lambda', \mu', \kappa'] = [\lambda + \lambda', \mu + \mu', \kappa + \kappa' + \lambda\mu' - \mu\lambda']$$

for $\lambda, \lambda', \mu, \mu', \kappa, \kappa' \in \mathbb{R}$. For $\left[\begin{smallmatrix} a & b \\ c & d \end{smallmatrix}\right] \in \mathrm{SL}(2, \mathbb{R})$, $[\lambda, \mu, \kappa] \in H(\mathbb{R})$, and $(\tau, z) \in \mathcal{H}_1 \times \mathbb{C}$ we define

$$(f\big|_{k,m} \left[\begin{smallmatrix} a & b \\ c & d \end{smallmatrix}\right])(\tau, z) = (c\tau + d)^{-k} e^{2\pi i m(\frac{-cz^2}{c\tau+d})} f(\frac{a\tau + b}{c\tau + d}, \frac{z}{c\tau + d}) \tag{10.19}$$

and

$$(f\big|_m [\lambda, \mu, \kappa])(\tau, z) = e^{2\pi i m(\lambda^2\tau + 2\lambda z + \lambda\mu + \kappa)} f(\tau, z + \lambda\tau + \mu). \tag{10.20}$$

Calculations show that $(f\big|_{k,m} g_1)\big|_{k,m} g_2 = f\big|_{k,m} g_1 g_2$ for $g_1, g_2 \in \mathrm{SL}(2, \mathbb{R})$ and $(f\big|_m h_1)\big|_m h_2 = f\big|_m h_1 h_2$ for $h_1, h_2 \in H(\mathbb{R})$. Also, if $g = \left[\begin{smallmatrix} a & b \\ c & d \end{smallmatrix}\right] \in \mathrm{SL}(2, \mathbb{R})$ and $[\lambda, \mu, \kappa] \in H(\mathbb{R})$, then we have the following commutation rule

$$(f\big|_m [\lambda, \mu, \kappa])\big|_{k,m} \left[\begin{smallmatrix} a & b \\ c & d \end{smallmatrix}\right] = (f\big|_{k,m} \left[\begin{smallmatrix} a & b \\ c & d \end{smallmatrix}\right])\big|_m [\lambda a + \mu c, \lambda b + \mu d, \kappa]. \tag{10.21}$$

We say that f is a *Jacobi form of weight k and index m on* $\mathrm{SL}(2, \mathbb{Z})$ if f is holomorphic, if $f\big|_{k,m} g = f$ for $g \in \mathrm{SL}(2, \mathbb{Z})$, $f\big|_m h = f$ for $h \in H(\mathbb{Z})$, and f has an absolutely convergent Fourier expansion of the form

$$f(\tau, z) = \sum_{n=0}^{\infty} \sum_{\substack{r \in \mathbb{Z} \\ r^2 \le 4nm}} c(n, r) e^{2\pi i(n\tau + rz)}.$$

We say that f is a *cusp form* if $c(n, r) = 0$ for $r^2 = 4nm$. See [38], p. 1 and p. 9. We denote the space of Jacobi forms of weight k and index m by $J_{k,m}$, and its subspace of cusp forms by $J_{k,m}^{\mathrm{cusp}}$. If f is a Jacobi form of weight k and index m on $\mathrm{SL}(2, \mathbb{Z})$, then

$$f(\tau + n_1, z + n_2) = f(\tau, z) \quad \text{and} \quad f(\tau, -z) = (-1)^k f(\tau, z) \tag{10.22}$$

for $n_1, n_2 \in \mathbb{Z}$, $\tau \in \mathcal{H}_1$, and $z \in \mathbb{C}$. Let $F \in M_k(\mathrm{K}_s(N))$ or $F \in M_k(\mathrm{K}(N))$. Then F has a *Fourier-Jacobi expansion*

$$F(Z) = \sum_{m=0}^{\infty} f_m(\tau, z) e^{2\pi i m\tau'}, \qquad Z = \left[\begin{smallmatrix} \tau & z \\ z & \tau' \end{smallmatrix}\right] \in \mathcal{H}_2, \tag{10.23}$$

where f_m is a Jacobi form of weight k and index m on $\mathrm{SL}(2, \mathbb{Z})$; if F is a cusp form, then $f_0 = 0$. Suppose that $F \in M_k(\mathrm{K}(N))$. Then it is easy to see that

$$f_m \neq 0 \implies N \mid m. \tag{10.24}$$

Suppose that $M \in M_k(\mathrm{K}_s(N))$. Then, similarly,

$$f_m \neq 0 \quad \Longrightarrow \quad N_s \mid m. \tag{10.25}$$

Here, N_s is as in (10.12).

Adelic Automorphic Forms

To connect the representation theory of Part I to Siegel modular forms we need to define a certain representation of $\mathrm{GSp}(4, \mathbb{A}_{\mathrm{fin}})$. Let

$$K_\infty = \{\left[\begin{smallmatrix} A & B \\ -B & A \end{smallmatrix}\right] \in \mathrm{GL}(4, \mathbb{R}) \mid {}^tAA + {}^tBB = 1, {}^tAB = {}^tBA\}.$$

Then K_∞ is a maximal compact subgroup of $\mathrm{Sp}(4, \mathbb{R})$ and is the stabilizer in $\mathrm{GSp}(4, \mathbb{R})^+$ of I. The function from K_∞ to $\mathbb{C}^\times$ that sends $k \in K_\infty$ to $j(k, I)$ is a character of K_∞ by the cocycle condition (10.5). Let k be an integer such that $k > 0$. We define $\mathcal{A}_k$ to be the set of all continuous functions $\Phi : \mathrm{GSp}(4, \mathbb{A}) \to \mathbb{C}$ such that

(1) $\Phi(\rho g) = \Phi(g)$ for all $\rho \in \mathrm{GSp}(4, \mathbb{Q})$ and $g \in \mathrm{GSp}(4, \mathbb{A})$;
(2) $\Phi(gz) = \Phi(g)$ for all $z \in \mathbb{I}$ and $g \in \mathrm{GSp}(4, \mathbb{A})$;
(3) for some compact, open subgroup K of $\mathrm{GSp}(4, \mathbb{A}_{\mathrm{fin}})$ we have $\Phi(g\kappa) = \Phi(g)$ for all $\kappa \in K$ and $g \in \mathrm{GSp}(4, \mathbb{A})$;
(4) $\Phi(g\kappa) = j(\kappa, I)^{-k}\Phi(g)$ for all $\kappa \in K_\infty$ and $g \in \mathrm{GSp}(4, \mathbb{A})$;
(5) For any $g_{\mathrm{fin}} \in \mathrm{GSp}(4, \mathbb{A}_{\mathrm{fin}})$, the function $\mathrm{GSp}(4, \mathbb{R})^+ \to \mathbb{C}$ defined by $g \mapsto \Phi(g_{\mathrm{fin}}g)$ is smooth and is annihilated by $\mathfrak{p}_\mathbb{C}^-$ (we refer to Sect. 3.5 of [6] for the definition of $\mathfrak{p}_\mathbb{C}^-$).

Then $\mathcal{A}_k$ is a complex vector space under addition of functions. Moreover, it is evident that $\mathcal{A}_k$ is a smooth representation of $\mathrm{GSp}(4, \mathbb{A}_{\mathrm{fin}})$ under the right translation action. We let $\mathcal{A}_k^\circ$ be the subspace of $\Phi \in \mathcal{A}_k$ that satisfy the following additional condition:

(6) For any proper parabolic subgroup P of $\mathrm{GSp}(4)$ and $g \in \mathrm{GSp}(4, \mathbb{A})$ we have

$$\int\limits_{N_P(\mathbb{Q})\backslash N_P(\mathbb{A})} \Phi(ng)\,dn = 0;$$

here, N_P is the unipotent radical of P.

We refer to elements of $\mathcal{A}_k^\circ$ as cusp forms. The subspace $\mathcal{A}_k^\circ$ is closed under the action of $\mathrm{GSp}(4, \mathbb{A}_{\mathrm{fin}})$.

Next, let $\{K_p\}_p$, where p runs over the primes of $\mathbb{Z}$, be a family of compact, open subgroups of $\mathrm{GSp}(4, \mathbb{Q}_p)$. We will say that $\{K_p\}_p$ is an *admissible family* if

$K_p = \mathrm{GSp}(4, \mathbb{Z}_p)$ for all but finitely many primes p of $\mathbb{Z}$ and $\lambda(K_p) = \mathbb{Z}_p^\times$ for all primes p of $\mathbb{Z}$. Assume that $\{K_p\}_p$ is admissible. Define

$$\mathcal{K} = \prod_{p<\infty} K_p. \tag{10.26}$$

Then $\mathcal{K}$ is a compact, open subgroup of $\mathrm{GSp}(4, \mathbb{A}_{\mathrm{fin}})$. We define

$$\Gamma = \mathrm{GSp}(4, \mathbb{Q}) \cap \mathrm{GSp}(4, \mathbb{R})^+ \mathcal{K}. \tag{10.27}$$

Then Γ is a subgroup of $\mathrm{Sp}(4, \mathbb{Q})$ commensurable with $\mathrm{Sp}(4, \mathbb{Z})$. Also, since strong approximation holds for $\mathrm{Sp}(4)$ (see Satz 2 of [77]), since $\mathbb{Q}$ has class number one, and since $\lambda(K_p) = \mathbb{Z}_p^\times$ for all primes p of $\mathbb{Z}$, we have

$$\mathrm{GSp}(4, \mathbb{A}) = \mathrm{GSp}(4, \mathbb{Q})\mathrm{GSp}(4, \mathbb{R})^+ \mathcal{K}.$$

We now define

$$\mathcal{A}_k(\mathcal{K}) = \{\Phi \in \mathcal{A}_k \mid \kappa \cdot \Phi = \Phi \text{ for } \kappa \in \mathcal{K}\};$$

here, the action of $\mathcal{K}$ on $\mathcal{A}_k(\mathcal{K})$ is defined by $(\kappa \cdot \Phi)(g) = \Phi(g\kappa)$ for $\kappa \in \mathcal{K}$ and $g \in \mathrm{GSp}(4, \mathbb{A})$. We let $\mathcal{A}_k^\circ(\mathcal{K})$ be the subspace of $\mathcal{A}_k(\mathcal{K})$ consisting of cusp forms. An element of $\mathcal{A}_k(\mathcal{K})$ is called an *adelic automorphic form of weight k with respect to* $\mathcal{K}$, and an element of $\mathcal{A}_k^\circ(\mathcal{K})$ is called a *cuspidal adelic automorphic form of weight k with respect to* $\mathcal{K}$. The next lemma proves that the vector space $M_k(\Gamma)$ of Siegel modular forms and the vector space $\mathcal{A}_k(\mathcal{K})$ of adelic automorphic forms are naturally isomorphic.

Lemma 10.2.1 *Let k be an integer such that $k > 0$, and let $\{K_p\}_p$, where p runs over the primes of $\mathbb{Z}$, be an admissible family of compact, open subgroups of* $\mathrm{GSp}(4, \mathbb{Q}_p)$. *Define $\mathcal{K}$ as in* (10.26) *and define Γ as in* (10.27). *For $F \in M_k(\Gamma)$, define $\Phi_F : \mathrm{GSp}(4, \mathbb{A}) \to \mathbb{C}$ by*

$$\Phi_F(\rho g \kappa) = \lambda(g)^k j(g, I)^{-k} F(g\langle I \rangle) \tag{10.28}$$

for $\rho \in \mathrm{GSp}(4, \mathbb{Q})$, $g \in \mathrm{GSp}(4, \mathbb{R})^+$, and $\kappa \in \mathcal{K}$. Then Φ_F is a well-defined element of $\mathcal{A}_k(\mathcal{K})$, so that there is a linear map

$$M_k(\Gamma) \longrightarrow \mathcal{A}_k(\mathcal{K}) \tag{10.29}$$

defined by $F \mapsto \Phi_F$. This map sends $S_k(\Gamma)$ into $\mathcal{A}_k^\circ(\mathcal{K})$. Conversely, for $\Phi \in \mathcal{A}_k(\mathcal{K})$ define $F_\Phi : \mathcal{H}_2 \to \mathbb{C}$ by

$$F_\Phi(Z) = \lambda(g)^{-k} j(g, I)^k \Phi(g) \tag{10.30}$$

for $Z \in \mathcal{H}_2$ and $g \in \mathrm{GSp}(4, \mathbb{R})^+$ such that $g\langle I \rangle = Z$. Then F_Φ is a well-defined element of $M_k(\Gamma)$, so that there is a linear map

$$\mathcal{A}_k(\mathcal{K}) \longrightarrow M_k(\Gamma) \tag{10.31}$$

defined by $\Phi \mapsto F_\Phi$. This map sends $\mathcal{A}_k^\circ(\mathcal{K})$ into $S_k(\Gamma)$. Moreover, the maps in (10.29) *and* (10.31) *are inverses of each other.*

Proof See the proof of Lemma 4.1 of [74] and also [6]. □

We may apply this theory to the paramodular and stable Klingen settings. Let N be an integer such that $N > 0$. One easily checks that

$$\{\mathrm{K}(\mathfrak{p}^{v_p(N)})\}_p \quad \text{and} \quad \{\mathrm{K}_s(\mathfrak{p}^{v_p(N)})\}_p$$

are admissible families of compact, open subgroups of $\mathrm{GSp}(4, \mathbb{Q}_p)$. We define

$$\mathcal{K}(N) = \prod_p \mathrm{K}(\mathfrak{p}^{v_p(N)}) \quad \text{and} \quad \mathcal{K}_s(N) = \prod_p \mathrm{K}_s(\mathfrak{p}^{v_p(N)}).$$

Here, $\mathrm{K}(\mathfrak{p}^{v_p(N)})$ is defined as in (2.25) and $\mathrm{K}_s(\mathfrak{p}^{v_p(N)})$ is defined as in (3.2); we note again that in this second part of this work we use (10.1), while in the first part we used (10.3); the conversion between formulas from Parts I and II is achieved by conjugation by the matrix (10.4). It is straightforward to verify that

$$\mathrm{K}(N) = \mathrm{GSp}(4, \mathbb{Q}) \cap \mathrm{GSp}(4, \mathbb{R})^+ \prod_p \mathrm{K}(\mathfrak{p}^{v_p(N)})$$

and

$$\mathrm{K}_s(N) = \mathrm{GSp}(4, \mathbb{Q}) \cap \mathrm{GSp}(4, \mathbb{R})^+ \prod_p \mathrm{K}_s(\mathfrak{p}^{v_p(N)}).$$

Applying Lemma 10.2.1, we see that there are natural isomorphisms

$$M_k(\mathrm{K}(N)) \overset{\sim}{\longleftrightarrow} \mathcal{A}_k(\mathcal{K}(N)) \quad \text{and} \quad M_k(\mathrm{K}_s(N)) \overset{\sim}{\longleftrightarrow} \mathcal{A}_k(\mathcal{K}_s(N)).$$

Under these isomorphisms cusp forms are mapped onto cusp forms.

Paramodular Old- and Newforms

To define paramodular old- and newforms we first recall three level raising operators from [104]. These level raising operators are obtained from the corresponding local operators from Sect. 2.3.

Lemma 10.2.2 *Let N and k be integers such that $N > 0$ and $k > 0$, and let p be a prime of $\mathbb{Z}$. Let $F \in M_k(\mathrm{K}(N))$. Define*

$$\eta_p F = F\big|_k \begin{bmatrix} 1 &&& \\ & p && \\ && 1 & \\ &&& p^{-1} \end{bmatrix}, \tag{10.32}$$

$$\theta_p F = F\big|_k \begin{bmatrix} 1 &&& \\ & 1 && \\ && p^{-1} & \\ &&& p^{-1} \end{bmatrix} + \sum_{x \in \mathbb{Z}/p\mathbb{Z}} F\big|_k \begin{bmatrix} p^{-1} &&& \\ & 1 && \\ && 1 & \\ &&& p^{-1} \end{bmatrix} \begin{bmatrix} 1 & x && \\ & 1 && \\ && 1 & \\ &&& 1 \end{bmatrix}, \tag{10.33}$$

$$\theta'_p F = \eta_p F + \sum_{c \in \mathbb{Z}/p\mathbb{Z}} F\big|_k \begin{bmatrix} 1 &&& \\ & 1 && cp^{-1}N^{-1} \\ && 1 & \\ &&& 1 \end{bmatrix}. \tag{10.34}$$

Then $\eta_p F \in M_k(\mathrm{K}(Np^2))$ and $\theta_p F, \theta'_p F \in M_k(\mathrm{K}(Np))$. If F is a cusp form, then $\eta_p F$, $\theta_p F$, and $\theta'_p F$ are cusp forms.

Proof We will prove the lemma for $\theta_p F$; the proofs for $\eta_p F$ and $\theta'_p F$ are similar. Let W be the subspace of $\mathcal{A}_k$ of all $\Phi \in \mathcal{A}_k$ such that $\kappa \cdot \Phi = \Phi$ for all $\kappa \in \prod_{q \neq p} \mathrm{K}(\mathfrak{q}^{v_q(N)})$. Then W is a smooth representation of $\mathrm{GSp}(4, \mathbb{Q}_p)$. Let $\theta : W(\mathfrak{p}^{v_p(N)}) \to W(\mathfrak{p}^{v_p(N)+1})$ be the level raising operator from Sect. 2.3; note that we are also using the adjusted notation from (10.7). Since $W(\mathfrak{p}^{v_p(N)}) = \mathcal{A}_k(\mathcal{K}(N))$ and $W(\mathfrak{p}^{v_p(N)+1}) = \mathcal{A}_k(\mathcal{K}(Np))$, we obtain an operator $\theta : \mathcal{A}_k(\mathcal{K}(N)) \to \mathcal{A}_k(\mathcal{K}(Np))$. To prove the lemma, it will suffice to prove that under the composition

$$M_k(\mathrm{K}(N)) \xrightarrow{\sim} \mathcal{A}_k(\mathcal{K}(N)) \xrightarrow{\theta} \mathcal{A}_k(\mathcal{K}(Np)) \xrightarrow{\sim} M_k(\mathrm{K}(Np))$$

the Siegel modular form F is mapped to $\theta_p F$; here the first and last maps are the isomorphisms from Lemma 10.2.1. Define $\Phi = \Phi_F$; the image of F under the composition is $F_{\theta\Phi}$. Let $Z \in \mathcal{H}_2$, and let $g \in \mathrm{GSp}(4, \mathbb{R})^+$ be such that $g\langle I \rangle = Z$. Then

$$\begin{aligned} F_{\theta\Phi}(Z) &= \lambda(g)^{-k} j(g, I)^k (\theta\Phi)(g) \\ &= \lambda(g)^{-k} j(g, I)^k \Big(\Phi(g \begin{bmatrix} 1 &&& \\ & 1 && \\ && p & \\ &&& p \end{bmatrix}_p) \\ &\quad + \sum_{x \in \mathbb{Z}/p\mathbb{Z}} \Phi(g \begin{bmatrix} 1 & x && \\ & 1 && \\ && 1 & \\ &&& 1 \end{bmatrix}_p \begin{bmatrix} p &&& \\ & 1 && \\ && 1 & \\ &&& p \end{bmatrix}_p)\Big) \qquad (10.35) \\ &= \lambda(g)^{-k} j(g, I)^k \Big(\Phi(\underbrace{\begin{bmatrix} 1 &&& \\ & 1 && \\ && p & \\ &&& p \end{bmatrix}^{-1}}_{\text{all places}} g \begin{bmatrix} 1 &&& \\ & 1 && \\ && p & \\ &&& p \end{bmatrix}_p) \end{aligned}$$

$$+\sum_{x\in\mathbb{Z}/p\mathbb{Z}}\Phi(\underbrace{\begin{bmatrix}p&&&\\&1&&\\&&1&\\&&&p\end{bmatrix}^{-1}\begin{bmatrix}1&x&&\\&1&&\\&&1&\\&&&1\end{bmatrix}^{-1}g}_{\text{all places}}\begin{bmatrix}1&x&&\\&1&&\\&&1&\\&&&1\end{bmatrix}_p\begin{bmatrix}p&&&\\&1&&\\&&1&\\&&&p\end{bmatrix}_p))$$

$$=\lambda(g)^{-k}j(g,I)^k\Big(\Phi(\begin{bmatrix}1&&&\\&1&&\\&&p^{-1}&\\&&&p^{-1}\end{bmatrix}_\infty g)$$

$$+\sum_{x\in\mathbb{Z}/p\mathbb{Z}}\Phi(\begin{bmatrix}p^{-1}&&&\\&1&&\\&&1&\\&&&p^{-1}\end{bmatrix}_\infty\begin{bmatrix}1&x&&\\&1&&\\&&1&\\&&&1\end{bmatrix}_\infty g)\Big)$$

$$=p^kF(\begin{bmatrix}1&&&\\&1&&\\&&p^{-1}&\\&&&p^{-1}\end{bmatrix}\langle Z\rangle)+\sum_{x\in\mathbb{Z}/p\mathbb{Z}}F((\begin{bmatrix}p^{-1}&&&\\&1&&\\&&1&\\&&&p^{-1}\end{bmatrix}\begin{bmatrix}1&x&&\\&1&&\\&&1&\\&&&1\end{bmatrix})\langle Z\rangle)$$

$$=(\theta_pF)(Z).$$

We have proved that $\theta_pF\in M_k(\mathrm{K}(Np))$. If F is a cusp form then the same argument, with $\mathcal{A}_k^\circ$ replacing $\mathcal{A}_k$, shows that θ_pF is a cusp form. □

Next, let M be an integer such that $M>0$. Let $\langle\cdot,\cdot\rangle$ be the Petersson inner product on $S_k(\mathrm{K}(M))$ from Chap. 2, Theorem 5.3 on p. 89 of [3] or Chap. IV, Hilfsatz 4.10 on p. 271 of [42]. We define the subspace $S_k(\mathrm{K}(M))_{\mathrm{old}}$ of *oldforms* in $S_k(\mathrm{K}(M))$ to be the linear span of the functions η_pF where p is a prime such that $p^2\mid M$ and $F\in S_k(\mathrm{K}(Mp^{-2}))$ and the functions θ_pF and θ'_pF where p is a prime such that $p\mid M$ and $F\in S_k(\mathrm{K}(Mp^{-1}))$. We define the subspace $S_k(\mathrm{K}(M))_{\mathrm{new}}$ of *newforms* in $S_k(\mathrm{K}(M))$ to be the orthogonal complement in $S_k(\mathrm{K}(M))$ of the subspace $S_k(\mathrm{K}(M))_{\mathrm{old}}$, so that $S_k(\mathrm{K}(M))_{\mathrm{new}}=S_k(\mathrm{K}(M))_{\mathrm{old}}^{\perp}$.

We remark that, as a consequence of local new- and oldforms theorems from [105], the spaces $S_k(\mathrm{K}(M))_{\mathrm{old}}$ and $S_k(\mathrm{K}(M))_{\mathrm{new}}$ defined as above coincide with the spaces of the same name defined in a more representation-theoretic way on p. 41.

Paramodular Hecke and Atkin-Lehner Operators

Let N and k be integers such that $N>0$ and $k>0$. For each prime of $\mathbb{Z}$, there exist three natural and important operators on $M_k(\mathrm{K}(N))$ and $S_k(\mathrm{K}(N))$. Let p be a prime of $\mathbb{Z}$ and let $n=v_p(N)$. There are finite disjoint decompositions

$$\mathrm{K}(N)\begin{bmatrix}1&&&\\&1&&\\&&p&\\&&&p\end{bmatrix}\mathrm{K}(N)=\sqcup_i\mathrm{K}(N)h_i,\qquad \mathrm{K}(N)\begin{bmatrix}p&&&\\&1&&\\&&p&\\&&&p^2\end{bmatrix}\mathrm{K}(N)=\sqcup_j\mathrm{K}(N)h'_j.$$

Define

$$T(1,1,p,p),\, T(p,1,p,p^2) : M_k(\mathrm{K}(N)) \longrightarrow M_k(\mathrm{K}(N)) \tag{10.36}$$

by

$$T(1,1,p,p)F = p^{k-3}\sum_i F\big|_k h_i, \qquad T(p,1,p,p^2)F = p^{2(k-3)}\sum_j F\big|_k h'_j \tag{10.37}$$

for $F \in M_k(\mathrm{K}(N))$. It is straightforward to verify that $T(1,1,p,p)$ and $T(p,1,p,p^2)$ are well-defined and map cusp forms to cusp forms. Note that we use the same normalization as in the case $N = 1$ (see, for example, (1.3.3) of [1]). We refer to $T(1,1,p,p)$ and $T(p,1,p,p^2)$ as the *paramodular Hecke operators*. Next, since the canonical map $\mathrm{Sp}(4,\mathbb{Z}) \to \mathrm{Sp}(4,\mathbb{Z}/N\mathbb{Z}) \cong \mathrm{Sp}(4,\mathbb{Z}/p^n\mathbb{Z}) \times \mathrm{Sp}(4,\mathbb{Z}/Np^{-n}\mathbb{Z})$ is surjective, there exists $\gamma \in \mathrm{Sp}(4,\mathbb{Z})$ such that

$$\gamma \equiv \begin{bmatrix} &&&-1\\ &&1&\\ &1&&\\ -1&&& \end{bmatrix} \pmod{p^n}, \qquad \gamma \equiv \begin{bmatrix} 1&&&\\ &1&&\\ &&1&\\ &&&1 \end{bmatrix} \pmod{Np^{-n}}.$$

Define

$$u = \begin{bmatrix} 1&&&\\ &1&&\\ &&p^n&\\ &&&p^n \end{bmatrix}\gamma.$$

Then arguments show that u normalizes $\mathrm{K}(N)$ and that $u^2 \in p^n\mathrm{K}(N)$. We now define

$$w_p : M_k(\mathrm{K}(N)) \longrightarrow M_k(\mathrm{K}(N)) \tag{10.38}$$

by

$$w_p F = F\big|_k u. \tag{10.39}$$

We see that w_p is a well-defined involution of $M_k(\mathrm{K}(N))$; it is also evident that w_p maps cusp forms to cusp forms. We refer to w_p as a *paramodular Atkin-Lehner operator*.

These operators are essentially the operators induced by the corresponding local operators from Sect. 2.3. To see this, let W be the smooth representation of $\mathrm{GSp}(4,\mathbb{Q}_p)$ from the proof of Lemma 10.2.2. Let

$$T_{0,1},\, T_{1,0},\, u_n : W(\mathfrak{p}^n) \longrightarrow W(\mathfrak{p}^n)$$

be the operators from (2.30) and (2.31). Since $W(\mathfrak{p}^n) = \mathcal{A}_k(\mathcal{K}(N))$, we obtain operators

$$T_{0,1}, T_{1,0}, u_n : \mathcal{A}_k(\mathcal{K}(N)) \longrightarrow \mathcal{A}_k(\mathcal{K}(N)).$$

Let $T_{0,1}(p)$, $T_{1,0}(p)$, and $u_{n,p}$ be the respective compositions

$$M_k(\mathrm{K}(N)) \xrightarrow{\sim} \mathcal{A}_k(\mathcal{K}(N)) \overset{T_{0,1},T_{1,0},u_n}{\longrightarrow} \mathcal{A}_k(\mathcal{K}(N)) \xrightarrow{\sim} M_k(\mathrm{K}(N)), \tag{10.40}$$

where the first and last maps are the isomorphisms from Lemma 10.2.1.

Lemma 10.2.3 *Let $T(1,1,p,p)$, $T(p,1,p,p^2)$, and w_p be as in* (10.36) *and* (10.38), *and let $T_{0,1}(p)$, $T_{1,0}(p)$, and $u_{n,p}$ be as in* (10.40). *Then*

$$T(1,1,p,p) = p^{k-3}T_{0,1}(p), \quad T(p,1,p,p^2) = p^{2(k-3)}T_{1,0}(p), \quad w_p = u_{n,p}.$$

Proof Using the decompositions in Sect. 6.1 of [105], along with some further refinements, one can prove that there exist finite disjoint decompositions

$$\mathrm{K}(\mathfrak{p}^n)\begin{bmatrix} p &&& \\ & p && \\ && 1 & \\ &&& 1 \end{bmatrix}\mathrm{K}(\mathfrak{p}^n) = \sqcup_i g_i \mathrm{K}(\mathfrak{p}^n), \qquad \mathrm{K}(\mathfrak{p}^n)\begin{bmatrix} p &&& \\ & p^2 && \\ && p & \\ &&& 1 \end{bmatrix}\mathrm{K}(\mathfrak{p}^n) = \sqcup_j g'_j \mathrm{K}(\mathfrak{p}^n)$$

such that $g_i, g'_j \in \mathrm{GSp}(4, \mathbb{Q})$ and

$$\mathrm{K}(N)\begin{bmatrix} 1 &&& \\ & 1 && \\ && p & \\ &&& p \end{bmatrix}\mathrm{K}(N) = \sqcup_i \mathrm{K}(N) p g_i^{-1},$$

$$\mathrm{K}(N)\begin{bmatrix} p &&& \\ & 1 && \\ && p & \\ &&& p^2 \end{bmatrix}\mathrm{K}(N) = \sqcup_j \mathrm{K}(N) p^2 g_j'^{-1}.$$

Since $\mathrm{K}(N) \subset \mathrm{K}(\mathfrak{q}^{v_q(N)})$ for all primes q of $\mathbb{Z}$, we have $g_i, g'_j \in \mathrm{K}(\mathfrak{q}^{v_q(N)})$ for all primes q of $\mathbb{Z}$ such that $q \neq p$. Let $F \in M_k(\mathrm{K}(N))$ and $Z \in \mathcal{H}_2$. Let $h \in \mathrm{GSp}(4, \mathbb{R})^+$ be such that $Z = h\langle I \rangle$. Then

$$\begin{aligned} p^{k-3}(T_{0,1}(p)F)(Z) &= p^{k-3}\lambda(h)^{-k} j(h, I)^k T_{0,1}\Phi_F(h) \\ &= p^{k-3}\lambda(h)^{-k} j(h, I)^k \sum_i \Phi_F(hg_{i,p}) \\ &= p^{k-3}\lambda(h)^{-k} j(h, I)^k \sum_i \Phi_F(\underbrace{g_i^{-1}}_{\text{all places}} hg_{i,p}) \\ &= p^{k-3}\lambda(h)^{-k} j(h, I)^k \sum_i \Phi_F(g_{i,\infty}^{-1} h) \end{aligned} \tag{10.41}$$

$$= p^{k-3}\Big(\sum_i F\big|_k g_i^{-1}\Big)(Z)$$

$$= p^{k-3}\Big(\sum_i F\big|_k p g_i^{-1}\Big)(Z)$$

$$= (T(1, 1, p, p)F)(Z).$$

The proofs of the remaining statements are similar. □

Lemma 10.2.4 *Let p be a prime of $\mathbb{Z}$. Let $T(1, 1, p, p)$, $T(p, 1, p, p^2)$, and w_p be the endomorphisms of $S_k(\mathrm{K}(N))$ as in* (10.36) *and* (10.38). *Then $T(1, 1, p, p)$, $T(p, 1, p, p^2)$, and w_p map $S_k(\mathrm{K}(N))_{\mathrm{old}}$ to $S_k(\mathrm{K}(N))_{\mathrm{old}}$ and map $S_k(\mathrm{K}(N))_{\mathrm{new}}$ to $S_k(\mathrm{K}(N))_{\mathrm{new}}$.*

Proof Let A be one of $T(1, 1, p, p)$, $T(p, 1, p, p^2)$, and w_p. We first prove that $S_k(\mathrm{K}(N))_{\mathrm{old}}$ is preserved by A. Let $F \in S_k(\mathrm{K}(N))_{\mathrm{old}}$. We may assume that there exists a prime q of $\mathbb{Z}$ such that $q \mid N$ and F is of the form $\theta_q F_0$ or $\theta'_q F_0$ for some $F_0 \in S_k(\mathrm{K}(Nq^{-1}))$, or that there exists a prime q of $\mathbb{Z}$ such that $q^2 \mid N$ and F is of the form $\eta_q F_0$ for some $F_0 \in S_k(\mathrm{K}(Nq^{-2}))$. Assume that $q \neq p$. If $F = \theta_q F_0$, then $AF = A\theta_q F_0 = \theta_q A' F_0 \in S_k(\mathrm{K}(N))$. Here, in $\theta_q A' F_0$, the map A' is the endomorphism of $S_k(\mathrm{K}(Nq^{-1}))$ defined in the same way as A; also, to verify $A\theta_q F_0 = \theta_q A' F_0$ we use the formulas (10.35) and (10.41). If $F = \theta'_q F_0$ or $F = \eta_q F_0$ then a similar argument proves that $AF \in S_k(\mathrm{K}(N))_{\mathrm{old}}$. Now assume that $q = p$. Then $AF \in S_k(\mathrm{K}(N))_{\mathrm{old}}$ by Lemma 2.3.6; for the connection to representation theory we use the space W from the proof of Lemma 10.2.2. Next we prove that $S_k(\mathrm{K}(N))_{\mathrm{new}}$ is preserved by A. Let $\langle \cdot, \cdot \rangle'$ be the inner product on $L^2(\mathbb{I} \cdot \mathrm{GSp}(4, \mathbb{Q})\backslash \mathrm{GSp}(4, \mathbb{A}))$ defined by

$$\langle \Phi_1, \Phi_2 \rangle' = \int\limits_{\mathbb{I}\cdot \mathrm{GSp}(4,\mathbb{Q})\backslash \mathrm{GSp}(4,\mathbb{A})} \Phi_1(g)\overline{\Phi_2(g)}\, dg$$

for $\Phi_1, \Phi_2 \in L^2(\mathbb{I} \cdot \mathrm{GSp}(4, \mathbb{Q})\backslash \mathrm{GSp}(4, \mathbb{A}))$ where dg is a fixed right $\mathrm{GSp}(4, \mathbb{A})$ invariant measure on $\mathbb{I} \cdot \mathrm{GSp}(4, \mathbb{Q})\backslash \mathrm{GSp}(4, \mathbb{A})$. There exists a positive real number c such that if $F_1, F_2 \in S_k(\mathrm{K}(N))$, then $\langle F_1, F_2 \rangle = c\langle \Phi_{F_1}, \Phi_{F_2} \rangle'$; see Lemma 6 of [6]. Using this formula, (3.22), and Lemma 10.2.3 one can verify that $\langle AF_1, F_2 \rangle = \langle F_1, AF_2 \rangle$ for $F_1, F_2 \in S_k(\mathrm{K}(N))$, i.e., the operator A is self-adjoint. This calculation repeatedly uses the fact that $L^2(\mathbb{I} \cdot \mathrm{GSp}(4, \mathbb{Q})\backslash \mathrm{GSp}(4, \mathbb{A}))$ with the above inner product is unitary with respect to the right translation action of $\mathrm{GSp}(4, \mathbb{A})$. Since $AS_k(\mathrm{K}(N))_{\mathrm{old}} \subset S_k(\mathrm{K}(N))_{\mathrm{old}}$, the definition of $S_k(\mathrm{K}(N))_{\mathrm{new}}$ now implies that A preserves $S_k(\mathrm{K}(N))_{\mathrm{new}}$. □

Chapter 11
Operators on Siegel Modular Forms

In this chapter we translate the upper block level changing operators from Chap. 3 to operators on Siegel modular forms defined with respect to the stable Klingen congruence subgroups. We give a slash formula for each such operator; since this formula involves only upper block matrices, we are able to calculate the Fourier and Fourier-Jacobi expansions of the resulting Siegel modular forms.

11.1 Overview

The local level changing operators from Chap. 3 naturally induce operators on Siegel modular forms. To describe this, let p be a prime of $\mathbb{Z}$, and let N', k, n_1, and n_2 be integers such that $N' > 0$ and N' is relatively prime to p, $k > 0$, and $n_1, n_2 \geq 0$. Define $N_1 = N' p^{n_1}$ and $N_2 = N' p^{n_2}$. Define $\mathcal{A}_k$ as in Sect. 10.2, and let W be the subspace of all $\Phi \in \mathcal{A}_k$ such that $\kappa \cdot \Phi = \Phi$ for all $\kappa \in \prod_{q \neq p} \mathrm{K}(\mathfrak{q}^{v_q(N')})$. Then W is a smooth representation of $\mathrm{GSp}(4, \mathbb{Q}_p)$. We may consider the subspaces $W_s(\mathfrak{p}^{n_1})$ and $W_s(\mathfrak{p}^{n_2})$ of stable Klingen vectors as defined in (3.16) (with the notational adjustment of (10.8)). Let

$$\delta : W_s(\mathfrak{p}^{n_1}) \longrightarrow W_s(\mathfrak{p}^{n_2})$$

be a linear map; a typical example for us will be one of the level changing operators from Chap. 3. From the involved definitions we have

$$W_s(\mathfrak{p}^{n_1}) = \mathcal{A}_k(\mathcal{K}_s(N_1)) \quad \text{and} \quad W_s(\mathfrak{p}^{n_2}) = \mathcal{A}_k(\mathcal{K}_s(N_2)),$$

so that we may view δ as map from $\mathcal{A}_k(\mathcal{K}_s(N_1))$ to $\mathcal{A}_k(\mathcal{K}_s(N_2))$:

$$\delta : \mathcal{A}_k(\mathcal{K}_s(N_1)) \longrightarrow \mathcal{A}_k(\mathcal{K}_s(N_2)).$$

J. Johnson-Leung et al., *Stable Klingen Vectors and Paramodular Newforms*,
Lecture Notes in Mathematics 2342, https://doi.org/10.1007/978-3-031-45177-5_11

By Lemma 10.2.1 we have natural isomorphisms

$$M_k(\mathrm{K}_s(N_1)) \xrightarrow{\sim} \mathcal{A}_k(\mathcal{K}_s(N_1)) \quad \text{and} \quad \mathcal{A}_k(\mathcal{K}_s(N_2)) \xrightarrow{\sim} M_k(\mathrm{K}_s(N_2)).$$

Therefore, we may compose and obtain a linear map

$$\delta_p : M_k(\mathrm{K}_s(N_1)) \longrightarrow M_k(\mathrm{K}_s(N_2)) \tag{11.1}$$

so that the diagram

$$\begin{array}{ccc} \mathcal{A}_k(\mathcal{K}_s(N_1)) & \xleftarrow{\sim} & M_k(\mathrm{K}_s(N_1)) \\ \delta \big\downarrow & & \big\downarrow \delta_p \\ \mathcal{A}_k(\mathcal{K}_s(N_2)) & \xrightarrow{\sim} & M_k(\mathrm{K}_s(N_2)) \end{array} \tag{11.2}$$

commutes. If $F \in M_k(\mathrm{K}_s(N_1))$, then in the notation of Lemma 10.2.1,

$$(\delta_p F)(Z) = \lambda(h)^{-k} j(h, I)^k (\delta \Phi_F)(h) \tag{11.3}$$

where $Z \in \mathcal{H}_2$ and $h \in \mathrm{GSp}(4, \mathbb{R})^+$ is such that $h\langle I \rangle = Z$. In this construction, if the subspace of cusp forms $\mathcal{A}_k^\circ$ is used in place of $\mathcal{A}_k$, then the result is a linear map $S_k(\mathrm{K}_s(N_1)) \to S_k(\mathrm{K}_s(N_2))$ that is the restriction of δ_p to $S_k(\mathrm{K}_s(N_1))$. Thus, δ_p maps cusp forms to cusp forms.

In the following sections we will calculate δ_p for each of the upper block level changing operators δ from Chap. 3. Before beginning, we note a general feature of the calculation of Fourier expansions. Let $F \in S_k(\mathrm{K}_s(N_1))$, and let

$$F(Z) = \sum_{S \in B(N_1)} a(S) e^{2\pi i \mathrm{Tr}(SZ)}$$

be the Fourier expansion of F. In calculating the Fourier expansion of $\delta_p F$, we will typically arrive at an expression of the form

$$(\delta_p F)(Z) = \sum_{S \in X} a(S) e^{2\pi i \mathrm{Tr}(t(S)Z)}.$$

Here, X is a subset of $B(N_1)$ and $t : X \xrightarrow{\sim} Y$ is a bijection, where Y is a subset of $B(N_2)$. We may rewrite this expression as

$$(\delta_p F)(Z) = \sum_{S \in Y} a(t^{-1}(S)) e^{2\pi i \mathrm{Tr}(SZ)}$$

and thus obtain the Fourier expansion of $\delta_p F$. We point out that it may not be immediately obvious from the expression for $t^{-1}(S)$ appearing in the statements of the lemmas below that this matrix lies in the domain $B(N_1)$ for $a(\cdot, F) : B(N_1) \to \mathbb{C}$. Nevertheless, it is apparent from the method that this is the case.

Let N and k be integers such that $N > 0$ and $k > 0$, and let $F \in M_k(\mathrm{K}_s(N))$. In this chapter we will write the Fourier expansion and Fourier-Jacobi expansion of F as in (10.17) and (10.23), respectively, so that

$$F(Z) = \sum_{S \in B(N)} a(S) e^{2\pi i \mathrm{Tr}(SZ)}, \tag{11.4}$$

$$F(Z) = \sum_{\substack{m=0 \\ N_s | m}}^{\infty} f_m(\tau, z) e^{2\pi i m \tau'}, \qquad Z = \left[\begin{smallmatrix} \tau & z \\ z & \tau' \end{smallmatrix}\right]. \tag{11.5}$$

11.2 Level Raising Operators

We begin by calculating formulas for the level raising operators τ_p, θ_p, and η_p obtained from the corresponding local operators defined in Sect. 3.5 via the procedure discussed in Sect. 11.1. For θ_p and η_p we find that the resulting formulas in terms of the Fourier-Jacobi expansion involve the well-known operators V_p and U_p on Jacobi forms defined in [38].

Lemma 11.2.1 *Let N and k be integers such that $N > 0$ and $k > 0$, and let p be a prime of $\mathbb{Z}$. Define $n = v_p(N)$. Let $\tau : \mathcal{A}_k(\mathcal{K}_s(N)) \to \mathcal{A}_k(\mathcal{K}_s(Np))$ be the linear map obtained as in Sect. 11.1 from the level raising operator* (3.34), *and let τ_p be as in* (11.1), *so that τ_p is a linear map*

$$\tau_p : M_k(\mathrm{K}_s(N)) \longrightarrow M_k(\mathrm{K}_s(Np)).$$

Let $F \in M_k(\mathrm{K}_s(N))$ with Fourier expansion as in (11.4). *Then*

$$\tau_p F = p^{-1} \sum_{x \in \mathbb{Z}/p\mathbb{Z}} F\Big|_k \begin{bmatrix} 1 & & & \\ & 1 & & xp^{-n} \\ & & 1 & \\ & & & 1 \end{bmatrix}. \tag{11.6}$$

If F is a cusp form, then $\tau_p F$ is a cusp form. We have

$$(\tau_p F)(Z) = \sum_{S \in B(Np)} a(S) e^{2\pi i \mathrm{Tr}(SZ)}. \tag{11.7}$$

The Fourier-Jacobi expansion of $\tau_p F$ is given by

$$(\tau_p F)(Z) = \sum_{\substack{m=0\\ N_s|m,\ p^n|m}}^{\infty} f_m(\tau, z) e^{2\pi i m\tau'}, \qquad Z = \left[\begin{smallmatrix}\tau & z\\ z & \tau'\end{smallmatrix}\right] \in \mathcal{H}_2. \tag{11.8}$$

Proof Define $\Phi = \Phi_F$. Let $Z \in \mathcal{H}_2$ and let $h \in \mathrm{GSp}(4, \mathbb{R})^+$ be such that $Z = h\langle I \rangle$. Then by (11.3),

$$\begin{aligned}
(\tau_p F)(Z) &= \lambda(h)^{-k} j(h, I)^k \tau\Phi(h)\\
&= \lambda(h)^{-k} j(h, I)^k \int\limits_{\mathbb{Z}_p} \Phi(h\begin{bmatrix}1&&&\\&1&xp^{-n}&\\&&1&\\&&&1\end{bmatrix}) \, dx\\
&= \lambda(h)^{-k} j(h, I)^k \int\limits_{\mathbb{Z}_p} \Phi(\begin{bmatrix}1&&&\\&1&xp^{-n}&\\&&1&\\&&&1\end{bmatrix} h) \, dx\\
&= \lambda(h)^{-k} j(h, I)^k \int\limits_{\mathbb{Z}_p/p^n\mathbb{Z}_p} \int\limits_{p^n\mathbb{Z}_p} \Phi(\begin{bmatrix}1&&&\\&1&(x_1+x_2)p^{-n}&\\&&1&\\&&&1\end{bmatrix} h) \, dx_1 \, dx_2\\
&= \lambda(h)^{-k} j(h, I)^k p^{-n} \sum_{x \in \mathbb{Z}/p^n\mathbb{Z}} \Phi(\begin{bmatrix}1&&&\\&1&xp^{-n}&\\&&1&\\&&&1\end{bmatrix}_p h)\\
&= \lambda(h)^{-k} j(h, I)^k p^{-n} \sum_{x \in \mathbb{Z}/p^n\mathbb{Z}} \Phi(\underbrace{\begin{bmatrix}1&&&\\&1&-xp^{-n}&\\&&1&\\&&&1\end{bmatrix}}_{\text{all places}} \begin{bmatrix}1&&&\\&1&xp^{-n}&\\&&1&\\&&&1\end{bmatrix}_p h)\\
&= \lambda(h)^{-k} j(h, I)^k p^{-n} \sum_{x \in \mathbb{Z}/p^n\mathbb{Z}} \Phi(\underbrace{\begin{bmatrix}1&&&\\&1&xp^{-n}&\\&&1&\\&&&1\end{bmatrix}}_{\text{all places but } p} h)\\
&= \lambda(h)^{-k} j(h, I)^k p^{-n} \sum_{x \in \mathbb{Z}/p^n\mathbb{Z}} \Phi(\begin{bmatrix}1&&&\\&1&xp^{-n}&\\&&1&\\&&&1\end{bmatrix}_\infty h \underbrace{\begin{bmatrix}1&&&\\&1&xp^{-n}&\\&&1&\\&&&1\end{bmatrix}}_{\text{all places but } p,\, \infty})\\
&= \lambda(h)^{-k} j(h, I)^k p^{-n} \sum_{x \in \mathbb{Z}/p^n\mathbb{Z}} \Phi(\begin{bmatrix}1&&&\\&1&xp^{-n}&\\&&1&\\&&&1\end{bmatrix}_\infty h)
\end{aligned}$$

$$= \big(p^{-n} \sum_{x \in \mathbb{Z}/p^n\mathbb{Z}} F\big|_k \begin{bmatrix} 1 & & & \\ & 1 & & xp^{-n} \\ & & 1 & \\ & & & 1 \end{bmatrix}\big)(Z)$$

$$= \big(p^{-1} \sum_{x \in \mathbb{Z}/p\mathbb{Z}} F\big|_k \begin{bmatrix} 1 & & & \\ & 1 & & xp^{-n} \\ & & 1 & \\ & & & 1 \end{bmatrix}\big)(Z).$$

This proves (11.6). To compute the Fourier expansion, we calculate as follows:

$$\begin{aligned}(\tau_p F)(Z) &= p^{-n} \sum_{x \in \mathbb{Z}/p^n\mathbb{Z}} \sum_{S \in B(N)} a(S) e^{2\pi i \mathrm{Tr}\left(S(Z + [\ \ xp^{-n}])\right)} \\ &= \sum_{S = \left[\begin{smallmatrix} \alpha & \beta \\ \beta & \gamma \end{smallmatrix}\right] \in B(N)} \big(p^{-n} \sum_{x \in \mathbb{Z}/p^n\mathbb{Z}} e^{2\pi i \gamma x p^{-n}}\big) a(S) e^{2\pi i \mathrm{Tr}(SZ)} \\ &= \sum_{\substack{S = \left[\begin{smallmatrix} \alpha & \beta \\ \beta & \gamma \end{smallmatrix}\right] \in B(N) \\ p^n | \gamma}} a(S) e^{2\pi i \mathrm{Tr}(SZ)} \\ &= \sum_{S \in B(Np)} a(S) e^{2\pi i \mathrm{Tr}(SZ)}.\end{aligned}$$

This proves (11.7). Finally,

$$\begin{aligned}(\tau_p F)(Z) &= p^{-n} \sum_{x \in \mathbb{Z}/p^n\mathbb{Z}} (F\big|_k \begin{bmatrix} 1 & & & \\ & 1 & & xp^{-n} \\ & & 1 & \\ & & & 1 \end{bmatrix})(Z) \\ &= p^{-n} \sum_{x \in \mathbb{Z}/p^n\mathbb{Z}} F(Z + [\ \ xp^{-n}]) \\ &= p^{-n} \sum_{x \in \mathbb{Z}/p^n\mathbb{Z}} \sum_{\substack{m=0 \\ N_s | m}}^{\infty} f_m(\tau, z) e^{2\pi i m(\tau' + xp^{-n})} \\ &= p^{-n} \sum_{\substack{m=0 \\ N_s | m}}^{\infty} f_m(\tau, z) e^{2\pi i m \tau'} \sum_{x \in \mathbb{Z}/p^n\mathbb{Z}} e^{2\pi i m(xp^{-n})} \\ &= \sum_{\substack{m=0 \\ N_s | m,\ p^n | m}}^{\infty} f_m(\tau, z) e^{2\pi i m \tau'}.\end{aligned}$$

This completes the proof. □

Lemma 11.2.2 *Let N and k be integers such that $N > 0$ and $k > 0$, and let p be a prime of $\mathbb{Z}$. Let $\theta : \mathcal{A}_k(\mathcal{K}_s(N)) \to \mathcal{A}_k(\mathcal{K}_s(Np))$ be the linear map obtained as in Sect. 11.1 from the level raising operator* (3.41), *and let θ_p be as in* (11.1), *so that θ_p is a linear map*

$$\theta_p : M_k(\mathrm{K}_s(N)) \longrightarrow M_k(\mathrm{K}_s(Np)).$$

Let $F \in M_k(\mathrm{K}_s(N))$ with Fourier expansion as in (11.4). *Then*

$$\theta_p F = F\big|_k \begin{bmatrix} 1 &&& \\ & 1 && \\ && p^{-1} & \\ &&& p^{-1} \end{bmatrix} + \sum_{x \in \mathbb{Z}/p\mathbb{Z}} F|_k(\begin{bmatrix} p^{-1} &&& \\ & 1 && \\ && 1 & \\ &&& p^{-1} \end{bmatrix}\begin{bmatrix} 1 && x & \\ & 1 && \\ && 1 & \\ &&& 1 \end{bmatrix}). \tag{11.9}$$

If F is a cusp form, then $\theta_p F$ is a cusp form. We have

$$\begin{aligned}(\theta_p F)(Z) = &\sum_{\substack{S \in B(N) \\ p|\alpha,\ p|2\beta,\ N_s p|\gamma}} p^k a(p^{-1}S) e^{2\pi i \mathrm{Tr}(SZ)} \\ &+ \sum_{\substack{S \in B(N) \\ N_s p|\gamma}} p\, a(p^{-1}S[\begin{bmatrix} p & \\ & 1 \end{bmatrix}]) e^{2\pi i \mathrm{Tr}(SZ)}.\end{aligned} \tag{11.10}$$

The Fourier-Jacobi expansion of $\theta_p F$ is

$$(\theta_p F)(Z) = p \sum_{\substack{m=0 \\ N_s p|m}}^{\infty} (f_{mp^{-1}}|_{k,mp^{-1}} V_p)(\tau, z) e^{2\pi i m \tau'}, \qquad Z = \begin{bmatrix} \tau & z \\ z & \tau' \end{bmatrix} \in \mathcal{H}_2. \tag{11.11}$$

Here, V_p is the operator from (2) of Chap. I, Sect. 4 on p. 41 of [38]. Explicitly,

$$(f_{mp^{-1}}\big|_{k,mp^{-1}} V_p)(\tau, z) = p^{k-1} f_{mp^{-1}}(p\tau, pz) + p^{-1} \sum_{x \in \mathbb{Z}/p\mathbb{Z}} f_{mp^{-1}}(p^{-1}(\tau + x), z)$$

for $\tau \in \mathcal{H}_1$ and $z \in \mathbb{C}$.

Proof The proof of (11.9) is very similar to the proof of Lemma 10.2.2 and will be omitted. To compute the Fourier expansion we use (11.9):

$$\begin{aligned}(\theta_p F)(Z) &= (F\big|_k \begin{bmatrix} 1 &&& \\ & 1 && \\ && p^{-1} & \\ &&& p^{-1} \end{bmatrix})(Z) + \sum_{x \in \mathbb{Z}/p\mathbb{Z}} F|_k(\begin{bmatrix} p^{-1} &&& \\ & 1 && \\ && 1 & \\ &&& p^{-1} \end{bmatrix}\begin{bmatrix} 1 && x & \\ & 1 && \\ && 1 & \\ &&& 1 \end{bmatrix})(Z) \\ &= p^k F(pZ) + \sum_{x \in \mathbb{Z}/p\mathbb{Z}} F(\begin{bmatrix} p^{-1} & \\ & 1 \end{bmatrix} Z \begin{bmatrix} 1 & \\ & p \end{bmatrix} + \begin{bmatrix} p^{-1}x & \\ & \end{bmatrix})\end{aligned}$$

$$= p^k \sum_{S\in B(N)} a(S)e^{2\pi i \mathrm{Tr}(pSZ)}$$

$$+ \sum_{x\in\mathbb{Z}/p\mathbb{Z}} \sum_{S\in B(N)} a(S)e^{2\pi i \mathrm{Tr}(S(\left[\begin{smallmatrix} p^{-1} & \\ & 1\end{smallmatrix}\right] Z \left[\begin{smallmatrix} 1 & \\ & p\end{smallmatrix}\right] + \left[\begin{smallmatrix} p^{-1}x & \\ & \end{smallmatrix}\right]))}$$

$$= p^k \sum_{S=\left[\begin{smallmatrix} \alpha & \beta \\ \beta & \gamma\end{smallmatrix}\right]\in B(N)} a(S)e^{2\pi i \mathrm{Tr}(pSZ)}$$

$$+ \sum_{S=\left[\begin{smallmatrix} \alpha & \beta \\ \beta & \gamma\end{smallmatrix}\right]\in B(N)} a(S)\Big(\sum_{x\in\mathbb{Z}/p\mathbb{Z}} e^{2\pi i p^{-1}x\alpha}\Big)e^{2\pi i \mathrm{Tr}(S\left[\begin{smallmatrix} p^{-1} & \\ & 1\end{smallmatrix}\right] Z \left[\begin{smallmatrix} 1 & \\ & p\end{smallmatrix}\right])}$$

$$= p^k \sum_{S=\left[\begin{smallmatrix} \alpha & \beta \\ \beta & \gamma\end{smallmatrix}\right]\in B(N)} a(S)e^{2\pi i \mathrm{Tr}(pSZ)}$$

$$+ p \sum_{\substack{S=\left[\begin{smallmatrix} \alpha & \beta \\ \beta & \gamma\end{smallmatrix}\right]\in B(N) \\ p\mid\alpha}} a(S)e^{2\pi i \mathrm{Tr}(\left[\begin{smallmatrix} 1 & \\ & p\end{smallmatrix}\right] S \left[\begin{smallmatrix} p^{-1} & \\ & 1\end{smallmatrix}\right] Z)}$$

$$= \sum_{\substack{S=\left[\begin{smallmatrix} \alpha & \beta \\ \beta & \gamma\end{smallmatrix}\right]\in B(N) \\ p\mid\alpha,\ p\mid 2\beta,\ N_s p\mid\gamma}} p^k a(p^{-1}S)e^{2\pi i \mathrm{Tr}(SZ)}$$

$$+ \sum_{\substack{S=\left[\begin{smallmatrix} \alpha & \beta \\ \beta & \gamma\end{smallmatrix}\right]\in B(N) \\ N_s p\mid\gamma}} pa(\left[\begin{smallmatrix} 1 & \\ & p^{-1}\end{smallmatrix}\right] S \left[\begin{smallmatrix} p & \\ & 1\end{smallmatrix}\right])e^{2\pi i \mathrm{Tr}(SZ)}.$$

This proves (11.10).

To obtain the Fourier-Jacobi expansion of $\theta_p F$ we proceed from the calculation for (11.10) and use (11.5):

$$(\theta_p F)(Z) = p^k F(pZ) + \sum_{x\in\mathbb{Z}/p\mathbb{Z}} F(\left[\begin{smallmatrix} p^{-1} & \\ & 1\end{smallmatrix}\right] Z \left[\begin{smallmatrix} 1 & \\ & p\end{smallmatrix}\right] + \left[\begin{smallmatrix} p^{-1}x & \\ & \end{smallmatrix}\right])$$

$$= p^k \sum_{\substack{m=0 \\ N_s\mid m}}^{\infty} f_m(p\tau, pz)e^{2\pi i mp\tau'} + \sum_{x\in\mathbb{Z}/p\mathbb{Z}} \sum_{\substack{m=0 \\ N_s\mid m}}^{\infty} f_m(p^{-1}\tau + p^{-1}x, z)e^{2\pi i mp\tau'}$$

$$= \sum_{\substack{m=0\\N_s p|m}}^{\infty} (p^k f_{mp^{-1}}(p\tau, pz) + \sum_{x\in\mathbb{Z}/p\mathbb{Z}} f_{mp^{-1}}(p^{-1}\tau + p^{-1}x, z))e^{2\pi i m\tau'}$$

$$= p \sum_{\substack{m=0\\N_s p|m}}^{\infty} (f_{mp^{-1}}\big|_k V_p)(\tau, z)e^{2\pi i m\tau'}.$$

This completes the proof. □

Lemma 11.2.3 *Let N and k be integers such that $N > 0$ and $k > 0$, and let p be a prime of $\mathbb{Z}$. Let $\eta : \mathcal{A}_k(\mathcal{K}_s(N)) \to \mathcal{A}_k(\mathcal{K}_s(Np^2))$ be the linear map obtained as in Sect. 11.1 from the level raising operator* (3.29), *and let η_p be as in* (11.1), *so that η_p is a linear map*

$$\eta_p : M_k(\mathrm{K}_s(N)) \longrightarrow M_k(\mathrm{K}_s(Np^2)).$$

Let $F \in M_k(\mathrm{K}_s(N))$ with Fourier expansion as in (11.4). *Then*

$$\eta_p F = F\big|_k \begin{bmatrix} 1 &&& \\ & p && \\ && 1 & \\ &&& p^{-1} \end{bmatrix}. \tag{11.12}$$

If F is a cusp form, then $\eta_p F$ is a cusp form. We have

$$(\eta_p F)(Z) = \sum_{\substack{S=\left[\begin{smallmatrix}\alpha & \beta\\ \beta & \gamma\end{smallmatrix}\right]\in B(Np^2)\\ p|2\beta,\ p^2|\gamma}} p^k a(S[\left[\begin{smallmatrix}1 & \\ & p^{-1}\end{smallmatrix}\right]])e^{2\pi i \mathrm{Tr}(SZ)}. \tag{11.13}$$

The Fourier-Jacobi expansion of $\eta_p F$ is

$$(\eta_p F)(Z) = \sum_{\substack{m=0\\ p^2|m,\ N_s|m}}^{\infty} p^k (f_{mp^{-2}}\big|_{k,mp^{-2}} U_p)(\tau, z)e^{2\pi i m\tau'} \tag{11.14}$$

for $Z = \left[\begin{smallmatrix}\tau & z\\ z & \tau'\end{smallmatrix}\right] \in \mathcal{H}_2$. Here, U_p is the operator from (1) of Chap. I, Sect. 4 on p. 41 of [38]. Explicitly, we have

$$(f_{mp^{-2}}\big|_{k,mp^{-2}} U_p)(\tau, z) = f_{mp^{-2}}(\tau, pz)$$

for integers m such that $m \geq 0$, $p^2 \mid m$, $\tau \in \mathcal{H}_1$, and $z \in \mathbb{C}$.

Proof Define $\Phi = \Phi_F$. Let $Z \in \mathcal{H}_2$ and let $h \in \mathrm{GSp}(4,\mathbb{R})^+$ be such that $Z = h\langle I\rangle$. Then by (11.3),

$$
\begin{aligned}
(\eta_p F)(Z) &= \lambda(h)^{-k} j(h,I)^k \eta\Phi(h) \\
&= \lambda(h)^{-k} j(h,I)^k \Phi(h \begin{bmatrix}1&&&\\&p^{-1}&&\\&&1&\\&&&p\end{bmatrix}_p) \\
&= \lambda(h)^{-k} j(h,I)^k \Phi(\begin{bmatrix}1&&&\\&p^{-1}&&\\&&1&\\&&&p\end{bmatrix}_p h) \\
&= \lambda(h)^{-k} j(h,I)^k \Phi(\underbrace{\begin{bmatrix}1&&&\\&p&&\\&&1&\\&&&p^{-1}\end{bmatrix}}_{\text{all places}}\begin{bmatrix}1&&&\\&p^{-1}&&\\&&1&\\&&&p\end{bmatrix}_p h) \\
&= \lambda(h)^{-k} j(h,I)^k \Phi(\begin{bmatrix}1&&&\\&p&&\\&&1&\\&&&p^{-1}\end{bmatrix}_\infty h) \\
&= (F\big|_k \begin{bmatrix}1&&&\\&p&&\\&&1&\\&&&p^{-1}\end{bmatrix})(Z).
\end{aligned}
$$

To prove (11.13) we use (11.12):

$$
\begin{aligned}
(\eta_p F)(Z) &= (F\big|_k \begin{bmatrix}1&&&\\&p&&\\&&1&\\&&&p^{-1}\end{bmatrix})(Z) \\
&= p^k F(\begin{bmatrix}1&\\&p\end{bmatrix} Z \begin{bmatrix}1&\\&p\end{bmatrix}) \\
&= \sum_{S\in B(N)} p^k a(S) e^{2\pi i \mathrm{Tr}(S\begin{bmatrix}1&\\&p\end{bmatrix} Z \begin{bmatrix}1&\\&p\end{bmatrix})} \\
&= \sum_{S\in B(N)} p^k a(S) e^{2\pi i \mathrm{Tr}(\begin{bmatrix}1&\\&p\end{bmatrix} S \begin{bmatrix}1&\\&p\end{bmatrix} Z)} \\
&= \sum_{\substack{S=\begin{bmatrix}\alpha&\beta\\\beta&\gamma\end{bmatrix}\in B(Np^2)\\ p|2\beta,\ p^2|\gamma}} p^k a(\begin{bmatrix}1&\\&p^{-1}\end{bmatrix} S \begin{bmatrix}1&\\&p^{-1}\end{bmatrix}) e^{2\pi i \mathrm{Tr}(SZ)} \\
&= \sum_{\substack{S=\begin{bmatrix}\alpha&\beta\\\beta&\gamma\end{bmatrix}\in B(Np^2)\\ p|2\beta,\ p^2|\gamma}} p^k a(S[\begin{bmatrix}1&\\&p^{-1}\end{bmatrix}]) e^{2\pi i \mathrm{Tr}(SZ)}.
\end{aligned}
$$

This proves (11.13).

To obtain the Fourier-Jacobi expansion, we calculate as follows:

$$\begin{aligned}
(\eta_p F)(Z) &= (F\big|_k \begin{bmatrix} 1 & & & \\ & p & & \\ & & 1 & \\ & & & p^{-1} \end{bmatrix})(Z) \\
&= p^k F(\begin{bmatrix} 1 & \\ & p \end{bmatrix}\begin{bmatrix} \tau & z \\ z & \tau' \end{bmatrix}\begin{bmatrix} 1 & \\ & p \end{bmatrix}) \\
&= p^k \sum_{\substack{m=0 \\ N_s|m}}^{\infty} f_m(\tau, pz) e^{2\pi i m p^2 \tau'} \\
&= p^k \sum_{\substack{m=0 \\ p^2|m,\ N_s|m}}^{\infty} f_{mp^{-2}}(\tau, pz) e^{2\pi i m \tau'} \\
&= p^k \sum_{\substack{m=0 \\ p^2|m,\ N_s|m}}^{\infty} (f_{mp^{-2}}\big|_k U_p)(\tau, z) e^{2\pi i m \tau'}.
\end{aligned}$$

This completes the proof. □

11.3 A Level Lowering Operator

In this section we calculate the formula for the level lowering operator σ_p obtained from the corresponding local operator defined in Sect. 3.7. The formula for the Fourier-Jacobi expansion of $\sigma_p F$ involves a certain index lowering operator L_{c^2} on Jacobi forms. This operator is described in the next lemma. Using the Petersson inner product as described in [38] and [79], one can define the adjoint of a linear operator between spaces of Jacobi cusp forms. One can show that, up to a constant, $L_{c^2} : J_{k,mc^2}^{\text{cusp}} \to J_{k,m}^{\text{cusp}}$ is the adjoint of the operator $U_c : J_{k,m}^{\text{cusp}} \to J_{k,mc^2}^{\text{cusp}}$ from Chap. I, Sect. 4 on p. 41 of [38].

Lemma 11.3.1 *Let k, m, and c be integers such that $k > 0$, $m \geq 0$ and $c > 0$. Let f be a Jacobi form of weight k and index mc^2 on* SL(2, $\mathbb{Z}$)*. Define $L_{c^2} f : \mathcal{H}_1 \times \mathbb{C} \to \mathbb{C}$ by*

$$(L_{c^2} f)(\tau, z) = c^{-k} \sum_{a,b \in \mathbb{Z}/c\mathbb{Z}} f(\tau, (b + a\tau + z)c^{-1}) e^{2\pi i m(2az + a^2\tau)} \qquad (11.15)$$

for $\tau \in \mathcal{H}_1$ and $z \in \mathbb{C}$. Then $L_{c^2} f$ is a well-defined Jacobi form of weight k and index m on SL(2, $\mathbb{Z}$)*. If f is a cusp form, then $L_{c^2} f$ is a cusp form.*

Proof If $g : \mathcal{H}_1 \times \mathbb{C} \to \mathbb{C}$ is a function, then we define $g\big|U : \mathcal{H}_1 \times \mathbb{C} \to \mathbb{C}$ by $g(\tau, z) = g(\tau, c^{-1}z)$ for $\tau \in \mathcal{H}_1$ and $z \in \mathbb{C}$. If $\gamma \in \mathrm{SL}(2, \mathbb{R})$, $[\lambda, \mu, \kappa] \in H(\mathbb{R})$, and $g : \mathcal{H}_1 \times \mathbb{C} \to \mathbb{C}$ is a function, then

$$(g\big|U)\big|_{k,m}\gamma = (g\big|_{k,mc^2}\gamma)\big|U, \tag{11.16}$$

$$(g\big|U)\big|_m[\lambda, \mu, \kappa] = (g\big|_{mc^2}[c^{-1}\lambda, c^{-1}\mu, c^{-2}\kappa])\big|U. \tag{11.17}$$

Using (11.17) we see that the sum

$$c^{-k} \sum_{a,b\in\mathbb{Z}/c\mathbb{Z}} (f\big|U)\big|_m[a, b, 0] = c^{-k} \sum_{a,b\in\mathbb{Z}/c\mathbb{Z}} f(\tau, (b + a\tau + z)c^{-1})e^{2\pi i m(2az+a^2\tau)}$$

is well-defined. It follows that $L_{c^2} f$ is well-defined. It is straightforward to verify that $(L_{c^2} f)\big|_m[\lambda, \mu, \kappa] = L_{c^2} f$ for $[\lambda, \mu, \kappa] \in H(\mathbb{Z})$. Let $\gamma = \left[\begin{smallmatrix} \gamma_1 & \gamma_2 \\ \gamma_3 & \gamma_4 \end{smallmatrix}\right] \in \mathrm{SL}(2, \mathbb{Z})$. Then

$$\begin{aligned}
(L_{c^2} f)\big|_{k,m}\gamma &= c^{-k} \sum_{a,b\in\mathbb{Z}/c\mathbb{Z}} ((f\big|U)\big|_m[a, b, 0])\big|_{k,m}\gamma \\
&= c^{-k} \sum_{a,b\in\mathbb{Z}/c\mathbb{Z}} ((f\big|U)\big|_{k,m}\gamma)\big|_m[a\gamma_1 + b\gamma_3, a\gamma_2 + b\gamma_4, 0] \\
&= c^{-k} \sum_{a,b\in\mathbb{Z}/c\mathbb{Z}} ((f\big|_{k,mc^2}\gamma)\big|U)\big|_m[a, b] \\
&= c^{-k} \sum_{a,b\in\mathbb{Z}/c\mathbb{Z}} (f\big|U)\big|_m[a, b] \\
&= L_{c^2} f.
\end{aligned}$$

Here, the second equality follows from the commutation rule (10.21) and the third equality is verified by a calculation. We next show that $L_{c^2} f$ has an absolutely convergent Fourier expansion of the form

$$L_{c^2} f(\tau, z) = \sum_{n'=0}^{\infty} \sum_{\substack{r\in\mathbb{Z} \\ r'^2\leq 4n'm}} c'(n', r')e^{2\pi i(n'\tau+r'z)}.$$

Let

$$f(\tau, z) = \sum_{n=0}^{\infty} \sum_{\substack{r\in\mathbb{Z} \\ r^2\leq 4nmc^2}} c(n, r)e^{2\pi i(n\tau+rz)}$$

be the Fourier expansion of f. Then

$$
\begin{aligned}
(L_{c^2}f)(\tau,z) &= c^{-k}\sum_{a,b\in\mathbb{Z}/c\mathbb{Z}} f(\tau,(b+a\tau+z)c^{-1})e^{2\pi i m(2az+a^2\tau)}\\
&= c^{-k}\sum_{a=0}^{c-1}\sum_{n=0}^{\infty}\sum_{\substack{r\in\mathbb{Z}\\ r^2\le 4nmc^2}}\Big(\sum_{b\in\mathbb{Z}/c\mathbb{Z}} e^{2\pi i rbc^{-1}}\Big)c(n,r)\\
&\quad\times e^{2\pi i(n+arc^{-1}+ma^2)\tau}e^{2\pi i(rc^{-1}+2ma)z}\\
&= c^{1-k}\sum_{a=0}^{c-1}\sum_{n=0}^{\infty}\sum_{\substack{r\in\mathbb{Z}\\ c\mid r,\ r^2\le 4nmc^2}} c(n,r)\\
&\quad\times e^{2\pi i(n+arc^{-1}+ma^2)\tau}e^{2\pi i(rc^{-1}+2ma)z}\\
&= c^{1-k}\sum_{a=0}^{c-1}\sum_{n=0}^{\infty}\sum_{\substack{r\in\mathbb{Z}\\ r^2\le 4nm}} c(n,rc)e^{2\pi i(n+ar+ma^2)\tau}e^{2\pi i(r+2ma)z}.
\end{aligned}
$$

Let $X=\{(n,r)\in\mathbb{Z}\times\mathbb{Z}\mid n\ge 0,\ r^2\le 4nm\}$. Let $a\in\{0,\dots,c-1\}$. Define $i:\mathbb{Z}\times\mathbb{Z}\to\mathbb{Z}\times\mathbb{Z}$ by $i(n,r)=(n+ra+ma^2,r+2ma)$ for $(n,r)\in\mathbb{Z}\times\mathbb{Z}$. Then i is injective. Set $X_a=i(X)$. Calculations show that if $(n',r')\in X_a$, then $n'\ge 0$ and $r'^2\le 4n'm$. It follows that $X_a\subset X$. We now have

$$
(L_{c^2}f)(\tau,z) = c^{1-k}\sum_{a=0}^{c-1}\sum_{(n',r')\in X_a} c(n,rc)e^{2\pi i(n'\tau+r'z)}
$$

where $(n,r)=i^{-1}(n',r')$. This proves that $L_{c^2}f$ has a Fourier expansion with the required properties. It is easy to see from the Fourier expansion that if f is a cusp form, then so is $L_{c^2}f$. □

Lemma 11.3.2 *Let N and k be integers such that $N>0$ and $k>0$, and let p be a prime of $\mathbb{Z}$. Assume that $v_p(N)\ge 2$. Let $\sigma:\mathcal{A}_k(\mathcal{K}_s(N))\to\mathcal{A}_k(\mathcal{K}_s(Np^{-1}))$ be the linear map obtained as in Sect. 11.1 from the level lowering operator* (3.60), *and let σ_p be as in* (11.1), *so that σ_p is a linear map*

$$
\sigma_p : M_k(\mathrm{K}_s(N)) \longrightarrow M_k(\mathrm{K}_s(Np^{-1})).
$$

Let $F\in M_k(\mathrm{K}_s(N))$ with Fourier expansion as in (11.4). *Then*

$$
\sigma_p F = p^{-3}\sum_{a,b,c\in\mathbb{Z}/p\mathbb{Z}} F\big|_k \begin{bmatrix} p&&&\\ &1&&\\ &&p&\\ &&&p^2\end{bmatrix}\begin{bmatrix}1&&&b\\ a&1&b&cp^{-n+2}\\ &&1&-a\\ &&&1\end{bmatrix}. \tag{11.18}
$$

If F is a cusp form, then $\sigma_p F$ is a cusp form. We have

$$(\sigma_p F)(Z) = \sum_{S\in B(Np^{-1})} \sum_{a\in\mathbb{Z}/p\mathbb{Z}} p^{-k-1} a(S[\left[\begin{smallmatrix}1&\\a&p\end{smallmatrix}\right]])e^{2\pi i \mathrm{Tr}(SZ)}. \tag{11.19}$$

The Fourier-Jacobi expansion of $\sigma_p F$ is given by

$$(\sigma_p F)(Z) = p^{-2} \sum_{\substack{m=0\\ p^{-1}N_s|m}}^{\infty} (L_{p^2} f_{mp^2})(\tau, z)e^{2\pi i m\tau'}, \qquad Z = \left[\begin{smallmatrix}\tau&z\\z&\tau'\end{smallmatrix}\right] \in \mathcal{H}_2. \tag{11.20}$$

Here L_{p^2} is as in Lemma 11.3.1.

Proof Define $\Phi = \Phi_F$. Let $Z \in \mathcal{H}_2$ and let $h \in \mathrm{GSp}(4, \mathbb{R})^+$ be such that $Z = h\langle I\rangle$. Let $n = v_p(N)$. Then by (11.3) and (3.61),

$$\begin{aligned}
(\sigma_p F)(Z) &= \lambda(h)^{-k} j(h, I)^k \sigma\Phi(h)\\
&= \lambda(h)^{-k} j(h, I)^k p^{-3} \sum_{a,b,c\in\mathbb{Z}/p\mathbb{Z}} \Phi(h \begin{bmatrix}1&&b&\\a&1&b&cp^{-n+2}\\&&1&-a\\&&&1\end{bmatrix}_p \begin{bmatrix}1&&&\\&p&&\\&&1&\\&&&p^{-1}\end{bmatrix}_p)\\
&= \lambda(h)^{-k} j(h, I)^k p^{-3} \sum_{a,b,c\in\mathbb{Z}/p\mathbb{Z}} \Phi(\underbrace{\begin{bmatrix}1&&&\\&p^{-1}&&\\&&1&\\&&&p\end{bmatrix}}_{\text{all places}}\\
&\qquad\times \underbrace{\begin{bmatrix}1&&-b&\\-a&1&-b&-cp^{-n+2}\\&&1&a\\&&&1\end{bmatrix}}_{\text{all places}} h \begin{bmatrix}1&&b&\\a&1&b&cp^{-n+2}\\&&1&-a\\&&&1\end{bmatrix}_p \begin{bmatrix}1&&&\\&p&&\\&&1&\\&&&p^{-1}\end{bmatrix}_p)\\
&= \lambda(h)^{-k} j(h, I)^k p^{-3} \sum_{a,b,c\in\mathbb{Z}/p\mathbb{Z}} \Phi(\begin{bmatrix}p&&&\\&1&&\\&&p&\\&&&p^2\end{bmatrix}_\infty \begin{bmatrix}1&&b&\\a&1&b&cp^{-n+2}\\&&1&-a\\&&&1\end{bmatrix}_\infty h)\\
&= p^{-3} \sum_{a,b,c\in\mathbb{Z}/p\mathbb{Z}} (F\big|_k \begin{bmatrix}p&&&\\&1&&\\&&p&\\&&&p^2\end{bmatrix}\begin{bmatrix}1&&b&\\a&1&b&cp^{-n+2}\\&&1&-a\\&&&1\end{bmatrix})(Z).
\end{aligned}$$

To prove (11.19) we use (11.18):

$$(\sigma_p F)(Z) = p^{-3} \sum_{a,b,c\in\mathbb{Z}/p\mathbb{Z}} (F\big|_k \begin{bmatrix}p&&&\\&1&&\\&&p&\\&&&p^2\end{bmatrix}\begin{bmatrix}1&&b&\\a&1&b&cp^{-n+2}\\&&1&-a\\&&&1\end{bmatrix})(Z)$$

$$= p^{-k-3} \sum_{a,b,c\in\mathbb{Z}/p\mathbb{Z}} F(\begin{bmatrix} p & \\ a & 1 \end{bmatrix} Z \begin{bmatrix} p^{-1} & ap^{-2} \\ & p^{-2} \end{bmatrix} + \begin{bmatrix} & bp^{-1} \\ bp^{-1} & abp^{-2}+cp^{-n} \end{bmatrix})$$

$$= p^{-k-3} \sum_{S=\left[\begin{smallmatrix} \alpha & \beta \\ \beta & \gamma \end{smallmatrix}\right]\in B(N)} \Big(\sum_{a,b\in\mathbb{Z}/p\mathbb{Z}} e^{2\pi i(\gamma abp^{-2}+2\beta bp^{-1})} \Big) \Big(\sum_{c\in\mathbb{Z}/p\mathbb{Z}} e^{2\pi i\gamma cp^{-n}} \Big)$$

$$\times a(S) e^{2\pi i \operatorname{Tr}(\left[\begin{smallmatrix} p^{-1} & ap^{-2} \\ & p^{-2} \end{smallmatrix}\right] S \left[\begin{smallmatrix} p & \\ a & 1 \end{smallmatrix}\right] Z)}$$

$$= p^{-k-2} \sum_{\substack{S=\left[\begin{smallmatrix} \alpha & \beta \\ \beta & \gamma \end{smallmatrix}\right]\in B(N) \\ p^n|\gamma}} \sum_{a\in\mathbb{Z}/p\mathbb{Z}} \Big(\sum_{b\in\mathbb{Z}/p\mathbb{Z}} e^{2\pi i(2\beta bp^{-1})} \Big)$$

$$\times a(S) e^{2\pi i \operatorname{Tr}(\left[\begin{smallmatrix} p^{-1} & ap^{-2} \\ & p^{-2} \end{smallmatrix}\right] S \left[\begin{smallmatrix} p & \\ a & 1 \end{smallmatrix}\right] Z)}$$

$$= p^{-k-1} \sum_{a\in\mathbb{Z}/p\mathbb{Z}} \sum_{\substack{S=\left[\begin{smallmatrix} \alpha & \beta \\ \beta & \gamma \end{smallmatrix}\right]\in B(N) \\ p^n|\gamma,\ p|2\beta}} a(S) e^{2\pi i \operatorname{Tr}(\left[\begin{smallmatrix} p^{-1} & ap^{-2} \\ & p^{-2} \end{smallmatrix}\right] S \left[\begin{smallmatrix} p & \\ a & 1 \end{smallmatrix}\right] Z)}$$

$$= p^{-k-1} \sum_{a\in\mathbb{Z}/p\mathbb{Z}} \sum_{S\in B(Np^{-1})} a(\left[\begin{smallmatrix} 1 & a \\ & p \end{smallmatrix}\right] S \left[\begin{smallmatrix} 1 & \\ a & p \end{smallmatrix}\right]) e^{2\pi i \operatorname{Tr}(SZ)}$$

$$= \sum_{S\in B(Np^{-1})} \sum_{a\in\mathbb{Z}/p\mathbb{Z}} p^{-k-1} a(S[\left[\begin{smallmatrix} 1 & \\ a & p \end{smallmatrix}\right]]) e^{2\pi i \operatorname{Tr}(SZ)}.$$

This proves (11.19).

Finally, beginning from (11.18), we have:

$$(\sigma_p F)(Z) = p^{-k-3} \sum_{a,b,c\in\mathbb{Z}/p\mathbb{Z}} F(\begin{bmatrix} p & \\ a & 1 \end{bmatrix} Z \begin{bmatrix} p^{-1} & ap^{-2} \\ & p^{-2} \end{bmatrix} + \begin{bmatrix} & bp^{-1} \\ bp^{-1} & abp^{-2}+cp^{-n} \end{bmatrix})$$

$$= p^{-k-3} \sum_{a,b,c\in\mathbb{Z}/p\mathbb{Z}} \sum_{\substack{m=0 \\ N_s|m}}^{\infty} f_m(\tau, (b+a\tau+z)p^{-1})$$

$$\times e^{2\pi i m(2ap^{-2}z+a^2p^{-2}\tau+cp^{-n}+abp^{-2}+p^{-2}\tau')}$$

$$= p^{-k-3} \sum_{a,b\in\mathbb{Z}/p\mathbb{Z}} \sum_{\substack{m=0 \\ N_s|m}}^{\infty} f_m(\tau, (b+a\tau+z)p^{-1})$$

$$\times \Big(\sum_{c\in\mathbb{Z}/p\mathbb{Z}} e^{2\pi i mcp^{-n}} \Big) e^{2\pi i m(2ap^{-2}z+a^2p^{-2}\tau+abp^{-2}+p^{-2}\tau')}$$

$$= p^{-k-2} \sum_{a,b\in\mathbb{Z}/p\mathbb{Z}} \sum_{\substack{m=0\\ pN_s|m}}^{\infty} f_m(\tau, (b+a\tau+z)p^{-1}) e^{2\pi i m p^{-2}(2az+a^2\tau+\tau')}$$

$$= p^{-k-2} \sum_{a,b\in\mathbb{Z}/p\mathbb{Z}} \sum_{\substack{m=0\\ pN_s|m}}^{\infty} f_m(\tau, (b+a\tau+z)p^{-1}) e^{2\pi i m p^{-2}(2az+a^2\tau)} e^{2\pi i \tau' m p^{-2}}$$

$$= p^{-k-2} \sum_{\substack{m=0\\ p^{-1}N_s|m}}^{\infty} \sum_{a,b\in\mathbb{Z}/p\mathbb{Z}} f_{mp^2}(\tau, (b+a\tau+z)p^{-1}) e^{2\pi i m(2az+a^2\tau)} e^{2\pi i m\tau'}$$

$$= p^{-2} \sum_{\substack{m=0\\ p^{-1}N_s|m}}^{\infty} (L_{p^2} f_{mp^2})(\tau, z) e^{2\pi i m\tau'}.$$

This completes the proof. □

11.4 Hecke Operators

In this section we calculate the formulas for the Hecke operators $T^s_{0,1}(p)$ and $T^s_{1,0}(p)$ obtained from the corresponding local operators defined in Sect. 3.8. The formula for the Fourier-Jacobi expansion of $T^s_{1,0}(p)F$ involves the index lowering operator L_{c^2} on Jacobi forms discussed in the previous section. For the formula for the Fourier-Jacobi expansion of $T^s_{0,1}(p)F$ we introduce in Lemma 11.4.1 below another index lowering operator L'_p on Jacobi forms. One can show that, up to a constant, $L'_p : J^{\text{cusp}}_{k,mp} \to J^{\text{cusp}}_{k,m}$ is the adjoint of the operator $V_p : J^{\text{cusp}}_{k,m} \to J^{\text{cusp}}_{k,mp}$ from Chap. I, Sect. 4 on p. 41 of [38]. The adjoint of $V_N : J^{\text{cusp}}_{k,1} \to J^{\text{cusp}}_{k,N}$ was also considered in [79].

Lemma 11.4.1 *Let k and m be integers such that $k > 0$ and $m \geq 0$. Let p be a prime of $\mathbb{Z}$. Let f be a Jacobi form of weight k and index pm on* SL(2, $\mathbb{Z}$). *Define $L'_p f : \mathcal{H}_1 \times \mathbb{C} \to \mathbb{C}$ by*

$$L'_p f(\tau, z) = p \sum_{a\in\mathbb{Z}/p\mathbb{Z}} f(p\tau, a\tau+z) e^{2\pi i m(a^2\tau+2az)}$$
$$+ p^{1-k} \sum_{a,b\in\mathbb{Z}/p\mathbb{Z}} f(p^{-1}(\tau+b), p^{-1}(z+a))$$

for $\tau \in \mathcal{H}_1$ and $z \in \mathbb{C}$. Then $L'_p f$ is a well-defined Jacobi form of weight k and index m on SL(2, $\mathbb{Z}$). *If f is a cusp form, then $L'_p f$ is a cusp form.*

Proof To prove that $L'_p f$ is well-defined it will suffice to prove that

$$f(p\tau, (a+\ell p)+z)\mathrm{e}^{2\pi i m((a+\ell p)^2\tau+2(a+\ell p)z)} = f(p\tau, a\tau+z)\mathrm{e}^{2\pi i m(a^2\tau+2az)} \tag{11.21}$$

and

$$f(p^{-1}(\tau+b+jp), p^{-1}(z+a+\ell p)) = f(p^{-1}(\tau+b), p^{-1}(z+a)) \tag{11.22}$$

for $a, b, \ell, j \in \mathbb{Z}$, $\tau \in \mathcal{H}_1$, and $z \in \mathbb{C}$. The equation (11.22) follows from the fact that $f(\tau+n_1, z+n_2) = f(\tau, z)$ for $\tau \in \mathcal{H}_1$, $z \in \mathbb{C}$, and $n_1, n_2 \in \mathbb{Z}$. For (11.21), we note first that

$$f(p\tau, a\tau+z)\mathrm{e}^{2\pi i m(a^2\tau+2az)} = (f\big|_{pm}[ap^{-1}, 0, 0])(p\tau, z)$$

for $a \in \mathbb{Z}$, $\tau \in \mathcal{H}_1$, and $z \in \mathbb{C}$; (11.21) now follows from $f\big|_{pm}[n, 0, 0] = f$ for $n \in \mathbb{Z}$.

To prove that $L'_p f\big|_{k,m} g = L'_p f$ for $g \in \mathrm{SL}(2, \mathbb{Z})$ it will suffice to prove that $(L'_p f)\big|_{k,m} \left[\begin{smallmatrix} & 1 \\ -1 & \end{smallmatrix}\right] = L'_p f$ and $(L'_p f)\big|_{k,m} \left[\begin{smallmatrix} 1 & 1 \\ & 1 \end{smallmatrix}\right] = L'_p f$. For each $b \in \mathbb{Z}$ relatively prime to p fix $e_b, f_b \in \mathbb{Z}$ such that $bf_b - e_b p = 1$. In the following calculation we will use that the map $(\mathbb{Z}/p\mathbb{Z})^\times \to (\mathbb{Z}/p\mathbb{Z})^\times$ defined by $b \mapsto -f_b$ is a bijection; we will also use (10.22). Let $\tau \in \mathcal{H}_1$ and $z \in \mathbb{C}$. Then

$$\begin{aligned}
&(L'_p f)\big|_{k,m} \left[\begin{smallmatrix} & 1 \\ -1 & \end{smallmatrix}\right](\tau, z) \\
&= (-\tau)^{-k}\mathrm{e}^{2\pi i m(-z^2\tau^{-1})} L'_p f(-\tau^{-1}, -z\tau^{-1}) \\
&= (-\tau)^{-k} e^{2\pi i m(-z^2\tau^{-1})} \\
&\quad \times \Big(p \sum_{a\in\mathbb{Z}/p\mathbb{Z}} f(-p\tau^{-1}, -a\tau^{-1} - z\tau^{-1})\mathrm{e}^{2\pi i m(-a^2\tau^{-1}-2az\tau^{-1})} \\
&\quad + p^{1-k} \sum_{a,b\in\mathbb{Z}/p\mathbb{Z}} f(p^{-1}(-\tau^{-1}+b), p^{-1}(-z\tau^{-1}+a))\Big) \\
&= (-\tau)^{-k} e^{2\pi i m(-z^2\tau^{-1})} \\
&\quad \times \Big(p \sum_{a\in\mathbb{Z}/p\mathbb{Z}} (f\big|_{k,pm} \left[\begin{smallmatrix} & -1 \\ 1 & \end{smallmatrix}\right])(-p\tau^{-1}, -a\tau^{-1} - z\tau^{-1})\mathrm{e}^{2\pi i m(-a^2\tau^{-1}-2az\tau^{-1})} \\
&\quad + p^{1-k} \sum_{a,b\in\mathbb{Z}/p\mathbb{Z}} (f\big|_{k,pm} \left[\begin{smallmatrix} & -1 \\ 1 & \end{smallmatrix}\right])(p^{-1}(-\tau^{-1}+b), p^{-1}(-z\tau^{-1}+a))\Big) \\
&= p^{1-k}(-\tau)^{-k}\mathrm{e}^{2\pi i m(-z^2\tau^{-1})}
\end{aligned}$$

$$\times \Big(\sum_{a\in\mathbb{Z}/p\mathbb{Z}} (-\tau)^k f(p^{-1}\tau, p^{-1}(z+a)) e^{2\pi i m (a+z)^2\tau^{-1}} e^{2\pi i m(-a^2\tau^{-1}-2az\tau^{-1})}$$

$$+ \sum_{a,b\in\mathbb{Z}/p\mathbb{Z}} (\frac{p\tau}{b\tau-1})^k e^{2\pi i m \frac{-(a\tau-z)^2}{\tau(b\tau-1)}} f(\frac{-p\tau}{b\tau-1}, \frac{a\tau-z}{b\tau-1})\Big)$$

$$= p^{1-k} \sum_{a\in\mathbb{Z}/p\mathbb{Z}} f(p^{-1}\tau, p^{-1}(z+a))$$

$$+ p \sum_{a,b\in\mathbb{Z}/p\mathbb{Z}} (\frac{-1}{b\tau-1})^k e^{2\pi i m(\frac{-a^2\tau+2az-z^2b}{b\tau-1})} f(\frac{-p\tau}{b\tau-1}, \frac{a\tau-z}{b\tau-1})$$

$$= p \sum_{a\in\mathbb{Z}/p\mathbb{Z}} e^{2\pi i m(a^2\tau-2az)} f(p\tau, -a\tau+z) + p^{1-k} \sum_{a\in\mathbb{Z}/p\mathbb{Z}} f(p^{-1}\tau, p^{-1}(z+a))$$

$$+ p \sum_{\substack{a,b\in\mathbb{Z}/p\mathbb{Z}\\ b\neq 0}} (\frac{-1}{b\tau-1})^k e^{2\pi i m(\frac{-a^2\tau+2az-z^2b}{b\tau-1})}$$

$$\times (f\big|_{k,pm} \begin{bmatrix} e_b & f_b \\ -b & -p \end{bmatrix})(\frac{-p\tau}{b\tau-1}, \frac{a\tau-z}{b\tau-1})$$

$$= p \sum_{a\in\mathbb{Z}/p\mathbb{Z}} e^{2\pi i m(a^2\tau+2az)} f(p\tau, a\tau+z) + p^{1-k} \sum_{a\in\mathbb{Z}/p\mathbb{Z}} f(p^{-1}\tau, p^{-1}(z+a))$$

$$+ p^{1-k}(-1)^k \sum_{\substack{a,b\in\mathbb{Z}/p\mathbb{Z}\\ b\neq 0}} e^{2\pi i m(a^2\tau-2az)} f(p^{-1}(\tau-f_b), p^{-1}(a\tau-z))$$

$$= p \sum_{a\in\mathbb{Z}/p\mathbb{Z}} e^{2\pi i m(a^2\tau+2az)} f(p\tau, a\tau+z) + p^{1-k} \sum_{a\in\mathbb{Z}/p\mathbb{Z}} f(p^{-1}\tau, p^{-1}(z+a))$$

$$+ p^{1-k}(-1)^k \sum_{\substack{a,b\in\mathbb{Z}/p\mathbb{Z}\\ b\neq 0}} e^{2\pi i m(a^2\tau-2az)} f(p^{-1}(\tau+b), p^{-1}(a\tau-z))$$

$$= p \sum_{a\in\mathbb{Z}/p\mathbb{Z}} e^{2\pi i m(a^2\tau+2az)} f(p\tau, a\tau+z) + p^{1-k} \sum_{a\in\mathbb{Z}/p\mathbb{Z}} f(p^{-1}\tau, p^{-1}(z+a))$$

$$+ p^{1-k}(-1)^k \sum_{\substack{a,b\in\mathbb{Z}/p\mathbb{Z}\\ b\neq 0}} e^{2\pi i m(a^2\tau-2az)}$$

$$\times (f\big|_{pm}[-a,0,0])(p^{-1}(\tau+b), p^{-1}(a\tau-z))$$

$$= p \sum_{a\in\mathbb{Z}/p\mathbb{Z}} e^{2\pi i m(a^2\tau+2az)} f(p\tau, a\tau+z) + p^{1-k} \sum_{a\in\mathbb{Z}/p\mathbb{Z}} f(p^{-1}\tau, p^{-1}(z+a))$$

$$
\begin{aligned}
&+ p^{1-k}(-1)^k \sum_{\substack{a,b\in\mathbb{Z}/p\mathbb{Z}\\ b\neq 0}} f(p^{-1}(\tau+b), -p^{-1}(z+ab))\\
&= p \sum_{a\in\mathbb{Z}/p\mathbb{Z}} e^{2\pi i m(a^2\tau+2az)} f(p\tau, a\tau+z) + p^{1-k} \sum_{a\in\mathbb{Z}/p\mathbb{Z}} f(p^{-1}\tau, p^{-1}(z+a))\\
&+ p^{1-k} \sum_{\substack{a,b\in\mathbb{Z}/p\mathbb{Z}\\ b\neq 0}} f(p^{-1}(\tau+b), p^{-1}(z+a))\\
&= p \sum_{a\in\mathbb{Z}/p\mathbb{Z}} e^{2\pi i m(a^2\tau+2az)} f(p\tau, a\tau+z)\\
&+ p^{1-k} \sum_{a,b\in\mathbb{Z}/p\mathbb{Z}} f(p^{-1}(\tau+b), p^{-1}(z+a))\\
&= L'_p(\tau, z).
\end{aligned}
$$

And:

$$
\begin{aligned}
(L'_p f)\big|_{k,m} \left[\begin{smallmatrix} 1 & 1\\ & 1\end{smallmatrix}\right](\tau, z) &= (L'_p f)(\tau+1, z)\\
&= p \sum_{a\in\mathbb{Z}/p\mathbb{Z}} f(p\tau+p, a\tau+z+a) e^{2\pi i m(a^2(\tau+1)+2az)}\\
&+ p^{1-k} \sum_{a,b\in\mathbb{Z}/p\mathbb{Z}} f(p^{-1}(\tau+b+1), p^{-1}(z+a))\\
&= p \sum_{a\in\mathbb{Z}/p\mathbb{Z}} f(p\tau, a\tau+z) e^{2\pi i m(a^2\tau+2az)}\\
&+ p^{1-k} \sum_{a,b\in\mathbb{Z}/p\mathbb{Z}} f(p^{-1}(\tau+b), p^{-1}(z+a))\\
&= L'_p(\tau, z),
\end{aligned}
$$

where the penultimate formula follows from (10.22). To prove that $L'_p f\big|_m h = L'_p f$ for $h \in H(\mathbb{Z})$, it suffices to prove that $L'_p f\big|_m h = L'_p f$ for $h = [1, 0, 0]$ and $h = [0, 1, 0]$. We have

$$
\begin{aligned}
(L'_p f\big|_m [1, 0, 0])(\tau, z) &= e^{2\pi i m(\tau+2z)} L'_p(\tau, z+\tau)\\
&= e^{2\pi i m(\tau+2z)} \Big(p \sum_{a\in\mathbb{Z}/p\mathbb{Z}} f(p\tau, a\tau+(z+\tau)) e^{2\pi i m(a^2\tau+2a(z+\tau))}\\
&+ p^{1-k} \sum_{a,b\in\mathbb{Z}/p\mathbb{Z}} f(p^{-1}(\tau+b), p^{-1}(z+\tau+a)) \Big)
\end{aligned}
$$

$$
\begin{aligned}
&= \mathrm{e}^{2\pi i m(\tau+2z)} \Big(p \sum_{a\in\mathbb{Z}/p\mathbb{Z}} f(p\tau, a\tau+z)\mathrm{e}^{2\pi i m((a-1)^2\tau+2(a-1)(z+\tau))} \\
&\quad + p^{1-k} \sum_{a,b\in\mathbb{Z}/p\mathbb{Z}} (f\big|_{pm}[-1,0,0])(p^{-1}(\tau+b), p^{-1}(z+\tau+a)) \Big) \\
&= p \sum_{a\in\mathbb{Z}/p\mathbb{Z}} f(p\tau, a\tau+z)\mathrm{e}^{2\pi i m(a^2\tau+2az)} \\
&\quad + p^{1-k} \sum_{a,b\in\mathbb{Z}/p\mathbb{Z}} \mathrm{e}^{2\pi i m(\tau+2z)} \mathrm{e}^{2\pi i m(-\tau-2z+b-2a)} \\
&\quad \times f(p^{-1}(\tau+b), p^{-1}(z+a-b)) \\
&= p \sum_{a\in\mathbb{Z}/p\mathbb{Z}} f(p\tau, a\tau+z)\mathrm{e}^{2\pi i m(a^2\tau+2az)} \\
&\quad + p^{1-k} \sum_{a,b\in\mathbb{Z}/p\mathbb{Z}} f(p^{-1}(\tau+b), p^{-1}(z+a)) \\
&= L'_p(\tau, z).
\end{aligned}
$$

And

$$
\begin{aligned}
(L'_p f\big|_m[0,1,0])(\tau, z) &= L'_p(\tau, z+1) \\
&= p \sum_{a\in\mathbb{Z}/p\mathbb{Z}} f(p\tau, a\tau+z+1)\mathrm{e}^{2\pi i m(a^2\tau+2a(z+1))} \\
&\quad + p^{1-k} \sum_{a,b\in\mathbb{Z}/p\mathbb{Z}} f(p^{-1}(\tau+b), p^{-1}(z+1+a)) \\
&= p \sum_{a\in\mathbb{Z}/p\mathbb{Z}} f(p\tau, a\tau+z)\mathrm{e}^{2\pi i m(a^2\tau+2az)} \\
&\quad + p^{1-k} \sum_{a,b\in\mathbb{Z}/p\mathbb{Z}} f(p^{-1}(\tau+b), p^{-1}(z+a)) \\
&= L'_p(\tau, z).
\end{aligned}
$$

We next show that $L'_p f$ has an absolutely convergent Fourier expansion of the form

$$
L'_p f(\tau, z) = \sum_{n'=0}^{\infty} \sum_{\substack{r'\in\mathbb{Z} \\ r'^2 \le 4n'm}} c'(n', r')\mathrm{e}^{2\pi i(n'\tau+r'z)}.
$$

We will show that each of the summands of $L'_p f$ has this property. Let

$$f(\tau, z) = \sum_{n=0}^{\infty} \sum_{\substack{r \in \mathbb{Z} \\ r^2 \leq 4npm}} c(n, r) e^{2\pi i(n\tau + rz)}$$

be the Fourier expansion of f. Let a be an integer such that $a \geq 0$. We have

$$e^{2\pi i m(a^2\tau + 2az)} f(p\tau, a\tau + z) = \sum_{n=0}^{\infty} \sum_{\substack{r \in \mathbb{Z} \\ r^2 \leq 4npm}} c(n, r) e^{2\pi i((np + ra + ma^2)\tau + (r + 2am)z)}.$$

Let $X = \{(n, r) \in \mathbb{Z} \times \mathbb{Z} : n \geq 0, r^2 \leq 4npm\}$. Define $i : \mathbb{Z} \times \mathbb{Z} \to \mathbb{Z} \times \mathbb{Z}$ by $i(n, r) = (np + ra + ma^2, r + 2am)$, and let $X' = i(X)$. The map i is an injection. Calculations show that if $(n', r') \in X'$, then $n' \geq 0$ and $r'^2 \leq 4n'm$. We now have

$$e^{2\pi i m(a^2\tau + 2az)} f(p\tau, a\tau + z) = \sum_{(n', r') \in X'} c(i^{-1}(n', r')) e^{2\pi i(n'\tau + r'z)}.$$

It follows that the first summand

$$p \sum_{a \in \mathbb{Z}/p\mathbb{Z}} f(p\tau, a\tau + z) e^{2\pi i m(a^2\tau + 2az)}$$

of $L'_p f$ has a Fourier expansion with the required properties. We may calculate the second summand of $L'_p f$ as follows:

$$\begin{aligned}
&p^{1-k} \sum_{a,b \in \mathbb{Z}/p\mathbb{Z}} f(p^{-1}(\tau + b), p^{-1}(z + a)) \\
&= p^{1-k} \sum_{a,b \in \mathbb{Z}/p\mathbb{Z}} \sum_{n=0}^{\infty} \sum_{\substack{r \in \mathbb{Z} \\ r^2 \leq 4npm}} c(n, r) e^{2\pi i(np^{-1}(\tau + b) + rp^{-1}(z + a))} \\
&= p^{1-k} \sum_{n=0}^{\infty} \sum_{\substack{r \in \mathbb{Z} \\ r^2 \leq 4npm}} \Big(\sum_{a,b \in \mathbb{Z}/p\mathbb{Z}} e^{2\pi i(np^{-1}b + rp^{-1}a)} \Big) c(n, r) e^{2\pi i(np^{-1}\tau + rp^{-1}z)} \\
&= p^{3-k} \sum_{\substack{n=0 \\ p \mid n}}^{\infty} \sum_{\substack{r \in \mathbb{Z} \\ r^2 \leq 4npm,\ p \mid r}} c(n, r) e^{2\pi i(np^{-1}\tau + rp^{-1}z)}
\end{aligned}$$

$$= p^{3-k} \sum_{n=0}^{\infty} \sum_{\substack{r \in \mathbb{Z} \\ r^2 \leq 4nm}} c(np, rp) e^{2\pi i(n\tau + rz)}.$$

This shows that the second summand also has a Fourier expansion with the required properties. It is easy to see from the Fourier expansion that if f is a cusp form, then so is $L'_p f$. □

Lemma 11.4.2 *Let N and k be integers such that $N > 0$ and $k > 0$, and let p be a prime of $\mathbb{Z}$. Assume that $v_p(N) \geq 1$. Let $T^s_{0,1} : \mathcal{A}_k(\mathcal{K}_s(N)) \to \mathcal{A}_k(\mathcal{K}_s(N))$ be the linear map obtained as in Sect. 11.1 from the Hecke operator* (3.69)*, and let $T^s_{0,1}(p) = (T^s_{0,1})_p$ be as in* (11.1)*, so that $T^s_{0,1}(p)$ is a linear map*

$$T^s_{0,1}(p) : M_k(\mathrm{K}_s(N)) \longrightarrow M_k(\mathrm{K}_s(N)).$$

Let $F \in M_k(\mathrm{K}_s(N))$ with Fourier expansion as in (11.4)*. Then*

$$T^s_{0,1}(p)F = \sum_{a,b \in \mathbb{Z}/p\mathbb{Z}} F\big|_k \begin{bmatrix} p &&& \\ & 1 && \\ && 1 & \\ &&& p \end{bmatrix} \begin{bmatrix} 1 &&& \\ a & 1 & & bp^{-n+1} \\ && 1 & -a \\ &&& 1 \end{bmatrix}$$
$$+ \sum_{a,b,c \in \mathbb{Z}/p\mathbb{Z}} F\big|_k \begin{bmatrix} 1 &&& \\ & 1 && \\ && p & \\ &&& p \end{bmatrix} \begin{bmatrix} 1 && c & a \\ & 1 & a & bp^{-n+1} \\ && 1 & \\ &&& 1 \end{bmatrix}. \tag{11.23}$$

If F is a cusp form, then $T^s_{0,1}(p)F$ is a cusp form. We have

$$(T^s_{0,1}(p)F)(Z) = \sum_{S = \left[\begin{smallmatrix} \alpha & \beta \\ \beta & \gamma \end{smallmatrix}\right] \in B(N)} \Big(p^{3-k} a(pS)$$
$$+ \sum_{\substack{y \in \mathbb{Z}/p\mathbb{Z} \\ p \mid (\alpha + 2\beta y + \gamma y^2)}} pa(p^{-1} S[\left[\begin{smallmatrix} 1 & \\ y & p \end{smallmatrix}\right]]) \Big) e^{2\pi i \mathrm{Tr}(SZ)}. \tag{11.24}$$

The Fourier-Jacobi expansion of $T^s_{0,1}(p)F$ is given by

$$(T^s_{0,1}(p)F)(Z) = \sum_{\substack{m=0 \\ N_s \mid m}}^{\infty} (L'_p f_{mp})(\tau, z) e^{2\pi i m\tau'}, \qquad Z = \left[\begin{smallmatrix} \tau & z \\ z & \tau' \end{smallmatrix}\right] \in \mathcal{H}_2. \tag{11.25}$$

Here L'_p is the operator on Jacobi forms from Lemma 11.4.1.

Proof Define $\Phi = \Phi_F$. Let $Z \in \mathcal{H}_2$ and let $h \in \mathrm{GSp}(4, \mathbb{R})^+$ be such that $Z = h\langle I \rangle$. Let $n = v_p(N)$. Then by (11.3) and (3.69) we have

$$
\begin{aligned}
(T^s_{0,1}(p)F)(Z) &= \lambda(h)^{-k} j(h,I)^k T^s_{0,1}\Phi(h)\\
&= \lambda(h)^{-k} j(h,I)^k \Big(\sum_{y,z\in\mathbb{Z}/p\mathbb{Z}} \Phi(h \begin{bmatrix} 1 &&& \\ y & 1 && zp^{-n+1} \\ && 1 & -y \\ &&& 1 \end{bmatrix}_p \begin{bmatrix} 1 &&& \\ & p && \\ && p & \\ &&& 1 \end{bmatrix}_p)\\
&\quad + \sum_{c,y,z\in\mathbb{Z}/p\mathbb{Z}} \Phi(h \begin{bmatrix} 1 && c & y \\ & 1 & y & zp^{-n+1} \\ && 1 & \\ &&& 1 \end{bmatrix}_p \begin{bmatrix} p &&& \\ & p && \\ && 1 & \\ &&& 1 \end{bmatrix}_p)\Big)\\
&= \lambda(h)^{-k} j(h,I)^k \Big(\sum_{y,z\in\mathbb{Z}/p\mathbb{Z}} \Phi(\underbrace{\begin{bmatrix} 1 &&& \\ & p^{-1} && \\ && p^{-1} & \\ &&& 1 \end{bmatrix} \begin{bmatrix} 1 &&& \\ -y & 1 && -zp^{-n+1} \\ && 1 & y \\ &&& 1 \end{bmatrix} h}_{\text{all places}}\\
&\quad \times \begin{bmatrix} 1 &&& \\ y & 1 && zp^{-n+1} \\ && 1 & -y \\ &&& 1 \end{bmatrix}_p \begin{bmatrix} 1 &&& \\ & p && \\ && p & \\ &&& 1 \end{bmatrix}_p)\\
&\quad + \sum_{c,y,z\in\mathbb{Z}/p\mathbb{Z}} \Phi(\underbrace{\begin{bmatrix} p^{-1} &&& \\ & p^{-1} && \\ && 1 & \\ &&& 1 \end{bmatrix} \begin{bmatrix} 1 && -c & -y \\ & 1 & -y & -zp^{-n+1} \\ && 1 & \\ &&& 1 \end{bmatrix} h}_{\text{all places}}\\
&\quad \times \begin{bmatrix} 1 && c & y \\ & 1 & y & zp^{-n+1} \\ && 1 & \\ &&& 1 \end{bmatrix}_p \begin{bmatrix} p &&& \\ & p && \\ && 1 & \\ &&& 1 \end{bmatrix}_p)\Big)\\
&= \lambda(h)^{-k} j(h,I)^k \Big(\sum_{y,z\in\mathbb{Z}/p\mathbb{Z}} \Phi(\begin{bmatrix} 1 &&& \\ & p^{-1} && \\ && p^{-1} & \\ &&& 1 \end{bmatrix}_\infty \begin{bmatrix} 1 &&& \\ -y & 1 && -zp^{-n+1} \\ && 1 & y \\ &&& 1 \end{bmatrix}_\infty h)\\
&\quad + \sum_{c,y,z\in\mathbb{Z}/p\mathbb{Z}} \Phi(\begin{bmatrix} p^{-1} &&& \\ & p^{-1} && \\ && 1 & \\ &&& 1 \end{bmatrix}_\infty \begin{bmatrix} 1 && -c & -y \\ & 1 & -y & -zp^{-n+1} \\ && 1 & \\ &&& 1 \end{bmatrix}_\infty h)\Big)\\
&= \lambda(h)^{-k} j(h,I)^k \Big(\sum_{y,z\in\mathbb{Z}/p\mathbb{Z}} \Phi(\begin{bmatrix} p &&& \\ & 1 && \\ && 1 & \\ &&& p \end{bmatrix}_\infty \begin{bmatrix} 1 &&& \\ y & 1 && zp^{-n+1} \\ && 1 & -y \\ &&& 1 \end{bmatrix}_\infty h)\\
&\quad + \sum_{c,y,z\in\mathbb{Z}/p\mathbb{Z}} \Phi(\begin{bmatrix} 1 &&& \\ & 1 && \\ && p & \\ &&& p \end{bmatrix}_\infty \begin{bmatrix} 1 && c & y \\ & 1 & y & zp^{-n+1} \\ && 1 & \\ &&& 1 \end{bmatrix}_\infty h)\Big)\\
&= \sum_{y,z\in\mathbb{Z}/p\mathbb{Z}} F\big|_k (\begin{bmatrix} p &&& \\ & 1 && \\ && 1 & \\ &&& p \end{bmatrix} \begin{bmatrix} 1 &&& \\ y & 1 && zp^{-n+1} \\ && 1 & -y \\ &&& 1 \end{bmatrix})(Z)\\
&\quad + \sum_{c,y,z\in\mathbb{Z}/p\mathbb{Z}} F\big|_k (\begin{bmatrix} 1 &&& \\ & 1 && \\ && p & \\ &&& p \end{bmatrix} \begin{bmatrix} 1 && c & y \\ & 1 & y & zp^{-n+1} \\ && 1 & \\ &&& 1 \end{bmatrix})(Z).
\end{aligned}
$$

This proves (11.23). To prove (11.24) we use (11.23):

$$
\begin{aligned}
&(T^s_{0,1}(p)F)(Z)\\
&= \sum_{y,z\in\mathbb{Z}/p\mathbb{Z}} F\big|_k(\begin{bmatrix} p &&& \\ & 1 && \\ && 1 & \\ &&& p \end{bmatrix}\begin{bmatrix} 1 &&& \\ y & 1 && zp^{-n+1} \\ && 1 & -y \\ &&& 1 \end{bmatrix})(Z)\\
&\quad + \sum_{c,y,z\in\mathbb{Z}/p\mathbb{Z}} F\big|_k(\begin{bmatrix} 1 &&& \\ & 1 && \\ && p & \\ &&& p \end{bmatrix}\begin{bmatrix} 1 && c & y \\ & 1 & y & zp^{-n+1} \\ && 1 & \\ &&& 1 \end{bmatrix})(Z)\\
&= \sum_{y,z\in\mathbb{Z}/p\mathbb{Z}} F\big|_k(\begin{bmatrix} p &&& \\ y & 1 && zp^{-n+1} \\ && 1 & -y \\ &&& p \end{bmatrix})(Z) + \sum_{c,y,z\in\mathbb{Z}/p\mathbb{Z}} F\big|_k(\begin{bmatrix} 1 && c & y \\ & 1 & y & zp^{-n+1} \\ && p & \\ &&& p \end{bmatrix})(Z)\\
&= \sum_{y,z\in\mathbb{Z}/p\mathbb{Z}} F(\begin{bmatrix} p &&& \\ y & 1 && zp^{-n+1} \\ && 1 & -y \\ &&& p \end{bmatrix}\langle Z\rangle) + p^{-k} \sum_{c,y,z\in\mathbb{Z}/p\mathbb{Z}} F(\begin{bmatrix} 1 && c & y \\ & 1 & y & zp^{-n+1} \\ && p & \\ &&& p \end{bmatrix}\langle Z\rangle)\\
&= \sum_{y,z\in\mathbb{Z}/p\mathbb{Z}} F(\begin{bmatrix} p & \\ y & 1 \end{bmatrix} Z \begin{bmatrix} 1 & yp^{-1} \\ & p^{-1} \end{bmatrix} + \begin{bmatrix} & \\ & zp^{-n} \end{bmatrix})\\
&\quad + p^{-k} \sum_{y,z,c\in\mathbb{Z}/p\mathbb{Z}} F(p^{-1}Z + \begin{bmatrix} cp^{-1} & yp^{-1} \\ yp^{-1} & zp^{-n} \end{bmatrix})\\
&= \sum_{y,z\in\mathbb{Z}/p\mathbb{Z}} \sum_{S\in B(N)} a(S) e^{2\pi i \mathrm{Tr}(S\begin{bmatrix} p & \\ y & 1 \end{bmatrix} Z \begin{bmatrix} 1 & yp^{-1} \\ & p^{-1} \end{bmatrix} + S\begin{bmatrix} & \\ & zp^{-n} \end{bmatrix})}\\
&\quad + p^{-k} \sum_{y,z,c\in\mathbb{Z}/p\mathbb{Z}} \sum_{S\in B(N)} a(S) e^{2\pi i \mathrm{Tr}(p^{-1}SZ + S\begin{bmatrix} cp^{-1} & yp^{-1} \\ yp^{-1} & zp^{-n} \end{bmatrix})}\\
&= \sum_{y\in\mathbb{Z}/p\mathbb{Z}} \sum_{S=\begin{bmatrix} \alpha & \beta \\ \beta & \gamma \end{bmatrix}\in B(N)} \sum_{z\in\mathbb{Z}/p\mathbb{Z}} a(S) e^{2\pi i\gamma zp^{-n}} e^{2\pi i \mathrm{Tr}(\begin{bmatrix} 1 & yp^{-1} \\ & p^{-1} \end{bmatrix} S \begin{bmatrix} p & \\ y & 1 \end{bmatrix} Z)}\\
&\quad + p^{-k} \sum_{S=\begin{bmatrix} \alpha & \beta \\ \beta & \gamma \end{bmatrix}\in B(N)} \sum_{y,z,c\in\mathbb{Z}/p\mathbb{Z}} a(S) e^{2\pi i(\alpha cp^{-1} + 2\beta yp^{-1} + \gamma zp^{-n})} e^{2\pi i \mathrm{Tr}(p^{-1}SZ)}\\
&= p \sum_{y\in\mathbb{Z}/p\mathbb{Z}} \sum_{\substack{S=\begin{bmatrix} \alpha & \beta \\ \beta & \gamma \end{bmatrix}\in B(N)\\ p^n\mid\gamma}} a(S) e^{2\pi i \mathrm{Tr}(\begin{bmatrix} 1 & yp^{-1} \\ & p^{-1} \end{bmatrix} S \begin{bmatrix} p & \\ y & 1 \end{bmatrix} Z)}\\
&\quad + p^{3-k} \sum_{\substack{S=\begin{bmatrix} \alpha & \beta \\ \beta & \gamma \end{bmatrix}\in B(N)\\ p^n\mid\gamma,\ p\mid\alpha,\ p\mid 2\beta}} a(S) e^{2\pi i \mathrm{Tr}(p^{-1}SZ)}
\end{aligned}
$$

$$= p \sum_{y \in \mathbb{Z}/p\mathbb{Z}} \sum_{\substack{S = \left[\begin{smallmatrix} \alpha & \beta \\ \beta & \gamma \end{smallmatrix}\right] \in B(N) \\ p \mid (\alpha - 2\beta y + \gamma y^2)}} a(p^{-1} S[\left[\begin{smallmatrix} 1 & \\ -y & p \end{smallmatrix}\right]]) e^{2\pi \mathrm{i} \mathrm{Tr}(SZ)}$$

$$+ p^{3-k} \sum_{S \in B(N)} a(pS) e^{2\pi \mathrm{i} \mathrm{Tr}(SZ)}$$

$$= \sum_{S = \left[\begin{smallmatrix} \alpha & \beta \\ \beta & \gamma \end{smallmatrix}\right] \in B(N)} \Big(p^{3-k} a(pS) + p \sum_{\substack{y \in \mathbb{Z}/p\mathbb{Z} \\ p \mid (\alpha + 2\beta y + \gamma y^2)}} a(p^{-1} S[\left[\begin{smallmatrix} 1 & \\ y & p \end{smallmatrix}\right]]) \Big) e^{2\pi \mathrm{i} \mathrm{Tr}(SZ)}.$$

This is (11.24).

Finally, to prove (11.25), we proceed from an equation from the proof of (11.24):

$$(T^s_{0,1}(p)F)(Z)$$

$$= \sum_{a,b \in \mathbb{Z}/p\mathbb{Z}} F(\left[\begin{smallmatrix} p & \\ a & 1 \end{smallmatrix}\right] Z \left[\begin{smallmatrix} 1 & ap^{-1} \\ & p^{-1} \end{smallmatrix}\right] + \left[\begin{smallmatrix} & \\ & bp^{-n} \end{smallmatrix}\right])$$

$$+ p^{-k} \sum_{a,b,c \in \mathbb{Z}/p\mathbb{Z}} F(p^{-1} Z + \left[\begin{smallmatrix} cp^{-1} & ap^{-1} \\ ap^{-1} & bp^{-n} \end{smallmatrix}\right])$$

$$= \sum_{a,b \in \mathbb{Z}/p\mathbb{Z}} F(\left[\begin{smallmatrix} p\tau & a\tau + z \\ a\tau + z & (a^2\tau + 2az + \tau' + bp^{-n+1}) p^{-1} \end{smallmatrix}\right])$$

$$+ p^{-k} \sum_{a,b,c \in \mathbb{Z}/p\mathbb{Z}} F(\left[\begin{smallmatrix} (\tau + c) p^{-1} & (z + a) p^{-1} \\ (z + a) p^{-1} & (\tau' + bp^{-n+1}) p^{-1} \end{smallmatrix}\right])$$

$$= \sum_{a,b \in \mathbb{Z}/p\mathbb{Z}} \sum_{\substack{m=0 \\ N_s \mid m}}^{\infty} f_m(p\tau, a\tau + z) e^{2\pi i m p^{-1} (a^2\tau + 2az + \tau' + bp^{-n+1})}$$

$$+ p^{-k} \sum_{a,b,c \in \mathbb{Z}/p\mathbb{Z}} \sum_{\substack{m=0 \\ N_s \mid m}}^{\infty} f_m(p^{-1}(\tau + c), p^{-1}(z + a)) e^{2\pi i m p^{-1} (\tau' + bp^{-n+1})}$$

$$= p \sum_{a \in \mathbb{Z}/p\mathbb{Z}} \sum_{\substack{m=0 \\ N_s p \mid m}}^{\infty} f_m(p\tau, a\tau + z) e^{2\pi i m p^{-1} (a^2\tau + 2az + \tau')}$$

$$+ p^{1-k} \sum_{a,c \in \mathbb{Z}/p\mathbb{Z}} \sum_{\substack{m=0 \\ N_s p \mid m}}^{\infty} f_m(p^{-1}(\tau + c), p^{-1}(z + a)) e^{2\pi i m p^{-1} \tau'}$$

$$= \sum_{\substack{m=0 \\ N_s p|m}}^{\infty} \Big(p \sum_{a \in \mathbb{Z}/p\mathbb{Z}} f_m(p\tau, a\tau + z) e^{2\pi i m p^{-1}(a^2\tau + 2az)}$$

$$+ p^{1-k} \sum_{a,c \in \mathbb{Z}/p\mathbb{Z}} f_m(p^{-1}(\tau + c), p^{-1}(z+a)) \Big) e^{2\pi i m p^{-1}\tau'}$$

$$= \sum_{\substack{m=0 \\ N_s|m}}^{\infty} \Big(p \sum_{a \in \mathbb{Z}/p\mathbb{Z}} f_{mp}(p\tau, a\tau + z) e^{2\pi i m(a^2\tau + 2az)}$$

$$+ p^{1-k} \sum_{a,c \in \mathbb{Z}/p\mathbb{Z}} f_{mp}(p^{-1}(\tau + c), p^{-1}(z+a)) \Big) e^{2\pi i m\tau'}$$

$$= \sum_{\substack{m=0 \\ N_s|m}}^{\infty} (L'_p f_{mp})(\tau, z) e^{2\pi i m\tau'}.$$

This completes the proof. □

Lemma 11.4.3 *Let N and k be integers such that $N > 0$ and $k > 0$, and let p be a prime of $\mathbb{Z}$. Assume that $v_p(N) \geq 1$. Let $T^s_{1,0} : \mathcal{A}_k(\mathcal{K}_s(N)) \to \mathcal{A}_k(\mathcal{K}_s(N))$ be the linear map obtained as in Sect. 11.1 from the Hecke operator* (3.70), *and let $T^s_{1,0}(p) = (T^s_{1,0})_p$ be as in* (11.1), *so that $T^s_{1,0}(p)$ is a linear map*

$$T^s_{1,0}(p) : M_k(\mathrm{K}_s(N)) \longrightarrow M_k(\mathrm{K}_s(N)).$$

Let $F \in M_k(\mathrm{K}_s(N))$ with Fourier expansion as in (11.4). *Then*

$$T^s_{1,0}(p)F = \sum_{\substack{x,y \in \mathbb{Z}/p\mathbb{Z} \\ z \in \mathbb{Z}/p^2\mathbb{Z}}} F\Big|_k \begin{bmatrix} p &&& \\ & 1 && \\ && p & \\ &&& p^2 \end{bmatrix} \begin{bmatrix} 1 &&& y \\ x & 1 & y & zp^{-n+1} \\ && 1 & -x \\ &&& 1 \end{bmatrix}. \tag{11.26}$$

If F is a cusp form, then $T^s_{1,0}(p)F$ is a cusp form. We have

$$(T^s_{1,0}(p)F)(Z) = \sum_{S \in B(N)} \sum_{a \in \mathbb{Z}/p\mathbb{Z}} p^{3-k} a(S[\left[\begin{smallmatrix} 1 & \\ a & p \end{smallmatrix}\right]]) e^{2\pi i \mathrm{Tr}(SZ)}. \tag{11.27}$$

The Fourier-Jacobi expansion of $T^s_{1,0}(p)F$ is given by

$$(T^s_{1,0}(p)F)(Z) = p^2 \sum_{\substack{m=0 \\ N_s|m}}^{\infty} (L_{p^2} f_{mp^2})(\tau, z) e^{2\pi i m\tau'}, \qquad Z = \left[\begin{smallmatrix} \tau & z \\ z & \tau' \end{smallmatrix}\right] \in \mathcal{H}_2. \tag{11.28}$$

Here L_{p^2} is as in Lemma 11.3.1.

Proof Define $\Phi = \Phi_F$. Let $Z \in \mathcal{H}_2$ and let $h \in \mathrm{GSp}(4, \mathbb{R})^+$ be such that $Z = h\langle I \rangle$. Let $n = v_p(N)$. By (11.3) and (3.70) we have

$$
\begin{aligned}
&T^s_{1,0}(p)F(Z)\\
&= \lambda(h)^{-k} j(h, I)^k T^s_{1,0}\Phi(h)\\
&= \lambda(h)^{-k} j(h, I)^k \sum_{\substack{x,y\in\mathbb{Z}/p\mathbb{Z}\\ z\in\mathbb{Z}/p^2\mathbb{Z}}} \Phi(h \begin{bmatrix} 1 &&& y\\ x & 1 & y & zp^{-n+1}\\ && 1 & -x\\ &&& 1\end{bmatrix}_p \begin{bmatrix} p &&&\\ & p^2 &&\\ && p &\\ &&& 1\end{bmatrix}_p)\\
&= \lambda(h)^{-k} j(h, I)^k \sum_{\substack{x,y\in\mathbb{Z}/p\mathbb{Z}\\ z\in\mathbb{Z}/p^2\mathbb{Z}}} \Phi(\underbrace{\begin{bmatrix} p &&&\\ & p^2 &&\\ && p &\\ &&& 1\end{bmatrix}^{-1}}_{\text{all places}} \underbrace{\begin{bmatrix} 1 &&& y\\ x & 1 & y & zp^{-n+1}\\ && 1 & -x\\ &&& 1\end{bmatrix}^{-1}}_{\text{all places}}\\
&\qquad \times h \begin{bmatrix} 1 &&& y\\ x & 1 & y & zp^{-n+1}\\ && 1 & -x\\ &&& 1\end{bmatrix}_p \begin{bmatrix} p &&&\\ & p^2 &&\\ && p &\\ &&& 1\end{bmatrix}_p)\\
&= \lambda(h)^{-k} j(h, I)^k \sum_{\substack{x,y\in\mathbb{Z}/p\mathbb{Z}\\ z\in\mathbb{Z}/p^2\mathbb{Z}}} \Phi(\begin{bmatrix} p^{-1} &&&\\ & p^{-2} &&\\ && p^{-1} &\\ &&& 1\end{bmatrix}_\infty \begin{bmatrix} 1 &&& -y\\ -x & 1 & -y & -zp^{-n+1}\\ && 1 & x\\ &&& 1\end{bmatrix}_\infty h)\\
&= \lambda(h)^{-k} j(h, I)^k \sum_{\substack{x,y\in\mathbb{Z}/p\mathbb{Z}\\ z\in\mathbb{Z}/p^2\mathbb{Z}}} \Phi(\begin{bmatrix} p &&&\\ & 1 &&\\ && p &\\ &&& p^2\end{bmatrix}_\infty \begin{bmatrix} 1 &&& y\\ x & 1 & y & zp^{-n+1}\\ && 1 & -x\\ &&& 1\end{bmatrix}_\infty h)\\
&= \sum_{\substack{x,y\in\mathbb{Z}/p\mathbb{Z}\\ z\in\mathbb{Z}/p^2\mathbb{Z}}} (F\big|_k \begin{bmatrix} p &&&\\ & 1 &&\\ && p &\\ &&& p^2\end{bmatrix}\begin{bmatrix} 1 &&& y\\ x & 1 & y & zp^{-n+1}\\ && 1 & -x\\ &&& 1\end{bmatrix})(Z).
\end{aligned}
$$

This proves (11.26). To prove (11.27) we use (11.26):

$$
\begin{aligned}
(T^s_{1,0}(p)F)(Z) &= \sum_{\substack{a,b\in\mathbb{Z}/p\mathbb{Z}\\ c\in\mathbb{Z}/p^2\mathbb{Z}}} (F\big|_k \begin{bmatrix} p &&&\\ & 1 &&\\ && p &\\ &&& p^2\end{bmatrix}\begin{bmatrix} 1 &&& b\\ a & 1 & b & cp^{-n+1}\\ && 1 & -a\\ &&& 1\end{bmatrix})(Z)\\
&= p^{-k} \sum_{\substack{a,b\in\mathbb{Z}/p\mathbb{Z}\\ c\in\mathbb{Z}/p^2\mathbb{Z}}} F(\begin{bmatrix} p & \\ a & 1\end{bmatrix} Z \begin{bmatrix} p^{-1} & ap^{-2}\\ & p^{-2}\end{bmatrix} + \begin{bmatrix} & p^{-1}b\\ bp^{-1} & abp^{-2}+cp^{-n-1}\end{bmatrix})\\
&= p^{-k} \sum_{\substack{a,b\in\mathbb{Z}/p\mathbb{Z}\\ c\in\mathbb{Z}/p^2\mathbb{Z}}} \sum_{S\in B(N)} a(S) e^{2\pi i \mathrm{Tr}(S(\begin{bmatrix} p & \\ a & 1\end{bmatrix} Z \begin{bmatrix} p^{-1} & ap^{-2}\\ & p^{-2}\end{bmatrix} + \begin{bmatrix} & p^{-1}b\\ p^{-1}b & abp^{-2}+cp^{-n-1}\end{bmatrix}))}
\end{aligned}
$$

$$= p^{-k} \sum_{S=\left[\begin{smallmatrix}\alpha & \beta\\ \beta & \gamma\end{smallmatrix}\right]\in B(N)} \sum_{\substack{a,b\in\mathbb{Z}/p\mathbb{Z}\\ c\in\mathbb{Z}/p^2\mathbb{Z}}} a(S)\mathrm{e}^{2\pi\mathrm{i}(2\beta bp^{-1}+\gamma abp^{-2}+\gamma cp^{-n-1})}$$

$$\times\, \mathrm{e}^{2\pi\mathrm{i}\mathrm{Tr}(\left[\begin{smallmatrix}p^{-1} & ap^{-2}\\ & p^{-2}\end{smallmatrix}\right]S\left[\begin{smallmatrix}p & \\ a & 1\end{smallmatrix}\right]Z)}$$

$$= p^{2-k} \sum_{\substack{S=\left[\begin{smallmatrix}\alpha & \beta\\ \beta & \gamma\end{smallmatrix}\right]\in B(N)\\ p^{n+1}|\gamma}} \sum_{a,b\in\mathbb{Z}/p\mathbb{Z}} a(S)\mathrm{e}^{2\pi\mathrm{i}(2\beta bp^{-1})}\mathrm{e}^{2\pi\mathrm{i}\mathrm{Tr}(\left[\begin{smallmatrix}p^{-1} & ap^{-2}\\ & p^{-2}\end{smallmatrix}\right]S\left[\begin{smallmatrix}p & \\ a & 1\end{smallmatrix}\right]Z)}$$

$$= p^{3-k} \sum_{a\in\mathbb{Z}/p\mathbb{Z}} \sum_{\substack{S=\left[\begin{smallmatrix}\alpha & \beta\\ \beta & \gamma\end{smallmatrix}\right]\in B(N)\\ p^{n+1}|\gamma,\ p|2\beta}} a(S)\mathrm{e}^{2\pi\mathrm{i}\mathrm{Tr}(\left[\begin{smallmatrix}p^{-1} & ap^{-2}\\ & p^{-2}\end{smallmatrix}\right]S\left[\begin{smallmatrix}p & \\ a & 1\end{smallmatrix}\right]Z)}$$

$$= \sum_{S\in B(N)} p^{3-k} \sum_{a\in\mathbb{Z}/p\mathbb{Z}} a(S[\left[\begin{smallmatrix}1 & \\ a & p\end{smallmatrix}\right]])\mathrm{e}^{2\pi\mathrm{i}\mathrm{Tr}(SZ)}.$$

This is (11.27).

Finally, to prove (11.28), we proceed from an equation from the proof of (11.27):

$$(T_{1,0}^s(p)F)(Z) = p^{-k} \sum_{\substack{a,b\in\mathbb{Z}/p\mathbb{Z}\\ c\in\mathbb{Z}/p^2\mathbb{Z}}} F(\left[\begin{smallmatrix}p & \\ a & 1\end{smallmatrix}\right] Z \left[\begin{smallmatrix}p^{-1} & ap^{-2}\\ & p^{-2}\end{smallmatrix}\right] + \left[\begin{smallmatrix} & p^{-1}b\\ bp^{-1} & abp^{-2}+cp^{-n-1}\end{smallmatrix}\right])$$

$$= p^{-k} \sum_{\substack{a,b\in\mathbb{Z}/p\mathbb{Z}\\ c\in\mathbb{Z}/p^2\mathbb{Z}}} F(\left[\begin{smallmatrix}\tau & (a\tau+b+z)p^{-1}\\ (a\tau+b+z)p^{-1} & (a^2\tau+2az+\tau'+ab+cp^{-n+1})p^{-2}\end{smallmatrix}\right])$$

$$= p^{-k} \sum_{\substack{a,b\in\mathbb{Z}/p\mathbb{Z}\\ c\in\mathbb{Z}/p^2\mathbb{Z}}} \sum_{\substack{m=0\\ N_s|m}}^{\infty} f_m(\tau, (a\tau+b+z)p^{-1})e^{2\pi im(a^2\tau+2az+\tau'+ab+cp^{-n+1})p^{-2}}$$

$$= p^{2-k} \sum_{a,b\in\mathbb{Z}/p\mathbb{Z}} \sum_{\substack{m=0\\ p^2N_s|m}}^{\infty} f_m(\tau, (a\tau+b+z)p^{-1})\mathrm{e}^{2\pi\mathrm{i}mp^{-2}(a^2\tau+2az)}\mathrm{e}^{2\pi\mathrm{i}mp^{-2}\tau'}$$

$$= p^{2-k} \sum_{\substack{m=0\\ N_s|m}}^{\infty} \sum_{a,b\in\mathbb{Z}/p\mathbb{Z}} f_{mp^2}(\tau, (b+a\tau+z)p^{-1})e^{2\pi im(2az+a^2\tau)}\mathrm{e}^{2\pi\mathrm{i}m\tau'}$$

$$= p^2 \sum_{\substack{m=0\\ N_s|m}}^{\infty} (L_{p^2} f_{mp^2})(\tau, z)\mathrm{e}^{2\pi\mathrm{i}m\tau'},$$

where L_{p^2} is defined in (11.15). This completes the proof. □

11.5 Some Relations Between Operators

We conclude this chapter by stating some relations between the operators on stable Klingen forms introduced above; these relations follow from local statements proved in Chap. 3.

Proposition 11.5.1 *Let N and k be integers such that $N > 0$ and $k > 0$, and let p be a prime of $\mathbb{Z}$. Then, for $F \in M_k(\mathrm{K}_s(N))$,*

$$\theta_p \tau_p F = \tau_p \theta_p F, \tag{11.29}$$

$$T^s_{0,1}(p) T^s_{1,0}(p) F = T^s_{1,0}(p) T^s_{0,1}(p) F \qquad \text{if } p \mid N, \tag{11.30}$$

$$T^s_{0,1}(p) F = p^2 \sigma_p \theta_p F \qquad \text{if } p \mid N, \tag{11.31}$$

$$T^s_{1,0}(p) F = p^4 \sigma_p \tau_p F \qquad \text{if } p \mid N, \tag{11.32}$$

$$T^s_{1,0}(p) F = p^4 \tau_p \sigma_p F \qquad \text{if } p^2 \mid N, \tag{11.33}$$

$$T^s_{0,1}(p) \tau_p F = \tau_p T^s_{0,1}(p) F \qquad \text{if } p \mid N, \tag{11.34}$$

$$T^s_{1,0}(p) \tau_p F = \tau_p T^s_{1,0}(p) F \qquad \text{if } p \mid N, \tag{11.35}$$

$$T^s_{1,0}(p) \theta_p F = p^2 T^s_{0,1}(p) \tau_p F, \tag{11.36}$$

$$T^s_{0,1}(p) \theta_p F = \theta_p T^s_{0,1}(p) F + p^3 \tau_p F - p^3 \eta_p \sigma_p F \qquad \text{if } p^2 \mid N. \tag{11.37}$$

Proof These statements follow from the corresponding local statements contained in Lemmas 3.8.2, 3.8.4, and 3.9.3. For the connection to representation theory we use the space W from Sect. 11.1. □

Chapter 12
Hecke Eigenvalues and Fourier Coefficients

In this final chapter we present some applications of the local theory of Part I to the Hecke eigenvalues and Fourier coefficients of Siegel modular newforms F in $S_k(\mathrm{K}(N))_{\mathrm{new}}$ of degree two with paramodular level N. Assuming that F is an eigenform for the Hecke operators $T(1, 1, p, p)$ and $T(p, 1, p, p^2)$ for all primes p of $\mathbb{Z}$,[1] we begin by proving in Sect. 12.1 that the local results from Part I imply identities involving F and its images under the upper block operators from Chap. 11 at p for $p^2 \mid N$. We then show in Corollary 12.1.3 that these identities yield relations between Fourier coefficients and Hecke eigenvalues as well as conditions which determine properties of the attached local representations at p. Corollary 12.1.3 may be regarded as a solution to the problem mentioned in the introduction to this work. In the second section we show that the formulas of Corollary 12.1.3 can be rewritten in terms of the action of the Hecke ring of $\Gamma_0(N)$ on the vector space of $\mathbb{C}$ valued functions on the set of positive semi-definite 2×2 matrices with rational entries; these results represent a more conceptual and compact statement of the formulas of Corollary 12.1.3. We conclude this chapter with two applications of Corollary 12.1.3. First, in Sects. 12.3 and 12.4 we show that the equations of Corollary 12.1.3 do indeed hold for the examples of [99] and [98], and we indicate how the equations could be used to calculate Hecke eigenvalues from Fourier coefficients in other instances. Finally, in Sect. 12.5 we apply Corollary 12.1.3 to prove that the radial Fourier coefficients $a(p^t S)$ for $t \geq 0$ and $p^2 \mid N$ satisfy a recurrence relation determined by the spin L-factor of F at p. This extends results known in other cases (see, for example, Sect. 4.3.2 of [2]).

[1] In fact, assuming that F is an eigenform for $T(1, 1, p, p)$ and $T(p, 1, p, p^2)$ for all but finitely many p is sufficient: see Theorem 12.1.1 below.

J. Johnson-Leung et al., *Stable Klingen Vectors and Paramodular Newforms*,
Lecture Notes in Mathematics 2342, https://doi.org/10.1007/978-3-031-45177-5_12

12.1 Applications

In this section we apply some of the local results of Part I to the Siegel modular newforms with paramodular level. To connect such newforms to representation theory we will use the following theorem. In this statement $\mathfrak{g}'$ is the Lie algebra of $\mathrm{GSp}(4, \mathbb{R})$ and $K^{\pm}$ is the maximal compact subgroup of $\mathrm{GSp}(4, \mathbb{R})$ as in [113]. See Chap. 10 for general definitions about Siegel modular forms.

Theorem 12.1.1 *Let N and k be integers such that $N > 0$ and $k > 0$, and let $F \in S_k(\mathrm{K}(N))$. Assume that F is a newform and is an eigenvector for the Hecke operators $T(1, 1, q, q)$ and $T(q, 1, q, q^2)$ for all but finitely many of the primes q of $\mathbb{Z}$ such that $q \nmid N$. Then F is an eigenvector for $T(1, 1, q, q)$, $T(q, 1, q, q^2)$, and w_q for all primes q of $\mathbb{Z}$. Moreover, let V be the $\mathrm{GSp}(4, \mathbb{A}_{\mathrm{fin}}) \times (\mathfrak{g}', K^{\pm})$ submodule of the space of cuspidal automorphic forms on $\mathrm{GSp}(4, \mathbb{A})$ generated by Φ_F, with Φ_F as in Lemma 10.2.1. Then V is irreducible, so that $V \simeq \otimes_{v \leq \infty} \pi_v$, and where, for each prime q of $\mathbb{Z}$, π_q is an irreducible, admissible representation of $\mathrm{GSp}(4, \mathbb{Q}_q)$ with trivial central character, and π_∞ is an irreducible, admissible $(\mathfrak{g}', K^{\pm})$ module. Under this isomorphism, Φ_F corresponds to a pure tensor $\otimes_{v \leq \infty} x_v$, where x_q is a newform in π_q for each prime q of $\mathbb{Z}$. Let q be a prime of $\mathbb{Z}$, and let $\lambda_q, \mu_q \in \mathbb{C}$ and $\varepsilon_q \in \{\pm 1\}$ be such that $T(1, 1, q, q)F = q^{k-3} \lambda_q F$, $T(q, 1, q, q^2)F = q^{2(k-3)} \mu_q F$, and $w_q F = \varepsilon_q F$. Then $T_{0,1} x_q = \lambda_q x_q$, $T_{1,0} x_q = \mu_q x_q$, and $u_{v_q(N)} x_q = \varepsilon_q x_q$. The representation π_q is non-generic for some prime q of $\mathbb{Z}$ if and only if π_q is a Saito-Kurokawa representation for all primes q of $\mathbb{Z}$.*

Proof See Theorem 5.5 and Corollary 5.4 of [115]. □

Let k and N be positive integers. Using this theorem and further results from [114] and [115], one may describe the space $S_k(\mathrm{K}(N))_{\mathrm{new}}$ as follows. The Hecke operators $T(1, 1, q, q)$ and $T(q, 1, q, q^2)$ for primes $q \nmid N$ act on this space and commute; see Lemma 10.2.4. We thus may diagonalize $S_k(\mathrm{K}(N))_{\mathrm{new}}$ according to this action. Each eigenspace is 1-dimensional, and any eigenform is also an eigenform for the Hecke operators at places $p \mid N$. Each eigenform corresponds to an automorphic representation as in Theorem 12.1.1. Such eigenforms come in two types, known as lifts and non-lifts. The lifts are modular forms of Saito-Kurokawa type (type **(P)**), and the non-lifts are modular forms of general type (type **(G)**) in the sense of [114]. The lifts can be characterized in a number of ways; in particular, F is a lift if and only if π_q is non-generic for one (then all) primes q of $\mathbb{Z}$. (See [40] for a survey of further characterizations.)

We now apply the local results from Part I to Siegel modular forms with paramodular level.

Theorem 12.1.2 *Let N and k be integers such that $N > 0$ and $k > 0$, and let $F \in S_k(\mathrm{K}(N))$. Assume that F is a newform and is an eigenvector for the Hecke operators $T(1, 1, q, q)$ and $T(q, 1, q, q^2)$ for all but finitely many of the primes q of $\mathbb{Z}$ such that $q \nmid N$; then by Theorem 12.1.1, F is an eigenvector for $T(1, 1, q, q)$ and $T(q, 1, q, q^2)$ for all primes q of $\mathbb{Z}$. Let $\otimes_{v \leq \infty} \pi_v$ be as in Theorem 12.1.1. For* *every prime q of $\mathbb{Z}$ let $\lambda_q, \mu_q \in \mathbb{C}$ be such that*

$$T(1,1,q,q)F = q^{k-3}\lambda_q F, \tag{12.1}$$

$$T(q,1,q,q^2)F = q^{2(k-3)}\mu_q F. \tag{12.2}$$

Let p be a prime of $\mathbb{Z}$ with $v_p(N) \geq 2$. Then

$$\mu_p = 0 \Longleftrightarrow \sigma_p F = 0 \Longleftrightarrow T^s_{1,0}(p)F = 0. \tag{12.3}$$

Moreover:

(1) *If $v_p(N) \geq 3$, then $\sigma_p^2 F = 0$.*
(2) *We have*

$$\mu_p F = p^4\tau_p^2\sigma_p F - p^2\eta_p\sigma_p F. \tag{12.4}$$

(3) *Assume that $\mu_p \neq 0$. Then $\sigma_p F \neq 0$, and*

$$T^s_{0,1}(p)(\sigma_p F) = \lambda_p(\sigma_p F), \tag{12.5}$$

$$T^s_{1,0}(p)(\sigma_p F) = (\mu_p + p^2)(\sigma_p F), \tag{12.6}$$

and the representation π_p is generic.
(4) *Assume that $\mu_p = 0$. Then*

$$T^s_{1,0}(p)F = 0. \tag{12.7}$$

If F is not an eigenvector for $T^s_{0,1}(p)$, then π_p is generic and

$$T^s_{0,1}(p)^2 F = -p^3 F + \lambda_p T^s_{0,1}(p)F. \tag{12.8}$$

The newform F is an eigenvector for $T^s_{0,1}(p)$ if and only if

$$T^s_{0,1}(p)F = (1+p^{-1})^{-1}\lambda_p F; \tag{12.9}$$

in this case π_p is non-generic.

Proof Let Φ_F and $V \cong \otimes_{v\leq\infty}\pi_v$ be as in Theorem 12.1.1. For each place v of $\mathbb{Q}$, let V_v be the space of π_v. By Theorem 12.1.1, Φ_F corresponds to a pure tensor $\otimes_v x_v$ under the fixed isomorphism $i : \otimes_v V_v \xrightarrow{\sim} V$. Define $V_p \hookrightarrow V$ by $x \mapsto i(x \otimes \otimes_{v\neq p} x_v)$ for $x \in V_p$; then $V_p \hookrightarrow V$ is a well-defined injective $\mathrm{GSp}(4,\mathbb{Q}_p)$

map. Moreover, the properties of x_∞, and of x_q for q a prime of $\mathbb{Z}$ such that $q \neq p$, imply that the image of this map is contained in

$$W = \{\, \Phi \in \mathcal{A}_k^\circ \mid \kappa \cdot \Phi = \Phi \text{ for } \kappa \in \prod_{q \neq p} \mathrm{K}_s(\mathfrak{q}^{v_q(N)}) \,\}, \tag{12.10}$$

which is also a $\mathrm{GSp}(4, \mathbb{Q}_p)$ space. We thus obtain an injective $\mathrm{GSp}(4, \mathbb{Q}_p)$ map

$$t : V_p \hookrightarrow W. \tag{12.11}$$

Let $N' = Np^{-v_p(N)}$ so that p and N' are relatively prime. Let n_1 and n_2 be integers such that $n_1, n_2 \geq 0$, and define $N_1 = N'p^{n_1}$ and $N_2 = N'p^{n_2}$. For $i = 1$ or 2, since the subspace of $\mathrm{K}_s(\mathfrak{p}^{n_i})$-invariant vectors in W is $\mathcal{A}_k^\circ(\mathcal{K}_s(N_i))$ and since t is a $\mathrm{GSp}(4, \mathbb{Q}_p)$ map, the map t restricts to give inclusions

$$t : V_{p,s}(\mathfrak{p}^{n_1}) \hookrightarrow \mathcal{A}_k^\circ(\mathcal{K}_s(N_1)) \quad \text{and} \quad t : V_{p,s}(\mathfrak{p}^{n_2}) \hookrightarrow \mathcal{A}_k^\circ(\mathcal{K}_s(N_2)).$$

Next, let $g \in \mathrm{GSp}(4, \mathbb{Q}_p)$, and let

$$\delta = T_g : V_{p,s}(\mathfrak{p}^{n_1}) \longrightarrow V_{p,s}(\mathfrak{p}^{n_2}) \quad \text{and} \quad \delta = T_g : W_{p,s}(\mathfrak{p}^{n_1}) \longrightarrow W_{p,s}(\mathfrak{p}^{n_2})$$

be the operators corresponding to g as in Sect. 3.4. Since t is a $\mathrm{GSp}(4, \mathbb{Q}_p)$ map, there is a commutative diagram

$$\begin{array}{ccc} V_{p,s}(\mathfrak{p}^{n_1}) & \xrightarrow{\ t\ } & \mathcal{A}_k^\circ(\mathcal{K}_s(N_1)) \\ \delta\big\downarrow & & \big\downarrow\delta \\ V_{p,s}(\mathfrak{p}^{n_2}) & \xrightarrow{\ t\ } & \mathcal{A}_k^\circ(\mathcal{K}_s(N_2)) \end{array} \tag{12.12}$$

Also, from Sect. 11.1 there is a commutative diagram

$$\begin{array}{ccc} \mathcal{A}_k^\circ(\mathcal{K}_s(N_1)) & \xrightarrow[\sim]{\ r\ } & S_k(\mathrm{K}_s(N_1)) \\ \delta\big\downarrow & & \big\downarrow\delta_p \\ \mathcal{A}_k^\circ(\mathcal{K}_s(N_2)) & \xrightarrow[\sim]{\ r\ } & S_k(\mathrm{K}_s(N_2)) \end{array} \tag{12.13}$$

Combining (12.12) with (12.13), we obtain a commutative diagram

$$\begin{array}{ccc} V_{p,s}(\mathfrak{p}^{n_1}) & \xrightarrow{\ rot\ } & S_k(\mathrm{K}_s(N_1)) \\ \delta\big\downarrow & & \big\downarrow\delta_p \\ V_{p,s}(\mathfrak{p}^{n_2}) & \xrightarrow{\ rot\ } & S_k(\mathrm{K}_s(N_2)) \end{array} \tag{12.14}$$

We note that the map $r \circ t$ is injective. Also, for the following arguments we remind the reader that $v_p(N)$ is the paramodular level N_{π_p} of π_p, x_p is a newform in $V_p(\mathfrak{p}^{v_p(N)}) \subset V_{p,s}(\mathfrak{p}^{v_p(N)})$, $\lambda_{\pi_p} = \lambda_p$, $\mu_{\pi_p} = \mu_p$, and $F = (r \circ t)x_p$ (these statements follow from Theorem 12.1.1).

To prove that $\mu_p = 0$ if and only if $\sigma_p F = 0$ we use the diagram (12.14) with $n_1 = v_p(N)$, $n_2 = v_p(N) - 1$ and

$$\delta = \sigma = \sigma_{v_p(N)-1} : V_{p,s}(\mathfrak{p}^{v_p(N)}) \longrightarrow V_{p,s}(\mathfrak{p}^{v_p(N)-1})$$

to obtain the commutative diagram

$$\begin{array}{ccc} V_{p,s}(\mathfrak{p}^{v_p(N)}) & \xrightarrow{rot} & S_k(\mathrm{K}_s(N)) \\ \sigma \downarrow & & \downarrow \sigma_p \\ V_{p,s}(\mathfrak{p}^{v_p(N)-1}) & \xrightarrow{rot} & S_k(\mathrm{K}_s(Np^{-1})) \end{array} \tag{12.15}$$

We have $x_p \in V_{p,s}(\mathfrak{p}^{v_p(N)})$ and $(r \circ t)(x_p) = F$. The commutativity of (12.15) implies that $\sigma_p F = (r \circ t)(\sigma x_p)$. Since $r \circ t$ is injective it follows that $\sigma_p F = 0$ if and only if $\sigma x_p = 0$. By the equivalence of (2) and (3) of Proposition 5.7.3 we have $\sigma x_p = 0$ if and only if $\mu_{\pi_p} = \mu_p = 0$, so that $\mu_p = 0$ if and only if $\sigma_p F = 0$. Next, by (12.14) with $n_1 = n_2 = v_p(N)$ and $\delta = T_{1,0}^s$ we have the commutative diagram

$$\begin{array}{ccc} V_{p,s}(\mathfrak{p}^{v_p(N)}) & \xrightarrow{rot} & S_k(\mathrm{K}_s(N)) \\ T_{1,0}^s \downarrow & & \downarrow T_{1,0}^s(p) \\ V_{p,s}(\mathfrak{p}^{v_p(N)}) & \xrightarrow{rot} & S_k(\mathrm{K}_s(N)) \end{array} \tag{12.16}$$

The commutativity of (12.16) implies that $T_{1,0}^s(p)F = (r \circ t)(T_{1,0}^s x_p)$. By the injectivity of $r \circ t$ we have $T_{1,0}^s(p)F = 0$ if and only if $T_{1,0}^s x_p = 0$. By the equivalence of (2) and (5) of Proposition 5.7.3 we have $\mu_{\pi_p} = \mu_p = 0$ if and only if $T_{1,0}^s x_p = 0$. Therefore, $\mu_p = 0$ if and only if $T_{1,0}^s(p)F = 0$.

(1) Assume that $v_p(N) \geq 3$. We combine diagram (12.14) with $n_1 = v_p(N)$, $n_2 = v_p(N) - 1$, and $\delta = \sigma = \sigma_{v_p(N)-1}$ and diagram (12.14) with $n_1 = v_p(N) - 1$, $n_2 = v_p(N) - 2$, and $\delta = \sigma = \sigma_{v_p(N)-2}$ to obtain the commutative diagram

$$\begin{array}{ccc} V_{p,s}(\mathfrak{p}^{v_p(N)}) & \xrightarrow{rot} & S_k(\mathrm{K}_s(N)) \\ \sigma^2 \downarrow & & \downarrow \sigma_p^2 \\ V_{p,s}(\mathfrak{p}^{v_p(N)-2}) & \xrightarrow{rot} & S_k(\mathrm{K}_s(Np^{-2})) \end{array}$$

Since $v_p(N) = N_{\pi_p}$ we have $V_{p,s}(\mathfrak{p}^{v_p(N)-2}) = 0$ by Corollary 5.1.6. This implies that $\sigma_p^2((r \circ t)(x_p)) = 0$; since $F = (r \circ t)(x_p)$, we obtain $\sigma_p^2 F = 0$.

(2). Letting $n_1 = v_p(N) - 1$, $n_2 = v_p(N)$, and

$$\delta = \tau = \tau_{v_p(N)-1} : V_{p,s}(\mathfrak{p}^{v_p(N)-1}) \longrightarrow V_{p,s}(\mathfrak{p}^{v_p(N)})$$

in (12.14), we obtain the commutative diagram

$$\begin{array}{ccc} V_{p,s}(\mathfrak{p}^{v_p(N)-1}) & \xrightarrow{rot} & S_k(\mathrm{K}_s(Np^{-1})) \\ \tau \big\downarrow & & \big\downarrow \tau_p \\ V_{p,s}(\mathfrak{p}^{v_p(N)}) & \xrightarrow{rot} & S_k(\mathrm{K}_s(N)) \end{array} \tag{12.17}$$

Similarly, there are commutative diagrams

$$\begin{array}{ccc} V_{p,s}(\mathfrak{p}^{v_p(N)}) & \xrightarrow{rot} & S_k(\mathrm{K}_s(N)) \\ \tau \big\downarrow & & \big\downarrow \tau_p \\ V_{p,s}(\mathfrak{p}^{v_p(N)+1}) & \xrightarrow{rot} & S_k(\mathrm{K}_s(Np)) \end{array} \tag{12.18}$$

and

$$\begin{array}{ccc} V_{p,s}(\mathfrak{p}^{v_p(N)-1}) & \xrightarrow{rot} & S_k(\mathrm{K}_s(Np^{-1})) \\ \eta \big\downarrow & & \big\downarrow \eta_p \\ V_{p,s}(\mathfrak{p}^{v_p(N)+1}) & \xrightarrow{rot} & S_k(\mathrm{K}_s(Np)) \end{array} \tag{12.19}$$

By (5.16) of Lemma 5.2.1 we have $q^{-1}\mu_p x_p = q^3\tau^2\sigma x_p - q\eta\sigma x_p$. Applying $r \circ t$ to this equation and using the commutativity of (12.17), (12.18), and (12.19) yields (12.4).

(3). Assume that $\mu_p \neq 0$. Then $\sigma_p F \neq 0$ by (12.3). By (12.14) with $n_1 = n_2 = v_p(N) - 1$ and $\delta = T_{0,1}^s, T_{1,0}^s$ we have the commutative diagrams

$$\begin{array}{ccc} V_{p,s}(\mathfrak{p}^{v_p(N)-1}) & \xrightarrow{rot} & S_k(\mathrm{K}_s(Np^{-1})) \\ T_{0,1}^s,\, T_{1,0}^s \big\downarrow & & \big\downarrow T_{0,1}^s(p),\, T_{1,0}^s(p) \\ V_{p,s}(\mathfrak{p}^{v_p(N)-1}) & \xrightarrow{rot} & S_k(\mathrm{K}_s(Np^{-1})) \end{array} \tag{12.20}$$

Since $\mu_{\pi_p} = \mu_p \neq 0$ we have that π_p is a category 1 representation by Proposition 5.7.3. By (5.12) and (1) of Theorem 6.1.1 we have that $T_{0,1}^s\sigma x_p =$

$\lambda_p \sigma x_p$ and $T^s_{1,0} \sigma x_p = (\mu_p + p^2) \sigma x_p$. Applying rot to these equations and using the commutativity of (12.15) and (12.20), we obtain (12.5) and (12.6). The remaining claim follows from Lemma 6.2.3 and the fact that representations of type IVb are non-unitary while π_p is necessarily unitary.

(4). Assume that $\mu_p F = 0$. By (12.3) we have $T^s_{1,0}(p)F = 0$ which is (12.7). Since $\mu_{\pi_p} = \mu_p = 0$ we have that π_p is a category 2 representation by Proposition 5.7.3. By (12.14) with $n_1 = n_2 = v_p(N)$ and $\delta = T^s_{0,1}$ we have the commutative diagram

$$\begin{array}{ccc} V_{p,s}(\mathfrak{p}^{v_p(N)}) & \xrightarrow{rot} & S_k(\mathrm{K}_s(N)) \\ {\scriptstyle T^s_{0,1}} \downarrow & & \downarrow {\scriptstyle T^s_{0,1}(p)} \\ V_{p,s}(\mathfrak{p}^{v_p(N)}) & \xrightarrow{rot} & S_k(\mathrm{K}_s(N)) \end{array} \tag{12.21}$$

Assume that F is not an eigenform for $T^s_{0,1}(p)$. Then by (2) of Theorem 6.1.1 the representation π_p is generic and $(T^s_{0,1})^2 x_p = -p^3 x_p + \lambda_p T^s_{0,1} x_p$. Applying $r \circ t$ to this equation and using the commutativity of (12.21) yields (12.8). Assume that F is an eigenform for $T^s_{0,1}(p)$. Then by (2) of Theorem 6.1.1 the representation π_p is non-generic and $T^s_{0,1} x_p = (1 + p^{-1})^{-1} \lambda_p x_p$. Again applying $r \circ t$ to this equation and using the commutativity of (12.21), we obtain (12.9). □

We now combine Theorem 12.1.2 and the formulas from Chap. 11 to obtain relations between the Fourier coefficients and the Hecke eigenvalues of a paramodular newform.

Corollary 12.1.3 *Let N and k be integers such that $N > 0$ and $k > 0$, and let $F \in S_k(\mathrm{K}(N))$. Assume that F is a newform and is an eigenvector for the Hecke operators $T(1,1,q,q)$ and $T(q,1,q,q^2)$ for all but finitely many of the primes q of $\mathbb{Z}$ such that $q \nmid N$; by Theorem 12.1.1, F is an eigenvector for $T(1,1,q,q)$, $T(q,1,q,q^2)$ and w_q for all primes q of $\mathbb{Z}$. Let $\lambda_q, \mu_q \in \mathbb{C}$ and $\varepsilon_q \in \{\pm 1\}$ be such that*

$$T(1,1,q,q)F = q^{k-3} \lambda_q F,$$
$$T(q,1,q,q^2)F = q^{2(k-3)} \mu_q F,$$
$$w_q F = \varepsilon_q F$$

for all primes q of $\mathbb{Z}$. Regard F as an element of $S_k(\mathrm{K}_s(N))$, and let

$$F(Z) = \sum_{S \in B(N)^+} a(S) e^{2\pi i \mathrm{Tr}(SZ)}$$

be the Fourier expansion of F. Let $\pi \cong \otimes_{v \le \infty} \pi_v$ be as in Theorem 12.1.2. Let p be a prime of $\mathbb{Z}$ with $v_p(N) \ge 2$. Then

$$\sum_{a \in \mathbb{Z}/p\mathbb{Z}} a(S[\left[\begin{smallmatrix} 1 & \\ a & p \end{smallmatrix}\right]]) = 0 \qquad \text{for } S \in B(Np^{-1})^+$$

$$\Updownarrow \tag{12.22}$$

$$\mu_p = 0$$

$$\Updownarrow \tag{12.23}$$

$$\sum_{a \in \mathbb{Z}/p\mathbb{Z}} a(S[\left[\begin{smallmatrix} 1 & \\ a & p \end{smallmatrix}\right]]) = 0 \qquad \text{for } S \in B(N)^+.$$

Moreover:

(1) *If $v_p(N) \ge 3$ and $S \in B(Np^{-2})^+$, then*

$$\sum_{z \in \mathbb{Z}/p^2\mathbb{Z}} a(S\Big[\begin{bmatrix} 1 & \\ z & p^2 \end{bmatrix}\Big]) = 0. \tag{12.24}$$

(2) *If $S = \begin{bmatrix} \alpha & \beta \\ \beta & \gamma \end{bmatrix} \in B(Np)^+$, then*

$$\mu_p a(S) = \begin{cases} \sum\limits_{x \in \mathbb{Z}/p\mathbb{Z}} p^{3-k} a(S[\left[\begin{smallmatrix} 1 & \\ x & p \end{smallmatrix}\right]]) & \text{if } p \nmid 2\beta, \\ \sum\limits_{x \in \mathbb{Z}/p\mathbb{Z}} p^{3-k} a(S[\left[\begin{smallmatrix} 1 & \\ x & p \end{smallmatrix}\right]]) - \sum\limits_{x \in \mathbb{Z}/p\mathbb{Z}} p a(S\Big[\begin{bmatrix} 1 & \\ xp^{-1} & 1 \end{bmatrix}\Big]) & \text{if } p \mid 2\beta. \end{cases} \tag{12.25}$$

(3) *Assume that $\mu_p \neq 0$. If $S = \begin{bmatrix} \alpha & \beta \\ \beta & \gamma \end{bmatrix} \in B(Np^{-1})^+$, then*

$$\lambda_p \sum_{x \in \mathbb{Z}/p\mathbb{Z}} a(S[\left[\begin{smallmatrix} 1 & \\ x & p \end{smallmatrix}\right]]) = \sum_{x \in \mathbb{Z}/p\mathbb{Z}} p^{3-k} a(pS[\left[\begin{smallmatrix} 1 & \\ x & p \end{smallmatrix}\right]]) + \sum_{\substack{z \in \mathbb{Z}/p^2\mathbb{Z} \\ p \mid (\alpha + 2\beta z + \gamma z^2)}} p a(p^{-1} S\Big[\begin{bmatrix} 1 & \\ z & p^2 \end{bmatrix}\Big]) \tag{12.26}$$

and

$$\sum_{y\in\mathbb{Z}/p^2\mathbb{Z}} a(S[\left[\begin{smallmatrix}1 & \\ y & p^2\end{smallmatrix}\right]]) = \begin{cases} \varepsilon_p \sum\limits_{x\in\mathbb{Z}/p\mathbb{Z}} p^{k-2}a(S[\left[\begin{smallmatrix}1 & \\ x & p\end{smallmatrix}\right]]) & \text{if } v_p(N)=2, \\ 0 & \text{if } v_p(N)>2. \end{cases} \tag{12.27}$$

The representation π_p is generic.

(4) *Assume that $\mu_p = 0$. Then*

$$\sum_{x\in\mathbb{Z}/p\mathbb{Z}} a(S[\left[\begin{smallmatrix}1 & \\ x & p\end{smallmatrix}\right]]) = 0 \qquad \text{for } S \in B(Np^{-1})^+. \tag{12.28}$$

We have $T^s_{0,1}(p)F = \sum_{S\in B(N)^+} c(S)e^{2\pi i\mathrm{Tr}(SZ)}$ where

$$c(S) = p^{3-k}a(pS) + \sum_{\substack{x\in\mathbb{Z}/p\mathbb{Z}\\ p|(\alpha+2\beta x)}} pa(p^{-1}S[\left[\begin{smallmatrix}1 & \\ x & p\end{smallmatrix}\right]]) \tag{12.29}$$

for $S = \left[\begin{smallmatrix}\alpha & \beta \\ \beta & \gamma\end{smallmatrix}\right] \in B(N)^+$. If F is not an eigenvector for $T^s_{0,1}(p)$, then π_p is generic,

$$\lambda_p c(S) = p^3 a(S) + p^{6-2k}a(p^2S) + \sum_{\substack{y\in\mathbb{Z}/p\mathbb{Z}\\ p|(\alpha+2\beta y)}} p^{4-k}a(S[\left[\begin{smallmatrix}1 & \\ y & p\end{smallmatrix}\right]])$$
$$+ \sum_{\substack{z\in\mathbb{Z}/p^2\mathbb{Z}\\ p^2|(\alpha+2\beta z+\gamma z^2)}} p^2a(p^{-2}S[\left[\begin{smallmatrix}1 & \\ z & p^2\end{smallmatrix}\right]]) \tag{12.30}$$

for $S = \left[\begin{smallmatrix}\alpha & \beta \\ \beta & \gamma\end{smallmatrix}\right] \in B(N)^+$, and $c(S) \neq 0$ for some $S \in B(N)^+$. The newform F is an eigenvector for $T^s_{0,1}(p)$ if and only if

$$\lambda_p a(S) = (1+p^{-1})c(S) = (1+p)p^{2-k}a(pS)$$
$$+ \sum_{\substack{x\in\mathbb{Z}/p\mathbb{Z}\\ p|(\alpha+2\beta x)}} (1+p)a(p^{-1}S[\left[\begin{smallmatrix}1 & \\ x & p\end{smallmatrix}\right]]) \tag{12.31}$$

for $S = \left[\begin{smallmatrix}\alpha & \beta \\ \beta & \gamma\end{smallmatrix}\right] \in B(N)^+$; in this case π_p is non-generic.

Proof The equivalences (12.22) and (12.23) follow from (12.3), (11.19), and (11.27).

(1). Assume that $v_p(N) \geq 3$. Then $\sigma_p^2 F = 0$ by (1) of Theorem 12.1.2. Two applications of (11.19) yield

$$(\sigma_p^2 F)(Z) = \sum_{S\in B(Np^{-2})^+} \sum_{y\in\mathbb{Z}/p\mathbb{Z}} \sum_{x\in\mathbb{Z}/p\mathbb{Z}} p^{-2k-2} a\big((S[\left[\begin{smallmatrix}1&\\x&p\end{smallmatrix}\right]])[\left[\begin{smallmatrix}1&\\y&p\end{smallmatrix}\right]]\big)e^{2\pi i \mathrm{Tr}(SZ)}$$
$$= \sum_{S\in B(Np^{-2})^+} \sum_{z\in\mathbb{Z}/p^2\mathbb{Z}} p^{-2k-2} a(S[\begin{bmatrix}1&\\z&p^2\end{bmatrix}])e^{2\pi i \mathrm{Tr}(SZ)}.$$

Since $\sigma_p^2 F = 0$ we obtain (12.24).

(2). By (2) of Theorem 12.1.2 we have $\mu_p F = p^4\tau_p^2\sigma_p F - p^2\eta_p\sigma_p F$. Each term of this equality is contained in $S_k(\mathrm{K}_s(Np))$. Let

$$(\sigma_p F)(Z) = \sum_{S\in B(Np^{-1})^+} b(S)e^{2\pi i \mathrm{Tr}(SZ)} \tag{12.32}$$

be the Fourier expansion of $\sigma_p F$. By (11.19), for $S \in B(Np^{-1})^+$,

$$b(S) = \sum_{x\in\mathbb{Z}/p\mathbb{Z}} p^{-k-1} a(S[\left[\begin{smallmatrix}1&\\x&p\end{smallmatrix}\right]]). \tag{12.33}$$

Therefore, by (11.7),

$$p^4\tau_p^2\sigma_p F(Z) = \sum_{S\in B(Np)^+} p^4 b(S)e^{2\pi i \mathrm{Tr}(SZ)}$$
$$= \sum_{S\in B(Np)^+} \sum_{x\in\mathbb{Z}/p\mathbb{Z}} p^{3-k} a(S[\left[\begin{smallmatrix}1&\\x&p\end{smallmatrix}\right]])e^{2\pi i \mathrm{Tr}(SZ)}.$$

And by (11.13),

$$(\eta_p\sigma_p F)(Z) = \sum_{\substack{S=\left[\begin{smallmatrix}\alpha&\beta\\\beta&\gamma\end{smallmatrix}\right]\in B(Np)^+\\ p|2\beta,\ p^2|\gamma}} p^k b(S[\begin{bmatrix}1&\\&p^{-1}\end{bmatrix}])e^{2\pi i \mathrm{Tr}(SZ)}$$
$$= \sum_{\substack{S=\left[\begin{smallmatrix}\alpha&\beta\\\beta&\gamma\end{smallmatrix}\right]\in B(Np)^+\\ p|2\beta}} p^k b(S[\begin{bmatrix}1&\\&p^{-1}\end{bmatrix}])e^{2\pi i \mathrm{Tr}(SZ)}$$
$$= \sum_{\substack{S=\left[\begin{smallmatrix}\alpha&\beta\\\beta&\gamma\end{smallmatrix}\right]\in B(Np)^+\\ p|2\beta}} \sum_{x\in\mathbb{Z}/p\mathbb{Z}} p^{-1} a\big((S[\begin{bmatrix}1&\\&p^{-1}\end{bmatrix}])[\left[\begin{smallmatrix}1&\\x&p\end{smallmatrix}\right]]\big)e^{2\pi i \mathrm{Tr}(SZ)}$$

$$= \sum_{\substack{S=\left[\begin{smallmatrix}\alpha & \beta \\ \beta & \gamma\end{smallmatrix}\right]\in B(Np)^+ \\ p|2\beta}} \sum_{x\in\mathbb{Z}/p\mathbb{Z}} p^{-1}a(S[\left[\begin{smallmatrix}1 & \\ xp^{-1} & 1\end{smallmatrix}\right]])e^{2\pi i \mathrm{Tr}(SZ)}.$$

The assertion (12.25) follows now from $\mu_p F = p^4\tau_p^2\sigma_p F - p^2\eta_p\sigma_p F$.

(3). Assume that $\mu_p \neq 0$. Again let the Fourier expansion of $\sigma_p F$ be as (12.32) with b as in (12.33). By (11.24),

$$\begin{aligned}
&(T_{0,1}^s(p)\sigma_p F)(Z) \\
&\quad = \sum_{S=\left[\begin{smallmatrix}\alpha & \beta \\ \beta & \gamma\end{smallmatrix}\right]\in B(Np^{-1})^+} \Big(p^{3-k}b(pS) \\
&\qquad + \sum_{\substack{y\in\mathbb{Z}/p\mathbb{Z} \\ p|(\alpha+2\beta y+\gamma y^2)}} pb(p^{-1}S[\left[\begin{smallmatrix}1 & \\ x & p\end{smallmatrix}\right]])\Big)e^{2\pi i \mathrm{Tr}(SZ)}.
\end{aligned}$$

By (12.5), $T_{0,1}^s(p)\sigma_p F = \lambda_p\sigma_p F$. Hence, for $S = \left[\begin{smallmatrix}\alpha & \beta \\ \beta & \gamma\end{smallmatrix}\right] \in B(Np^{-1})^+$,

$$\lambda_p b(S) = p^{3-k}b(pS) + \sum_{\substack{y\in\mathbb{Z}/p\mathbb{Z} \\ p|(\alpha+2\beta y+\gamma y^2)}} pb(p^{-1}S[\left[\begin{smallmatrix}1 & \\ y & p\end{smallmatrix}\right]]).$$

Substituting (12.33) we obtain, for $S = \left[\begin{smallmatrix}\alpha & \beta \\ \beta & \gamma\end{smallmatrix}\right] \in B(Np^{-1})^+$,

$$\begin{aligned}
&\lambda_p \sum_{x\in\mathbb{Z}/p\mathbb{Z}} a(S[\left[\begin{smallmatrix}1 & \\ x & p\end{smallmatrix}\right]]) \\
&\quad = \sum_{x\in\mathbb{Z}/p\mathbb{Z}} p^{3-k}a(pS[\left[\begin{smallmatrix}1 & \\ x & p\end{smallmatrix}\right]]) + \sum_{\substack{z\in\mathbb{Z}/p^2\mathbb{Z} \\ p|(\alpha+2\beta z+\gamma z^2)}} pa(p^{-1}S[\left[\begin{smallmatrix}1 & \\ z & p^2\end{smallmatrix}\right]]).
\end{aligned}$$

This is (12.26). Next we prove (12.27). By (11.27) and (11.19) we have

$$(T_{1,0}^s(p)\sigma_p F)(Z) = \sum_{S\in B(Np^{-1})^+} \sum_{y\in\mathbb{Z}/p^2\mathbb{Z}} p^{2-2k}a(S[\left[\begin{smallmatrix}1 & \\ y & p^2\end{smallmatrix}\right]])e^{2\pi i \mathrm{Tr}(SZ)}.$$

By (11.19) again,

$$(\sigma_p F)(Z) = \sum_{S\in B(Np^{-1})^+} \sum_{x\in\mathbb{Z}/p\mathbb{Z}} p^{-k-1}a(S[\left[\begin{smallmatrix}1 & \\ x & p\end{smallmatrix}\right]])e^{2\pi i \mathrm{Tr}(SZ)}.$$

Hence, (12.6) implies that if $S \in B(Np^{-1})^+$, then

$$\sum_{y \in \mathbb{Z}/p^2\mathbb{Z}} a(S[\begin{bmatrix} 1 & \\ y & p^2 \end{bmatrix}]) = (\mu_p + p^2)p^{k-3} \sum_{x \in \mathbb{Z}/p\mathbb{Z}} a(S[\begin{bmatrix} 1 & \\ x & p \end{bmatrix}]). \tag{12.34}$$

Let $V \cong \otimes_v \pi_v$ be as in Theorem 12.1.2. Applying Corollary 6.1.2 to π_p, we find that if $v_p(N) = 2$, then $\mu_p = -p^2 + \varepsilon_p p$, and if $v_p(N) > 2$, then $\mu_p = -p^2$. This implies (12.27). The remaining claim follows from (3) of Theorem 12.1.2.

(4). Assume that $\mu_p = 0$. The assertion (12.28) follows immediately from (12.23). Let

$$T^s_{0,1}(p)F = \sum_{S \in B(N)^+} c(S) e^{2\pi i \mathrm{Tr}(SZ)}$$

be the Fourier expansion of $T^s_{0,1}(p)F$. By (11.24), for $S = \begin{bmatrix} \alpha & \beta \\ \beta & \gamma \end{bmatrix} \in B(N)^+$,

$$c(S) = p^{3-k} a(pS) + \sum_{\substack{x \in \mathbb{Z}/p\mathbb{Z} \\ p \mid (\alpha + 2\beta x)}} p a(p^{-1} S[\begin{bmatrix} 1 & \\ x & p \end{bmatrix}]).$$

For this, we note that $p \mid \gamma$ because $p \mid N_s$ and $N_s \mid \gamma$; hence $\alpha + 2\beta x + \gamma x^2 \equiv \alpha + 2\beta x \pmod{p}$ for $x \in \mathbb{Z}$, which simplifies (11.24). This proves (12.29). Assume that F is not an eigenvector for $T^s_{0,1}(p)$. Then π_p is generic by (4) of Theorem 12.1.2. Since F is not an eigenvector for $T^s_{0,1}(p)$, $c(S)$ is non-zero for some $S \in B(N)^+$. Applying (11.24) again, and also using (12.23), we obtain

$$\begin{aligned}
&(T^s_{0,1}(p)^2 F)(Z) \\
&= \sum_{S = \begin{bmatrix} \alpha & \beta \\ \beta & \gamma \end{bmatrix} \in B(N)^+} \Big(p^{3-k} c(pS) + \sum_{\substack{y \in \mathbb{Z}/p\mathbb{Z} \\ p \mid (\alpha + 2\beta y)}} p c(p^{-1} S[\begin{bmatrix} 1 & \\ y & p \end{bmatrix}])\Big) e^{2\pi i \mathrm{Tr}(SZ)} \\
&= \sum_{S = \begin{bmatrix} \alpha & \beta \\ \beta & \gamma \end{bmatrix} \in B(N)^+} \Big(p^{3-k}\big(p^{3-k} a(p^2 S) + \sum_{x \in \mathbb{Z}/p\mathbb{Z}} p a(S[\begin{bmatrix} 1 & \\ x & p \end{bmatrix}])\big) \\
&\quad + \sum_{\substack{y \in \mathbb{Z}/p\mathbb{Z} \\ p \mid (\alpha + 2\beta y)}} p\big(p^{3-k} a(S[\begin{bmatrix} 1 & \\ y & p \end{bmatrix}]) \\
&\quad + \sum_{\substack{x \in \mathbb{Z}/p\mathbb{Z} \\ p \mid (\alpha' + 2\beta x) \\ \alpha' = p^{-1}(\alpha + 2\beta y + \gamma y^2)}} p a(p^{-2} S[\begin{bmatrix} 1 & \\ y + px & p^2 \end{bmatrix}])\big)\Big) e^{2\pi i \mathrm{Tr}(SZ)}
\end{aligned}$$

$$= \sum_{S=\left[\begin{smallmatrix}\alpha & \beta\\ \beta & \gamma\end{smallmatrix}\right]\in B(N)^+} \Big(p^{6-2k}a(p^2S) + \sum_{\substack{y\in\mathbb{Z}/p\mathbb{Z}\\ p|(\alpha+2\beta y)}} p^{4-k}a(S[\left[\begin{smallmatrix}1 & \\ y & p\end{smallmatrix}\right]])$$

$$+ \sum_{\substack{y\in\mathbb{Z}/p\mathbb{Z}\\ p|(\alpha+2\beta y)}} \sum_{\substack{x\in\mathbb{Z}/p\mathbb{Z}\\ p|(\alpha'+2\beta x)\\ \alpha'=p^{-1}(\alpha+2\beta y+\gamma y^2)}} p^2a(p^{-2}S[\left[\begin{smallmatrix}1 & \\ y+px & p^2\end{smallmatrix}\right]])\Big)e^{2\pi i\mathrm{Tr}(SZ)}$$

$$= \sum_{S=\left[\begin{smallmatrix}\alpha & \beta\\ \beta & \gamma\end{smallmatrix}\right]\in B(N)^+} \Big(p^{6-2k}a(p^2S) + \sum_{\substack{y\in\mathbb{Z}/p\mathbb{Z}\\ p|(\alpha+2\beta y)}} p^{4-k}a(S[\left[\begin{smallmatrix}1 & \\ y & p\end{smallmatrix}\right]])$$

$$+ \sum_{\substack{z\in\mathbb{Z}/p^2\mathbb{Z}\\ p^2|(\alpha+2\beta z+\gamma z^2)}} p^2a(p^{-2}S[\left[\begin{smallmatrix}1 & \\ z & p^2\end{smallmatrix}\right]])\Big)e^{2\pi i\mathrm{Tr}(SZ)}.$$

By (12.8), $F = p^{-3}\lambda_p T^s_{0,1}(p)F - p^{-3}T^s_{0,1}(p)^2F$. Hence, for $S = \left[\begin{smallmatrix}\alpha & \beta\\ \beta & \gamma\end{smallmatrix}\right] \in B(N)^+$,

$$a(S) = p^{-3}\lambda_p\Big(p^{3-k}a(pS) + \sum_{\substack{x\in\mathbb{Z}/p\mathbb{Z}\\ p|(\alpha+2\beta x)}} pa(p^{-1}S[\left[\begin{smallmatrix}1 & \\ x & p\end{smallmatrix}\right]])\Big)$$

$$- p^{-3}\Big(p^{6-2k}a(p^2S) + \sum_{\substack{y\in\mathbb{Z}/p\mathbb{Z}\\ p|(\alpha+2\beta y)}} p^{4-k}a(S[\left[\begin{smallmatrix}1 & \\ y & p\end{smallmatrix}\right]])$$

$$+ \sum_{\substack{z\in\mathbb{Z}/p^2\mathbb{Z}\\ p^2|(\alpha+2\beta z+\gamma z^2)}} p^2a(p^{-2}S[\left[\begin{smallmatrix}1 & \\ z & p^2\end{smallmatrix}\right]])\Big)$$

$$= \lambda_p p^{-k}a(pS) + \lambda_p \sum_{\substack{x\in\mathbb{Z}/p\mathbb{Z}\\ p|(\alpha+2\beta x)}} p^{-2}a(p^{-1}S[\left[\begin{smallmatrix}1 & \\ x & p\end{smallmatrix}\right]])$$

$$- p^{3-2k}a(p^2S) - \sum_{\substack{y\in\mathbb{Z}/p\mathbb{Z}\\ p|(\alpha+2\beta y)}} p^{1-k}a(S[\left[\begin{smallmatrix}1 & \\ y & p\end{smallmatrix}\right]])$$

$$- \sum_{\substack{z\in\mathbb{Z}/p^2\mathbb{Z}\\ p^2|(\alpha+2\beta z+\gamma z^2)}} p^{-1}a(p^{-2}S[\left[\begin{smallmatrix}1 & \\ z & p^2\end{smallmatrix}\right]]).$$

This proves (12.30). Finally, if F is an eigenvector for $T^s_{0,1}(p)$, then π_p is non-generic by (4) of Theorem 12.1.2 and (12.31) follows from (12.9) and (12.29). □

We mention that the identities of Theorem 12.1.2 can also be expressed in terms of the Fourier-Jacobi coefficients of the newform F, using the formulas computed in Chap. 11.

12.2 Another Formulation

In this section we present another formulation of the consequences of Theorem 12.1.2 for Fourier coefficients. This approach is based on the observation that the formulas from Corollary 12.1.3 involve right coset representatives for elements of the classical Hecke ring of $\Gamma_0(N) \subset \mathrm{SL}(2, \mathbb{Z})$, suggesting a more compact and abstract description.

To describe the setting, let $A(\mathbb{Q})$ be the set of all positive semi-definite symmetric 2×2 matrices with entries from $\mathbb{Q}$; if X is a subset of $A(\mathbb{Q})$, then let X^+ be the subset of elements of X that are positive definite. We define an action of $\mathrm{GL}(2, \mathbb{Q})$ on $A(\mathbb{Q})$ by

$$g \cdot S = gS\,{}^{\mathrm{t}}g = S[{}^{\mathrm{t}}g] \tag{12.35}$$

for $g \in \mathrm{GL}(2, \mathbb{Q})$ and $S \in A(\mathbb{Q})$; this definition extends the action (10.15). Let $\mathcal{F}$ be the set of all functions $a : A(\mathbb{Q}) \to \mathbb{C}$. Then $\mathcal{F}$ is a $\mathbb{C}$-algebra under pointwise addition and multiplication of functions. Characteristic functions lying in $\mathcal{F}$ will be useful. Let X be a subset of $A(\mathbb{Q})$. We define $\mathrm{char}_X : A(\mathbb{Q}) \to \mathbb{C}$ to be the characteristic function of X, so that

$$\mathrm{char}_X(S) = \begin{cases} 1 & \text{if } S \in X, \\ 0 & \text{if } S \notin X. \end{cases}$$

We define a right action of $\mathrm{GL}(2, \mathbb{Q})$ on $\mathcal{F}$ by

$$(a|g)(S) = a(g \cdot S) = a(gS\,{}^{\mathrm{t}}g) = a(S[{}^{\mathrm{t}}g])$$

for $a \in \mathcal{F}$, $g \in \mathrm{GL}(2, \mathbb{Q})$ and $S \in A(\mathbb{Q})$. Let N be an integer such that $N > 0$, and let $\Gamma_0(N)$ be the subgroup of $\left[\begin{smallmatrix} a & b \\ c & d \end{smallmatrix}\right] \in \mathrm{SL}(2, \mathbb{Z})$ such that $c \equiv 0 \pmod N$. We define $\mathcal{F}^{\Gamma_0(N)}$ to be the subspace of $a \in \mathcal{F}$ such that $a|k = a$ for $k \in \Gamma_0(N)$. We note that $\mathrm{char}_{B(N)}$ and $\mathrm{char}_{B(N)^+}$ are in $\mathcal{F}^{\Gamma_0(N)}$. Fourier coefficients of Siegel modular forms are also contained in $\mathcal{F}^{\Gamma_0(N)}$. More precisely, let k be an integer such that $k > 0$, and let $F \in M_k(\mathrm{K}_s(N))$. Let

$$F(Z) = \sum_{S \in B(N)} a(S) e^{2\pi i \mathrm{Tr}(SZ)}$$

be the Fourier expansion of F. To F we associate the function

$$\hat{a}(F) : A(\mathbb{Q}) \longrightarrow \mathbb{C} \tag{12.36}$$

defined by

$$\hat{a}(F)(S) = \begin{cases} a(S) & \text{if } S \in B(N), \\ 0 & \text{if } S \notin B(N). \end{cases}$$

By (10.18) the function $\hat{a}(F)$ is contained in $\mathcal{F}^{\Gamma_0(N)}$. Next, let $\Delta_0(N)$ be the subset of $\left[\begin{smallmatrix} a & b \\ c & d \end{smallmatrix}\right] \in \mathrm{M}(2, \mathbb{Z})$ such that $c \equiv 0 \pmod{N}$, $(a, N) = 1$, and $ad - bc > 0$. Then $\Delta_0(N)$ is a semigroup under multiplication of matrices. We let $\mathcal{R}(N) = \mathcal{R}(\Gamma_0(N), \Delta_0(N))$ be the Hecke ring with respect to $\Gamma_0(N)$ and $\Delta_0(N)$; see, for example, (4.5.4) of [86]. The Hecke ring $\mathcal{R}(N)$ is commutative (see Theorem 4.5.3 of [86]), and we define an action of this ring on $\mathcal{F}^{\Gamma_0(N)}$ via the formula

$$a\big|\Gamma_0(N)g\Gamma_0(N) = \sum_i a\big|g_i$$

where $a \in \mathcal{F}^{\Gamma_0(N)}$, $g \in \Delta_0(N)$ and $\Gamma_0(N)g\Gamma_0(N) = \sqcup_i \Gamma_0(N)g_i$ is a disjoint decomposition. Following [86], we define

$$T(l, m) = \Gamma_0(N)\left[\begin{smallmatrix} l & \\ & m \end{smallmatrix}\right]\Gamma_0(N), \tag{12.37}$$

$$T(n) = \sum_{\det(g)=n} \Gamma_0(N)g\Gamma_0(N), \tag{12.38}$$

where $l \mid m$ and $(l, N) = 1$, and in the second equality the summation is over all double cosets $\Gamma_0(N)g\Gamma_0(N)$ where $g \in \Delta_0(N)$ with $\det(g) = n$. Let p be a prime of $\mathbb{Z}$ dividing N. By (4.5.17) and Lemma 4.5.7 of [86] we have

$$T(p) = T(1, p), \tag{12.39}$$

$$T(p)^2 = T(p^2) = T(1, p^2). \tag{12.40}$$

By Lemma 4.5.6 of [86] we have

$$\Gamma_0(N)\left[\begin{smallmatrix} 1 & \\ & p \end{smallmatrix}\right]\Gamma_0(N) = \bigsqcup_{a \in \mathbb{Z}/p\mathbb{Z}} \Gamma_0(N)\left[\begin{smallmatrix} 1 & a \\ & p \end{smallmatrix}\right], \tag{12.41}$$

$$\Gamma_0(N)\begin{bmatrix} 1 & \\ & p^2 \end{bmatrix}\Gamma_0(N) = \bigsqcup_{a \in \mathbb{Z}/p^2\mathbb{Z}} \Gamma_0(N)\begin{bmatrix} 1 & a \\ & p^2 \end{bmatrix}. \tag{12.42}$$

Finally, we will need three other operators on $\mathcal{F}$. Let $t \in \mathbb{Z}$ with $t > 0$. We define

$$\Delta_t^+,\ \Delta_t^-,\ \nabla_t : \mathcal{F} \longrightarrow \mathcal{F}$$

by

$$(\Delta_t^+ a)(S) = a(tS),$$
$$(\Delta_t^- a)(S) = a(t^{-1}S),$$
$$(\nabla_t a)(S) = (a\big|\begin{bmatrix} 1 & \\ & t^{-1}\end{bmatrix})(S) = a(\begin{bmatrix} 1 & \\ & t^{-1}\end{bmatrix} S \begin{bmatrix} 1 & \\ & t^{-1}\end{bmatrix})$$

for $a \in \mathcal{F}$ and $S \in A(\mathbb{Q})$. The operators Δ_t^+, Δ_t^-, and ∇_t mutually commute, and the action of the Hecke algebra $\mathcal{R}(\Gamma_0(N), \Delta_0(N))$ commutes with Δ_t^+ and Δ_t^-. Also, if $a \in \mathcal{F}$ and $g \in \mathrm{GL}(2, \mathbb{Q})$, then

$$(\mathrm{char}_{A(\mathbb{Q})^+} \cdot a)\big|g = \mathrm{char}_{A(\mathbb{Q})^+} \cdot a\big|g, \tag{12.43}$$
$$\Delta_t^+(\mathrm{char}_{A(\mathbb{Q})^+} \cdot a) = \mathrm{char}_{A(\mathbb{Q})^+} \cdot \Delta_t^+ a, \tag{12.44}$$
$$\Delta_t^-(\mathrm{char}_{A(\mathbb{Q})^+} \cdot a) = \mathrm{char}_{A(\mathbb{Q})^+} \cdot \Delta_t^- a, \tag{12.45}$$
$$\nabla_t(\mathrm{char}_{A(\mathbb{Q})^+} \cdot a) = \mathrm{char}_{A(\mathbb{Q})^+} \cdot \nabla_t a. \tag{12.46}$$

The next lemma translates the Fourier coefficient calculations of Chap. 11 into the language of the last paragraph.

Lemma 12.2.1 *Let N and k be integers such that $N > 0$ and $k > 0$, and let p be a prime of $\mathbb{Z}$. Let $F \in M_k(\mathrm{K}_{\text{ş}}(N))$, and set $\hat{a} = \hat{a}(F)$. Then:*

$$\hat{a}(\tau_p F) = \mathrm{char}_{B(Np)} \cdot \hat{a}, \tag{12.47}$$
$$\hat{a}(\theta_p F) = p^k \Delta_p^- \hat{a} + p \cdot \begin{Bmatrix} \mathrm{char}_{B(Np^2)} & \text{if } v_p(N) = 0 \\ \mathrm{char}_{B(Np)} & \text{if } v_p(N) > 0 \end{Bmatrix} \cdot \Delta_p^+ \nabla_p \hat{a}, \tag{12.48}$$
$$\hat{a}(\eta_p F) = p^k \nabla_p \hat{a}, \tag{12.49}$$
$$\hat{a}(\sigma_p F) = p^{-k-1} \mathrm{char}_{B(Np^{-1})} \cdot (\hat{a}\big|T(p)) \qquad \text{if } v_p(N) \geq 2, \tag{12.50}$$
$$\hat{a}(\sigma_p^2 F) = p^{-2k-2} \mathrm{char}_{B(Np^{-2})} \cdot (\hat{a}\big|T(p)^2) \qquad \text{if } v_p(N) \geq 3, \tag{12.51}$$
$$\begin{aligned}\hat{a}(T_{0,1}^s(p)F) = {} & p^{3-k} \mathrm{char}_{B(N)} \cdot \Delta_p^+ \hat{a} \\ & + p\, \mathrm{char}_{B(N)} \cdot (\Delta_p^- \hat{a}\big|T(p)) \qquad \text{if } v_p(N) \geq 1,\end{aligned} \tag{12.52}$$
$$\hat{a}(T_{1,0}^s(p)F) = p^{3-k} \mathrm{char}_{B(N)} \cdot (\hat{a}\big|T(p)) \qquad \text{if } v_p(N) \geq 1, \tag{12.53}$$
$$\hat{a}(T_{1,0}^s(p)\sigma_p F) = p^{2-2k} \mathrm{char}_{B(Np^{-1})} \cdot (\hat{a}\big|T(p)^2) \qquad \text{if } v_p(N) \geq 2. \tag{12.54}$$

If $F \in M_k(\mathrm{K}(N))$ and $v_p(N) \geq 2$, then

$$\begin{aligned}\hat{a}(T^s_{0,1}(p)\sigma_p F) &= p^{2-2k}\mathrm{char}_{B(Np^{-1})}\cdot(\Delta^+_p\hat{a}\big|T(p))\\ &\quad + p^{-k}\mathrm{char}_{B(Np^{-1})}\cdot(\Delta^-_p\hat{a}\big|T(p)^2), \qquad (12.55)\\ \hat{a}(T^s_{0,1}(p)^2 F) &= \mathrm{char}_{B(N)}\cdot\big(p^{6-2k}\Delta^+_{p^2}\hat{a}+p^{4-k}\hat{a}\big|T(p)\\ &\quad + p^{4-k}\Delta^-_p(\mathrm{char}_{B(N)}\cdot\hat{a})\big|T(p)+p^2\Delta^-_{p^2}\hat{a}\big|T(p)^2\big). \qquad (12.56)\end{aligned}$$

If F is a cusp form, then the same equations with all appearing sets of the form $B(M)$ replaced by $B(M)^+$, where M an integer such that $M > 0$, hold.

Proof (12.47). This follows immediately from (11.7).

(12.48). Let

$$X_1 = \{\left[\begin{smallmatrix}\alpha & \beta\\ \beta & \gamma\end{smallmatrix}\right] \in B(N) \mid p \mid \alpha,\ p \mid 2\beta,\ N_s p \mid \gamma\},$$

$$X_2 = \{\left[\begin{smallmatrix}\alpha & \beta\\ \beta & \gamma\end{smallmatrix}\right] \in B(N) \mid N_s p \mid \gamma\}.$$

By (11.10), since $\hat{a}(\theta_p F)$ is supported in $B(Np)$ and $B(Np) \subset B(N)$,

$$\begin{aligned}\hat{a}(\theta_p F)(S) &= p^k\mathrm{char}_{B(Np)}(S)\,\mathrm{char}_{X_1}(S)\hat{a}(p^{-1}S)\\ &\quad + p\,\mathrm{char}_{B(Np)}(S)\,\mathrm{char}_{X_2}(S)\hat{a}(p^{-1}S[\left[\begin{smallmatrix}p & \\ & 1\end{smallmatrix}\right]])\end{aligned}$$

for $S \in A(\mathbb{Q})$. Since $X_1, X_2 \subset B(Np)$ we obtain

$$\hat{a}(\theta_p F)(S) = p^k\mathrm{char}_{X_1}(S)\hat{a}(p^{-1}S) + p\,\mathrm{char}_{X_2}(S)\hat{a}(p^{-1}S[\left[\begin{smallmatrix}p & \\ & 1\end{smallmatrix}\right]])$$

for $S \in A(\mathbb{Q})$. Next, we note that

$$X_1 = pB(N) \quad\text{and}\quad X_2 = \begin{cases} B(Np^2) & \text{if } v_p(N) = 0,\\ B(Np) & \text{if } v_p(N) > 0.\end{cases}$$

Therefore, if $S \in A(\mathbb{Q})$, then

$$\begin{aligned}p^k\mathrm{char}_{X_1}(S)\hat{a}(p^{-1}S) &= p^k\mathrm{char}_{pB(N)}(S)\hat{a}(p^{-1}S)\\ &= p^k\mathrm{char}_{B(N)}(p^{-1}S)\hat{a}(p^{-1}S)\\ &= p^k\hat{a}(p^{-1}S)\\ &= p^k(\Delta^-_p\hat{a})(S).\end{aligned}$$

For the third equality we used that the support of $\hat{a}$ lies in $B(N)$. Also, if $S \in A(\mathbb{Q})$, then

$$p\,\mathrm{char}_{X_2}(S)\hat{a}(p^{-1}S[\left[\begin{smallmatrix} p & \\ & 1\end{smallmatrix}\right]]) = p\begin{Bmatrix} \mathrm{char}_{B(Np^2)}(S) \text{ if } v_p(N)=0 \\ \mathrm{char}_{B(Np)}(S) \text{ if } v_p(N)>0 \end{Bmatrix}(\Delta_p^{+}\nabla_p\hat{a})(S).$$

This proves (12.48).

(12.49). Let

$$X = \{\left[\begin{smallmatrix} \alpha & \beta \\ \beta & \gamma\end{smallmatrix}\right] \in B(Np^2) \mid p \mid 2\beta,\ p^2 \mid \gamma\}.$$

Then an argument shows that $X = \left[\begin{smallmatrix} 1 & \\ & p\end{smallmatrix}\right] B(N) \left[\begin{smallmatrix} 1 & \\ & p\end{smallmatrix}\right]$. By (11.13), for $S \in A(\mathbb{Q})$,

$$\begin{aligned}
\hat{a}(\eta_p F)(S) &= p^k \mathrm{char}_X(S)(\nabla_p\hat{a})(S)\\
&= p^k \mathrm{char}_{\left[\begin{smallmatrix} 1 & \\ & p\end{smallmatrix}\right] B(N) \left[\begin{smallmatrix} 1 & \\ & p\end{smallmatrix}\right]}(S)(\nabla_p\hat{a})(S)\\
&= p^k(\nabla_p \mathrm{char}_{B(N)})(S)(\nabla_p\hat{a})(S)\\
&= p^k\nabla_p(\mathrm{char}_{B(N)}\cdot\hat{a})(S)\\
&= p^k(\nabla_p\hat{a})(S).
\end{aligned}$$

This is (12.49).

(12.50). This follows immediately from (11.19).

(12.51). To prove this, assume that $v_p(N) \geq 3$. Let $S \in A(\mathbb{Q})$. If $S \notin B(Np^{-2})$, then both sides of (12.51) evaluated at S are zero. Assume that $S = \left[\begin{smallmatrix} \alpha & \beta \\ \beta & \gamma\end{smallmatrix}\right] \in B(Np^{-2})$. By (12.50),

$$\begin{aligned}
\hat{a}(\sigma_p^2 F)(S) &= p^{-k-1}(\hat{a}(\sigma_p F)\big|T(p))(S)\\
&= p^{-k-1}\sum_{x\in\mathbb{Z}/p\mathbb{Z}} \hat{a}(\sigma_p F)(S[\left[\begin{smallmatrix} 1 & \\ x & p\end{smallmatrix}\right]]).
\end{aligned}$$

If $x \in \mathbb{Z}$ then $S[\left[\begin{smallmatrix} 1 & \\ x & p\end{smallmatrix}\right]] = \left[\begin{smallmatrix} \alpha+2\beta x+\gamma x^2 & \beta p+\gamma p x \\ \beta p+\gamma p x & \gamma p^2\end{smallmatrix}\right] \in B(Np^{-1})$. Using (12.50) again, we have:

$$\begin{aligned}
&p^{-k-1}\sum_{x\in\mathbb{Z}/p\mathbb{Z}} \hat{a}(\sigma_p F)(S[\left[\begin{smallmatrix} 1 & \\ x & p\end{smallmatrix}\right]])\\
&= p^{-2k-2}\sum_{x\in\mathbb{Z}/p\mathbb{Z}}\sum_{y\in\mathbb{Z}/p\mathbb{Z}} \hat{a}(F)((S[\left[\begin{smallmatrix} 1 & \\ x & p\end{smallmatrix}\right]])[\left[\begin{smallmatrix} 1 & \\ y & p\end{smallmatrix}\right]])\\
&= p^{-2k-2}\mathrm{char}_{B(Np^{-2})}(S)\cdot(\hat{a}\big|T(p)^2)(S).
\end{aligned}$$

This proves (12.51).

(12.52). Assume that $v_p(N) \geq 1$, and let $S = \begin{bmatrix} \alpha & \beta \\ \beta & \gamma \end{bmatrix} \in A(\mathbb{Q})$. If $S \notin B(N)$, then both sides of (12.52) evaluated at S are trivially zero. Assume that $S \in B(N)$. Then by (12.41),

$$
\begin{aligned}
&p^{3-k}\mathrm{char}_{B(N)}(S)\,(\Delta_p^+\hat{a})(S) + p\,\mathrm{char}_{B(N)}(S)\,(\Delta_p^-\hat{a}\big|T(p))(S) \\
&\quad = p^{3-k}\mathrm{char}_{B(N)}(S)\,\hat{a}(pS) + p\,\mathrm{char}_{B(N)}(S) \sum_{y\in\mathbb{Z}/p\mathbb{Z}} \hat{a}(p^{-1}S[\left[\begin{smallmatrix} 1 & \\ y & p \end{smallmatrix}\right]]) \\
&\quad = p^{3-k}\hat{a}(pS) + p \sum_{y\in\mathbb{Z}/p\mathbb{Z}} \hat{a}(p^{-1}S[\left[\begin{smallmatrix} 1 & \\ y & p \end{smallmatrix}\right]]) \\
&\quad = p^{3-k}a(pS) + p \sum_{y\in\mathbb{Z}/p\mathbb{Z}} \mathrm{char}_{B(N)}(p^{-1}S[\left[\begin{smallmatrix} 1 & \\ y & p \end{smallmatrix}\right]])\hat{a}(p^{-1}S[\left[\begin{smallmatrix} 1 & \\ y & p \end{smallmatrix}\right]]) \\
&\quad = p^{3-k}a(pS) + p \sum_{y\in\mathbb{Z}/p\mathbb{Z}} \mathrm{char}_{B(N)}(\left[\begin{smallmatrix} (\alpha+2\beta y+\gamma y^2)p^{-1} & \beta+y\gamma \\ \beta+y\gamma & p\gamma \end{smallmatrix}\right])\hat{a}(p^{-1}S[\left[\begin{smallmatrix} 1 & \\ y & p \end{smallmatrix}\right]]) \\
&\quad = p^{3-k}a(pS) + p \sum_{\substack{y\in\mathbb{Z}/p\mathbb{Z} \\ p|(\alpha+2\beta y+\gamma y^2)}} a(p^{-1}S[\left[\begin{smallmatrix} 1 & \\ y & p \end{smallmatrix}\right]]) \\
&\quad = a(T_{0,1}^s(p)F)(S) \qquad \text{(by (11.24))} \\
&\quad = \hat{a}(T_{0,1}^s(p)F)(S).
\end{aligned}
$$

This proves (12.52).

(12.53). This follows immediately from (11.27).

(12.54). The proof of this is similar to the proof of (12.51).

(12.55). Assume that $v_p(N) \geq 2$ and $F \in M_k(\mathrm{K}(N))$, and let $S \in A(\mathbb{Q})$. Then using (12.52) and (12.50), we have:

$$
\begin{aligned}
&\hat{a}(T_{0,1}^s(p)\sigma_p F)(S) \\
&\qquad = p^{3-k}\mathrm{char}_{B(Np^{-1})}(S)\hat{a}(\sigma_p F)(pS) \\
&\qquad\quad + p\,\mathrm{char}_{B(Np^{-1})}(S) \sum_{x\in\mathbb{Z}/p\mathbb{Z}} \hat{a}(\sigma_p F)(p^{-1}S[\left[\begin{smallmatrix} 1 & \\ x & p \end{smallmatrix}\right]]) \\
&\qquad = p^{3-k}\mathrm{char}_{B(Np^{-1})}(S)\Big(p^{-k-1}\mathrm{char}_{B(Np^{-1})}(pS) \sum_{x\in\mathbb{Z}/p\mathbb{Z}} \hat{a}(pS[\left[\begin{smallmatrix} 1 & \\ x & p \end{smallmatrix}\right]])\Big) \\
&\qquad\quad + p\,\mathrm{char}_{B(Np^{-1})}(S) \sum_{x\in\mathbb{Z}/p\mathbb{Z}} \Big(p^{-k-1}\mathrm{char}_{B(Np^{-1})}(p^{-1}S[\left[\begin{smallmatrix} 1 & \\ x & p \end{smallmatrix}\right]]) \\
&\qquad\quad \times \sum_{y\in\mathbb{Z}/p\mathbb{Z}} \hat{a}(p^{-1}S[\left[\begin{smallmatrix} 1 & \\ x & p \end{smallmatrix}\right]][\left[\begin{smallmatrix} 1 & \\ y & p \end{smallmatrix}\right]])\Big)
\end{aligned}
$$

$$
\begin{aligned}
&= p^{2-2k}\mathrm{char}_{B(Np^{-1})}(S) \sum_{x\in\mathbb{Z}/p\mathbb{Z}} \hat{a}(pS[\left[\begin{smallmatrix}1&\\x&p\end{smallmatrix}\right]]) \\
&\quad + p^{-k}\,\mathrm{char}_{B(Np^{-1})}(S) \sum_{x\in\mathbb{Z}/p\mathbb{Z}} \Big(\mathrm{char}_{B(Np^{-1})}(p^{-1}S[\left[\begin{smallmatrix}1&\\x&p\end{smallmatrix}\right]]) \\
&\quad \times \sum_{y\in\mathbb{Z}/p\mathbb{Z}} \hat{a}(p^{-1}S[\left[\begin{smallmatrix}1&\\x&p\end{smallmatrix}\right]\left[\begin{smallmatrix}1&\\y&p\end{smallmatrix}\right]])\Big).
\end{aligned}
$$

We now observe the following. Let $x, y \in \mathbb{Z}$, and assume that

$$
\mathrm{char}_{B(Np^{-1})}(S) \neq 0 \quad \text{and} \quad \hat{a}(p^{-1}S[\left[\begin{smallmatrix}1&\\x&p\end{smallmatrix}\right]\left[\begin{smallmatrix}1&\\y&p\end{smallmatrix}\right]]) \neq 0. \tag{12.57}
$$

Let $S = \left[\begin{smallmatrix}\alpha&\beta\\\beta&\gamma\end{smallmatrix}\right]$, and write $N = Mp^{v_p(N)}$ with $M \in \mathbb{Z}$ and p and M relatively prime. Then $(Np^{-1})_s = M_s p^{v_p(N)-2}$. Since $\mathrm{char}_{B(Np^{-1})}(S) \neq 0$ we have $\alpha, 2\beta, \gamma \in \mathbb{Z}$ and $M_s p^{v_p(N)-2} \mid \gamma$. Since $F \in M_k(\mathrm{K}(N))$ the second statement in (12.57) implies that

$$
p^{-1}S[\left[\begin{smallmatrix}1&\\x&p\end{smallmatrix}\right]\left[\begin{smallmatrix}1&\\y&p\end{smallmatrix}\right]] = \left[\begin{smallmatrix}\gamma y^2p+2\gamma xy+2\beta y+(\alpha+2\beta x+\gamma x^2)p^{-1} & \gamma yp^2+\beta p+\gamma xp\\ \gamma yp^2+\beta p+\gamma xp & \gamma p^3\end{smallmatrix}\right] \in A(N).
$$

This implies that $p \mid (\alpha + 2\beta x + \gamma x^2)$ and $N \mid \gamma p^3$. Now

$$
p^{-1}S[\left[\begin{smallmatrix}1&\\x&p\end{smallmatrix}\right]] = \left[\begin{smallmatrix}(\alpha+2\beta x+\gamma x^2)p^{-1} & \beta+\gamma x\\ \beta+\gamma x & \gamma p\end{smallmatrix}\right]. \tag{12.58}
$$

Since $N \mid \gamma p^3$, the integer $(Np^{-1})_s = M_s p^{v_p(N)-2}$ divides γp; therefore, since also $p \mid (\alpha + 2\beta x + \gamma x^2)$, the matrix in (12.58) is in $B(Np^{-1})$. Thus,

$$
\mathrm{char}_{B(Np^{-1})}(p^{-1}S[\left[\begin{smallmatrix}1&\\x&p\end{smallmatrix}\right]]) = 1.
$$

It follows that:

$$
\begin{aligned}
\hat{a}(T^s_{0,1}(p)\sigma_p F)(S) &= p^{2-2k}\mathrm{char}_{B(Np^{-1})}(S) \sum_{x\in\mathbb{Z}/p\mathbb{Z}} \hat{a}(pS[\left[\begin{smallmatrix}1&\\x&p\end{smallmatrix}\right]]) \\
&\quad + p^{-k}\,\mathrm{char}_{B(Np^{-1})}(S) \sum_{x\in\mathbb{Z}/p\mathbb{Z}}\sum_{y\in\mathbb{Z}/p\mathbb{Z}} \hat{a}(p^{-1}S[\left[\begin{smallmatrix}1&\\x&p\end{smallmatrix}\right]\left[\begin{smallmatrix}1&\\y&p\end{smallmatrix}\right]]) \\
&= p^{2-2k}\mathrm{char}_{B(Np^{-1})}(S)\cdot(\Delta_p\hat{a}\big|T(p))(S) \\
&\quad + p^{-k}\,\mathrm{char}_{B(Np^{-1})}(S)\cdot(\Delta_p^-\hat{a}\big|T(p)^2)(S).
\end{aligned}
$$

This proves (12.55).

(12.56). Assume again that $v_p(N) \geq 2$ and $F \in M_k(\mathrm{K}(N))$, and let $S \in A(\mathbb{Q})$. Then:

$$
\begin{aligned}
&\hat{a}(T^s_{0,1}(p)T^s_{0,1}(p)F)(S)\\
&= p^{3-k}\mathrm{char}_{B(N)}(S)\,\hat{a}(T^s_{0,1}(p)F)(pS)\\
&\quad + p\,\mathrm{char}_{B(N)}(S) \sum_{x\in\mathbb{Z}/p\mathbb{Z}} \hat{a}(T^s_{0,1}(p)F)(p^{-1}S[\left[\begin{smallmatrix}1&\\x&p\end{smallmatrix}\right]])\\
&= p^{3-k}\mathrm{char}_{B(N)}(S)\Big(p^{3-k}\mathrm{char}_{B(N)}(pS)\,\hat{a}(p^2S)\\
&\quad + p\,\mathrm{char}_{B(N)}(pS) \sum_{x\in\mathbb{Z}/p\mathbb{Z}} \hat{a}(S[\left[\begin{smallmatrix}1&\\x&p\end{smallmatrix}\right]])\Big)\\
&\quad + p\,\mathrm{char}_{B(N)}(S) \sum_{x\in\mathbb{Z}/p\mathbb{Z}} \Big(p^{3-k}\mathrm{char}_{B(N)}(p^{-1}S[\left[\begin{smallmatrix}1&\\x&p\end{smallmatrix}\right]])\,\hat{a}(S[\left[\begin{smallmatrix}1&\\x&p\end{smallmatrix}\right]])\\
&\quad + p\,\mathrm{char}_{B(N)}(p^{-1}S[\left[\begin{smallmatrix}1&\\x&p\end{smallmatrix}\right]]) \sum_{y\in\mathbb{Z}/p\mathbb{Z}} \hat{a}(p^{-2}S[\left[\begin{smallmatrix}1&\\x&p\end{smallmatrix}\right]][\left[\begin{smallmatrix}1&\\y&p\end{smallmatrix}\right]])\Big)\\
&= p^{3-k}\mathrm{char}_{B(N)}(S)\Big(p^{3-k}\mathrm{char}_{B(N)}(pS)\,\hat{a}(p^2S)\\
&\quad + p\,\mathrm{char}_{B(N)}(pS) \sum_{x\in\mathbb{Z}/p\mathbb{Z}} \hat{a}(S[\left[\begin{smallmatrix}1&\\x&p\end{smallmatrix}\right]])\Big)\\
&\quad + p^{4-k}\,\mathrm{char}_{B(N)}(S) \sum_{x\in\mathbb{Z}/p\mathbb{Z}} \mathrm{char}_{B(N)}(p^{-1}S[\left[\begin{smallmatrix}1&\\x&p\end{smallmatrix}\right]])\,\hat{a}(S[\left[\begin{smallmatrix}1&\\x&p\end{smallmatrix}\right]])\\
&\quad + p^2\,\mathrm{char}_{B(N)}(S)\\
&\quad \times \sum_{x\in\mathbb{Z}/p\mathbb{Z}} \sum_{y\in\mathbb{Z}/p\mathbb{Z}} \mathrm{char}_{B(N)}(p^{-1}S[\left[\begin{smallmatrix}1&\\x&p\end{smallmatrix}\right]])\hat{a}(p^{-2}S[\left[\begin{smallmatrix}1&\\x&p\end{smallmatrix}\right]][\left[\begin{smallmatrix}1&\\y&p\end{smallmatrix}\right]]).
\end{aligned}
$$

Let x, $y \in \mathbb{Z}$, and assume that

$$\mathrm{char}_{B(N)}(S) \neq 0 \quad \text{and} \quad \hat{a}(p^{-2}S[\left[\begin{smallmatrix}1&\\x&p\end{smallmatrix}\right]][\left[\begin{smallmatrix}1&\\y&p\end{smallmatrix}\right]], F) \neq 0. \tag{12.59}$$

Let $S = \left[\begin{smallmatrix}\alpha&\beta\\\beta&\gamma\end{smallmatrix}\right]$, and write $N = Mp^{v_p(N)}$ with $M \in \mathbb{Z}$ and p and M relatively prime. Then $N_s = M_s p^{v_p(N)-1}$. Since $\mathrm{char}_{B(N)}(S) \neq 0$ we have $\alpha, 2\beta, \gamma \in \mathbb{Z}$ and $M_s p^{v_p(N)-1} \mid \gamma$. Since $F \in M_k(\mathrm{K}(N))$ the second equation in (12.59) implies that

$$p^{-2}S[\left[\begin{smallmatrix}1&\\x&p\end{smallmatrix}\right]][\left[\begin{smallmatrix}1&\\y&p\end{smallmatrix}\right]] = \begin{bmatrix}(\alpha+2\beta x+\gamma x^2)p^{-2}+(2\beta y+2\gamma xy)p^{-1}+\gamma y^2 & \gamma x+\gamma yp+\beta\\ x\gamma+\gamma yp+\beta & \gamma p^2\end{bmatrix} \in A(N).$$

This implies that $p \mid (\alpha + 2\beta x + \gamma x^2)$ and $N \mid \gamma p^2$. Now

$$p^{-1} S[\begin{bmatrix} 1 & \\ x & p \end{bmatrix}] = \begin{bmatrix} (\alpha+2\beta x+\gamma x^2)p^{-1} & \beta+\gamma x \\ \beta+\gamma x & \gamma p \end{bmatrix}. \tag{12.60}$$

Since $N \mid \gamma p^2$, the integer $N_s = M_s p^{v_p(N)-1}$ divides γp; therefore, since we also have $p \mid (\alpha + 2\beta x + \gamma x^2)$, the matrix in (12.60) is in $B(N)$. Thus,

$$\mathrm{char}_{B(N)}(p^{-1} S[\begin{bmatrix} 1 & \\ x & p \end{bmatrix}]) = 1.$$

It follows that

$$\begin{aligned}
&\hat{a}(T^s_{0,1}(p) T^s_{0,1}(p) F)(S) \\
&= p^{3-k} \mathrm{char}_{B(N)}(S) \big(p^{3-k} \mathrm{char}_{B(N)}(pS)\, \hat{a}(p^2 S) \\
&\quad + p\, \mathrm{char}_{B(N)}(pS) \sum_{x \in \mathbb{Z}/p\mathbb{Z}} \hat{a}(S[\begin{bmatrix} 1 & \\ x & p \end{bmatrix}]) \big) \\
&\quad + p^{4-k}\, \mathrm{char}_{B(N)}(S) \sum_{x \in \mathbb{Z}/p\mathbb{Z}} \mathrm{char}_{B(N)}(p^{-1} S[\begin{bmatrix} 1 & \\ x & p \end{bmatrix}])\, \hat{a}(S[\begin{bmatrix} 1 & \\ x & p \end{bmatrix}]) \\
&\quad + p^2\, \mathrm{char}_{B(N)}(S) \sum_{x \in \mathbb{Z}/p\mathbb{Z}} \sum_{y \in \mathbb{Z}/p\mathbb{Z}} \hat{a}(p^{-2} S[\begin{bmatrix} 1 & \\ x & p \end{bmatrix}][\begin{bmatrix} 1 & \\ y & p \end{bmatrix}]) \\
&= p^{3-k} \mathrm{char}_{B(N)}(S) \big(p^{3-k} \mathrm{char}_{B(N)}(pS)\, \hat{a}(p^2 S) \\
&\quad + p\, \mathrm{char}_{B(N)}(pS) \sum_{x \in \mathbb{Z}/p\mathbb{Z}} \hat{a}(S[\begin{bmatrix} 1 & \\ x & p \end{bmatrix}]) \big) \\
&\quad + p^{4-k}\, \mathrm{char}_{B(N)}(S) \sum_{x \in \mathbb{Z}/p\mathbb{Z}} \Delta^-_p(\mathrm{char}_{B(N)})(S[\begin{bmatrix} 1 & \\ x & p \end{bmatrix}])\, \hat{a}(S[\begin{bmatrix} 1 & \\ x & p \end{bmatrix}]) \\
&\quad + p^2\, \mathrm{char}_{B(N)}(S) \cdot (\Delta^-_{p^2} \hat{a} \big| T(p)^2)(S) \\
&= p^{3-k} \mathrm{char}_{B(N)}(S) \big(p^{3-k}\, (\Delta^+_{p^2} \hat{a})(S) + p\, (\hat{a} \big| T(p))(S) \big) \\
&\quad + p^{4-k}\, \mathrm{char}_{B(N)}(S) \cdot \big((\Delta^-_p(\mathrm{char}_{B(N)}) \cdot \hat{a}) \big| T(p) \big)(S) \\
&\quad + p^2\, \mathrm{char}_{B(N)}(S) \cdot \big(\Delta^-_{p^2} \hat{a} \big| T(p)^2 \big) \big)(S) \\
&= \mathrm{char}_{B(N)}(S) \big(p^{6-2k}\, (\Delta^+_{p^2} \hat{a})(S) + p^{4-k}\, (\hat{a} \big| T(p))(S) \\
&\quad + p^{4-k} ((\Delta^-_p(\mathrm{char}_{B(N)}) \cdot \hat{a}) \big| T(p))(S) + p^2 (\Delta^-_{p^2} \hat{a} \big| T(p^2))(S) \big).
\end{aligned}$$

This is (12.56).

Finally, assume that F is a cusp form. Straightforward calculations using $\hat{a} = \mathrm{char}_{A(\mathbb{Q})} \cdot \hat{a}$, (12.43), (12.44), (12.45), and (12.46) then prove that the same equations hold with all sets of the form $B(M)$ replaced by $B(M)^+$, where M is an integer such that $M > 0$. □

We can now present another formulation of the consequences of Theorem 12.1.2 for Fourier coefficients.

Theorem 12.2.2 *Let N and k be integers such that $N > 0$ and $k > 0$, and let $F \in S_k(\mathrm{K}(N))$. Assume that F is a newform and is an eigenvector for the Hecke operators $T(1,1,q,q)$ and $T(q,1,q,q^2)$ for all but finitely many of the primes q of $\mathbb{Z}$ such that $q \nmid N$; by Theorem 12.1.1, F is an eigenvector for $T(1,1,q,q)$, $T(q,1,q,q^2)$ and w_q for all primes q of $\mathbb{Z}$. Let $\lambda_q, \mu_q \in \mathbb{C}$ be such that*

$$T(1,1,q,q)F = q^{k-3}\lambda_q F,$$
$$T(q,1,q,q^2)F = q^{2(k-3)}\mu_q F.$$

for all primes q of $\mathbb{Z}$. Regard F as an element of $S_k(\mathrm{K}_s(N))$ and define $\hat{a} = \hat{a}(F)$ as in (12.36). Let p be a prime of $\mathbb{Z}$ with $v_p(N) \geq 2$. Then:

(1) If $v_p(N) \geq 3$, then

$$\mathrm{char}_{B(Np^{-2})} \cdot (\hat{a}|T(p^2)) = 0. \tag{12.61}$$

(2) We have

$$\mu_p\,\hat{a} = p^{-k+3}\,\mathrm{char}_{B(Np)^+} \cdot (\hat{a}|T(p)) - p\,\nabla_p\big(\mathrm{char}_{B(Np^{-1})^+} \cdot (\hat{a}|T(p))\big). \tag{12.62}$$

(3) Assume that $\mu_p \neq 0$. Then

$$0 = \mathrm{char}_{B(Np^{-1})^+} \cdot \big(\lambda_p\,\hat{a}|T(p) - p^{3-k}\,\Delta_p^+\hat{a}|T(p) - p\,\Delta_p^-\hat{a}|T(p^2)\big), \tag{12.63}$$

$$0 = \mathrm{char}_{B(Np^{-1})^+} \cdot \big(\hat{a}|T(p^2) - (\mu_p + p^2)p^{k-3}\,\hat{a}|T(p)\big). \tag{12.64}$$

(4) Assume that $\mu_p = 0$. Then

$$0 = \mathrm{char}_{B(Np^{-1})^+} \cdot (\hat{a}|T(p)). \tag{12.65}$$

If F is not an eigenform for $T^s_{0,1}(p)$, then

$$0 = \mathrm{char}_{B(N)^+} \cdot \Big(p^2\hat{a} + p^{5-2k}\,\Delta^+_{p^2}\hat{a} + p^{3-k}\,\hat{a}|T(p) + p\,\Delta^-_{p^2}\hat{a}|T(p)^2$$
$$+ p^{3-k}\,\big(\Delta_p^-(\mathrm{char}_{B(N)^+}) \cdot \hat{a}\big)|T(p) - \lambda_p\big(p^{2-k}\,\Delta_p^+\hat{a} + \Delta_p^-\hat{a}|T(p)\big)\Big). \tag{12.66}$$

If F is an eigenform for $T^s_{0,1}(p)$, then

$$0 = \mathrm{char}_{B(N)^+} \cdot \left((1+p^{-1})^{-1}\lambda_p \hat{a} - p^{3-k}\,\Delta_p^+\hat{a} - p\,\Delta_p^-\hat{a}\big|T(p)\right). \qquad (12.67)$$

Proof Throughout the proof we will use the formulas from Lemma 12.2.1, taking into account the last statement of this lemma.

(1). This follows from (12.51).

(2). By (12.4) we have

$$\mu_p F = p^4\tau_p^2\sigma_p F - p^2\eta_p\sigma_p F.$$

Therefore,

$$\begin{aligned}
\mu_p\hat{a}(F) &= p^4\hat{a}(\tau_p^2\sigma_p F) - p^2\hat{a}(\eta_p\sigma_p F)\\
&= p^4\mathrm{char}_{B(N)^+}\cdot\hat{a}(\tau_p\sigma_p F) - p^{k+2}\nabla_p\hat{a}(\sigma_p F) \quad \text{(by (12.47) and (12.49))}\\
&= p^4\mathrm{char}_{B(N)^+}\cdot\mathrm{char}_{B(Np)^+}\cdot\hat{a}(\sigma_p F)\\
&\quad - p\nabla_p(\mathrm{char}_{B(Np^{-1})^+}\cdot\hat{a}\big|T(p)) \qquad \text{(by (12.47) and (12.50))}\\
&= p^{-k+3}\mathrm{char}_{B(N)^+}\cdot\mathrm{char}_{B(Np)^+}\cdot\mathrm{char}_{B(Np^{-1})^+}\cdot\hat{a}\big|T(p)\\
&\quad - p\nabla_p(\mathrm{char}_{B(Np^{-1})^+}\cdot\hat{a}\big|T(p)) \qquad \text{(by (12.50))}\\
\mu_p\hat{a} &= p^{-k+3}\mathrm{char}_{B(Np)^+}\cdot\hat{a}\big|T(p) - p\nabla_p(\mathrm{char}_{B(Np^{-1})^+}\cdot\hat{a}\big|T(p)).
\end{aligned}$$

This proves (12.62).

(3). Assume that $\mu_p \neq 0$. By (12.5) we have $T^s_{0,1}(p)(\sigma_p F) = \lambda_p(\sigma_p F)$. Hence, by (12.50) and (12.55),

$$\begin{aligned}
\lambda_p\hat{a}(\sigma_p F) &= \hat{a}(T^s_{0,1}(p)\sigma_p F)\\
\lambda_p p^{-k-1}\mathrm{char}_{B(Np^{-1})^+}\cdot(\hat{a}\big|T(p)) &= p^{2-2k}\mathrm{char}_{B(Np^{-1})^+}\cdot(\Delta_p^+\hat{a}\big|T(p))\\
&\quad + p^{-k}\mathrm{char}_{B(Np^{-1})^+}\cdot(\Delta_p^-\hat{a}\big|T(p)^2)\\
\lambda_p\mathrm{char}_{B(Np^{-1})^+}\cdot(\hat{a}\big|T(p)) &= p^{3-k}\mathrm{char}_{B(Np^{-1})^+}\cdot(\Delta_p^+\hat{a}\big|T(p))\\
&\quad + p\,\mathrm{char}_{B(Np^{-1})^+}\cdot(\Delta_p^-\hat{a}\big|T(p)^2).
\end{aligned}$$

This is (12.63). Next, by (12.6) we have $T^s_{1,0}(p)(\sigma_p F) = (\mu_p + p^2)(\sigma_p F)$. Therefore, by (12.50) and (12.54),

$$\begin{aligned}
\hat{a}(T^s_{1,0}(p)\sigma_p F) &= (\mu_p + p^2)\hat{a}(\sigma_p F)\\
p^{2-2k}\mathrm{char}_{B(Np^{-1})^+}\cdot\hat{a}\big|T(p)^2 &= (\mu_p + p^2)p^{-k-1}\mathrm{char}_{B(Np^{-1})^+}\cdot\hat{a}\big|T(p).
\end{aligned}$$

This is (12.64).

(4). Assume that $\mu_p = 0$. The assertion (12.65) follows directly from (12.3) and (12.50). Assume that F is not an eigenform for $T^s_{0,1}(p)$. By (12.8) we have $T^s_{0,1}(p)^2 F = -p^3 F + \lambda_p T^s_{0,1}(p)F$. Hence,

$$\hat{a}(T^s_{0,1}(p)^2 F) = -p^3\hat{a} + \lambda_p \hat{a}(T^s_{0,1}(p)F). \tag{12.68}$$

By (12.56)

$$\begin{aligned}\hat{a}(T^s_{0,1}(p)^2 F) &= \mathrm{char}_{B(N)^+} \cdot \big(p^{6-2k}\Delta^+_{p^2}\hat{a} + p^{4-k}\hat{a}\big|T(p) \\ &\quad + p^{4-k}\Delta^-_p(\mathrm{char}_{B(N)^+}\cdot\hat{a})\big|T(p) + p^2\Delta^-_{p^2}\hat{a}\big|T(p)^2\big) \\ &= p\,\mathrm{char}_{B(N)^+}\cdot\big(p^{5-2k}\Delta^+_{p^2}\hat{a} + p^{3-k}\hat{a}\big|T(p) \\ &\quad + p^{3-k}\Delta^-_p(\mathrm{char}_{B(N)^+}\cdot\hat{a})\big|T(p) + p\Delta^-_{p^2}\hat{a}\big|T(p)^2\big)\end{aligned}$$

and by (12.52)

$$\begin{aligned}&-p^3\hat{a}(F) + \lambda_p\hat{a}(T^s_{0,1}(p)F) \\ &\quad = -p^3\hat{a} + \lambda_p\big(p^{3-k}\mathrm{char}_{B(N)^+}\cdot\Delta^+_p\hat{a} + p\,\mathrm{char}_{B(N)^+}\cdot(\Delta^-_p\hat{a}\big|T(p))\big) \\ &\quad = p\,\mathrm{char}_{B(N)^+}\cdot\big(-p^2\hat{a} + \lambda_p(p^{2-k}\Delta^+_p\hat{a} + \Delta^-_p\hat{a}\big|T(p))\big).\end{aligned}$$

Substituting into (12.68) now proves (12.66); note that we have used that $\hat{a}(F) = \mathrm{char}_{B(N)^+}\cdot\hat{a}(F)$.

Finally, assume that F is an eigenform for $T^s_{0,1}(p)$. By (12.9) we have $T^s_{0,1}(p)F = (1+p^{-1})^{-1}\lambda_p F$. Therefore, by (12.52),

$$(1+p^{-1})^{-1}\lambda_p\hat{a} = \hat{a}(T^s_{0,1}(p)F)$$

$$(1+p^{-1})^{-1}\lambda_p\mathrm{char}_{B(N)^+}\cdot\hat{a} = p^{3-k}\mathrm{char}_{B(N)^+}\cdot\Delta^+_p\hat{a} + p\,\mathrm{char}_{B(N)^+}\cdot(\Delta^-_p\hat{a}\big|T(p)).$$

This is (12.67). □

12.3 Examples

The work [99] and the website [98] present examples of 126 distinct newforms in $S_k(\mathrm{K}(16))$ for $k = 6, 7, 8, 9, 10, 11, 12, 13$ and 14 that are eigenvectors for $T(1,1,q,q)$, $T(q,1,q,q^2)$, and w_q for all primes q of $\mathbb{Z}$. These newforms are non-lifts and form 58 Galois orbits. Altogether, about 67,500 Fourier coefficients appear in [98]. These works also determine the Hecke eigenvalues λ_2 and μ_2 and provide information about π_2; in particular, π_2 is always generic. In this section we will describe how we verified that the seven Eqs. (12.24)–(12.31) from Corollary 12.1.3 hold for this data. This verification is a welcome extra check of Corollary 12.1.3 (in addition to the proof!).

Indices and Fourier Coefficients

We begin by reviewing some background concerning Fourier coefficients and their index set. Let $N, k \in \mathbb{Z}$ with $N > 0$ and $k > 0$, and let $F \in S_k(\mathrm{K}(N))$. Let

$$F(Z) = \sum_{S \in A(N)^+} a(S) e^{2\pi i \mathrm{Tr}(SZ)}$$

be the Fourier expansion of F. As defined in (10.15), the group $\Gamma_0(N)_\pm$ acts on $A(N)^+$. Thus, the set $A(N)^+$ is partitioned into $\Gamma_0(N)_\pm$ orbits. By (10.18), if $S \in A(N)^+$ and $a(S)$ is known, then $a(T)$ is known for all T in the orbit $\Gamma_0(N)_\pm \cdot S$. Also, if $d \in \mathbb{Z}$ with $d > 0$, then the number of orbits $\Gamma_0(N)_\pm \cdot S$ for $S \in A(N)^+$ such that $4\det(S) = d$ is finite. This is a consequence of a result of Gauss; see also Lemma 12.3.2 below. We may thus effectively list all the Fourier coefficients of F by first making a list of orbit representatives S for $\Gamma_0(N)_\pm \backslash A(N)^+$ with $4\det(S)$ in increasing order, and then specifying for each such orbit representative S the Fourier coefficient $a(S)$. Below, we will describe a method for obtaining such a list of orbit representatives for $\Gamma_0(N)_\pm \backslash A(N)^+$; as we will see, this technique can also be used to generate other lists of indices needed for the verification. To obtain orbit representatives we need three lemmas.

Lemma 12.3.1 *Let $d \in \mathbb{Z}$ be such that $d > 0$, and let $Y(d)$ be the set of all $(a, b, c) \in \mathbb{Z}^3$ such that*

(1) $d = 4ac - b^2$;
(2) $0 < a \le c$;
(3) $|b| \le a$.

If $(a, b, c) \in Y(d)$, then

$$1 \le a \le \sqrt{d/3}, \quad -a \le b \le a \le c, \quad \text{and} \quad c = (d + b^2)/(4a).$$

In particular, the set $Y(d)$ is finite.

Proof Let $(a, b, c) \in Y(d)$. Then

$$\begin{aligned} d &= 4ac - b^2 \\ &\geq 4a^2 - a^2 \qquad \text{(by (2) and (3))} \\ &= 3a^2. \end{aligned}$$

This implies that $\sqrt{d/3} \geq a$. It is immediate that $-a \leq b \leq a \leq c$ and that $c = (d + b^2)/(4a)$. Since the function

$$Y(d) \longrightarrow \{1, \ldots, \lfloor \sqrt{d/3} \rfloor\} \times \{-\lfloor \sqrt{d/3} \rfloor, \ldots, \lfloor \sqrt{d/3} \rfloor\}$$

defined by $(a, b, c) \mapsto (a, b)$ for $(a, b, c) \in Y(d)$ is well-defined and injective, the set $Y(d)$ is finite. □

Let N and d be integers such that $N > 0$ and $d > 0$. We define

$$A(N)^+(d) = \{S \in A(N)^+ \mid 4 \det(S) = d\}. \tag{12.69}$$

We are interested in obtaining orbit representatives for $\Gamma_0(N)_\pm \backslash A(N)^+(d)$. We note that the set $A(1)^+(d)$ contains $A(N)^+(d)$.

Lemma 12.3.2 *Let $d \in \mathbb{Z}$ be such that $d > 0$. If $S \in A(1)^+(d)$, then there exists an element $k \in \mathrm{SL}(2, \mathbb{Z})$ such that*

$$k \cdot S \in \{\left[\begin{smallmatrix} a & b/2 \\ b/2 & c \end{smallmatrix}\right] \in A(1)^+(d) \mid (a, b, c) \in Y(d)\}.$$

Proof We follow the argument in [133], p. 59. Let $S \in A(1)^+(d)$, and write

$$S = \left[\begin{smallmatrix} a & b/2 \\ b/2 & c \end{smallmatrix}\right]$$

with $a, b, c \in \mathbb{Z}$. Since S is positive-definite, we have $ax^2 + bxy + cy^2 > 0$ for $(x, y) \in \mathbb{Z}^2$ with $(x, y) \neq 0$. Let a' be the smallest integer in the set $\{ax^2 + bxy + cy^2 \mid (x, y) \in \mathbb{Z}^2, (x, y) \neq 0\}$. The integer a' is positive; let $(x, y) \in \mathbb{Z}^2$ be such that $a' = ax^2 + bxy + cy^2$. The definition of a' implies that x and y are relatively prime. Let $w, z \in \mathbb{Z}$ be such that $xz - yw = 1$. Then

$$k = \left[\begin{smallmatrix} x & y \\ w & z \end{smallmatrix}\right] \in \mathrm{SL}(2, \mathbb{Z}).$$

Moreover,

$$k \cdot S = \left[\begin{smallmatrix} ax^2+bxy+cy^2 & b''/2 \\ b''/2 & aw^2+bwz+cz^2 \end{smallmatrix}\right] = \left[\begin{smallmatrix} a' & b''/2 \\ b''/2 & c'' \end{smallmatrix}\right]$$

where $b'' \in \mathbb{Z}$ and $c'' \in \mathbb{Z}$. Next, there exists an integer n such that

$$2n - 1 \le b''/a' \le 2n + 1.$$

This implies that

$$-a' \le b'' - 2a'n \le a'.$$

Let

$$h = \begin{bmatrix} 1 & \\ -n & 1 \end{bmatrix} \quad \text{and} \quad b' = b'' - 2a'n.$$

Then $h \in \mathrm{SL}(2, \mathbb{Z})$ and

$$h \cdot (k \cdot S) = \begin{bmatrix} a' & b'/2 \\ b'/2 & c' \end{bmatrix}$$

for some $c' \in \mathbb{Z}$. Since $-a' \le b'' - 2a'n \le a'$ we have $|b'| \le a'$. Concerning c', we note that $h \cdot (k \cdot S) = (hk) \cdot S$; since

$$hk = \begin{bmatrix} x & y \\ w-nx & -ny+z \end{bmatrix}$$

we obtain

$$c' = a(w - nx)^2 + b(w - nx)(-ny + z) + c(-ny + z)^2.$$

By the definition of a', we must have $a' \le c'$. □

Lemma 12.3.3 *Let N and d be integers such that $N > 0$ and $d > 0$, and let Γ be a subgroup of $\Gamma_0(N)$. Let $Y(d) = \{(a_1, b_1, c_1), \dots, (a_t, b_t, c_t)\}$, and let $\mathrm{SL}(2, \mathbb{Z}) = \sqcup_{i \in I} \Gamma g_i$ be a disjoint decomposition. Define $X(\Gamma, N, d)$ to be the set of all the $g_i \cdot \begin{bmatrix} a_j & b_j/2 \\ b_j/2 & c_j \end{bmatrix}$ that are contained in $A(N)^+(d)$, where $i \in I$ and $j \in \{1, \dots, t\}$. Then $A(N)^+(d) = \cup_{S \in X(\Gamma,N,d)} \Gamma \cdot S$.*

Proof Clearly, $\cup_{S \in X(\Gamma,N,d)} \Gamma \cdot S \subset A(N)^+(d)$. Let $T \in A(N)^+(d)$. By Lemma 12.3.2 there exists $g \in \mathrm{SL}(2, \mathbb{Z})$ such that $g \cdot T = \begin{bmatrix} a_j & b_j/2 \\ b_j/2 & c_j \end{bmatrix}$ for some $j \in \{1, \dots, t\}$. Let $i \in I$ and $h \in \Gamma$ be such that $g^{-1} = hg_i$. Then $T = h \cdot (g_i \cdot \begin{bmatrix} a_j & b_j/2 \\ b_j/2 & c_j \end{bmatrix})$, so that $T \in \cup_{S \in X(\Gamma,N,d)} \Gamma \cdot S$. □

We can now state the steps for determining a set of orbit representatives for $\Gamma_0(N)_\pm \backslash A(N)^+(d)$ when $N, d \in \mathbb{Z}$ with $N > 0$ and $d > 0$:

(1) Determine the finite set $Y(d)$.
(2) Find a disjoint decomposition $\mathrm{SL}(2, \mathbb{Z}) = \sqcup_{i \in I} \Gamma_0(N) g_i$.
(3) Determine the set $X(\Gamma_0(N), N, d)$.

(4) By Lemma 12.3.3 the set $X(\Gamma_0(N), N, d)$ contains a set of orbit representatives for $\Gamma_0(N)\backslash A(N)^+(d)$, and hence a set of orbit representatives for $\Gamma_0(N)_\pm\backslash A(N)^+(d)$. Refine the set $X(\Gamma_0(N), N, d)$ to obtain a set of orbit representatives for $\Gamma_0(N)_\pm\backslash A(N)^+(d)$.

We will not describe an explicit implementation of this algorithm, but do mention two points. First, if $N = p^t$ is a power of a prime p, e.g., 16, then there is a convenient disjoint decomposition

$$\mathrm{SL}(2,\mathbb{Z}) = \bigsqcup_{x\in\mathbb{Z}/p^t\mathbb{Z}} \Gamma_0(p^t)\left[\begin{smallmatrix} 1 & \\ x & 1\end{smallmatrix}\right] \sqcup \bigsqcup_{y\in p\mathbb{Z}/p^t\mathbb{Z}} \Gamma_0(p^t)\left[\begin{smallmatrix} 1 & \\ y & 1\end{smallmatrix}\right]\left[\begin{smallmatrix} & 1 \\ -1 & \end{smallmatrix}\right].$$

Second, for the implementation of (4), given $S_1, S_2 \in A(N)^+(d)$, one needs a method for determining whether or not there exists $k \in \Gamma_0(N)_\pm$ such that $k \cdot S_1 = S_2$. One such method is described in Lemma 2.6 of [132]. Alternatively, one may use the meta-algorithm function `Solve` from Mathematica. For this, writing $S_1 = \left[\begin{smallmatrix} x_1 & y_1 \\ y_1 & z_1\end{smallmatrix}\right]$ and $S_2 = \left[\begin{smallmatrix} x_2 & y_2 \\ y_2 & z_2\end{smallmatrix}\right]$, it is convenient to note that there exists $k \in \Gamma_0(N)_\pm$ such that $k \cdot S_1 = S_2$ if and only if there exist $a, b, c, d \in \mathbb{Z}$ such that

$$\left[\begin{smallmatrix} a & b \\ Nc & d\end{smallmatrix}\right]\left[\begin{smallmatrix} x_1 & y_1 \\ y_1 & z_1\end{smallmatrix}\right] = \left[\begin{smallmatrix} x_2 & y_2 \\ y_2 & z_2\end{smallmatrix}\right]\left[\begin{smallmatrix} d & -Nc \\ -b & a\end{smallmatrix}\right] \quad \text{and} \quad ad - bNc = 1$$

or there exist $a, b, c, d \in \mathbb{Z}$ such that

$$\left[\begin{smallmatrix} a & b \\ Nc & d\end{smallmatrix}\right]\left[\begin{smallmatrix} x_1 & y_1 \\ y_1 & z_1\end{smallmatrix}\right] = \left[\begin{smallmatrix} x_2 & y_2 \\ y_2 & z_2\end{smallmatrix}\right]\left[\begin{smallmatrix} -d & Nc \\ b & -a\end{smallmatrix}\right] \quad \text{and} \quad ad - bNc = -1;$$

moreover, the number of $k \in \Gamma_0(N)_\pm$ such that $k \cdot S_1 = S_2$ is finite.

The Data from [98]

Next, we review the data from [98]. Let O be one of the 58 Galois orbits from [98]. Let $p_O(X) \in \mathbb{Z}[X]$ be the irreducible monic polynomial corresponding to O; also, let k be the weight corresponding to O. Fix a root ρ of $p_O(X)$. Then corresponding to ρ there exists a newform $F_\rho \in S_k(\mathrm{K}(16))$ that is an eigenform for the Hecke operators $T(1,1,p,p)$ and $T(p,1,p,p^2)$ for all primes p of $\mathbb{Z}$. Let

$$F_\rho(Z) = \sum_{S\in A(16)^+} a(S)\mathrm{e}^{2\pi\mathrm{i}\mathrm{Tr}(SZ)}$$

be the Fourier expansion of F_ρ. Some of the Fourier coefficients of F_ρ are given in [98] via a three-column table. A row of the table has the form

$$\boxed{(\alpha,\beta,\gamma)\,|\,d\,|\,z} \tag{12.70}$$

Here, $\alpha, 2\beta$ and γ are integers, and the triple (α, β, γ) is an abbreviation for the element S of $A(16)^+$ given by

$$S = \begin{bmatrix} \alpha & \beta \\ \beta & \gamma \end{bmatrix}. \tag{12.71}$$

Also,

$$d = 4\det(S) = 4\det(\begin{bmatrix} \alpha & \beta \\ \beta & \gamma \end{bmatrix}) = 4(\alpha\gamma - \beta^2).$$

The integer d is necessarily positive and satisfies $d \equiv 0$ or $3 \pmod 4$. Finally, z is an element of $\mathbb{Q}(\rho)$ such that

$$z = a(S) = a(\begin{bmatrix} \alpha & \beta \\ \beta & \gamma \end{bmatrix}),$$

The tables from [98] have several further properties. First, the rows are ordered by the second entry, $d = 4\det(S)$, and d is such that $d < 500$; most tables have approximately 1200 entries. Second, if $S_1 \in A(16)^+$ and $S_2 \in A(16)^+$ correspond to the first entries in two distinct rows, then S_1 and S_2 are not in the same $\Gamma_0(16)_\pm$ orbit (the action is defined in (10.15)). Thus, each row corresponds to a unique orbit of $\Gamma_0(16)_\pm$ acting on $A(16)^+$. Note that if $S \in A(16)^+$ corresponds to the first entry of a table, so that $a(S)$ is known, then $a(S')$ is also known for all S' in the $\Gamma_0(16)_\pm$ orbit of S by (10.18). Finally, not every orbit of $\Gamma_0(16)_\pm$ acting on $A(16)^+(d)$ with $d < 500$ is represented in every table. (In fact, our summary of the tables from [98] is not completely accurate: in the tables from [98] the first entry of each row is given as $(2\alpha, 2\beta, 2\gamma)$ rather than (α, β, γ).) As an illustration, Table 12.1 shows the first few Fourier coefficients of $F_{7\text{-}16\text{-}2}$ and $F_{10\text{-}16\text{-}2}$. Here, 7 and 10 are the weights of $F_{7\text{-}16\text{-}2}$

Table 12.1 Some Fourier coefficients of $F_{7\text{-}16\text{-}2}$ and $F_{10\text{-}16\text{-}2}$

$F_{7\text{-}16\text{-}2}$			$F_{10\text{-}16\text{-}2}$		
S	d	$a(S)$	S	d	$a(S)$
$(2, -53/2, 352)$	7	1	$(2, -53/2, 352)$	7	-1
$(4, -89/2, 496)$	15	-45	$(4, -89/2, 496)$	15	145
$(2, -25/2, 80)$	15	27	$(2, -25/2, 80)$	15	-95
$(6, -109/2, 496)$	23	-131	$(6, -109/2, 496)$	23	-133
$(3, -13/2, 16)$	23	-131	$(3, -13/2, 16)$	23	-133
$(3, -19/2, 32)$	23	-229	$(3, -19/2, 32)$	23	779
$(7, -21, 64)$	28	112	$(7, -21, 64)$	28	-256
$(7, -35, 176)$	28	112	$(7, -35, 176)$	28	-1280
$(4, -21, 112)$	28	112	$(4, -21, 112)$	28	-256
$(4, -53, 704)$	28	-112	$(4, -53, 704)$	28	768

For $S = \begin{bmatrix} \alpha & \beta \\ \beta & \gamma \end{bmatrix} = (\alpha, \beta, \gamma) \in A(16)^+$ we write $d = 4\det(S)$

Table 12.2 Conditions on S as in Corollary 12.1.3 needed for application of the data from [98]

For	S must lie in
(12.24)	$Z_1 = \{ S \in B(Np^{-2})^+ = A(2)^+ \mid 4\det(S) < \lfloor 500/16 \rfloor = 31 \}$
(12.25)	$Z_2 = \{ S \in B(Np)^+ = A(16)^+ \mid 4\det(S) < \lfloor 500/4 \rfloor = 125 \}$
(12.26)	$Z_3 = \{ S \in B(Np^{-1})^+ = A(4)^+ \mid 4\det(S) < \lfloor 500/16 \rfloor = 31 \}$
(12.27)	$Z_4 = \{ S \in B(Np^{-1})^+ = A(4)^+ \mid 4\det(S) < \lfloor 500/16 \rfloor = 31 \}$
(12.28)	$Z_5 = \{ S \in B(Np^{-1})^+ = A(4)^+ \mid 4\det(S) < \lfloor 500/4 \rfloor = 125 \}$
(12.30)	$Z_6 = \{ S \in B(N)^+ = A(8)^+ \mid 4\det(S) < \lfloor 500/16 \rfloor = 31 \}$
(12.31)	$Z_7 = \{ S \in B(N)^+ = A(8)^+ \mid 4\det(S) < \lfloor 500/4 \rfloor = 125 \}$

and $F_{10\text{-}16\text{-}2}$, respectively, 16 refers to the common level of $F_{7\text{-}16\text{-}2}$ and $F_{10\text{-}16\text{-}2}$, and 2 refers to the positions of $F_{7\text{-}16\text{-}2}$ and $F_{10\text{-}16\text{-}2}$ in the ordering from [98].

Verification

In this subsection we explain how we verified that the seven equations of Corollary 12.1.3 hold for the newforms from [99] and [98] when the data applies.

Our first step was to calculate appropriate finite sets of indices at which to evaluate the seven equations of Corollary 12.1.3. More precisely, let $i \in \{1, \ldots, 7\}$. By the summary of the data from [98] and the statement of Corollary 12.1.3, we see that, for the i-th equation of Corollary 12.1.3, S must lie in the set Z_i from Table 12.2. We calculated a finite subset $X_i \subset Z_i$ such that the i-th equation holds for $S \in X_i$ if and only if the i-th equation holds for all $S \in Z_i$. The idea is similar to the method for obtaining orbit representatives for $\Gamma_0(16)_\pm \backslash A(16)^+$ described above.

To explain this, let the notation be as in Corollary 12.1.3, and consider, say, (12.24) for the examples from [98]. This is the assertion that

$$\sum_{z \in \mathbb{Z}/4\mathbb{Z}} a(S[\begin{bmatrix} 1 & \\ z & 4 \end{bmatrix}]) = 0 \tag{12.72}$$

for $S \in Z_1$. To arrive at the set X_1 we will take advantage of the invariance property of the Fourier coefficient function $a : A(16)^+ \to \mathbb{C}$. We recall that the group $\mathrm{GL}(2, \mathbb{Q})^+$ acts on $A(\mathbb{Q})^+$ via the formula $g \cdot S = gS\,{}^t g$ for $g \in \mathrm{GL}(2, \mathbb{Q})$ and $S \in A(\mathbb{Q})^+$, and that with this action, $\Gamma_0(16)$ preserves the subset $A(16)^+$. By (10.18) we have

$$a(k \cdot S) = a(S) \tag{12.73}$$

for $k \in \Gamma_0(16)$ and $S \in A(16)^+$. The action also satisfies the identity

$$(g \cdot S)[B] = ({}^t B g\, {}^t B^{-1}) \cdot S[B] \tag{12.74}$$

for $g, B \in \mathrm{GL}(2, \mathbb{Q})$ and $S \in A(\mathbb{Q})^+$. Since (12.74) holds, we see that if H is a subgroup of $\Gamma_0(2)$ (so that H acts on $B(Np^{-2})^+ = A(2)^+$) and

$$g \in H \implies {}^t\begin{bmatrix} 1 & \\ z & 4 \end{bmatrix} g\, {}^t\begin{bmatrix} 1 & \\ z & 4 \end{bmatrix}^{-1} \in \Gamma_0(16)$$

for $z \in \mathbb{Z}$, then (12.72) will hold for $S \in Z_1$ if and only if (12.72) holds for all the elements in the orbit $H \cdot S$. Similar observations apply to the other equations of Corollary 12.1.3. We will define such a group H for each of the Eqs. (12.24)–(12.31). We first define

$$\begin{aligned}
G_1 &= \{\begin{bmatrix} a & b \\ c & d \end{bmatrix} \in \mathrm{SL}(2, \mathbb{Z}) \mid b \equiv c \equiv 0 \ (\mathrm{mod}\ 4),\ a \equiv d \ (\mathrm{mod}\ 4)\} \\
&= \{\begin{bmatrix} a & b \\ c & d \end{bmatrix} \in \mathrm{SL}(2, \mathbb{Z}) \mid b \equiv c \equiv 0 \ (\mathrm{mod}\ 4)\}, \\
G_2 &= \{\begin{bmatrix} a & b \\ c & d \end{bmatrix} \in \mathrm{SL}(2, \mathbb{Z}) \mid b \equiv 0 \ (\mathrm{mod}\ 2),\ c \equiv 0 \ (\mathrm{mod}\ 16),\ a \equiv d \ (\mathrm{mod}\ 2)\} \\
&= \{\begin{bmatrix} a & b \\ c & d \end{bmatrix} \in \mathrm{SL}(2, \mathbb{Z}) \mid b \equiv 0 \ (\mathrm{mod}\ 2),\ c \equiv 0 \ (\mathrm{mod}\ 16)\}, \\
G_3 &= \{\begin{bmatrix} a & b \\ c & d \end{bmatrix} \in \mathrm{SL}(2, \mathbb{Z}) \mid b \equiv 0 \ (\mathrm{mod}\ 4),\ c \equiv 0 \ (\mathrm{mod}\ 8),\ a \equiv d \ (\mathrm{mod}\ 4)\}, \\
&= \{\begin{bmatrix} a & b \\ c & d \end{bmatrix} \in \mathrm{SL}(2, \mathbb{Z}) \mid b \equiv 0 \ (\mathrm{mod}\ 4),\ c \equiv 0 \ (\mathrm{mod}\ 8)\}.
\end{aligned}$$

We have $G_1 \subset \Gamma_0(4) \subset \Gamma_0(2)$, $G_2 \subset \Gamma_0(16)$, and $G_3 \subset \Gamma_0(8) \subset \Gamma_0(4)$. Also, let $\Gamma(16)$ be the subgroup of $g \in \mathrm{SL}(2, \mathbb{Z})$ such that $g \equiv 1 \ (\mathrm{mod}\ 16)$.

Lemma 12.3.4 *If* $x \in \mathbb{Z}$, *then*

$$\begin{aligned}
g \in G_1 &\implies {}^t\begin{bmatrix} 1 & \\ x & 4 \end{bmatrix} g\, {}^t\begin{bmatrix} 1 & \\ x & 4 \end{bmatrix}^{-1} \in \Gamma_0(16), \\
g \in G_2 &\implies {}^t\begin{bmatrix} 1 & \\ x & 2 \end{bmatrix} g\, {}^t\begin{bmatrix} 1 & \\ x & 2 \end{bmatrix}^{-1},\ {}^t\begin{bmatrix} 1 & \\ x/2 & 1 \end{bmatrix} g\, {}^t\begin{bmatrix} 1 & \\ x/2 & 1 \end{bmatrix}^{-1} \in \Gamma_0(16), \\
g \in G_3 &\implies {}^t\begin{bmatrix} 1 & \\ x & 2 \end{bmatrix} g\, {}^t\begin{bmatrix} 1 & \\ x & 2 \end{bmatrix}^{-1},\ {}^t\begin{bmatrix} 1 & \\ x & 4 \end{bmatrix} g\, {}^t\begin{bmatrix} 1 & \\ x & 4 \end{bmatrix}^{-1} \in \Gamma_0(16).
\end{aligned}$$

We have $[\mathrm{SL}(2, \mathbb{Z}) : \Gamma(16)] = 3072$, $[\mathrm{SL}(2, \mathbb{Z}) : G_1] = 24$, $[\mathrm{SL}(2, \mathbb{Z}) : G_2] = 48$, *and* $[\mathrm{SL}(2, \mathbb{Z}) : G_3] = 48$.

Proof Let $g = \begin{bmatrix} a & b \\ c & d \end{bmatrix}$. Then

$$\begin{aligned}
{}^t\begin{bmatrix} 1 & \\ x & 4 \end{bmatrix} g\, {}^t\begin{bmatrix} 1 & \\ x & 4 \end{bmatrix}^{-1} &= \begin{bmatrix} a+cx & b/4-cx^2/4+(d-a)x/4 \\ 4c & d-cx \end{bmatrix}, \\
{}^t\begin{bmatrix} 1 & \\ x & 2 \end{bmatrix} g\, {}^t\begin{bmatrix} 1 & \\ x & 2 \end{bmatrix}^{-1} &= \begin{bmatrix} a+cx & b/2-cx^2/2+(d-a)x/2 \\ 2c & d-cx \end{bmatrix}, \\
{}^t\begin{bmatrix} 1 & \\ x/2 & 1 \end{bmatrix} g\, {}^t\begin{bmatrix} 1 & \\ x/2 & 1 \end{bmatrix}^{-1} &= \begin{bmatrix} a+cx/2 & b-cx^2/4+(d-a)x/2 \\ c & d-cx/2 \end{bmatrix}.
\end{aligned}$$

The inclusion statements of the lemma follow from these identities. The equality $[\mathrm{SL}(2,\mathbb{Z}) : \Gamma(16)] = 3072$ follows from p. 22 of [118]. To prove $[\mathrm{SL}(2,\mathbb{Z}) : G_1] = 24$, let $u = \begin{bmatrix} & 1 \\ -4 & \end{bmatrix}$. Then u normalizes $\Gamma_0(4)$, and $uG_1u^{-1} = \Gamma_0(16)$. It follows that $[\Gamma_0(4) : G_1] = [\Gamma_0(4) : \Gamma_0(16)]$. Hence,

$$\begin{aligned}[\mathrm{SL}(2,\mathbb{Z}) : G_1] &= [\mathrm{SL}(2,\mathbb{Z}) : \Gamma_0(4)][\Gamma_0(4) : G_1] \\ &= [\mathrm{SL}(2,\mathbb{Z}) : \Gamma_0(4)][\Gamma_0(4) : \Gamma_0(16)] \\ &= [\mathrm{SL}(2,\mathbb{Z}) : \Gamma_0(16)] \\ &= 24 \qquad \text{(see p. 24 of [118]).}\end{aligned}$$

Next, let $g = \begin{bmatrix} 1 & \\ & 2 \end{bmatrix}$. We have $G_2 \subset {}^t\Gamma_0(2)$. Conjugating this inclusion by g, we obtain $\Gamma_0(32) = gG_2g^{-1} \subset \Gamma_0(2)$. Hence, $[{}^t\Gamma_0(2) : G_2] = [\Gamma_0(2) : \Gamma_0(32)]$ and:

$$\begin{aligned}[\mathrm{SL}(2,\mathbb{Z}) : G_2] &= [\mathrm{SL}(2,\mathbb{Z}) : {}^t\Gamma_0(2)][{}^t\Gamma_0(2) : G_2] \\ &= [\mathrm{SL}(2,\mathbb{Z}) : \Gamma_0(2)][\Gamma_0(2) : \Gamma_0(32)] \\ &= [\mathrm{SL}(2,\mathbb{Z}) : \Gamma_0(32)] \\ &= 48 \qquad \text{(see p. 24 of [118]).}\end{aligned}$$

Finally, let u be as before; conjugating the inclusion $G_3 \subset \Gamma_0(4)$ we have $G_2 = uG_3u^{-1} \subset \Gamma_0(4)$. Hence, $[\mathrm{SL}(2,\mathbb{Z}) : G_3] = [\mathrm{SL}(2,\mathbb{Z}) : G_2] = 48$. □

We define the desired subgroups as follows.

Lemma 12.3.5 *Define $H_1 = G_1$, $H_2 = G_2$, $H_3 = G_3$, $H_4 = G_3$, $H_5 = G_2$, $H_6 = G_3$, and $H_7 = G_2$. Let $i \in \{1, \dots, 7\}$. The group H_i acts on Z_i. If $S \in Z_i$, then the corresponding statement from Corollary 12.1.3 as in Table 12.2 holds for S if and only if it holds for all the elements of the orbit $H_i \cdot S$.*

Proof This follows from the definitions of the groups H_i for $i \in \{1, \dots, 7\}$ and Lemma 12.3.4. □

By Lemma 12.3.3 the sets $\cup_{d=1}^{30} X(H_1, 2, d)$, $\cup_{d=1}^{124} X(H_2, 16, d)$, $\cup_{d=1}^{30} X(H_3, 4, d)$, $\cup_{d=1}^{30} X(H_4, 4, d)$, $\cup_{d=1}^{124} X(H_5, 4, d)$, $\cup_{d=1}^{30} X(H_6, 8, d)$, and $\cup_{d=1}^{124} X(H_7, 8, d)$ contain orbit representatives for $H_1\backslash Z_1$, $H_2\backslash Z_2$, $H_3\backslash Z_3$, $H_4\backslash Z_4$, $H_5\backslash Z_5$, $H_6\backslash Z_6$, and $H_7\backslash Z_7$, respectively. However, each of the aforementioned sets contains multiple pairs of elements that define the same orbit. To avoid repetitive calculations, for each $i \in \{1, \dots, 7\}$ we refined each set to obtain a set X_i of orbit representatives for $H_i\backslash Z_i$. The orders of these sets appear in Table 12.3. By our discussion, verifying the i-th equation for $i \in \{1, \dots, 7\}$ for $S \in Z_i$ is now reduced to verifying this equation for $S \in X_i$.

Table 12.3 The number of orbit representatives

X_i	H_i	Order of $X_i \cong H_i \backslash Z_i$
X_1	G_1	208
X_2	G_2	536
X_3	G_3	184
X_4	G_3	88
X_5	G_2	2320
X_6	G_3	88
X_7	G_2	1120

For every newform F from [98] and S in the appropriate set X_i we verified the applicable equations from the Corollary 12.1.3. We note that for some equations and some $S \in X_i$ it may happen that not all the Fourier coefficients were available; as pointed out above, the data from [98] is not quite complete. The calculations had several steps. First, we determined the list L of all indices S that appear in the tables defining the newforms from [98] as discussed above. As mentioned, each element S of L is a distinct orbit representative for $\Gamma_0(16)_\pm \backslash A(16)^+$ with $4\det(S) < 500$. To explain the second step, consider, for example, the first Eq. (12.24). In this equation there appear the indices $S[\left[\begin{smallmatrix} 1 & \\ z & 4 \end{smallmatrix}\right]]$ for $S \in X_1$ and $z = 0, 1, 2, 3$; evidently, each of these indices is contained in $A(16)^+$. For each $z = 0, 1, 2, 3$ we explicitly computed a function $r_z : X_1 \to \{\pm 1\} \times L$ such that $r_z(S) = (\varepsilon_z(S), T_z(S))$ if and only if there exists $k \in \Gamma_0(16)_\pm$ such that $k \cdot T_z(S) = S[\left[\begin{smallmatrix} 1 & \\ z & 4 \end{smallmatrix}\right]]$ and $\det(k) = \varepsilon_z(S)$. Note that the functions r_z for $z \in \{0, 1, 2, 3\}$ do not involve any particular newform: this is algebra that applies to all the examples. We computed similar functions for the remaining six equations of Corollary 12.1.3. Finally, we calculated the equations. For example, for a particular newform F with weight k and $S \in X_1$, the first Eq. (12.24) now has the form

$$\varepsilon_0(S)^k a(T_0(S)) + \varepsilon_1(S)^k a(T_1(S)) + \varepsilon_2(S)^k a(T_2(S)) + \varepsilon_3(S)^k a(T_3(S)) = 0$$

where we have used (10.18). As an illustration, in Table 12.4 we present the data for some of the verifications of (12.24) for the earlier presented newforms $F_{7\text{-}16\text{-}2}$ and $F_{10\text{-}16\text{-}2}$. By [98], the μ_2 eigenvalue of $F_{7\text{-}16\text{-}2}$ is 0, the μ_2 eigenvalue of $F_{10\text{-}16\text{-}2}$ is -4, the λ_2 eigenvalue of $F_{7\text{-}16\text{-}2}$ is -3, and the λ_2 eigenvalue of $F_{10\text{-}16\text{-}2}$ is -2; also, π_2 is generic for both of these newforms (as is the case for all of newforms from [98] described earlier in this section). Interestingly, for $F_{7\text{-}16\text{-}2}$, the first and third terms already sum to zero, as do the second and fourth terms. This is a consequence of $\mu_2 = 0$ for $F_{7\text{-}16\text{-}2}$; for this, use (12.28). Some examples of (12.28) appear in Table 12.5. The example $F_{10\text{-}16\text{-}2}$ shows that this canceling pairs phenomenon does not hold if $\mu_2 \neq 0$.

Table 12.4 Examples of (12.24) for $F_{7\text{-}16\text{-}2}$ and $F_{10\text{-}16\text{-}2}$

$S \in X_1$	d	A_0	A_1	A_2	A_3	$\sum_{i=0}^{3} A_i$
$F_{7\text{-}16\text{-}2}$:						
(257, 241, 226)	4	2304	2304	−2304	−2304	0
(242, 453/2, 212)	7	4352	9216	−4352	−9216	0
(1 , 0 , 2)	8	0	4608	0	−4608	0
(259, 243, 228)	12	−27648	−55296	27648	55296	0
(244, 457/2, 214)	15	25344	−27648	−25344	27648	0
(1, 0, 4)	16	0	−55296	0	55296	0
(261 , 245 , 230)	20	129024	27648	−129024	−27648	0
(246, 461/2 , 216)	23	−109312	193536	109312	−193536	0
(1, 0, 6)	24	0	248832	0	−248832	0
(263, 247, 232)	28	−258048	442368	258048	−442368	0
$F_{10\text{-}16\text{-}2}$:						
(257, 241, 226)	4	−24576	24576	24576	−24576	0
(242, 453/2, 212)	7	0	131072	−262144	131072	0
(1 , 0 , 2)	8	−81920	524288	−966656	524288	0
(259, 243, 228)	12	5799936	−2949120	−5799936	2949120	0
(244, 457/2, 214)	15	0	3276800	−6553600	3276800	0
(1, 0, 4)	16	−5898240	6291456	−6684672	6291456	0
(261 , 245 , 230)	20	−48168960	22118400	48168960	−22118400	0
(246, 461/2 , 216)	23	0	−42336256	84672512	−42336256	0
(1, 0, 6)	24	12419072	−30408704	48398336	−30408704	0
(263, 247, 232)	28	−168689664	−53477376	168689664	53477376	0

For $S = \begin{bmatrix} \alpha & \beta \\ \beta & \gamma \end{bmatrix} = (\alpha, \beta, \gamma) \in X_1$, $d = 4\det(S)$; $A_0 = a(S[\begin{smallmatrix} 1 & \\ & 4 \end{smallmatrix}])$, $A_1 = a(S[\begin{smallmatrix} 1 & \\ 1 & 4 \end{smallmatrix}])$, $A_2 = a(S[\begin{smallmatrix} 1 & \\ 2 & 4 \end{smallmatrix}])$, $A_3 = a(S[\begin{smallmatrix} 1 & \\ 3 & 4 \end{smallmatrix}])$.

Table 12.5 Examples of (12.28) for $F_{7\text{-}16\text{-}2}$

$S \in X_5$	d	A_0	A_1	$A_0 + A_1$
$F_{7\text{-}16\text{-}2}$				
(242, 453/2, 212)	7	−112	112	0
(259, 243, 228)	12	−192	192	0
(199156, 373529/2, 175144)	15	−1584	1584	0
(1, 0, 4)	16	0	0	0
(246, 461/2, 216)	23	−3152	3152	0
(263, 247, 232)	28	9216	−9216	0
(401656, 753329/2, 353228)	31	800	−800	0
(1, 0, 8)	32	0	0	0
(250, 469/2, 220)	39	−3504	3504	0
(267, 251, 236)	44	−8896	8896	0

For $S = \begin{bmatrix} \alpha & \beta \\ \beta & \gamma \end{bmatrix} = (\alpha, \beta, \gamma) \in X_5$, $d = 4\det(S)$; $A_i = a(S[\begin{smallmatrix} 1 & \\ i & 2 \end{smallmatrix}])$

For all of these calculations, it is important to note that in Corollary 12.1.3 the Siegel modular form F is contained in $S_k(\mathrm{K}(16))$ but is regarded as an element of the larger space $S_k(\mathrm{K}_s(16))$; as such, F has the Fourier expansion

$$F(Z) = \sum_{S \in B(16)^+} a(S) e^{2\pi i \mathrm{Tr}(SZ)}.$$

Thus, we trivially have $a(S) = 0$ for $S \in B(16)^+$ with $S \notin A(16)^+$.

12.4 Computing Eigenvalues

A natural application of Corollary 12.1.3 is to the computation of the Hecke eigenvalues λ_p and μ_p of a paramodular newform F. As mentioned in the introduction, knowledge of λ_p and μ_p is equivalent to knowing the L-factor at p of the spin L-function attached to F. This new method of calculating λ_p and μ_p is easier than the method of "restriction to a modular curve" used in [99].

Let the notation be as in Corollary 12.1.3. The following algorithm then determines the Hecke eigenvalues λ_p and μ_p and determines whether or not π_p is generic for $p^2 \mid N$ in terms of the Fourier coefficients of F.

(1) Find $S \in A(N)^+ \subset B(Np)^+$ such that $a(S) \neq 0$. Solve (12.25) for μ_p. If $\mu_p \neq 0$, then go to (2). If $\mu_p = 0$, then go to (3).
(2) Find $S \in B(Np^{-1})^+$ such that $\sum_{x \in \mathbb{Z}/p\mathbb{Z}} a(S[\begin{bmatrix} 1 & \\ x & p \end{bmatrix}]) \neq 0$; such an S exists by (12.22). Solve (12.26) for λ_p. By Corollary 12.1.3 the representation π_p is generic.
(3) Using (12.29) determine whether or not F is an eigenvector for $T^s_{0,1}(p)$. If F is an eigenvector for $T^s_{0,1}(p)$, then find $S \in A(N)^+ \subset B(N)^+$ such that $a(S) \neq 0$, and use (12.31) to solve for λ_p; in this case π_p is non-generic. If F is not an eigenvector for $T^s_{0,1}(p)$, then find $S \in B(N)$ such that $c(S) \neq 0$ (such an S exists by (4) of Corollary 12.1.3), and use (12.30) to solve for λ_p; in this case π_p is generic.

We note that, in case it is known that F is a non-lift, then π_p is generic, and hence F is not an eigenform for $T^s_{0,1}(p)$. Thus, in this situation, (3) of the algorithm simplifies.

For the examples discussed in Sect. 12.3 we found that the above algorithm quickly determines λ_2 and μ_2 from a few Fourier coefficients. As an illustration, we consider the newforms $F_{7\text{-}16\text{-}2}$ and $F_{10\text{-}16\text{-}2}$ of [98] that were recalled in Sect. 12.3. In Table 12.6 we show some examples of the determination of μ_2 for $F_{7\text{-}16\text{-}2}$ and $F_{10\text{-}16\text{-}2}$ as in (1) of the algorithm. We see that $\mu_2(F_{10\text{-}16\text{-}2}) = -4$ and $\mu_2(F_{7\text{-}16\text{-}2}) = 0$. Since $\mu_2(F_{10\text{-}16\text{-}2}) \neq 0$, we proceed to (2) for $F_{10\text{-}16\text{-}2}$ to determine $\lambda_2(F_{10\text{-}16\text{-}2})$.

Table 12.6 Determining μ_2 for $F_{7\text{-}16\text{-}2}$ and $F_{10\text{-}16\text{-}2}$

$F_{7\text{-}16\text{-}2}$:

$S \in X_2$	d	$a(S)$	A_0	A_1	B_0	B_1	μ_2
(7666, 14389/2, 6752)	7	−1	112	−112	–	–	0
(36340, 68185/2, 31984)	15	45	−1584	1584	–	–	0
(98038, 183917/2, 86256)	23	131	−3152	3152	–	–	0
(354631, 332565, 311872)	28	−112	9216	−9216	−112	112	0
(205048, 384625/2, 180368)	31	−178	−800	800	–	–	0
(369658, 693349/2, 325120)	39	123	3504	−3504	–	–	0
(604156, 1133129/2, 531312)	47	290	3232	−3232	–	–	0
(1, 2, 16)	48	−192	−55296	55296	−192	192	0
(254, 477/2, 224)	55	−1871	44816	−44816	–	–	0
(271, 255, 240)	60	1584	−27648	27648	1584	−1584	0

$F_{10\text{-}16\text{-}2}$:

$S \in X_2$	d	$a(S)$	A_0	A_1	B_0	B_1	μ_2
(7666, 14389/2, 6752)	7	−1	768	−256	–	–	−4
(36340, 68185/2, 31984)	15	145	−49920	−24320	–	–	−4
(98038, 183917/2, 86256)	23	−133	−131328	199424	–	–	−4
(354631, 332565, 311872)	28	−256	131072	131072	−256	768	−4
(205048, 384625/2, 180368)	31	−1402	327168	390656	–	–	−4
(369658, 693349/2, 325120)	39	−1341	1226496	−539904	–	–	−4
(604156, 1133129/2, 531312)	47	−2558	1853952	−544256	–	–	−4
(1, 2, 16)	48	0	2949120	−2949120	0	0	*
(254, 477/2, 224)	55	−14575	3452160	4010240	–	–	−4
(271, 255, 240)	60	1280	−3276800	−3276800	1280	−24320	−4

For $S = \begin{bmatrix} \alpha & \beta \\ \beta & \gamma \end{bmatrix} = (\alpha, \beta, \gamma) \in X_2$, $d = 4\det(S)$; $A_i = a(S[\begin{bmatrix} 1 & \\ i & p \end{bmatrix}])$; $B_i = a(S[\begin{bmatrix} 1 & \\ i/2 & 1 \end{bmatrix}])$ if $2 \mid 2\beta$; the μ_2 column lists the numbers obtained by solving (12.25) for μ_2; a * indicates an equation with both sides equal to zero

In Table 12.7 we show some examples of the determination of $\lambda_2(F_{10\text{-}16\text{-}2})$ as in (2) of the algorithm. Since $\mu_2(F_{7\text{-}16\text{-}2}) = 0$, we proceed to (3) for $F_{7\text{-}16\text{-}2}$ to determine $\lambda_2(F_{7\text{-}16\text{-}2})$. In Table 12.8 we present some examples showing that $F_{7\text{-}16\text{-}2}$ cannot be an eigenvector for $T^s_{0,1}(2)$. The argument is by contradiction. If $F_{7\text{-}16\text{-}2}$ were an eigenvector for $T^s_{0,1}(2)$, then (12.31) would have to hold for all $S \in B(16)^+ = A(8)^+$; however, in Table 12.8 we see that this impossible, either because it leads to equations of the form $0 = C$ where $C \neq 0$, or because it leads to multiple distinct values for $\lambda_2(F_{7\text{-}16\text{-}2})$. Since $F_{7\text{-}16\text{-}2}$ is not an eigenvector for $T^s_{0,1}(2)$, we may use (12.30) to solve for λ_2. This is carried out for some examples in Table 12.9.

Table 12.7 Determining λ_2 for $F_{10\text{-}16\text{-}2}$

$F_{10\text{-}16\text{-}2}$

$S \in X_3$	d	A_0	A_1	B_0	B_1	C_0	C_1	C_2	C_3	λ_2
(242, 453/2, 212)	7	−256	−1280	0	262144	−256	–	768	–	−2
(259, 243, 228)	12	0	0	0	0	–	–	–	–	*
(199156, 373529/2, 175144)	15	−49920	−24320	6553600	0	11520	–	37120	–	−2
(1, 0, 4)	16	49152	0	−12582912	0	–	–	–	–	−2
(246, 461/2, 216)	23	−131328	199424	24903680	−59768832	199424	–	−131328	–	−2
(263, 247, 232)	28	−131072	−131072	33554432	33554432	–	–	–	–	−2

For $S = \left[\begin{smallmatrix} \alpha & \beta \\ \beta & \gamma \end{smallmatrix}\right] = (\alpha, \beta, \gamma) \in X_3$, $d = 4\det(S)$; $A_i = a(S[\left[\begin{smallmatrix} 1 \\ i & 2 \end{smallmatrix}\right]])$; $B_i = a(2S[\left[\begin{smallmatrix} 1 \\ i & 2 \end{smallmatrix}\right]])$ $C_i = a(2^{-1}S[\left[\begin{smallmatrix} 1 \\ i & 4 \end{smallmatrix}\right]])$ if $p \mid (\alpha + 2\beta i + \gamma i^2)$ and is undefined if $p \nmid (\alpha + 2\beta i + \gamma i^2)$; the λ_2 column lists the numbers obtained by solving (12.26) for λ_2; a * indicates an equation with both sides equal to zero

Table 12.8 Showing that $F_{7\text{-}16\text{-}2}$ is not an eigenvector for $T^s_{0,1}(2)$

$F_{10\text{-}16\text{-}2}$

$S \in X_7$	d	$A(S)$	$A(2S)$	A_0	A_1	A
(71666, 134421/2, 63032)	7	0	−112	−1	–	⇒⇐
(199156, 373529/2, 175144)	15	0	432	−27	–	⇒⇐
(1, 2, 8)	16	0	2304	–	–	⇒⇐
(246, 461/2, 216)	23	0	3152	131	–	⇒⇐
(263, 247, 232)	28	0	−9216	–	–	⇒⇐
(205048, 384625/2, 180368)	31	−178	−800	−178	–	609/178
(1, 0, 8)	32	0	0	–	–	0 = 0
(369658, 693349/2, 325120)	39	123	−18384	195	–	−759/82
(604156, 1133129/2, 531312)	47	290	−11744	−358	–	−15/2
(1, 2, 16)	48	−192	33792	–	–	−33/2

For $S = \begin{bmatrix} \alpha & \beta \\ \beta & \gamma \end{bmatrix} = (\alpha, \beta, \gamma) \in X_7$, $d = 4\det(S)$; $A_i = a(2^{-1}S[\begin{bmatrix} 1 & \\ i & 2 \end{bmatrix}])$ if $2 \mid (\alpha + 2\beta i)$ and is undefined if $2 \nmid (\alpha + 2\beta i)$; A is ⇒⇐ if this instance of (12.31) produces a contradiction of the form $0 = C$ where $C \neq 0$, A is $0 = 0$ if both sides of (12.31) are zero, and A is a number, putatively equal to λ_2, if (12.31) can be solved for λ_2

12.5 A Recurrence Relation

In this final section, assuming that F is as in Corollary 12.1.3, we prove that the radial Fourier coefficients $a(p^t S)$ of F for $t \geq 0$, $p^2 \mid N$, satisfy a recurrence relation determined by the spin L-factor $L_p(s, F)$ of F at p. This theorem extends results known in other situations (e.g., Sect. 4.3.2 of [2]). We begin with the following lemma.

Lemma 12.5.1 *Let N and k be integers such that $N > 0$ and $k > 0$, and let $F \in S_k(\mathrm{K}(N))$. Assume that F is a newform and is an eigenvector for the Hecke operators $T(1, 1, q, q)$ and $T(q, 1, q, q^2)$ for all but finitely many of the primes q of $\mathbb{Z}$ such that $q \nmid N$; by Theorem 12.1.1, F is an eigenvector for $T(1, 1, q, q)$, $T(q, 1, q, q^2)$ and w_q for all primes q of $\mathbb{Z}$. Let $\lambda_q, \mu_q \in \mathbb{C}$ be such that $T(1, 1, q, q)F = q^{k-3}\lambda_q F$ and $T(q, 1, q, q^2)F = q^{2(k-3)}\mu_q F$ for all primes q of $\mathbb{Z}$. Regard F as an element of $S_k(\mathrm{K}_s(N))$, and let*

$$F(Z) = \sum_{S \in B(N)^+} a(S) e^{2\pi i \mathrm{Tr}(SZ)}$$

be the Fourier expansion of F. Let $\pi \cong \otimes_{v \leq \infty} \pi_v$ be as in Theorem 12.1.2. Let p be a prime of $\mathbb{Z}$ with $v_p(N) \geq 2$.

Table 12.9 Determining λ_2 for $F_{7\text{-}16\text{-}2}$

$F_{7\text{-}16\text{-}2}$													
$S \in X_6$	d	$A(S)$	$A(2S)$	A_0	A_1	$A(4S)$	B_0	B_1	C_0	C_1	C_2	C_3	λ_2
(71666, 134421/2, 63032)	7	0	−112	−1	–	4352	112	0	–	–	−1	–	−3
(199156, 373529/2, 175144)	15	0	432	−27	–	25344	−1584	0	45	–	–	–	−3
(1, 2, 8)	16	0	2304	–	–	−110592	–	0	–	–	–	–	−3
(246, 461/2, 216)	23	0	3152	131	–	−17152	−3152	0	–	–	−229	–	−3
(263, 247, 232)	28	0	−9216	–	–	442368	–	0	–	–	–	–	−3

For $S = \left[\begin{smallmatrix} \alpha & \beta \\ \beta & \gamma \end{smallmatrix}\right] = (\alpha, \beta, \gamma) \in X_6$, $d = 4\det(S)$; $A_i = a(2^{-1}S[\left[\begin{smallmatrix} 1 & \\ i & 2 \end{smallmatrix}\right]])$ if $2 \mid (\alpha + 2\beta i)$ and is undefined if $2 \nmid (\alpha + 2\beta i)$; $B_i = a(S[\left[\begin{smallmatrix} 1 & \\ i & 2 \end{smallmatrix}\right]])$ if $2 \mid (\alpha + 2\beta i)$ and is undefined if $2 \nmid (\alpha + 2\beta i)$; $C_i = a(2^{-2}S[\left[\begin{smallmatrix} 1 & \\ i & 4 \end{smallmatrix}\right]])$ if $4 \mid (\alpha + 2\beta i + \gamma i^2)$ and is undefined if $p \nmid (\alpha + 2\beta i + \gamma i^2)$; the λ_2 column lists the numbers obtained by solving (12.30) for λ_2

(1) *Assume that* $\mu_p \neq 0$. *If* $S \in B(N)^+$, *then there is a formal identity*

$$\sum_{t=0}^{\infty} a(p^t S) X^t = \frac{N_1(X,S)}{D_1(X)} \tag{12.75}$$

where

$$\begin{aligned} N_1(X,S) = a(S) &+ \big(a(pS) - p^{k-3}\lambda_p a(S)\big)X \\ &+ \big(a(p^2S) - p^{k-3}\lambda_p a(pS) + p^{2k-5}(\mu_p + p^2)a(S)\big)X^2, \end{aligned} \tag{12.76}$$

$$D_1(X) = 1 - p^{k-3}\lambda_p X + p^{2k-5}(\mu_p + p^2)X^2. \tag{12.77}$$

(2) *Assume that* $\mu_p = 0$. *If* $S \in B(N)^+$ *and* F *is not an eigenvector for* $T^s_{0,1}(p)$ *(so that* π_p *is generic by Theorem 12.1.2), then there is a formal identity*

$$\sum_{t=0}^{\infty} a(p^t S) X^t = \frac{N_2(X,S)}{D_2(X)} \tag{12.78}$$

where

$$\begin{aligned} N_2(X,S) = a(S) &+ \big(a(pS) - p^{k-3}\lambda_p a(S)\big)X \\ &+ \big(a(p^2S) - p^{k-3}\lambda_p a(pS) + p^{2k-3}a(S)\big)X^2, \end{aligned} \tag{12.79}$$

$$D_2(X) = 1 - p^{k-3}\lambda_p X + p^{2k-3}X^2. \tag{12.80}$$

If $S \in B(N)^+$ *and* F *is an eigenvector for* $T^s_{0,1}(p)$ *(so that* π_p *is non-generic by Theorem 12.1.2), then there is a formal identity*

$$\sum_{t=0}^{\infty} a(p^t S) X^t = \frac{N_3(X,S)}{D_3(X)} \tag{12.81}$$

where

$$N_3(X,S) = a(S) + \big(a(pS) - p^{k-2}(1+p)^{-1}\lambda_p a(S)\big)X, \tag{12.82}$$

$$D_3(X) = 1 - p^{k-2}(1+p)^{-1}\lambda_p X. \tag{12.83}$$

Proof (1). For $S \in B(Np^{-1})^+$ define

$$a'(S) = \sum_{x \in \mathbb{Z}/p\mathbb{Z}} a(S[\begin{bmatrix} 1 & \\ x & p \end{bmatrix}]). \tag{12.84}$$

We will first prove that if $S \in B(Np^{-1})^+$, then there is a formal identity

$$\sum_{t=0}^{\infty} a'(p^t S)X^t = \frac{N_0(X, S)}{D_1(X)} \tag{12.85}$$

where

$$N_0(X, S) = a'(S) + \big(a'(pS) - p^{k-3}\lambda_p a'(S)\big)X, \tag{12.86}$$

$$D_1(X) = 1 - p^{k-3}\lambda_p X + p^{2k-5}(\mu_p + p^2)X^2. \tag{12.87}$$

Let $S \in B(Np^{-1})^+$. Let $i \in \mathbb{Z}$ be such that $i \geq 1$ and define $S' = p^i S$. Then $S' \in B(Np^{-1})^+$. Write $S' = \left[\begin{smallmatrix} \alpha & \beta \\ \beta & \gamma \end{smallmatrix}\right]$. By (12.26) we have

$$\lambda_p \sum_{x \in \mathbb{Z}/p\mathbb{Z}} a(S'[\left[\begin{smallmatrix} 1 & \\ x & p \end{smallmatrix}\right]]) = \sum_{x \in \mathbb{Z}/p\mathbb{Z}} p^{3-k} a(pS'[\left[\begin{smallmatrix} 1 & \\ x & p \end{smallmatrix}\right]])$$

$$+ \sum_{\substack{z \in \mathbb{Z}/p^2\mathbb{Z} \\ p|(\alpha+2\beta z+\gamma z^2)}} pa(p^{-1}S'[\left[\begin{smallmatrix} 1 & \\ z & p^2 \end{smallmatrix}\right]]).$$

Since p divides α, 2β, and γ, by (12.84) this is

$$\lambda_p a'(p^i S) = p^{3-k} a'(p^{i+1}S) + \sum_{z \in \mathbb{Z}/p^2\mathbb{Z}} pa(p^{-1}S'[\left[\begin{smallmatrix} 1 & \\ z & p^2 \end{smallmatrix}\right]]). \tag{12.88}$$

Applying (12.34) to $p^{-1}S' \in B(Np^{-1})^+$, we also have

$$\sum_{z \in \mathbb{Z}/p^2\mathbb{Z}} pa(p^{-1}S'[\left[\begin{smallmatrix} 1 & \\ z & p^2 \end{smallmatrix}\right]]) = (\mu_p + p^2)p^{k-2} \sum_{x \in \mathbb{Z}/p\mathbb{Z}} a(p^{-1}S'[\left[\begin{smallmatrix} 1 & \\ x & p \end{smallmatrix}\right]])$$

$$= (\mu_p + p^2)p^{k-2} a'(p^{i-1}S). \tag{12.89}$$

Substituting (12.89) into (12.88), we now have

$$\lambda_p a'(p^i S) = p^{3-k} a'(p^{i+1}S) + p^{k-2}(\mu_p + p^2) a'(p^{i-1}S).$$

Rewriting, we have proven that for $t \in \mathbb{Z}$, $t \geq 2$, and $S \in B(Np^{-1})^+$,

$$a'(p^t S) - p^{k-3}\lambda_p a'(p^{t-1}S) + p^{2k-5}(\mu_p + p^2) a'(p^{t-2}S) = 0. \tag{12.90}$$

It is straightforward to derive (12.85) from this recurrence relation. Let $S \in B(N)^+$. Then

$$\mu_p \sum_{t=0}^{\infty} a(p^t S)X^t = \mu_p a(S) + \mu_p X \sum_{t=1}^{\infty} a(p^t S)X^{t-1}$$

$$= \mu_p a(S) + X \sum_{t=0}^{\infty} \mu_p a(p^{t+1} S)X^t$$

$$= \mu_p a(S) + X \sum_{t=0}^{\infty} \sum_{x \in \mathbb{Z}/p\mathbb{Z}} p^{3-k} a(p^{t+1} S[\left[\begin{smallmatrix} 1 & \\ x & p \end{smallmatrix}\right]])X^t$$

$$- X \sum_{t=0}^{\infty} \sum_{x \in \mathbb{Z}/p\mathbb{Z}} pa(p^{t+1} S[\begin{bmatrix} & 1 \\ xp^{-1} & 1 \end{bmatrix}])X^t \qquad \text{(by (12.25))}$$

$$= \mu_p a(S) + X \sum_{t=0}^{\infty} \sum_{x \in \mathbb{Z}/p\mathbb{Z}} p^{3-k} a(p^{t+1} S[\left[\begin{smallmatrix} 1 & \\ x & p \end{smallmatrix}\right]])X^t$$

$$- X \sum_{t=0}^{\infty} \sum_{x \in \mathbb{Z}/p\mathbb{Z}} pa(p^{t+1} S[\begin{bmatrix} 1 & \\ & p^{-1} \end{bmatrix}][\begin{bmatrix} 1 & \\ x & p \end{bmatrix}])X^t$$

$$= \mu_p a(S) + p^{3-k} X \sum_{t=0}^{\infty} a'(p^{t+1} S)X^t$$

$$- pX \sum_{t=0}^{\infty} a'(p^{t+1} S[\begin{bmatrix} 1 & \\ & p^{-1} \end{bmatrix}])X^t$$

$$= \mu_p a(S) + p^{3-k} X N_0(X, pS) D_1(X)^{-1}$$

$$- pX N_0(X, pS[\begin{bmatrix} 1 & \\ & p^{-1} \end{bmatrix}]) D_1(X)^{-1} \qquad \text{(use (12.85))}$$

$$= \big(\mu_p a(S) D_1(X) + p^{3-k} a'(pS)X + p^{3-k} a'(p^2 S)X^2 - \lambda_p a'(pS)X^2$$

$$- pa'(pS[\begin{bmatrix} 1 & \\ & p^{-1} \end{bmatrix}])X - pa'(p^2 S[\begin{bmatrix} 1 & \\ & p^{-1} \end{bmatrix}])X^2$$

$$+ p^{k-2} \lambda_p a'(pS[\begin{bmatrix} 1 & \\ & p^{-1} \end{bmatrix}])X^2\big) D_1(X)^{-1} \qquad \text{(by (12.86))}$$

$$= (\mu_p a(S) D_1(X) + (p^{3-k} a'(pS) - pa'(pS[\begin{bmatrix} 1 & \\ & p^{-1} \end{bmatrix}]))X$$

$$+ (p^{3-k} a'(p^2 S) - pa'(p^2 S[\begin{bmatrix} 1 & \\ & p^{-1} \end{bmatrix}]))X^2$$

$$- p^{k-3} \lambda_p (p^{3-k} a'(pS) - pa'(pS[\begin{bmatrix} 1 & \\ & p^{-1} \end{bmatrix}]))X^2) D_1(X)^{-1}$$

$$= (\mu_p a(S) D_1(X) + \mu_p a(pS) X + \mu_p a(p^2 S) X^2$$
$$- p^{k-3} \lambda_p \mu_p a(pS) X^2) D_1(X)^{-1} \qquad \text{(by (12.25))}$$
$$= \mu_p (a(S) + (a(pS) - p^{k-3} \lambda_p a(S)) X$$
$$+ (a(p^2 S) - p^{k-3} \lambda_p a(pS) + p^{2k-5} (\mu_p + p^2) a(S)) X^2) D_1(X)^{-1}.$$

For the last equality we used (12.87). Canceling μ_p from both sides of the last equation now yields (12.75).

(2). Assume first that F is not an eigenvector for $T^s_{0,1}(p)$. Let $S \in B(N)^+$. Let $i \in \mathbb{Z}$ be such that $i \geq 1$ and define $S' = p^i S$. Then $S' \in B(N)^+$; let $S' = \left[\begin{smallmatrix} \alpha & \beta \\ \beta & \gamma \end{smallmatrix}\right]$. By (12.30) we have

$$p^{3-k} \lambda_p a(pS') + \lambda_p \sum_{\substack{x \in \mathbb{Z}/p\mathbb{Z} \\ p \mid (\alpha + 2\beta x)}} p a(p^{-1} S'[\left[\begin{smallmatrix} 1 & \\ x & p \end{smallmatrix}\right]])$$
$$= p^3 a(S') + p^{6-2k} a(p^2 S') + \sum_{\substack{y \in \mathbb{Z}/p\mathbb{Z} \\ p \mid (\alpha + 2\beta y)}} p^{4-k} a(S'[\left[\begin{smallmatrix} 1 & \\ y & p \end{smallmatrix}\right]])$$
$$+ \sum_{\substack{z \in \mathbb{Z}/p^2\mathbb{Z} \\ p^2 \mid (\alpha + 2\beta z + \gamma z^2)}} p^2 a(p^{-2} S'[\left[\begin{smallmatrix} 1 & \\ z & p^2 \end{smallmatrix}\right]]).$$

Since p divides α and 2β, and since p^2 divides γ, this equation is

$$p^{3-k} \lambda_p a(pS') + \lambda_p \sum_{x \in \mathbb{Z}/p\mathbb{Z}} p a(p^{-1} S'[\left[\begin{smallmatrix} 1 & \\ x & p \end{smallmatrix}\right]])$$
$$= p^3 a(S') + p^{6-2k} a(p^2 S') + \sum_{y \in \mathbb{Z}/p\mathbb{Z}} p^{4-k} a(S'[\left[\begin{smallmatrix} 1 & \\ y & p \end{smallmatrix}\right]])$$
$$+ \sum_{\substack{z \in \mathbb{Z}/p^2\mathbb{Z} \\ p^2 \mid (\alpha + 2\beta z)}} p^2 a(p^{-2} S'[\left[\begin{smallmatrix} 1 & \\ z & p^2 \end{smallmatrix}\right]]). \qquad (12.91)$$

Since $p^{-1}S', S' \in B(N)^+ \subset B(Np^{-1})^+$ the identity (12.28) implies that

$$\sum_{x \in \mathbb{Z}/p\mathbb{Z}} a(p^{-1} S'[\left[\begin{smallmatrix} 1 & \\ x & p \end{smallmatrix}\right]]) = 0 \quad \text{and} \quad \sum_{y \in \mathbb{Z}/p\mathbb{Z}} a(S'[\left[\begin{smallmatrix} 1 & \\ y & p \end{smallmatrix}\right]]) = 0. \qquad (12.92)$$

We also have

$$\begin{aligned}
\sum_{\substack{z\in\mathbb{Z}/p^2\mathbb{Z}\\ p^2|(\alpha+2\beta z)}} a(p^{-2}S'[\left[\begin{smallmatrix}1&\\ z&p^2\end{smallmatrix}\right]]) &= \sum_{x\in\mathbb{Z}/p\mathbb{Z}} \sum_{\substack{y\in\mathbb{Z}/p\mathbb{Z}\\ p^2|(\alpha+2\beta(x+yp))}} a(p^{-2}S'[\left[\begin{smallmatrix}1&\\ x+yp&p^2\end{smallmatrix}\right]])\\
&= \sum_{x\in\mathbb{Z}/p\mathbb{Z}} \sum_{\substack{y\in\mathbb{Z}/p\mathbb{Z}\\ p^2|(\alpha+2\beta x)}} a((p^{-2}S'[\left[\begin{smallmatrix}1&\\ x&p\end{smallmatrix}\right]])[\left[\begin{smallmatrix}1&\\ y&p\end{smallmatrix}\right]]) \qquad \text{(note that } p \mid 2\beta)\\
&= \sum_{\substack{x\in\mathbb{Z}/p\mathbb{Z}\\ p^2|(\alpha+2\beta x)}} \sum_{y\in\mathbb{Z}/p\mathbb{Z}} a((p^{-2}S'[\left[\begin{smallmatrix}1&\\ x&p\end{smallmatrix}\right]])[\left[\begin{smallmatrix}1&\\ y&p\end{smallmatrix}\right]])\\
&= \sum_{\substack{x\in\mathbb{Z}/p\mathbb{Z}\\ p^2|(\alpha+2\beta x)}} 0 \qquad \text{(by (12.28))}\\
&= 0.
\end{aligned} \tag{12.93}$$

Substituting (12.92) and (12.93) into (12.91) now gives

$$\begin{aligned}
p^{3-k}\lambda_p a(pS') &= p^3 a(S') + p^{6-2k}a(p^2S')\\
p^{3-k}\lambda_p a(p^{i+1}S) &= p^3 a(p^iS) + p^{6-2k}a(p^{i+2}S).
\end{aligned}$$

Rewriting, we have proven that if $t \in \mathbb{Z}$ with $t \geq 3$, then

$$a(p^tS) - p^{k-3}\lambda_p a(p^{t-1}S) + p^{2k-3}a(p^{t-2}S) = 0.$$

This recurrence relation implies (12.78). Finally, assume that F is an eigenvector for $T^s_{0,1}(p)$. Let $S \in B(N)^+$. Let $i \in \mathbb{Z}$ be such that $i \geq 1$ and define $S' = p^iS$. Then $S' \in B(N)^+$; let $S' = \left[\begin{smallmatrix}\alpha&\beta\\ \beta&\gamma\end{smallmatrix}\right]$. By (12.31) we have

$$\lambda_p a(S') = (1+p)p^{2-k}a(pS') + \sum_{\substack{x\in\mathbb{Z}/p\mathbb{Z}\\ p|(\alpha+2\beta x)}} (1+p)a(p^{-1}S'[\left[\begin{smallmatrix}1&\\ x&p\end{smallmatrix}\right]]).$$

Since p divides α and 2β, this is

$$\lambda_p a(S') = (1+p)p^{2-k}a(pS') + \sum_{x\in\mathbb{Z}/p\mathbb{Z}} (1+p)a(p^{-1}S'[\left[\begin{smallmatrix}1&\\ x&p\end{smallmatrix}\right]]).$$

Also, since $p^{-1}S' \in B(N)^+ \subset B(Np^{-1})^+$, (12.28) implies that

$$\sum_{x\in\mathbb{Z}/p\mathbb{Z}} a(p^{-1}S'[\left[\begin{smallmatrix}1&\\ x&p\end{smallmatrix}\right]]) = 0.$$

Therefore,

$$\lambda_p a(S') = (1+p)p^{2-k}a(pS')$$
$$\lambda_p a(p^i S) = (1+p)p^{2-k}a(p^{i+1}S).$$

Rewriting, we have proven that if $t \in \mathbb{Z}$ and $t \geq 2$, then

$$a(p^t S) - (1+p)^{-1}p^{k-2}\lambda_p a(p^{t-1}S) = 0. \tag{12.94}$$

This recurrence relation implies (12.81). □

We can now prove the main result of this section.

Theorem 12.5.2 *Let N and k be integers such that $N > 0$ and $k > 0$, and let $F \in S_k(\mathrm{K}(N))$. Assume that F is a newform and is an eigenvector for the Hecke operators $T(1,1,q,q)$ and $T(q,1,q,q^2)$ for all but finitely many of the primes q of $\mathbb{Z}$ such that $q \nmid N$; by Theorem 12.1.1, F is an eigenvector for $T(1,1,q,q)$, $T(q,1,q,q^2)$ and w_q for all primes q of $\mathbb{Z}$. Let $\lambda_q, \mu_q \in \mathbb{C}$ be such that*

$$T(1,1,q,q)F = q^{k-3}\lambda_q F,$$
$$T(q,1,q,q^2)F = q^{2(k-3)}\mu_q F$$

for all primes q of $\mathbb{Z}$. Regard F as an element of $S_k(\mathrm{K}_s(N))$, and let

$$F(Z) = \sum_{S \in B(N)^+} a(S)e^{2\pi i \mathrm{Tr}(SZ)} \tag{12.95}$$

be the Fourier expansion[2] of F. Let p be a prime of $\mathbb{Z}$ with $v_p(N) \geq 2$. If $S \in B(N)^+$, then there is a formal identity of power series in p^{-s}

$$\sum_{t=0}^{\infty} \frac{a(p^t S)}{p^{ts}} = N(p^{-s}, S)L_p(s, F) \tag{12.96}$$

where

$$N(p^{-s}, S) = a(S) + \big(a(pS) - p^{k-3}\lambda_p a(S)\big)p^{-s}$$
$$+ \big(a(p^2S) - p^{k-3}\lambda_p a(pS) + p^{2k-5}(\mu_p + p^2)a(S)\big)p^{-2s} \tag{12.97}$$

[2] Note that, since $F \in S_k(\mathrm{K}(N))$, the Fourier coefficient $a(S)$ is zero if $S \notin A(N)^+$.

and

$$L_p(s, F) = \frac{1}{1 - p^{k-3}\lambda_p p^{-s} + p^{2k-5}(\mu_p + p^2)p^{-2s}} \tag{12.98}$$

is the spin L-factor of F at p (e.g., see [71], p. 547).

Proof The identity (12.96) follows from Lemma 12.5.1 if $\mu_p \neq 0$ or if $\mu_p = 0$ and F is not an eigenvector for $T^s_{0,1}(p)$. Assume that $\mu_p = 0$ and F is an eigenvector for $T^s_{0,1}(p)$. Let $\pi \cong \otimes_{v\leq\infty}\pi_v$ be as in Theorem 12.1.1. By (4) of Theorem 12.1.2 π_p is non-generic and by Theorem 12.1.1 π_p is a Saito-Kurokawa representation. An inspection of Table A.4 shows that $\lambda_p^2 = p^2(p+1)^2$. Let $S \in B(N)^+$. By (12.81) there is a formal identity

$$\sum_{t=0}^{\infty} \frac{a(p^t S)}{p^{ts}} = \frac{a(S) + \big(a(pS) - p^{k-2}(1+p)^{-1}\lambda_p a(S)\big)p^{-s}}{1 - p^{k-2}(1+p)^{-1}\lambda_p p^{-s}}. \tag{12.99}$$

To prove (12.96) we will multiply the numerator and denominator in (12.99) by $1 - p^{k-2}(1+p)^{-1}\lambda_p p^{-(s+1)}$. Using $\lambda_p^2 = p^2(p+1)^2$ and $\mu_p = 0$ it is easy to verify that

$$(1 - p^{k-2}(1+p)^{-1}\lambda_p p^{-s})(1 - p^{k-2}(1+p)^{-1}\lambda_p p^{-(s+1)}) = L_p(s, F)^{-1}.$$

And:

$$\begin{aligned}
&(a(S) + (a(pS) - p^{k-2}(1+p)^{-1}\lambda_p a(S))p^{-s})(1 - p^{k-2}(1+p)^{-1}\lambda_p p^{-(s+1)})\\
&\quad = a(S) + (a(pS) - p^{k-3}\lambda_p a(S))p^{-s}\\
&\qquad + (-p^{k-3}(1+p)^{-1}\lambda_p a(pS) + p^{2k-3}a(S))p^{-2s}\\
&\quad = a(S) + (a(pS) - p^{k-3}\lambda_p a(S))p^{-s}\\
&\qquad + (a(p^2S) - a(p^2S) - p^{k-3}(1+p)^{-1}\lambda_p a(pS) + p^{2k-3}a(S))p^{-2s}\\
&\quad = a(S) + (a(pS) - p^{k-3}\lambda_p a(S))p^{-s}\\
&\qquad + (a(p^2S) - p^{k-2}(1+p)^{-1}\lambda_p a(pS)\\
&\qquad - p^{k-3}(1+p)^{-1}\lambda_p a(pS) + p^{2k-3}a(S))p^{-2s} \qquad \text{(by (12.94))}\\
&\quad = a(S) + (a(pS) - p^{k-3}\lambda_p a(S))p^{-s}\\
&\qquad + (a(p^2S) - p^{k-3}\lambda_p a(pS) + p^{2k-3}a(S))p^{-2s}\\
&\quad = N(p^{-s}, S).
\end{aligned}$$

This completes the proof. □

We remark that the examples of [99] and [98] can be used to exhibit F and S such that the coefficients of p^{-ts} for $t = 0, 1$ and 2 in $N(p^{-s}, S)$ are non-zero.

Examining the radial Fourier coefficient formula (12.96) further, one can derive interesting identities involving Hecke eigenvalues and Fourier coefficients. We thank an anonymous referee for the following consequence.

Corollary 12.5.3 *With the notations and assumptions as in Theorem 12.5.2,*

$$\lambda_p \Delta(S, p) = p^{3-k} \left(a(pS)a(p^4S) - a(p^2S)a(p^3S) \right), \tag{12.100}$$

$$(\mu_p + p^2)\Delta(S, p) = p^{5-2k} \left(a(p^2S)a(p^4S) - a(p^3S)^2 \right), \tag{12.101}$$

where $\Delta(S, p) = a(pS)a(p^3S) - a(p^2S)^2$.

Appendix A
Tables

In this appendix we present a number of tables summarizing some of the results of the text. The following is a guide to these tables.

Table A.1: Non-Supercuspidal Representations of GSp(4, F)

This table lists all the non-supercuspidal, irreducible, admissible representations π of GSp(4, F); see Sect. 2.2 for more details. In addition, this table shows in the "tempered" column the conditions required for π to be tempered; indicates with a • in the "ess. L^2" column when π is essentially square-integrable, i.e., square-integrable after an appropriate twisting; and indicates with a • in the "generic" column if π is generic.

Table A.2: Non-Paramodular Representations

This table lists all the irreducible, admissible representations π of GSp(4, F) with trivial central character that are not paramodular. In this table, π is not paramodular if and only if the listed condition on the defining data for π is satisfied. This table follows from Theorem 3.4.3 of [105] and Table A.12 of [105].

Table A.3: Stable Klingen Dimensions

This table lists, for every irreducible, admissible representation (π, V) of GSp(4, F) with trivial central character, the paramodular level N_π, the stable Klingen level

J. Johnson-Leung et al., *Stable Klingen Vectors and Paramodular Newforms*, Lecture Notes in Mathematics 2342, https://doi.org/10.1007/978-3-031-45177-5

$N_{\pi,s}$, the dimensions of the spaces $V_s(n)$ for integers $n \geq N_{\pi,s}$, the quotient stable Klingen level $\bar{N}_{\pi,s}$, the dimensions of the spaces $\bar{V}_s(n)$ for integers $n \geq \bar{N}_{\pi,s}$, the category of π, and some additional information for some π. The symbols and abbreviations $\boxed{1}$, $\boxed{2}$, "SK", non-unit., and one-dim., stand for category 1 and category 2 (see p. 151), Saito-Kurokawa (see p. 81), non-unitary, and one-dimensional, respectively. The entry for $\bar{N}_{\pi,s}$ is – if $V_s(n) = V(n)$ for all integers $n \geq 0$, so that $\bar{N}_{\pi,s}$ is not defined. See Theorem 5.6.1.

Table A.4: Hecke Eigenvalues

For every irreducible, admissible representation (π, V) of GSp(4, F) with trivial central character, this table lists the paramodular level N_π, the Atkin-Lehner eigenvalue ε_π, and the paramodular Hecke eigenvalues λ_π and μ_π from Table A.14 of [105]; also listed are the category of π, and some additional information for some π. The symbols and abbreviations $\boxed{1}$, $\boxed{2}$, "SK", non-unit., and one-dim., stand for category 1 and category 2 (see p. 151), Saito-Kurokawa (see p. 81), non-unitary, and one-dimensional, respectively. For typesetting reasons, some of the eigenvalues are given as below.

$$(\text{A1}) = q^{3/2}\sigma(\varpi)\big(1 + \chi_1(\varpi) + \chi_2(\varpi) + \chi_1(\varpi)\chi_2(\varpi)\big),$$

$$(\text{A2}) = q^2\big(\chi_1(\varpi) + \chi_2(\varpi) + \chi_1(\varpi)^{-1} + \chi_2(\varpi)^{-1} + 1 - q^{-2}\big),$$

$$(\text{B1}) = q^{3/2}(\sigma(\varpi) + \sigma(\varpi)^{-1}) + (q+1)(\sigma\chi)(\varpi),$$

$$(\text{B2}) = q^{3/2}(\chi(\varpi) + \chi(\varpi)^{-1}),$$

$$(\text{C1}) = q^{3/2}(\sigma(\varpi) + \sigma(\varpi)^{-1}) + q(q+1)(\sigma\chi)(\varpi),$$

$$(\text{C2}) = q^{3/2}(q+1)(\chi(\varpi) + \chi^{-1}(\varpi)) + q^2 - 1,$$

$$(\text{D}) = q^2(\chi(\varpi) + \chi^{-1}(\varpi) + q + 1) + q - 1.$$

The last column provides additional information about the representation.

Table A.5: Characteristic Polynomials of $T^s_{0,1}$ and $T^s_{1,0}$ on $V_s(1)$ in Terms of Inducing Data

This table lists, for every irreducible, admissible representation (π, V) of GSp(4, F) with trivial central character such that $V_s(1) \neq 0$, the characteristic polynomials of $T^s_{0,1}$ and $T^s_{1,0}$ in terms of the inducing data for π. See Theorem 9.4.6.

Table A.6: Characteristic Polynomials of $T^s_{0,1}$ and $T^s_{1,0}$ on $V_s(1)$ in Terms of Paramodular Eigenvalues

This table lists, for every irreducible, admissible representation (π, V) of $\mathrm{GSp}(4, F)$ with trivial central character such that $V_s(1) \neq 0$, the characteristic polynomials of $T^s_{0,1}$ and $T^s_{1,0}$ in terms of the paramodular eigenvalues ε_π, λ_π, and μ_π for π. See Theorem 9.4.6.

Table A.1 Non-supercuspidal representations of GSp(4, *F*)

Constituent of	Group		Representation	Tempered	ess. L^2	Generic
$\chi_1 \times \chi_2 \rtimes \sigma$ (irreducible)	I		$\chi_1 \times \chi_2 \rtimes \sigma$	χ_i, σ unit.		•
$\nu^{1/2}\chi \times \nu^{-1/2}\chi \rtimes \sigma$	II	a	$\chi \mathrm{St}_{\mathrm{GL}(2)} \rtimes \sigma$	χ, σ unit.		•
$(\chi^2 \neq \nu^{\pm 1}, \chi \neq \nu^{\pm 3/2})$		b	$\chi 1_{\mathrm{GL}(2)} \rtimes \sigma$			
$\chi \times \nu \rtimes \nu^{-1/2}\sigma$	III	a	$\chi \rtimes \sigma \mathrm{St}_{\mathrm{GSp}(2)}$	χ, σ unit.		•
$(\chi \notin \{1, \nu^{\pm 2}\})$		b	$\chi \rtimes \sigma 1_{\mathrm{GSp}(2)}$			
$\nu^2 \times \nu \rtimes \nu^{-3/2}\sigma$	IV	a	$\sigma \mathrm{St}_{\mathrm{GSp}(4)}$	σ unit.	•	•
		b	$L(\nu^2, \nu^{-1}\sigma \mathrm{St}_{\mathrm{GSp}(2)})$			
		c	$L(\nu^{3/2}\mathrm{St}_{\mathrm{GL}(2)}, \nu^{-3/2}\sigma)$			
		d	$\sigma 1_{\mathrm{GSp}(4)}$			
$\nu\xi \times \xi \rtimes \nu^{-1/2}\sigma$	V	a	$\delta([\xi, \nu\xi], \nu^{-1/2}\sigma)$	σ unit.	•	•
$(\xi^2 = 1,\ \xi \neq 1)$		b	$L(\nu^{1/2}\xi \mathrm{St}_{\mathrm{GL}(2)}, \nu^{-1/2}\sigma)$			
		c	$L(\nu^{1/2}\xi \mathrm{St}_{\mathrm{GL}(2)}, \xi\nu^{-1/2}\sigma)$			
		d	$L(\nu\xi, \xi \rtimes \nu^{-1/2}\sigma)$			
$\nu \times 1_{F^\times} \rtimes \nu^{-1/2}\sigma$	VI	a	$\tau(S, \nu^{-1/2}\sigma)$	σ unit.		•
		b	$\tau(T, \nu^{-1/2}\sigma)$	σ unit.		
		c	$L(\nu^{1/2}\mathrm{St}_{\mathrm{GL}(2)}, \nu^{-1/2}\sigma)$			
		d	$L(\nu, 1_{F^\times} \rtimes \nu^{-1/2}\sigma)$			
$\chi \rtimes \pi$ (irreducible)	VII		$\chi \rtimes \pi$	χ, π unit.		•
$1_{F^\times} \rtimes \pi$	VIII	a	$\tau(S, \pi)$	π unit.		•
		b	$\tau(T, \pi)$	π unit.		
$\nu\xi \rtimes \nu^{-1/2}\pi$	IX	a	$\delta(\nu\xi, \nu^{-1/2}\pi)$	π unit.	•	•
$(\xi \neq 1,\ \xi\pi = \pi)$		b	$L(\nu\xi, \nu^{-1/2}\pi)$			
$\pi \rtimes \sigma$ (irreducible)	X		$\pi \rtimes \sigma$	π, σ unit.		•
$\nu^{1/2}\pi \rtimes \nu^{-1/2}\sigma$	XI	a	$\delta(\nu^{1/2}\pi, \nu^{-1/2}\sigma)$	π, σ unit.	•	•
$(\omega_\pi = 1)$		b	$L(\nu^{1/2}\pi, \nu^{-1/2}\sigma)$			

Table A.2 Non-paramodular representations

		Representation	Condition on defining data
II	b	$\chi 1_{\mathrm{GL}(2)} \rtimes \sigma$	$\chi\sigma$ ramified
III	b	$\chi \rtimes \sigma 1_{\mathrm{GSp}(2)}$	σ ramified
IV	b	$L(\nu^2, \nu^{-1}\sigma \mathrm{St}_{\mathrm{GSp}(2)})$	σ ramified
	c	$L(\nu^{3/2}\mathrm{St}_{\mathrm{GL}(2)}, \nu^{-3/2}\sigma)$	σ ramified
	d	$\sigma 1_{\mathrm{GSp}(4)}$	σ ramified
V	b	$L(\nu^{1/2}\xi \mathrm{St}_{\mathrm{GL}(2)}, \nu^{-1/2}\sigma)$	σ ramified
	c	$L(\nu^{1/2}\xi \mathrm{St}_{\mathrm{GL}(2)}, \xi\nu^{-1/2}\sigma)$	$\xi\sigma$ ramified
	d	$L(\nu\xi, \xi \rtimes \nu^{-1/2}\sigma)$	σ or ξ ramified
VI	b	$\tau(T, \nu^{-1/2}\sigma)$	None
	c	$L(\nu^{1/2}\mathrm{St}_{\mathrm{GL}(2)}, \nu^{-1/2}\sigma)$	σ ramified
	d	$L(\nu, 1_{F^\times} \rtimes \nu^{-1/2}\sigma)$	σ ramified
VIII	b	$\tau(T, \pi)$	None
IX	b	$L(\nu\xi, \nu^{-1/2}\pi)$	None
XI	b	$L(\nu^{1/2}\pi, \nu^{-1/2}\sigma)$	σ ramified
		π supercuspidal	Non-generic

Table A.3 Stable Klingen dimensions

		Inducing data	N_π	$N_{\pi,s}$	$\dim V_s(n)$	$\bar{N}_{\pi,s}$	$\dim \bar{V}_s(n)$	Cat.	Cmt.
I		χ_1, χ_2, σ unr.	0	0	$\frac{n^2+5n+2}{2}$	1	n	2	
		χ_1, χ_2 ram., σ unr.	$a := a(\chi_1) + a(\chi_2)$	a	$\frac{(n-a+1)(n-a+4)}{2}$	a	$n-a+1$	2	
		$\chi_i\sigma$ unr., σ ram.	$a := 2a(\sigma)$	a	$\frac{(n-a+1)(n-a+4)}{2}$	a	$n-a+1$	2	
		$\chi_i\sigma$ ram., σ ram.	$a:=a(\chi_1\sigma)+a(\chi_2\sigma)+2a(\sigma)$	$a-1$	$\frac{(n-a+2)(n-a+3)}{2}$	$a-1$	$n-a+2$	1	
II	a	σ, χ unr.	1	1	$\frac{n(n+3)}{2}$	1	n	2	
		σ ram., $\chi\sigma$ unr.	$a := 2a(\sigma) + 1$	$a-1$	$\frac{(n-a+2)(n-a+3)}{2}$	$a-1$	$n-a+2$	1	
		σ unr., $\chi\sigma$ ram.	$a := 2a(\sigma\chi)$	a	$\frac{(n-a+1)(n-a+4)}{2}$	a	$n-a+1$	2	
		σ, $\chi\sigma$ ram.	$a := 2a(\chi\sigma) + 2a(\sigma)$	$a-1$	$\frac{(n-a+2)(n-a+3)}{2}$	$a-1$	$n-a+2$	1	
	b	$\chi\sigma$ unr., σ unr.	0	0	$n+1$	–	0	2	SK
		$\chi\sigma$ unr., σ ram.	$a := 2a(\sigma)$	a	$n-a+1$	–	0	2	SK
		$\chi\sigma$ ram.	Not paramodular						SK
III	a	σ unr.	2	1	$\frac{n(n+1)}{2}$	1	n	1	
		σ ram.	$a := 4a(\sigma)$	$a-1$	$\frac{(n-a+2)(n-a+3)}{2}$	$a-1$	$n-a+2$	1	
	b	σ unr.	0	0	$2n+1$	–	0	2	
		σ ram.	Not paramodular						

(continued)

Table A.3 (continued)

		Inducing data	N_π	$N_{\pi,s}$	$\dim V_s(n)$	$\bar{N}_{\pi,s}$	$\dim \bar{V}_s(n)$	Cat.	Cmt.
IV	a	σ unr.	3	2	$\frac{(n-1)n}{2}$	2	$n-1$	$\boxed{1}$	
		σ ram.	$a := 4a(\sigma)$	$a-1$	$\frac{(n-a+2)(n-a+3)}{2}$	$a-1$	$n-a+2$	$\boxed{1}$	
	b	σ unr.	2	1	n	1	1	$\boxed{1}$	Non-unit.
		σ ram.	Not paramodular						Non-unit.
	c	σ unr.	1	1	$2n$	1	1	$\boxed{2}$	Non-unit.
		σ ram.	Not paramodular						Non-unit.
	d	σ unr.	0	0	1	–	0	$\boxed{2}$	One-dim.
		σ ram.	Not paramodular						One-dim.
V	a	σ, ξ unr.	2	1	$\frac{n(n+1)}{2}$	1	n	$\boxed{1}$	
		σ unr., ξ ram.	$a := 2a(\xi)+1$	$a-1$	$\frac{(n-a+2)(n-a+3)}{2}$	$a-1$	$n-a+2$	$\boxed{1}$	
		σ ram., $\sigma\xi$ unr.	$a := 2a(\sigma)+1$	$a-1$	$\frac{(n-a+2)(n-a+3)}{2}$	$a-1$	$n-a+2$	$\boxed{1}$	
		$\sigma, \sigma\xi$ ram.	$a := 2a(\xi\sigma)+2a(\sigma)$	$a-1$	$\frac{(n-a+2)(n-a+3)}{2}$	$a-1$	$n-a+2$	$\boxed{1}$	
	b	σ, ξ unr.	1	1	n	–	0	$\boxed{2}$	SK
		σ unr., ξ ram.	$2a(\xi)$	$2a(\xi)$	$n-2a(\xi)+1$	–	0	$\boxed{2}$	SK
		σ ram., $\sigma\xi$ unr.	Not paramodular						SK
		$\sigma, \sigma\xi$ ram.	Not paramodular						SK

V	c	σ, ξ unr.	1	1	n	–	0	$\boxed{2}$	SK
		σ unr., ξ ram.	Not paramodular						SK
		σ ram., $\sigma\xi$ unr.	$2a(\sigma)$	$2a(\sigma)$	$n-2a(\sigma)+1$	–	0	$\boxed{2}$	SK
		$\sigma, \sigma\xi$ ram.	Not paramodular						SK
	d	σ, ξ unr.	0	0	1	–	0	$\boxed{2}$	
		σ or ξ ram.	Not paramodular						
VI	a	σ unr.	2	1	$\frac{n(n+1)}{2}$	1	n	$\boxed{1}$	
		σ ram.	$a := 4a(\sigma)$	$a-1$	$\frac{(n-a+2)(n-a+3)}{2}$	$a-1$	$n-a+2$	$\boxed{1}$	
	b	σ unr.	Not paramodular						
		σ ram.	Not paramodular						
	c	σ unr.	1	1	n	–	0	$\boxed{2}$	SK
		σ ram.	Not paramodular						SK
	d	σ unr.	0	0	$n+1$	–	0	$\boxed{2}$	
		σ ram.	Not paramodular						
VII			$a := 2a(\pi)$	$a-1$	$\frac{(n-a+2)(n-a+3)}{2}$	$a-1$	$n-a+2$	$\boxed{1}$	
VIII	a		$a := 2a(\pi)$	$a-1$	$\frac{(n-a+2)(n-a+3)}{2}$	$a-1$	$n-a+2$	$\boxed{1}$	
	b		Not paramodular						
IX	a		$a := 2a(\pi)$	$a-1$	$\frac{(n-a+2)(n-a+3)}{2}$	$a-1$	$n-a+2$	$\boxed{1}$	
	b		Not paramodular						

(continued)

Table A.3 (continued)

		Inducing data	N_π	$N_{\pi,s}$	$\dim V_s(n)$	$\bar{N}_{\pi,s}$	$\dim \bar{V}_s(n)$	Cat.	Cmt.
X		σ unr.	$a := a(\pi)$	a	$\frac{(n-a+1)(n-a+4)}{2}$	a	$n-a+1$	$\boxed{2}$	
		σ ram.	$a := a(\sigma\pi) + 2a(\sigma)$	$a-1$	$\frac{(n-a+2)(n-a+3)}{2}$	$a-1$	$n-a+2$	$\boxed{1}$	
XI	a	σ unr.	$a := a(\sigma\pi) + 1$	$a-1$	$\frac{(n-a+2)(n-a+3)}{2}$	$a-1$	$n-a+2$	$\boxed{1}$	
		σ ram.	$a := a(\sigma\pi) + 2a(\sigma)$	$a-1$	$\frac{(n-a+2)(n-a+3)}{2}$	$a-1$	$n-a+2$	$\boxed{1}$	
	b	σ unr.	$a := a(\pi)$	a	$n-a+1$	–	0	$\boxed{2}$	SK
		σ ram.	Not paramodular						SK
s.c.		generic	$a \geq 4$	$a-1$	$\frac{(n-a+2)(n-a+3)}{2}$	$a-1$	$n-a+2$	$\boxed{1}$	
		non-generic	Not paramodular						

Table A.4 Hecke eigenvalues

		Inducing data	N_π	ε_π	λ_π	μ_π	Cat.	Cmt.
I		χ_1, χ_2, σ unr.	0	1	(A1)	(A2)	$\boxed{2}$	
		χ_1, χ_2 ram., σ unr.	$a(\chi_1)+a(\chi_2)$	$\chi_1(-1)$	$q^{3/2}(\sigma(\varpi)+\sigma(\varpi^{-1}))$	0	$\boxed{2}$	
		$\chi_i\sigma$ unr., σ ram.	$2a(\sigma)$	$\chi_1(-1)$	$q^{3/2}((\chi_1\sigma)(\varpi)+(\chi_2\sigma)(\varpi))$	0	$\boxed{2}$	
		$\chi_i\sigma$ ram., σ ram.	$a(\chi_1\sigma)+a(\chi_2\sigma)+2a(\sigma)$	$\chi_1(-1)$	0	$-q^2$	$\boxed{1}$	
II	a	σ, χ unr.	1	$-(\chi\sigma)(\varpi)$	(B1)	(B2)	$\boxed{2}$	
		σ ram., $\chi\sigma$ unr.	$2a(\sigma)+1$	$-\sigma(-1)(\chi\sigma)(\varpi)$	$q(\chi\sigma)(\varpi)$	$-q^2$	$\boxed{1}$	
		σ unr., $\chi\sigma$ ram.	$2a(\sigma\chi)$	$\chi(-1)$	$q^{3/2}(\sigma(\varpi)+\sigma(\varpi)^{-1})$	0	$\boxed{2}$	
		$\sigma, \chi\sigma$ ram.	$2a(\chi\sigma)+2a(\sigma)$	$\chi(-1)$	0	$-q^2$	$\boxed{1}$	
	b	$\chi\sigma$ unr., σ unr.	0	1	(C1)	(C2)	$\boxed{2}$	SK
		$\chi\sigma$ unr., σ ram.	$2a(\sigma)$	$\chi(-1)$	$q(q+1)(\sigma\chi)(\varpi)$	0	$\boxed{2}$	SK
		$\chi\sigma$ ram.	Not paramodular					SK
III	a	σ unr.	2	1	$q(\sigma(\varpi)+\sigma(\varpi)^{-1})$	$-q^2+q$	$\boxed{1}$	
		σ ram.	$4a(\sigma)$	1	0	$-q^2$	$\boxed{1}$	
	b	σ unr.	0	1	$q(q+1)\sigma(\varpi)(1+\chi(\varpi))$	(D)	$\boxed{2}$	
		σ ram.	Not paramodular					

(continued)

Table A.4 (continued)

		Inducing data	N_π	ε_π	λ_π	μ_π	Cat.	Cmt.
IV	a	σ unr.	3	$-\sigma(\varpi)$	$\sigma(\varpi)$	$-q^2$	$\boxed{1}$	
		σ ram.	$4a(\sigma)$	1	0	$-q^2$	$\boxed{1}$	
	b	σ unr.	2	1	$\sigma(\varpi)(1+q^2)$	$-q^2+q$	$\boxed{1}$	Non-unit.
		σ ram.	not paramodular					Non-unit.
	c	σ unr.	1	$-\sigma(\varpi)$	$\sigma(\varpi)(q^3+q+2)$	q^3+1	$\boxed{2}$	Non-unit.
		σ ram.	not paramodular					Non-unit.
	d	σ unr.	0	1	$\sigma(\varpi)(q+1)(q^2+1)$	$q(q+1)(q^2+1)$	$\boxed{2}$	One-dim.
		σ ram.	not paramodular					One-dim.
V	a	σ, ξ unr.	2	-1	0	$-q^2-q$	$\boxed{1}$	
		σ unr., ξ ram.	$2a(\xi)+1$	$-\sigma(\varpi)\xi(-1)$	$\sigma(\varpi)q$	$-q^2$	$\boxed{1}$	
		σ ram., $\sigma\xi$ unr.	$2a(\sigma)+1$	$-\sigma(-1)(\xi\sigma)(\varpi)$	$(\xi\sigma)(\varpi)q$	$-q^2$	$\boxed{1}$	
		$\sigma, \sigma\xi$ ram.	$2a(\xi\sigma)+2a(\sigma)$	$\xi(-1)$	0	$-q^2$	$\boxed{1}$	
	b	σ, ξ unr.	1	$\sigma(\varpi)$	$\sigma(\varpi)(q^2-1)$	$-q^2-q$	$\boxed{2}$	SK
		σ unr., ξ ram.	$2a(\xi)$	$\xi(-1)$	$\sigma(\varpi)(q^2+q)$	0	$\boxed{2}$	SK
		σ ram., $\sigma\xi$ unr.	not paramodular					SK
		$\sigma, \sigma\xi$ ram.	not paramodular					SK

V	c	σ, ξ unr.	1	$-\sigma(\varpi)$	$-\sigma(\varpi)(q^2-1)$	$-q^2-q$	$\boxed{2}$	SK
		σ unr., ξ ram	Not paramodular					SK
		σ ram., $\sigma\xi$ unr.	$2a(\sigma)$	$\xi(-1)$	$(\xi\sigma)(\varpi)(q^2+q)$	0	$\boxed{2}$	SK
		$\sigma, \sigma\xi$ ram.	Not paramodular					SK
	d	σ, ξ unr.	0	1	0	$-(q+1)(q^2+1)$	$\boxed{2}$	
		σ or ξ ram.	Not paramodular					
VI	a	σ unr.	2	1	$2q\sigma(\varpi)$	$-q^2+q$	$\boxed{1}$	
		σ ram.	$4a(\sigma)$	1	0	$-q^2$	$\boxed{1}$	
	b	σ unr.	Not paramodular					
		σ ram.	Not paramodular					
	c	σ unr.	1	$-\sigma(\varpi)$	$\sigma(\varpi)(q+1)^2$	$q(q+1)$	$\boxed{2}$	SK
		σ ram.	Not paramodular					SK
	d	σ unr.	0	1	$2q(q+1)\sigma(\varpi)$	$(q+1)(q^2+2q-1)$	$\boxed{2}$	
		σ ram.	Not paramodular					
VII			$2a(\pi)$	$\chi(-1)$	0	$-q^2$	$\boxed{1}$	
VIII	a		$2a(\pi)$	1	0	$-q^2$	$\boxed{1}$	
	b		Not paramodular					

(continued)

Table A.4 (continued)

		Inducing data	N_π	ε_π	λ_π	μ_π	Cat.	Cmt.
IX	a		$2a(\pi)$	$\xi(-1)$	0	$-q^2$	$\boxed{1}$	
	b		Not paramodular					
X		σ unr.	$a(\pi)$	$\varepsilon(1/2, \sigma\pi)$	$q^{3/2}(\sigma(\varpi)+\sigma(\varpi)^{-1})$	0	$\boxed{2}$	
		σ ram.	$a(\sigma\pi)+2a(\sigma)$	$\sigma(-1)\varepsilon(1/2,\sigma\pi)$	0	$-q^2$	$\boxed{1}$	
XI	a	σ unr.	$a(\sigma\pi)+1$	$-\sigma(\varpi)\varepsilon(1/2,\sigma\pi)$	$q\sigma(\varpi)$	$-q^2$	$\boxed{1}$	
		σ ram.	$a(\sigma\pi)+2a(\sigma)$	$\sigma(-1)\varepsilon(1/2,\sigma\pi)$	0	$-q^2$	$\boxed{1}$	
	b	σ unr.	$a(\pi)$	$\varepsilon(1/2, \sigma\pi)$	$(q^2+q)\sigma(\varpi)$	0	$\boxed{2}$	SK
		σ ram.	Not paramodular					SK
s.c.		Generic	$N_\pi \geq 4$	ε_π	0	$-q^2$	$\boxed{1}$	
		Non-generic	Not paramodular					

Table A.5 Characteristic polynomials of $T^s_{0,1}$ and $T^s_{1,0}$ on $V_s(1)$ in terms of inducing data

type		π	$p(T^s_{0,1}, V_s(1), X)$	$p(T^s_{1,0}, V_s(1), X)$
I χ_1, χ_2, σ unram. $\chi_1\chi_2\sigma^2 = 1$		$\chi_1 \times \chi_2 \rtimes \sigma$	$\left(X - \chi_1(\varpi)(1+\chi_2(\varpi))\sigma(\varpi)q^{\frac{3}{2}}\right)$ $\times\left(X - \chi_2(\varpi)(1+\chi_1(\varpi))\sigma(\varpi)q^{\frac{3}{2}}\right)$ $\times\left(X - (1+\chi_1(\varpi))\sigma(\varpi)q^{\frac{3}{2}}\right)$ $\times\left(X - (1+\chi_2(\varpi))\sigma(\varpi)q^{\frac{3}{2}}\right)$	$(X - \chi_1(\varpi)q^2)$ $\times(X - \chi_2(\varpi)q^2)$ $\times(X - \chi_2(\varpi)^{-1}q^2)$ $\times(X - \chi_1(\varpi)^{-1}q^2)$
II χ, σ unram. $\chi^2\sigma^2 = 1$	a	$\chi\mathrm{St}_{\mathrm{GL}(2)} \rtimes \sigma$	X^2 $-\left(2(\chi\sigma)(\varpi)q + \left(\sigma(\varpi)+\sigma(\varpi)^{-1}\right)q^{\frac{3}{2}}\right)X$ $+q^2+q^3+\left(\chi(\varpi)+\chi(\varpi)^{-1}\right)q^{\frac{5}{2}}$	$X^2 - \left(\chi(\varpi)+\chi(\varpi)^{-1}\right)q^{\frac{3}{2}}X + q^3$
	b	$\chi 1_{\mathrm{GL}(2)} \rtimes \sigma$	X^2 $-\left(2(\chi\sigma)(\varpi)q^2 + \left(\sigma(\varpi)+\sigma(\varpi)^{-1}\right)q^{\frac{3}{2}}\right)X$ $+q^3+q^4+\left(\chi(\varpi)+\chi(\varpi)^{-1}\right)q^{\frac{7}{2}}$	X^2 $-\left(\chi(\varpi)+\chi(\varpi)^{-1}\right)q^{5/2}X + q^5$
III χ, σ unram. $\chi\sigma^2 = 1$	a	$\chi \rtimes \sigma\mathrm{St}_{\mathrm{GSp}(2)}$	$X - q(\sigma(\varpi)+\sigma(\varpi)^{-1})$	$X - q$
	b	$\chi \rtimes \sigma 1_{\mathrm{GSp}(2)}$	X^3 $-\left(\sigma(\varpi)+\sigma(\varpi)^{-1}\right)q(1+2q)X^2$ $+\left(1+3q+\left(\chi(\varpi)+\chi(\varpi)^{-1}\right)q\right)q^2(q+1)X$ $-\left(\sigma(\varpi)+\sigma(\varpi)^{-1}\right)q^4(q+1)^2$	X^3 $-\left(\left(\chi(\varpi)+\chi(\varpi)^{-1}\right)q^2+q^3\right)X^2$ $+\left(\left(\chi(\varpi)+\chi(\varpi)^{-1}\right)q^5+q^4\right)X$ $-q^7$

(continued)

Table A.5 (continued)

type		π	$p(T^s_{0,1}, V_s(1), X)$	$p(T^s_{1,0}, V_s(1), X)$
IV	b	$L(\nu^2, \nu^{-1}\sigma \mathrm{St}_{\mathrm{GSp}(2)})$	$X - \sigma(\varpi)(1+q^2)$	$X - q$
σ unram. $\sigma^2 = 1$	c	$L(\nu^{\frac{3}{2}} \mathrm{St}_{\mathrm{GL}(2)}, \nu^{-\frac{3}{2}}\sigma)$	$X^2 - \sigma(\varpi)(1+2q+q^3)X$ $+q(1+q)(1+q^2)$	$X^2 - (1+q^3)X + q^3$
	d	$\sigma 1_{\mathrm{GSp}(4)}$	$X - \sigma(\varpi)(q^2+q^3)$	$X - q^4$
V	a	$\delta([\xi, \nu\xi], \nu^{-\frac{1}{2}}\sigma)$	X	$X + q$
σ, ξ unram.	b	$L(\nu^{\frac{1}{2}}\xi \mathrm{St}_{\mathrm{GL}(2)}, \nu^{-\frac{1}{2}}\sigma)$	$X - \sigma(\varpi)(q^2 - q)$	$X + q^2$
$\sigma^2 = \xi^2 = 1$	c	$L(\nu^{\frac{1}{2}}\xi \mathrm{St}_{\mathrm{GL}(2)}, \xi\nu^{-\frac{1}{2}}\sigma)$	$X + \sigma(\varpi)(q^2 - q)$	$X + q^2$
$\xi \neq 1$	d	$L(\nu\xi, \xi \rtimes \nu^{-\frac{1}{2}}\sigma)$	X	$X + q^3$
VI	a	$\tau(S, \nu^{-\frac{1}{2}}\sigma)$	$X - 2q\sigma(\varpi)$	$X - q$
σ unram.	c	$L(\nu^{\frac{1}{2}} \mathrm{St}_{\mathrm{GL}(2)}, \nu^{-\frac{1}{2}}\sigma)$	$X - \sigma(\varpi)(q^2 + q)$	$X - q^2$
$\sigma^2 = 1$	d	$L(\nu, 1_{F^\times} \rtimes \nu^{-\frac{1}{2}}\sigma)$	$X^2 - \sigma(\varpi)(q + 3q^2)X$ $+2q^3(q+1)$	$(X - q^2)(X - q^3)$

Table A.6 Characteristic polynomials of $T^s_{0,1}$ and $T^s_{1,0}$ on $V_s(1)$ in terms of paramodular eigenvalues

Type		π	$p(T^s_{0,1}, V_s(1), X)$	$p(T^s_{1,0}, V_s(1), X)$
I χ_1, χ_2, σ unram. $\chi_1\chi_2\sigma^2 = 1$		$\chi_1 \times \chi_2 \rtimes \sigma$	$X^4 - 2\lambda_\pi X^3$ $+(\lambda_\pi^2 + q\mu_\pi + 3q^3 + q)X^2$ $-q\lambda_\pi(\mu_\pi + 3q^2 + 1)X + q^3\lambda_\pi^2$	$X^4 - (\mu_\pi - q^2 + 1)X^3$ $+(q\lambda_\pi^2 - 2q^2\mu_\pi - 2q^2)X^2$ $-q^4(\mu_\pi - q^2 + 1)X + q^8$
II χ, σ unram. $\chi^2\sigma^2 = 1$	a	$\chi \mathrm{St}_{\mathrm{GL}(2)} \rtimes \sigma$	X^2 $+((q-1)\varepsilon_\pi - \lambda_\pi)X$ $+q(q^2 + q + \mu)$	$X^2 - \mu_\pi X + q^3$
	b	$\chi 1_{\mathrm{GL}(2)} \rtimes \sigma$	X^2 $-\lambda_\pi \dfrac{\mu_\pi + 1 + q^2 + 2q^3}{\mu_\pi + 1 + q + q^2 + q^3} X$ $+q^2\mu_\pi(q+1)^{-1} + q^2(q^2+1)$	X^2 $-q(\mu_\pi(q+1)^{-1} - q + 1)X$ $+q^5$
III	a	$\chi \rtimes \sigma \mathrm{St}_{\mathrm{GSp}(2)}$	$X - \lambda_\pi$	$X - q$
χ, σ unram. $\chi\sigma^2 = 1$	b	$\chi \rtimes \sigma 1_{\mathrm{GSp}(2)}$	X^3 $-\lambda_\pi(1 + q(1+q)^{-1})X^2$ $+(1 + 2q^2 - q^3 + \mu_\pi)q(q+1)X$ $-\lambda_\pi q^3(q+1)$	X^3 $-(\mu_\pi + 1 - q - q^2)X^2$ $+q^3(\mu_\pi + 1 - q^2 - q^3)X$ $-q^7$
IV	b	$L(\nu^2, \nu^{-1}\sigma \mathrm{St}_{\mathrm{GSp}(2)})$	$X - \lambda_\pi$	$X - q$
σ unram. $\sigma^2 = 1$	c	$L(\nu^{\frac{3}{2}} \mathrm{St}_{\mathrm{GL}(2)}, \nu^{-\frac{3}{2}}\sigma)$	X^2 $-\lambda_\pi \dfrac{1 + 2q + q^3}{2 + q + q^3} X$ $+q(1+q)(1+q^2)$	$X^2 - (1+q^3)X + q^3$
	d	$\sigma 1_{\mathrm{GSp}(4)}$	$X - \lambda_\pi q^2(q^2+1)^{-1}$	$X - q^4$

(continued)

Table A.6 (continued)

Type		π	$p(T^s_{0,1}, V_s(1), X)$	$p(T^s_{1,0}, V_s(1), X)$
V	a	$\delta([\xi, \nu\xi], \nu^{-\frac{1}{2}}\sigma)$	X	$X+q$
σ, ξ unram.	b	$L(\nu^{\frac{1}{2}}\xi \mathrm{St}_{\mathrm{GL}(2)}, \nu^{-\frac{1}{2}}\sigma)$	$X-\lambda_\pi q(q+1)^{-1}$	$X+q^2$
$\sigma^2=\xi^2=1$	c	$L(\nu^{\frac{1}{2}}\xi \mathrm{St}_{\mathrm{GL}(2)}, \xi\nu^{-\frac{1}{2}}\sigma)$	$X-\lambda_\pi q(q+1)^{-1}$	$X+q^2$
$\xi \neq 1$	d	$L(\nu\xi, \xi \rtimes \nu^{-\frac{1}{2}}\sigma)$	X	$X+q^3$
VI	a	$\tau(S, \nu^{-\frac{1}{2}}\sigma)$	$X-\lambda_\pi$	$X-q$
σ unram.	c	$L(\nu^{\frac{1}{2}} \mathrm{St}_{\mathrm{GL}(2)}, \nu^{-\frac{1}{2}}\sigma)$	$X-\lambda_\pi q(q+1)^{-1}$	$X-q^2$
$\sigma^2=1$	d	$L(\nu, 1_{F^\times} \rtimes \nu^{-\frac{1}{2}}\sigma)$	$X^2-\frac{\lambda_\pi}{2}(1+2q(1+q)^{-1})X$ $+2q^3(q+1)$	$(X-q^2)(X-q^3)$

References

1. Andrianov, A.N.: Euler products that correspond to Siegel's modular forms of genus 2. Russian Math. Surveys **29**(3 (177)), 45–116 (1974)
2. Andrianov, A.N.: Quadratic forms and Hecke operators, Grundlehren der mathematischen Wissenschaften [Fundamental Principles of Mathematical Sciences], vol. 286. Springer-Verlag, Berlin (1987)
3. Andrianov, A.N., Zhuravlev, V.G.: Modular forms and Hecke operators, Translations of Mathematical Monographs, vol. 145. American Mathematical Society, Providence (1995). Translated from the 1990 Russian original by Neal Koblitz
4. Arthur, J.: Automorphic representations of GSp(4). In: Contributions to Automorphic Forms, Geometry, and Number Theory, pp. 65–81. Johns Hopkins University Press, Baltimore (2004)
5. Arthur, J.: The endoscopic classification of representations, American Mathematical Society Colloquium Publications, vol. 61. American Mathematical Society, Providence (2013). Orthogonal and symplectic groups
6. Asgari, M., Schmidt, R.: Siegel modular forms and representations. Manuscripta Math. **104**(2), 173–200 (2001)
7. Assaf, E., Ladd, W., Rama, G., Tornaría, G., Voight, J.: A database of paramodular forms from quinary orthogonal modular forms (2023). arXiv:2308.09824
8. Atobe, H., Kondo, S., Yasuda, S.: Local newforms for the general linear groups over a non-archimedean local field. Forum Math. Pi **10**, Paper No. e24, 56 (2022)
9. Bastian, B., Hohenegger, S.: Symmetries in A-type little string theories. Part I. Reduced free energy and paramodular groups. J. High Energy Phys. (3), 062, 28 (2020)
10. Böcherer, S., Schulze-Pillot, R.: Paramodular groups and theta series (2021). arXiv:2011.09597
11. Belin, A., Castro, A., Gomes, J.A., Keller, C.A.: Siegel modular forms and black hole entropy. J. High Energy Phys. (4), 057, front matter + 48 (2017)
12. Belin, A., Castro, A., Gomes, J.A., Keller, C.A.: Siegel paramodular forms and sparseness in AdS_3/CFT_2. J. High Energy Phys. (11), 037, front matter+51 (2018)
13. Berger, T., Dembélé, L., Pacetti, A., Sengün, M.H.: Theta lifts of Bianchi modular forms and applications to paramodularity. J. Lond. Math. Soc. (2) **92**(2), 353–370 (2015)
14. Berger, T., Klosin, K.: Deformations of Saito-Kurokawa type and the paramodular conjecture. Am. J. Math. **142**(6), 1821–1875 (2020). With an appendix by Chris Poor, Jerry Shurman, and David S. Yuen
15. Bernstein, J.N., Zelevinsky, A.V.: Representations of the group $GL(n, F)$, where F is a local non-Archimedean field. Uspehi Mat. Nauk **31**(3(189)), 5–70 (1976)

J. Johnson-Leung et al., *Stable Klingen Vectors and Paramodular Newforms*,
Lecture Notes in Mathematics 2342, https://doi.org/10.1007/978-3-031-45177-5

16. Björner, A., Brenti, F.: Combinatorics of Coxeter groups, Graduate Texts in Mathematics, vol. 231. Springer, New York (2005)
17. Borel, A., Jacquet, H.: Automorphic forms and automorphic representations. In: Automorphic Forms, Representations and L-Functions (Proc. Sympos. Pure Math., Oregon State Univ., Corvallis, Ore., 1977), Part 1, Proceedings of Symposia in Pure Mathematics, vol. XXXIII, pp. 189–207. American Mathematical Society, Providence (1979). With a supplement "On the notion of an automorphic representation" by R. P. Langlands
18. Bourbaki, N.: Lie groups and Lie algebras. Chapters 4–6. Elements of Mathematics (Berlin). Springer-Verlag, Berlin (2002). Translated from the 1968 French original by Andrew Pressley
19. Breeding, II, J., Poor, C., Yuen, D.S.: Computations of spaces of paramodular forms of general level. J. Korean Math. Soc. **53**(3), 645–689 (2016)
20. Brown, J., Li, H.: Congruence primes for Siegel modular forms of paramodular level and applications to the Bloch-Kato conjecture. Glasg. Math. J. **63**(3), 660–681 (2021)
21. Brumer, A., Kramer, K.: Paramodular abelian varieties of odd conductor. Trans. Am. Math. Soc. **366**(5), 2463–2516 (2014)
22. Brumer, A., Kramer, K.: Certain abelian varieties bad at only one prime. Algebra Number Theory **12**(5), 1027–1071 (2018)
23. Brumer, A., Kramer, K.: Corrigendum to "Paramodular abelian varieties of odd conductor". Trans. Am. Math. Soc. **372**(3), 2251–2254 (2019)
24. Brumer, A., Pacetti, A., Poor, C., Tornaría, G., Voight, J., Yuen, D.S.: On the paramodularity of typical abelian surfaces. Algebra Number Theory **13**(5), 1145–1195 (2019)
25. Calegari, F., Chidambaram, S., Ghitza, A.: Some modular abelian surfaces. Math. Comput. **89**(321), 387–394 (2020)
26. Cartier, P.: Representations of p-adic groups: a survey. In: Automorphic forms, representations and L-functions (Proc. Sympos. Pure Math., Oregon State Univ., Corvallis, Ore., 1977), Part 1, Proceedings of Symposia in Pure Mathematics, vol. XXXIII, pp. 111–155. American Mathematical Society, Providence (1979)
27. Casselman, W.: On some results of Atkin and Lehner. Math. Ann. **201**, 301–314 (1973)
28. Casselman, W.: Introduction to the theory of admissible representations of p-adic reductive groups. Unpublished Notes (1974)
29. Christian, U.: Einführung in die Theorie der paramodularen Gruppen. Math. Ann. **168**, 59–104 (1967)
30. Cohen, H.: Computing L-functions: a survey. J. Théor. Nombres Bordeaux **27**(3), 699–726 (2015)
31. Conforto, F.: Funzioni abeliane modulari. Vol. 1. Preliminari e parte gruppale. Geometria simplettica. Edizioni Universitarie "Docet", Roma (1952). Lezioni raccolte dal dott. Mario Rosati
32. Deligne, P.: La conjecture de Weil. I. Inst. Hautes Études Sci. Publ. Math. (43), 273–307 (1974)
33. Dembélé, L., Kumar, A.: Examples of abelian surfaces with everywhere good reduction. Math. Ann. **364**(3–4), 1365–1392 (2016)
34. Dern, T.: Paramodular forms of degree 2 and level 3. Comment. Math. Univ. St. Paul. **51**(2), 157–194 (2002)
35. Dirichlet, P.G.L.: Beweis des Satzes, daß jede unbegrenzte arithmetische Progression, deren erstes Glied und Differenz ganze Zahlen ohne gemeinschaftlichen Factor sind, unendlich viele Primzahlen enthält. Abhandlungen der Königlich Preussischen Akademie der Wissenschaften **1837**, 45–81
36. Dummigan, N.: Congruences of Saito-Kurokawa lifts and denominators of central spinor L-values. Glasg. Math. J. **64**(2), 504–525 (2022)
37. Dummigan, N., Pacetti, A., Rama, G., Tornaría, G.: Quinary forms and paramodular forms (2021). arXiv:2112.03797
38. Eichler, M., Zagier, D.: The theory of Jacobi forms. Progress in Mathematics, vol. 55. Birkhäuser Boston, Boston (1985)

39. Evdokimov, S.A.: Characterization of the Maass space of Siegel modular cusp forms of genus 2. Mat. Sb. (N.S.) **112(154)**(1(5)), 133–142,144 (1980)
40. Farmer, D.W., Pitale, A., Ryan, N.C., Schmidt, R.: Survey article: Characterizations of the Saito-Kurokawa lifting. Rocky Mountain J. Math. **43**(6), 1747–1757 (2013)
41. Flath, D.: Decomposition of representations into tensor products. In: Automorphic forms, representations and L-functions (Proc. Sympos. Pure Math., Oregon State Univ., Corvallis, Ore., 1977), Part 1, Proceedings of Symposia in Pure Mathematics, vol. XXXIII, pp. 179–183. American Mathematical Society, Providence (1979)
42. Freitag, E.: Siegelsche Modulfunktionen, Grundlehren der Mathematischen Wissenschaften, vol. 254. Springer-Verlag, Berlin (1983)
43. Fretwell, D.: Genus 2 paramodular Eisenstein congruences. Ramanujan J. **46**(2), 447–473 (2018)
44. Gallenkämper, J., Krieg, A.: The Hecke algebras for the orthogonal group SO(2, 3) and the paramodular group of degree 2. Int. J. Number Theory **14**(9), 2409–2423 (2018)
45. Gan, W.T., Takeda, S.: The local Langlands conjecture for GSp(4). Ann. Math. (2) **173**(3), 1841–1882 (2011)
46. Gritsenko, V.: Irrationality of the moduli spaces of polarized abelian surfaces. Internat. Math. Res. Notices (6), 235 ff., approx. 9 pp. (1994)
47. Gritsenko, V.: Arithmetical lifting and its applications. In: Number theory (Paris, 1992–1993), London Mathematical Society. Lecture Note Series, vol. 215, pp. 103–126. Cambridge University Press, Cambridge (1995)
48. Gritsenko, V.: Exponential lifting and Hecke correspondence. 1002, pp. 119–136 (1997). Research on automorphic forms and zeta functions (Japanese) (Kyoto, 1997)
49. Gritsenko, V.A., Vang, K.: Weight 3 antisymmetric paramodular forms. Mat. Sb. **210**(12), 43–66 (2019)
50. Gritsenko, V., Poor, C., Yuen, D.S.: Borcherds products everywhere. J. Number Theory **148**, 164–195 (2015)
51. Gritsenko, V., Poor, C., Yuen, D.S.: Antisymmetric paramodular forms of weights 2 and 3. Int. Math. Res. Not. IMRN (20), 6926–6946 (2020)
52. Gross, B.K.: On the Langlands correspondence for symplectic motives. Izv. Ross. Akad. Nauk Ser. Mat. **80**(4), 49–64 (2016)
53. Hecke, E.: Über die Bestimmung Dirichletscher Reihen durch ihre Funktionalgleichung. Math. Ann. **112**(1), 664–699 (1936)
54. Hecke, E.: Über Modulfunktionen und die Dirichletschen Reihen mit Eulerscher Produktentwicklung. I. Math. Ann. **114**(1), 1–28 (1937)
55. Hecke, E.: Über Modulfunktionen und die Dirichletschen Reihen mit Eulerscher Produktentwicklung. II. Math. Ann. **114**(1), 316–351 (1937)
56. Heim, B., Krieg, A.: The Maaßspace for paramodular groups. Kyoto J. Math. **60**(4), 1191–1207 (2020)
57. van Hoften, P.: A geometric Jacquet-Langlands correspondence for paramodular Siegel threefolds. Math. Z. **299**(3–4), 2029–2061 (2021)
58. Howe, R., Piatetski-Shapiro, I.I.: A counterexample to the "generalized Ramanujan conjecture" for (quasi-) split groups. In: Automorphic forms, representations and L-functions (Proc. Sympos. Pure Math., Oregon State University, Corvallis, Ore., 1977), Part 1, Proceedings of Symposia in Pure Mathematics, vol. XXXIII, pp. 315–322. American Mathematical Society, Providence (1979)
59. Humphreys, J.E.: Linear algebraic groups. Graduate Texts in Mathematics, vol. 21. Springer-Verlag, New York-Heidelberg (1975)
60. Ibukiyama, T.: On symplectic Euler factors of genus two. Proc. Jpn. Acad. Ser. A Math. Sci. **57**(5), 271–275 (1981)
61. Ibukiyama, T.: On relations of dimensions of automorphic forms of Sp(2, **R**) and its compact twist Sp(2). I. In: Automorphic forms and number theory (Sendai, 1983), Advanced Studies in Pure Mathematics, vol. 7, pp. 7–30. North-Holland, Amsterdam (1985)

62. Ibukiyama, T.: Paramodular forms and compact twist. In: Furusawa, M. (Ed.) Automorphic Forms on GSp(4), Proceedings of the 9th Autumn Workshop on Number Theory, pp. 37–48 (2007)
63. Ibukiyama, T.: Siegel modular forms of weight three and conjectural correspondence of Shimura type and Langlands type. In: The Conference on L-Functions, pp. 55–69. World Scientific Publication, Hackensack (2007)
64. Ibukiyama, T., Kitayama, H.: Dimension formulas of paramodular forms of squarefree level and comparison with inner twist. J. Math. Soc. Jpn. **69**(2), 597–671 (2017)
65. Igusa, J.I.: On Siegel modular forms of genus two. Am. J. Math. **84**, 175–200 (1962)
66. Igusa, J.I.: Theta functions. Die Grundlehren der mathematischen Wissenschaften, Band 194. Springer-Verlag, New York-Heidelberg (1972)
67. Iwahori, N.: Generalized Tits system (Bruhat decompostition) on p-adic semisimple groups. In: Algebraic Groups and Discontinuous Subgroups (Proceedings of Symposia in Pure Mathematics, Boulder, Colo., 1965), pp. 71–83. American Mathematical Society, Providence (1966)
68. Jacquet, H., Langlands, R.P.: Automorphic forms on GL(2). Lecture Notes in Mathematics, vol. 114. Springer-Verlag, Berlin-New York (1970)
69. Jacquet, H., Piatetski-Shapiro, I.I., Shalika, J.: Conducteur des représentations du groupe linéaire. Math. Ann. **256**(2), 199–214 (1981)
70. Jiang, D., Soudry, D.: Generic representations and local Langlands reciprocity law for p-adic SO_{2n+1}. In: Contributions to Automorphic Forms, Geometry, and Number Theory, pp. 457–519. Johns Hopkins University Press, Baltimore (2004)
71. Johnson-Leung, J., Roberts, B.: Siegel modular forms of degree two attached to Hilbert modular forms. J. Number Theory **132**(4), 543–564 (2012)
72. Johnson-Leung, J., Roberts, B.: Twisting of paramodular vectors. Int. J. Number Theory **10**(4), 1043–1065 (2014)
73. Johnson-Leung, J., Roberts, B.: Fourier coefficients for twists of Siegel paramodular forms. J. Ramanujan Math. Soc. **32**(2), 101–119 (2017)
74. Johnson-Leung, J., Roberts, B.: Twisting of Siegel paramodular forms. Int. J. Number Theory **13**(7), 1755–1854 (2017)
75. Johnson-Leung, J., Parker, J., Roberts, B.: The paramodular Hecke algebra (2023). arXiv:2310.13179
76. Klingen, H.: Introductory Lectures on Siegel Modular Forms. Cambridge Studies in Advanced Mathematics, vol. 20. Cambridge University Press, Cambridge (1990)
77. Kneser, M.: Starke Approximation in algebraischen Gruppen. I. J. Reine Angew. Math. **218**, 190–203 (1965)
78. Koecher, M.: Zur Theorie der Modulformen n-ten Grades. I. Math. Z. **59**, 399–416 (1954)
79. Kohnen, W., Skoruppa, N.P.: A certain Dirichlet series attached to Siegel modular forms of degree two. Invent. Math. **95**(3), 541–558 (1989)
80. Kreuzer, J.: Borcherds lift on the paramodular group of level 3. In: Automorphic Forms, Springer Proceedings in Mathematics and Statistics, vol. 115, pp. 151–161. Springer, Cham (2014)
81. Kurokawa, N.: Examples of eigenvalues of Hecke operators on Siegel cusp forms of degree two. Invent. Math. **49**(2), 149–165 (1978)
82. Ladd, W.B.: Algebraic modular forms on $SO_5(\mathbb{Q})$ and the computation of paramodular forms. Ph.D. Thesis, University of California, Berkeley, 2018
83. Marzec, J.: On Bessel models for GSp_4 and Fourier coefficients of Siegel modular forms of degree 2. Ph.D. Thesis, University of Bristol, 2016
84. Marzec, J.: Non-vanishing of fundamental Fourier coefficients of paramodular forms. J. Number Theory **182**, 311–324 (2018)
85. Marzec, J.: Maass relations for Saito-Kurokawa lifts of higher levels. Ramanujan J. **55**(1), 25–51 (2021)
86. Miyake, T.: Modular Forms. Springer-Verlag, Berlin (1989). Translated from the Japanese by Yoshitaka Maeda

87. Miyauchi, M.: On local newforms for unramified U(2, 1). Manuscripta Math. **141**(1–2), 149–169 (2013)
88. Miyazaki, T.: On Siegel paramodular forms corresponding to skew-holomorphic Jacobi cusp forms. Int. J. Math. **31**(8), 2050064, 34 (2020)
89. Nazaroglu, C.: Jacobi forms of higher index and paramodular groups in $N = 2$, $D = 4$ compactifications of string theory. J. High Energy Phys. (12), 074, front matter + 49 (2013)
90. Oda, T.: On the poles of Andrianov L-functions. Math. Ann. **256**(3), 323–340 (1981)
91. Okazaki, T.: Local Whittaker-newforms for GSp(4) matching to Langlands parameters (2019). arXiv:1902.07801
92. Piatetski-Shapiro, I.I.: Multiplicity one theorems. In: Automorphic forms, representations and L-functions (Proceedings of Symposia in Pure Mathematics, Oregon State University, Corvallis, Ore., 1977). Part 1, Proceedings of Symposia in Pure Mathematics, vol. XXXIII, pp. 209–212. American Mathematical Society, Providence (1979)
93. Piatetski-Shapiro, I.I.: On the Saito-Kurokawa lifting. Invent. Math. **71**(2), 309–338 (1983)
94. Poor, C., Yuen, D.S.: Paramodular cusp forms. Math. Comp. **84**(293), 1401–1438 (2015)
95. Poor, C., Shurman, J., Yuen, D.S.: Siegel paramodular forms of weight 2 and squarefree level. Int. J. Number Theory **13**(10), 2627–2652 (2017)
96. Poor, C., Schmidt, R., Yuen, D.S.: Paramodular forms of level 8 and weights 10 and 12. Int. J. Number Theory **14**(2), 417–467 (2018)
97. Poor, C., Shurman, J., Yuen, D.S.: Theta block Fourier expansions, Borcherds products and a sequence of Newman and Shanks. Bull. Aust. Math. Soc. **98**(1), 48–59 (2018)
98. Poor, C., Schmidt, R., Yuen, D.S.: Degree 2 Siegel paramodular forms of level 16 and weights up to 14. http://www.siegelmodularforms.org/pages/degree2/paramodular-level-16/ (2019) [Online; accessed 18 June 2023]
99. Poor, C., Schmidt, R., Yuen, D.S.: Paramodular forms of level 16 and supercuspidal representations. Mosc. J. Comb. Number Theory **8**(4), 289–324 (2019)
100. Poor, C., Shurman, J., Yuen, D.S.: Finding all Borcherds product paramodular cusp forms of a given weight and level. Math. Comp. **89**(325), 2435–2480 (2020)
101. Poor, C., Shurman, J., Yuen, D.S.: Nonlift weight two paramodular eigenform constructions. J. Korean Math. Soc. **57**(2), 507–522 (2020)
102. Rama, G., Tornaría, G.: Computation of paramodular forms. In: ANTS XIV—Proceedings of the Fourteenth Algorithmic Number Theory Symposium, Open Book Series, vol. 4, pp. 353–373. Mathematical Sciences Publishers, Berkeley (2020)
103. Roberts, B., Schmidt, R.: An alternative proof of a theorem about local newforms for GSp(4). In: Furusawa, M. (Ed.) Automorphic Forms on GSp(4). In: Proceedings of the 9th Autumn Workshop on Number Theory, pp. 227–255 (2006)
104. Roberts, B., Schmidt, R.: On modular forms for the paramodular groups. In: Automorphic Forms and Zeta Functions, pp. 334–364. World Scientific Publication, Hackensack (2006)
105. Roberts, B., Schmidt, R.: Local newforms for GSp(4), Lecture Notes in Mathematics, vol. 1918. Springer, Berlin (2007)
106. Roberts, B., Schmidt, R.: Some results on Bessel functionals for GSp(4). Doc. Math. **21**, 467–553 (2016)
107. Ryan, N.C., Tornaría, G.: A Böcherer-type conjecture for paramodular forms. Int. J. Number Theory **7**(5), 1395–1411 (2011)
108. Ryan, N.C., Tornaría, G.: Formulas for central values of twisted spin L-functions attached to paramodular forms. Math. Comp. **85**(298), 907–929 (2016). With an appendix by Ralf Schmidt
109. Sally, Jr., P.J., Tadić, M.: Induced representations and classifications for GSp(2, F) and Sp(2, F). Mém. Soc. Math. Fr. (N.S.) (52), 75–133 (1993)
110. Schmidt, R.: Some remarks on local newforms for GL(2). J. Ramanujan Math. Soc. **17**(2), 115–147 (2002)
111. Schmidt, R.: Iwahori-spherical representations of GSp(4) and Siegel modular forms of degree 2 with square-free level. J. Math. Soc. Jpn. **57**(1), 259–293 (2005)

112. Schmidt, R.: On classical Saito-Kurokawa liftings. J. Reine Angew. Math. **604**, 211–236 (2007)
113. Schmidt, R.: Archimedean aspects of Siegel modular forms of degree 2. Rocky Mt. J. Math. **47**(7), 2381–2422 (2017)
114. Schmidt, R.: Packet structure and paramodular forms. Trans. Am. Math. Soc. **370**(5), 3085–3112 (2018)
115. Schmidt, R.: Paramodular forms in CAP representations of GSp(4). Acta Arith. **194**(4), 319–340 (2020)
116. Schmidt, R., Shukla, A.: On Klingen Eisenstein series with level in degree two. J. Ramanujan Math. Soc. **34**(4), 373–388 (2019)
117. Shimura, G.: Modules des variétés abéliennes polarisées et fonctions modulaires. In: Séminaire Henri Cartan, vol. 18–20 (1957–1958)
118. Shimura, G.: Introduction to the arithmetic theory of automorphic functions. Kanô Memorial Lectures. Publications of the Mathematical Society of Japan, vol. 11. Princeton University Press/Iwanami Shoten Publishers, Princeton/Tokyo (1971)
119. Shukla, A.: Pullback of Klingen Eisenstein series and certain critical L-values identities. Ramanujan J. **55**(2), 471–495 (2021)
120. Siegel, C.L.: Symplectic geometry. Am. J. Math. **65**, 1–86 (1943)
121. Skoruppa, N.P.: Computations of Siegel modular forms of genus two. Math. Comp. **58**(197), 381–398 (1992)
122. Soudry, D.: The CAP representations of GSp(4, **A**). J. Reine Angew. Math. **383**, 87–108 (1988)
123. Soudry, D.: Rankin-Selberg convolutions for $SO_{2l+1} \times GL_n$: local theory. Mem. Am. Math. Soc. **105**(500), vi+100 (1993)
124. Takase, K.: An extension for the generalized Poisson summation formula of Weil and its applications. Comment. Math. Univ. St. Paul. **50**(1), 29–51 (2001)
125. Takloo-Bighash, R.: *L*-functions for the *p*-adic group GSp(4). Am. J. Math. **122**(6), 1085–1120 (2000)
126. Tate, J.T.: Fourier analysis in number fields, and Hecke's zeta-functions. In: Algebraic Number Theory (Proceedings of the Instructional Conference, Brighton, 1965), pp. 305–347. Thompson, Washington (1967)
127. Tsai, P.Y.: On Newforms for Split Special Odd Orthogonal Groups. Ph.D. Thesis, Harvard University, Ann Arbor, 2013
128. Tsai, P.Y.: Newforms for odd orthogonal groups. J. Number Theory **161**, 75–87 (2016)
129. Williams, B.: Graded rings of paramodular forms of levels 5 and 7. J. Number Theory **209**, 483–515 (2020)
130. Yi, S.: Klingen $\mathfrak{p}^2$ vectors for GSp(4). Ramanujan J. **54**(3), 511–554 (2021)
131. Yoshida, H.: On an explicit construction of Siegel modular forms of genus 2. Proc. Jpn. Acad. Ser. A Math. Sci. **55**(8), 297–300 (1979)
132. Zacarías, E.F.: Abelian surfaces, Siegel modular forms, and the Paramodularity Conjecture. Master's Thesis, Universitat de Barcelona, 2021
133. Zagier, D.B.: Zetafunktionen und quadratische Körper. Hochschultext [University Text]. Springer-Verlag, Berlin-New York (1981). Eine Einführung in die höhere Zahlentheorie [An introduction to higher number theory]

Index

J. Johnson-Leung et al., *Stable Klingen Vectors and Paramodular Newforms*,
Lecture Notes in Mathematics 2342, https://doi.org/10.1007/978-3-031-45177-5

LECTURE NOTES IN MATHEMATICS **Springer**

Editors in Chief: J.-M. Morel, B. Teissier;

Editorial Policy

1. Lecture Notes aim to report new developments in all areas of mathematics and their applications – quickly, informally and at a high level. Mathematical texts analysing new developments in modelling and numerical simulation are welcome.

   Manuscripts should be reasonably self-contained and rounded off. Thus they may, and often will, present not only results of the author but also related work by other people. They may be based on specialised lecture courses. Furthermore, the manuscripts should provide sufficient motivation, examples and applications. This clearly distinguishes Lecture Notes from journal articles or technical reports which normally are very concise. Articles intended for a journal but too long to be accepted by most journals, usually do not have this "lecture notes" character. For similar reasons it is unusual for doctoral theses to be accepted for the Lecture Notes series, though habilitation theses may be appropriate.

2. Besides monographs, multi-author manuscripts resulting from SUMMER SCHOOLS or similar INTENSIVE COURSES are welcome, provided their objective was held to present an active mathematical topic to an audience at the beginning or intermediate graduate level (a list of participants should be provided).

   The resulting manuscript should not be just a collection of course notes, but should require advance planning and coordination among the main lecturers. The subject matter should dictate the structure of the book. This structure should be motivated and explained in a scientific introduction, and the notation, references, index and formulation of results should be, if possible, unified by the editors. Each contribution should have an abstract and an introduction referring to the other contributions. In other words, more preparatory work must go into a multi-authored volume than simply assembling a disparate collection of papers, communicated at the event.

3. Manuscripts should be submitted either online at www.editorialmanager.com/lnm to Springer's mathematics editorial in Heidelberg, or electronically to one of the series editors. Authors should be aware that incomplete or insufficiently close-to-final manuscripts almost always result in longer refereeing times and nevertheless unclear referees' recommendations, making further refereeing of a final draft necessary. The strict minimum amount of material that will be considered should include a detailed outline describing the planned contents of each chapter, a bibliography and several sample chapters. Parallel submission of a manuscript to another publisher while under consideration for LNM is not acceptable and can lead to rejection.

4. In general, **monographs** will be sent out to at least 2 external referees for evaluation.

   A final decision to publish can be made only on the basis of the complete manuscript, however a refereeing process leading to a preliminary decision can be based on a pre-final or incomplete manuscript.

   Volume Editors of **multi-author works** are expected to arrange for the refereeing, to the usual scientific standards, of the individual contributions. If the resulting reports can be forwarded to the LNM Editorial Board, this is very helpful. If no reports are forwarded or if other questions remain unclear in respect of homogeneity etc, the series editors may wish to consult external referees for an overall evaluation of the volume.

5. Manuscripts should in general be submitted in English. Final manuscripts should contain at least 100 pages of mathematical text and should always include

   - a table of contents;
   - an informative introduction, with adequate motivation and perhaps some historical remarks: it should be accessible to a reader not intimately familiar with the topic treated;
   - a subject index: as a rule this is genuinely helpful for the reader.
   - For evaluation purposes, manuscripts should be submitted as pdf files.

6. Careful preparation of the manuscripts will help keep production time short besides ensuring satisfactory appearance of the finished book in print and online. After acceptance of the manuscript authors will be asked to prepare the final LaTeX source files (see LaTeX templates online: https://www.springer.com/gb/authors-editors/book-authors-editors/manuscriptpreparation/5636) plus the corresponding pdf- or zipped ps-file. The LaTeX source files are essential for producing the full-text online version of the book, see http://link.springer.com/bookseries/304 for the existing online volumes of LNM). The technical production of a Lecture Notes volume takes approximately 12 weeks. Additional instructions, if necessary, are available on request from lnm@springer.com.

7. Authors receive a total of 30 free copies of their volume and free access to their book on SpringerLink, but no royalties. They are entitled to a discount of 33.3 % on the price of Springer books purchased for their personal use, if ordering directly from Springer.

8. Commitment to publish is made by a *Publishing Agreement*; contributing authors of multiauthor books are requested to sign a *Consent to Publish form*. Springer-Verlag registers the copyright for each volume. Authors are free to reuse material contained in their LNM volumes in later publications: a brief written (or e-mail) request for formal permission is sufficient.

Addresses:

Professor Jean-Michel Morel, CMLA, École Normale Supérieure de Cachan, France
E-mail: moreljeanmichel@gmail.com

Professor Bernard Teissier, Equipe Géométrie et Dynamique,
Institut de Mathématiques de Jussieu – Paris Rive Gauche, Paris, France
E-mail: bernard.teissier@imj-prg.fr

Springer: Ute McCrory, Mathematics, Heidelberg, Germany,
E-mail: lnm@springer.com

Printed in the United States
by Baker & Taylor Publisher Services